BIOINFORMATICS FOR GENETICISTS

BIOINFORMATICS FOR GENETICISTS

Edited by

Michael R. Barnes
Genetic Bioinformatics
GlaxoSmithKline Pharmaceuticals, UK

and

Ian C. Gray
Discovery Genetics
GlaxoSmithKline Pharmaceuticals, UK

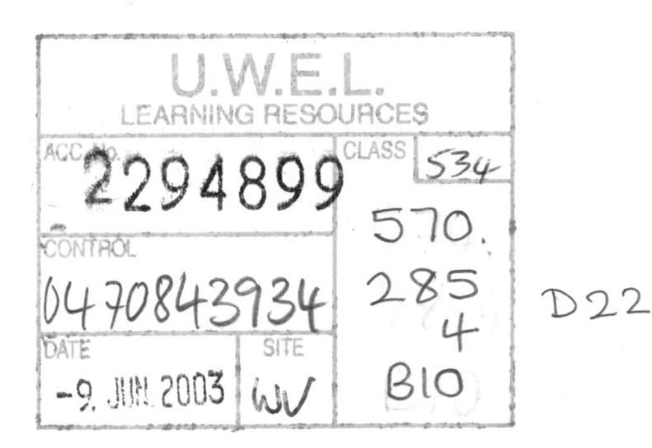

Telephone (+44) 1243 779777

Email (for orders and customer service enquiries): cs-books@wiley.co.uk
Visit our Home Page on www.wileyeurope.com or www.wiley.com

Other Wiley Editorial Offices

John Wiley & Sons Inc., 111 River Street, Hoboken, NJ 07030, USA

Jossey-Bass, 989 Market Street, San Francisco, CA 94103-1741, USA

Wiley-VCH Verlag GmbH, Boschstr. 12, D-69469 Weinheim, Germany

John Wiley & Sons Australia Ltd, 33 Park Road, Milton, Queensland 4064, Australia

John Wiley & Sons (Asia) Pte Ltd, 2 Clementi Loop #02-01, Jin Xing Distripark, Singapore 129809

John Wiley & Sons Canada Ltd, 22 Worcester Road, Etobicoke, Ontario, Canada M9W 1L1

Wiley also publishes its books in a variety of electronic formats. Some content that appears in print may not be available in electronic books.

British Library Cataloguing in Publication Data

A catalogue record for this book is available from the British Library

ISBN 0-470-84393-4
ISBN 0-470-84394-2 (PB)

Typeset in 9/11pt Times by Laserwords Private Limited, Chennai, India
Printed and bound in Great Britain by TJ International, Padstow, Cornwall
This book is printed on acid-free paper responsibly manufactured from sustainable forestry in which at least two trees are planted for each one used for paper production.

CONTENTS

LIST OF CONTRIBUTORS

Aruna Bansal
Population Genetics
GlaxoSmithKline Pharmaceuticals
New Frontiers Science Park (North)
Third Avenue
Harlow, Essex CM19 5AW, UK

Michael R. Barnes
Genetic Bioinformatics
GlaxoSmithKline Pharmaceuticals
New Frontiers Science Park (North)
Third Avenue
Harlow, Essex CM19 5AW, UK

Matthew J. Betts
Bioinformatics
DeCODE Genetics
Sturlugötu 8
101 Reykjavík, Iceland

Judith A. Blake
The Jackson Laboratory
600 Main Street
Bar Harbor, ME 04609, USA

Peter R. Boyd
Population Genetics
GlaxoSmithKline Pharmaceuticals
Medicines Research Centre
Gunnels Wood Road,
Stevenage, Herts, SG1 2NY

Carol J. Bult
The Jackson Laboratory
600 Main Street
Bar Harbor, ME 04609, USA

H. N. Caron
Department of Pediatric Oncology,
Academic Medical Center,
University of Amsterdam
Meibergdreef 9, 1105 AZ Amsterdam
The Netherlands

Janan Eppig
The Jackson Laboratory
600 Main Street
Bar Harbor, ME 04609, USA

Ian C. Gray
Discovery Genetics
GlaxoSmithKline Pharmaceuticals
New Frontiers Science Park (North)
Third Avenue
Harlow, Essex CM19 5AW, UK

Alexandre Hamburger
Hybrigenics
3–5 impasse Reille
F75014 Paris, France

Pui-Yan Kwok
University of California, San Francisco
505 Parnassus Ave,
Long 1332A, Box 0130
San Francisco
CA 94143-0130, USA

Gabor Marth
National Center for Biotechnology Information, NLM, NIH
8600 Rockville Pike
Building 45, Room 5AS29
Bethesda, Maryland 20894, USA

Ralph McGinnis
Population Genetics
GlaxoSmithKline Pharmaceuticals
New Frontiers Science Park (North)
Third Avenue
Harlow, Essex CM19 5AW, UK

J. M. Ruijter
Department of Anatomy and Embryology,
Academic Medical Center,
University of Amsterdam
Meibergdreef 9, 1105 AZ Amsterdam
The Netherlands

Robert B. Russell
Structural & Computational Biology
Programme
EMBL, Meyerhofstrasse 1
69117 Heidelberg, Germany

Colin A. Semple
Bioinformatics
MRC Human Genetics Unit
Edinburgh EH4 2XU, UK

Christopher Southan
Oxford GlycoSciences UK Ltd.
The Forum, 86 Milton Science Park
Abingdon OX14 4RY, UK

B. D. C. van Schaik
Bioinformatics Laboratory
Academic Medical Center,
University of Amsterdam
Meibergdreef 9, 1105 AZ Amsterdam
The Netherlands

Antoine H. C. van Kampen
Bioinformatics Laboratory
Academic Medical Center,
University of Amsterdam
Meibergdreef 9, 1105 AZ Amsterdam
The Netherlands

R. Versteeg
Department of Human Genetics
Academic Medical Center,
University of Amsterdam
Meibergdreef 9, 1105 AZ Amsterdam
The Netherlands

Ellen Vieux
Washington University School of Medicine
660 S. Euclid Ave
Box 8123
St Louis
MO 63110, USA

Thomas Werner
Genomatix Software GmbH
Landsberger Str. 6
D-80339 Muenchen, Germany

Jérôme Wojcik
Hybrigenics
3–5 impasse Reille
F75014 Paris, France

FOREWORD

Despite a relatively short existence, bioinformatics has always seemed an unusually multidisciplinary field. Fifteen years ago, when sequence data were still scarce and only a small fraction of the power of today's pizza-box supercomputers was available, bioinformatics was already entrenched in a diverse array of topics. Database development, sequence alignment, protein structure prediction, coding and promoter site identification, RNA folding, and evolutionary tree construction were all within the remit of the early bioinformaticist.[1,2]. To address these problems, the field drew from the foundations of statistics, mathematics, physics, computer science, and of course, molecular biology. Today, predictably, bioinformatics still reflects the broad base on which it started, comprising an eclectic collection of scientific specialists.

As a result of its inherent diversity, it is difficult to define the scope of bioinformatics as a discipline. It may be even fruitless to try to draw hard boundaries around the field. It is ironic, therefore, that even now, if one were to compile an intentionally broad list of research areas within the bioinformatics purview, it would often exclude one biological discipline with which it shares a fundamental basis: Genetics. On one hand, this seems difficult to believe, since the fields share a strong common grounding in statistical methodology, dependence on efficient computational algorithms, rapidly growing biological data, and shared principles of molecular biology. On the other hand, this is completely understandable, since a large part of bioinformatics has spent the last few years helping to sequence a number of genomes, including that of man. In many cases, these sequencing projects have focused on constructing a single representative sequence — the consensus — a concept that is completely foreign to the core genetics principles of variability and individual differences. Despite a growing awareness of each other, and with a few clear exceptions, genetics and bioinformatics have managed to maintain separate identities.

Geneticists needs bioinformatics. This is particularly true of those trying to identify and understand genes that influence complex phenotypes. In the realm of human genetics, this need has become particularly clear, so that most large laboratories now have one or two bioinformatics 'specialists' to whom other lab members turn for computing matters. These specialists are required to support a dauntingly wide assortment of applications: typical queries for such people might range from how to find instructions for accessing the internet, to how to disentangle a complex database schema, to how to optimize numerically intensive algorithms on parallel computing farms. These people, though somewhat scarce, are essential to the success of the laboratory.

With the ever-increasing volume of sequence data, expression information and well-characterized structures, as well as the imminent genotype and haplotype data on large and diverse human populations, genetics laboratories now must move beyond singular dependence on the bioinformatics handyman. Some level of understanding and ability to use bioinformatics applications is becoming necessary by everyone in the lab. Fortunately, bioinformaticians have been particularly successful in developing user-friendly software that renders complex statistical methods accessible to the bench scientists who generated and should know most about the data being analysed. To further these analyses, ingenious software applications have been constructed to display the outcomes and integrate them with a host of useful annotation features such as chromosome characteristics, sequence signatures, disease correlates and species comparisons[3]. With these tools freely available and undergoing continued development, mapping projects that make effective use of genetic and genomic information will naturally enjoy greater success than those less equipped to do so. Simply put, genetics groups that cannot capitalize on bioinformatics applications will be increasingly scooped by those who can.

The emerging requirement of broader understanding of bioinformatics within genetics is the focus of this text, as easily appreciated by a quick glance at the title. Equally obvious is that geneticists are the editors' target audience. Still, one might ask 'toward what specific group of geneticists is this text aimed?' The software and computational backbone of bioinformatics is shared most noticeably with the areas of statistical and population genetics, so the statistical specialists would seem a plausible audience. By design, however, this text is not aimed at these specialists so much as at those with broader backgrounds in molecular and medical genetics, including both human and model organism research. The content should be accessible by skilled bench scientists, clinical researchers and even laboratory heads. Computationally, one needs only basic computing skills to work through most of the material. Biologically, appreciation of the problems described requires general familiarity with genetics research and recognition of the inherent value in careful use of in silico genetic and genomic information.

By necessity, the bioinformatics topics covered in this text reflect the diversity of the field. In order to obtain some order in this broad area, the editors have focused on computer applications and effective use of available databases. This concentration on applications means that descriptions of the statistical theory, numerical algorithms and database organization are left to other texts. The editors have intentionally bypassed much of this material to emphasize applications in widespread use—the focus is on efficient use, rather than development, of bioinformatics methods and tools.

The data behind many of the bioinformatics tools described here are rapidly changing and expanding. In response, the software tools and databases themselves tend to be (infuriatingly) dynamic. A consequence of this fluid state is that learning to use existing programs by no means guarantees a knack for using those in the future. Thus, one cannot expect long-term consistency in the tools and data-types described here (or in most any other contemporary bioinformatics text). By learning to use current tools more effectively, however, geneticists can not only capitalize on technology available, but perhaps engage more bioinformaticians in the excitement of genetics research. Bringing bioinformatics to geneticists is a crucial first step towards integrating the kindred fields and characterizing the frustratingly elusive genes that influence complex phenotypes.

Lon R. Cardon
Professor of Bioinformatics
Wellcome Trust Centre for Human Genetics
University of Oxford

1. Doolittle, R. F. *Of URFs and ORFs: A primer on how to analyze derived amino acid sequences* (University Science Books, Mill Valley, California, 1987).
2. von Heijne, G. *Sequence analysis in molecular biology: Treasure trove or trivial pursuit* (Academic Press, London, 1987).
3. Wolfsberg, T. G., Wetterstrand, K. A., Guyer, M. S., Collins, F. S. & Baxevanis, A. D. A user's guide to the human genome. *Nature Genetics* **32 (suppl)** (2002).

SECTION 1

AN INTRODUCTION TO BIOINFORMATICS FOR THE GENETICIST

CHAPTER 1

Introduction: The Role of Genetic Bioinformatics

MICHAEL R. BARNES[1] and IAN C. GRAY[2]

[1]*Genetic Bioinformatics and* [2]*Discovery Genetics*
Genetics Research Division
GlaxoSmithKline Pharmaceuticals, Harlow, Essex, UK

1.1 INTRODUCTION

In February 2000, scientists announced the draft completion of the human genome. If media reports were accepted at face value, then it might be reasonable to predict that most geneticists would be unemployed within a decade of this announcement and human disease would become a distant memory. As we all know this is very far from the truth, the human genome is many things but it is not in itself a panacea for all human ailments, nor is it a revelation akin to the elucidation of the DNA double helix or the theory of evolution. The human genome is simply a resource borne out of technical prowess, perhaps with a little human inspiration. One thing that is certain is that we do not yet understand the functional significance of the majority of our genome, but what we do know is finally put into context. Over the past 200 years mankind has developed an

Bioinformatics for Geneticists. Edited by M.R. Barnes and I.C. Gray
 ISBNs: 0 470 84393 4; 0 470 84394 2 (PB)

ever increasing understanding of genetics; Darwin and Mendel provided the 19th century theories of evolution and inheritance, while Bateson, Morgan and others established a framework for the mechanisms of genetics at the beginning of the 20th century. The tentative identification of DNA as the genetic material by Avery and colleagues in the 1940s preceded the elucidation of the structure of the DNA molecule in 1953 by Watson and Crick, which in turn provided a mechanism for DNA replication and ushered in the era of modern molecular genetics. In 2003, precisely 50 years after this landmark discovery it is anticipated that the entire human genome sequence will be completed in a final, polished form; a fully indexed but currently only semi-intelligible 'book of life'. Here lies the most overlooked property of the genome — its value as a framework for data integration, a central index for biology and genetics. Almost any form of biological data can be mapped to a genomic region based on the genes or regulatory elements that mediate it. So the sequencing of the human genome means new order for biology. This order is perhaps comparable to the order the periodic table brought to chemistry in the 19th century. Where elements were placed in an ordered chemical landscape, biological elements will be grouped and ordered on the new landscape of the human genome. This presents excellent opportunities to draw together very diverse biological data; only then will the 'book of life' begin to make sense.

The human genome and peripheral data associated with and generated as a result of it require increasingly sophisticated data storage, retrieval and handling systems. With the promises and challenges that lie ahead, bioinformatics can no longer be the exclusive realm of the Unix guru or the Perl hacker and in recent years web browsers have made tools accessible and user friendly to the average biologist or geneticist. Bioinformatics is now both custodian and gatekeeper of the new genome data and with it most other biological data. This makes bioinformatics expertise a prerequisite for the effective geneticist. This expertise is no mystery; modern bioinformatics tools coupled with an inquiring mind and a willingness to experiment (key requirements for any scientist, bioinformatician or not) can yield confidence and competence in bioinformatic data handling in a very short space of time. The objective of this book is not to act as an exhaustive guide to bioinformatics, other texts are available to fulfil this role, but instead is intended as a specialist guide to help the typical geneticist navigate the internet jungle to some of the best tools and databases for the job, that is, associating genes, polymorphisms and mutations with diseases and genetic traits. In this chapter we give a flavour of the many processes in modern genetics where bioinformatics has a major impact and refer to subsequent chapters for greater detail.

At the risk of over simplifying a very complex issue, the process of understanding genetic disease typically proceeds through three stages. First, recognition of the disease state or syndrome including an assessment of its hereditary character; second, discovery and mapping of the related polymorphism(s) or mutation(s) and third, elucidation of the biochemical/biophysical mechanism leading to the disease phenotype. Each of these stages proceeds with a variable degree of laboratory investigation and bioinformatics. Both activities are complementary, bioinformatics without laboratory work is a sterile activity as much as laboratory work without bioinformatics can be a futile and inefficient one. In fact these two sciences are really one, genetics and genomics generate data and computational systems allow efficient storage, access and analysis of the data — together, they constitute bioinformatics. Almost every laboratory process has a complementary bioinformatics process, Table 1.1 lists a few of these — building on these basic applications will maximize the effect of bioinformatics on workflow efficiency.

TABLE 1.1 Examples of Bioinformatics Applications in Genetics Research

Data	Related Laboratory Techniques	Associated Bioinformatics Applications
Human genome sequence	DNA sequencing PCR Novel gene identification by expression analysis *In vitro* characterization of regulatory elements	Gene and regulatory region prediction BLAST homology searching Electronic PCR PCR primer design Electronic translation and protein secondary structure prediction *In silico* design of expression constructs
Genetic markers	Genotyping	Identification of optimal marker sets Genotyping assay design QC checking and statistical analysis of genotype data
Model organism genome sequence	Comparative genetics (e.g. linkage) and genomics (e.g. transgenics and gene knock-outs)	Linkage analysis of models of human diseases Comparative genetic and physical maps for cross-species analysis of linkage regions Functional assessment of gene regulatory regions by cross-species comparison *In silico* drafting of gene knock-out and transgenic constructs
Expression RNA and protein	Microarrays Serial analysis of gene expression (SAGE) Proteomics	Gene regulatory analysis Tumour and other disease tissue classification Elucidation of gene–gene interactions and disease pathway expansion
Three-dimensional protein structure	Crystallography/NMR	Prediction and visualization of molecular structures related to disease and mutation

1.2 GENETICS IN THE POST-GENOME ERA – THE ROLE OF BIOINFORMATICS

In the role of genome data custodian and gatekeeper, bioinformatics is an integral part of almost every field of biology, including of course, genetics. In the broadest sense it covers the following main aspects of biological research:

- Knowledge management and expansion
- Data management and mining
- Study design and support
- Data analysis
- Determination of function

These categories are quite generic and could apply to any field of biology, but are clearly applicable to genetics. Both genetics and bioinformatics are essentially concerned with asking the right questions, generating and testing hypotheses and organizing and interpreting large amounts of data to detect biological patterns.

1.3 KNOWLEDGE MANAGEMENT AND EXPANSION

Few areas of biological research call for a broader background in biology than the modern approach to genetics. This background is tested to the extreme in the selection of candidate genes to test for involvement with a disease process, where genes need to be chosen and prioritized based on many criteria. Often biological links may be very subtle, for example a candidate gene may regulate a gene which regulates a gene that in turn may act upon the target disease pathway. Faced with the complexity of relationships between genes, geneticists need to be able to expand pathways and identify complex cross talk between pathways. As this process can extend almost interminably to a point where virtually every gene is a candidate for every disease, knowledge management is important to help to weigh up evidence to prioritize genes. The geneticist may not be an authority in the disease area under study, and in today's climate of reductionist biology an expert with a global picture of the disease process at the molecular level may be hard to find. Therefore effective tools are needed to quickly evaluate the role of each candidate and its related pathways with respect to the target phenotype.

Literature is the most powerful resource to support this process, but it is also the most complex and confounding data source to search. To expedite this process, some databases have been constructed which attempt to encapsulate the available literature, e.g. On-line Mendelian Inheritance in Man (OMIM). These centralized data resources can often be very helpful for gaining a quick overview of an unfamiliar pathway or gene, but inevitably one needs to re-enter the literature to build up a fuller picture and to answer the questions that are most relevant to the target phenotype or gene. The internet is also an excellent resource to help in this process. In Chapter 2, we offer some pointers to help the reader with effective literature searching strategies and give suggestions as to some of the best disease databases and related resources on the internet.

1.4 DATA MANAGEMENT AND MINING

Efficient application of knowledge relies on well organized data and genetics is highly dependent upon good data, often in very large volumes. Accessing available data, particularly in large volumes is often the biggest informatic frustration for geneticists. Here we focus on aspects of accessing data from public databases; solutions for in-house data collection, either in the form of 'off the shelf' or custom-built laboratory information management systems (LIMS) belong to a specialist area that lies beyond the scope of this book.

Genetic data have grown exponentially over the last few years, fuelled by the expressed sequence tag (EST) cDNA sequence resources generated largely during the 1990s and more recently the increasing genomic sequence data from the human genome and other genome sequencing projects. Genetic database evolution has matched this growth in some areas, with some resources leading the efforts towards whole genome integration of genetic data, particularly the combined human genome sequence, genetic map, EST and SNP databases exemplified by the Golden Path. Curiously, development in many of the older more established genetic resources (for example, GDB and HGMD) has been somewhat stagnant. This may be partly due to the difficulties involved in data integration with the draft genome sequence, which is effectively a moving target as the data are updated on a regular basis. Many of the traditional genetic databases have not seized the opportunity to integrate genetic data with the human genome sequence. The future survival of these databases will certainly depend on this taking place and there is no question that the role of these databases will change. One might question the value of some of the older genetic datasets, for example, why would we need radiation hybrid maps of the human genome, when we have the ultimate physical map — the human genome sequence? These painstakingly collected datasets have already played a critical role in the process of generating the maps that allowed the sequencing of the human genome and they may still have some value as an aid for QC of new data and perhaps more importantly as a point of reference for all the studies that have previously taken place.

A key problem that frequently hinders effective genetic data mining is the localization of data in many independent databases rather than a few centralized repositories. A clear exception to this is SNP data which has now coalesced around a single central database — dbSNP at NCBI (Sherry *et al.*, 2001). By contrast human mutation data, which has been collected over many years, is still stored in disparate sources, although moves are afoot to move to a similar central database — Hobbies (Fredman *et al.*, 2002). These developments are timely; human mutation and polymorphism data both hold complementary keys to a better understanding of how genes function and malfunction in disease. The availability of a complete human genome presents us with an ideal framework to integrate both sets of data, as our understanding of the mechanisms of complex disease increase, the full genomic context of variation will become increasingly significant.

With the exception of dbSNP most recent database development has not been implicitly designed for geneticists, instead genomic databases and genome viewers have developed to aid the annotation of the human genome. Of course this data is vital for genetics, but this explains why the available tools often appear to lack important functionality. One has to make use of what functionality is available, although sometimes this means using tools in ways that were not originally intended (for example many geneticists use BLAST to identify sequence primer homology in the human genome, but few realize that the default parameters of this tool are entirely unsuited for this task). We will attempt to address these issues throughout this book and offer practical solutions for obtaining the most value from existing tools wherever possible. In Chapter 5 we examine the use of human genome browsers for genetic research. Tools such as Ensembl and the UCSC human genome browser annotate important genetic information on the human genome, including SNPs, some microsatellites and of course, genes and regulatory regions. User-defined queries place genes and genetic variants in their full genomic context, giving very detailed information on nearby genes, promoters or regions conserved between species, including mouse and fish. It is difficult to overstate the value of this information for genetics. For example, cross-species genome comparison is invaluable for the analysis of function, as inter-species sequence conservation is generally thought to be restricted to a functionally important gene or regulatory regions and so this is one of the most powerful tools for identifying potential regulatory elements or undetected genes (Aparicio *et al.*, 1995). Several chapters in this book cover tools and databases to support these approaches (see Chapters 12 and 13).

As technology developments have scaled up the throughput of genotyping to enable studies of tens (and possibly hundreds) of thousands of polymorphisms and provided the capability to generate equally impressive amounts of microarray transcript data to name just two examples, the need for more effective data management has intensified. This reveals the major drawback of the ultra user-friendly 'point and click' interfaces to most genetics and genomics tools — they often do not allow retrieval of bulk datasets; instead data often has to be retrieved on a point by point basis. For many applications this is highly inefficient at best or simply non-viable at worst. One solution to this problem is to query the database directly at a UNIX or SQL level, but this may not be a trivial process for the occasional user with no or limited knowledge of command lines and in many cases it will not be possible to access the data directly in this manner. If the raw data are available, it may be possible to build custom databases, using database tools such as Microsoft ACCESS. However, the authors accept that this is not a straightforward option nor the method of choice of most users and instead this book will focus on web-based methods for data access. Where there is no web-based method to achieve a data mining goal, geneticists should consider contacting the developers of databases to request new functionality, such requests are generally welcomed by database developers, many of whom would be very pleased to know that their tools are being used! Several developers have already improved their methods for bulk data retrieval (probably as a result of requests from users), but interfaces are still lacking in some critical areas for genetics. For example, several tools allow the user to generate a list of SNPs across a locus (e.g. dbSNP, Ensembl and UCSC), but only one allows the user to retrieve the flanking sequence of each SNP in one batch to allow primer design (SNPper — see Chapter 3). We will try to tackle these problems as they arise throughout the book.

1.5 GENETIC STUDY DESIGNS

There are a number of approaches to disease gene hunting and many arguments to support the merits of one approach over another. Whatever the method, comprehensive informatics input at the study design stage can contribute greatly to the quality, efficiency and speed of the study. It can help to define a locus clearly in terms of the genes and markers that it contains and supports a logical and systematic approach to marker and gene selection and subsequent genetic analysis, simultaneously reducing the cost of a project and improving the chances of successfully discovering a phenotype–genotype correlation.

Despite the recent improvements in the throughput of genetic and genomic techniques and the increased availability of gene and marker data, genes which contribute to the most common human diseases are still very elusive. By contrast, the identification of genes mutated in relatively rare single gene disorders (so-called Mendelian or monogenic disorders) is now straightforward if suitable kindreds are available. The identification of the genes responsible for a plethora of monogenic disorders is one of the genetics success stories of the late 1980s and the 1990s; genes identified include, to name but a few — *CFTR* (cystic fibrosis; Riordan *et al.*, 1989), Huntingtin (Huntington's disease; Huntington's Disease Collaborative Research Group, 1993), Frataxin (Friedreich's ataxia; Campuzano *et al.*, 1996) and *BRCA1* in breast and ovarian cancer (Miki *et al.*, 1994).

Unfortunately, success in the identification of genes with a role in complex (i.e. multigenic) disease has been far less successful. Notable examples are the involvement of *APOE* in late-onset Alzheimer's disease and cardiovascular disease and the role of *NOD2* in Crohn's disease (Hugot *et al.*, 2001; Saunders *et al.*, 1993). However, genes for most of the common complex diseases remain elusive. Our ability to detect disease genes is often dependent on the analysis method applied. Methods for the identification of disease genes can be divided neatly into two broad categories, linkage and association. Although many common principles apply to both of these study types, each approach has distinct informatics demands.

1.5.1 The Linkage Approach

The vast majority of Mendelian disease genes have been identified by linkage analysis. This involves identifying a correlation between the inheritance pattern of the phenotypic trait (usually a disease state) with that of a genetic marker, or a series of adjacent markers. Because of the relatively low number of recombination events observed in the 2–5 generation families typically used for linkage analyses (around one per Morgan, which is roughly equivalent to 100 megabases, per meiosis), these marker/disease correlations extend over many megabases (Mb), allowing adequate coverage of the entire human genome with a linkage scan of only 300–600 simple tandem repeat (STR) markers giving an average spacing of 10 or 5 cM respectively. STRs are the markers of choice for linkage analysis, due to the fact that they show a high degree of heterozygosity. Markers with a heterozygosity level of >70% are typically selected for linkage panels (i.e. from 100 individuals selected at random, at least 70 would have two different alleles for a given marker; clearly the higher the heterozygosity the greater the chance of following the inheritance pattern from parent to offspring). Such marker panels are well characterized and can be accessed from several public sources at various densities (see Chapter 7). Just over 16,000 STR markers have been characterized in humans, which represents a small fraction of the estimated total numbers of polymorphic STRs. Analysis of the December 2001 human genome draft sequence suggests that there may be somewhere in the order of 200,000 potentially polymorphic STRs in the human genome (Viknaraja *et al.*, unpublished data). Software tools are now available to assist in the sequence-based identification of these potentially polymorphic STR markers across a given locus, should additional markers be required to narrow a linkage region (see Chapter 9 for details).

Clearly the limited degree of recombination that facilitates linkage analysis with sparse marker panels is a double-edged sword; the investigator may be left with several megabases of DNA containing a large number of potential candidate genes. However, combining data from several different families often results in reduction of the genetic interval under study, and the high-throughput sequencing capabilities available in many modern genetics laboratories coupled with complete genome sequence render the systematic screening of a large number of candidate genes a far less daunting task than it was 10 years ago.

Unlike single gene Mendelian diseases, complex genetic diseases are caused by the combined effect of multiple polymorphisms in a number of genes, often coupled with environmental factors. The successes of linkage analysis in the rapid identification of

Mendelian disease genes has spawned large-scale efforts to track down genes involved in the more common complex disease phenotypes. This approach is not restricted to academic research groups; many pharmaceutical and biotechnology companies have joined what many would perceive to be a 'genetic gold-rush', in an attempt to identify new drug targets for common diseases such as asthma, diabetes and schizophrenia, in a manner reminiscent of the rush to mine drug targets from expressed sequence tags (ESTs) in the late 1990s (Debouck and Metcalf, 2000). The application of a linkage approach to complex disease typically involves combining data from a large number of affected sib-pairs. Publicly available software for linkage analysis of sib-pairs is described in detail in Chapter 11.

Unfortunately the identification of genes involved in common diseases using a linkage strategy has been largely unsuccessful to date, mainly because each gene with phenotypic relevance is thought to make a relatively small contribution to disease susceptibility. These small effects are likely to be below the threshold of detection by linkage analysis in the absence of unfeasibly large sample sizes (Risch, 2000). In an attempt to circumvent this problem researchers using linkage approaches to identify genes involved in complex disease typically relax the threshold of acceptable 'log of the odds' (LOD) score (see Chapter 11) from 3, the traditionally accepted threshold of evidence for linkage in monogenic disease to 2, or sometimes even lower (Pericak-Vance *et al.*, 1998). However we would expect to see a number of hits due to chance alone with a comprehensive genome scan at this threshold. The rationale for lowering the threshold for detection of linkage, i.e. the effect of each contributing gene in a complex disease is smaller than would be expected for a monogenic disease, can result in a situation where a true signal is indistinguishable from background noise. In order to distinguish true linkage from false positives, many investigators are now using a combination of both linkage and association, relying on linkage analysis to reveal tentative, broad map positions which are subsequently confirmed and narrowed with an association study (see Chapter 8).

1.5.2 The Association Approach

In its simplest form, the aim of a genetic association study is to compare an allele frequency in a disease population with that in a matched control population. A significant difference may be indicative that the locus under test is in some way related to the disease phenotype. This association could be direct, i.e. the polymorphism being tested may have functional consequences that have a direct bearing on the disease state. Alternatively, the relationship between a genetic marker and phenotype may be indirect, reflecting proximity of the marker under test to a polymorphism predisposing to disease. The phenomenon of co-occurrence of alleles (in this case a disease-conferring allele and a surrogate marker allele) more often than would be expected by chance is termed linkage disequilibrium (LD). Suitable population structures for genetic association studies and statistical methods and software tools for the analysis of data resulting from such studies are discussed in detail in Chapters 8 and 11. Our aim here is to give the reader the briefest of introductions.

Association studies have three main advantages over linkage studies for the analysis of complex disease: (i) case–control cohorts are generally easier to collect than extended pedigrees; (ii) association studies have greater power to detect small genetic effects than linkage studies; a clear example is the insulin gene, which shows extremely strong association with type 2 diabetes, but very weak linkage (Speilman *et al.*, 1993); (iii) LD typically stretches over tens of kilobases rather than several megabases (Reich *et al.*, 2001), allowing focus on much smaller and more manageable loci. Among other reasons (discussed in

Chapter 8), this is because an association-based approach exploits recombination in the context of the entire population, rather than within the local confines of a family structure.

Of course, this last point is the other side of the double-edged sword of marker density and resolution mentioned in the context of linkage analysis above. The trade-off is reduced range over which each marker can detect an effect, resulting in a need for increased marker density. The required marker density for an association-based genome scan is unknown at present as we do not have enough information regarding human genome diversity in terms of polymorphic variability and genome-wide patterns of LD. However, typical guesses are in the range of 30,000–300,000 markers (Collins *et al.*, 1999; Kruglyak, 1999); orders of magnitude higher than the numbers required for linkage analysis. The high cost of generating the several million genotypes for such an experiment has prevented any such undertaking at the time of writing, although several proof of concept studies have demonstrated that high-density SNP maps can be efficiently generated using existing technologies and should be achievable in a reasonable time-frame (Antonellis *et al.*, 2002; Lai *et al.*, 1998). In the meantime, it is likely that research groups will continue to test individual genes for association with disease (the 'candidate gene' approach — see Section 1.7 below).

Once the genomic landscape, in terms of polymorphism and LD, is known with some degree of accuracy, it is highly likely that the number of markers required for a whole genome association study can be reduced by an intelligent study design with heavy reliance on bioinformatics input. Testing all available markers in a given region for association with a disease is expensive, laborious and frequently unnecessary; a simple example to illustrate this would be two adjacent markers which always demonstrate co-segregation; in other words, the genotypic status of one can always be predicted by genotyping the other — there is no point in genotyping both. Although this example is simple in the extreme, as adjacent markers typically show varying degrees of (rather than absolute) co-segregation, there is a trade-off between minimizing the amount of required genotyping whilst minimizing loss of information. Selection of optimal non-redundant marker sets, coupled with an initial focus on gene-rich regions, is the key to providing lower overall genotyping costs whilst retaining high power to detect association. This will require extensive knowledge of the blocks of preserved marker patterns (haplotypes) in the population under study; bioinformatics tools for constructing and analysing haplotypes and selecting optimal marker sets based on haplotypic information are discussed in detail in Chapters 8 and 11.

1.5.3 Markers for Association Studies

STRs were (and still are) the vanguard of linkage analysis, mainly because of their high levels of heterozygosity and hence increased informativeness when compared to an earlier marker system, the restriction fragment length polymorphism (RFLP); the majority of RFLPs are the result of a single nucleotide polymorphism (SNP) which creates or destroys a restriction site. SNPs have made a comeback worthy of Lazarus in recent years and are now the marker of choice for genetic association studies. The main reasons for the return to favour of SNPs are their abundance (an estimated 7 million with a minor allele frequency of greater than 5% in the human genome; Kruglyak and Nickerson, 2001) and binary nature which renders them well suited to automated, high-throughput genotyping. As mentioned above, tens or hundreds of thousands of SNPs will be required for whole genome association scans (even with optimized marker sets). Until very recently, studies on this scale were unfeasible, not only as a result of unacceptably high genotyping costs, but also due to the lack of available markers. Large-scale SNP discovery projects such as the SNP consortium (TSC; Altshuler *et al.*, 2000a) have increased the number of known SNPs dramatically. We now have a great deal of SNP data (3.4 million non-redundant SNPs deposited in dbSNP at the time of going to press), however it is becoming apparent that even this number of markers will be insufficient for comprehensive association studies (note that the figure of 3.4 million includes a considerable number of SNPs with a minor allele frequency of less than 5%, which may be of limited use in association studies; this is discussed in Chapter 8).

We have already touched on the importance and potential impact of defining haplotypes as the basis for identifying optimal marker sets. This method has already been applied in small-scale studies with striking results. For example, in a study of nine genes spanning a total of 135 kb, Johnson *et al.* (2001) found that just 34 SNPs from a total of 122 could be used to define all common haplotypes (those with a frequency of greater than 5%) across the nine genes, an impressive validation of the approach of defining maximally informative minimal marker sets based on haplotypic data. However this study also highlighted the inadequacy of the current public SNP resource; only 10% of the SNPs identified by Johnson *et al.* were found to be present in dbSNP. Using dbSNP data alone, it was impossible to capture comprehensive haplotype data; in fact for four of the nine genes, no SNPs whatsoever were registered in dbSNP. Unfortunately it appears that our current public SNP resource represents the tip of the iceberg in terms of requisite information for the proper implementation of modest candidate gene association studies, let alone whole genome scans. However, given the burgeoning nature of dbSNP, we are optimistic that this situation is transient.

As a footnote to this section, it should be noted that although STRs have been largely swept aside by the wave of SNP euphoria, STRs may still be useful for association studies; indeed, it is possible that LD can be detected over far greater distances with STRs than SNPs under some circumstances, as discussed in Chapter 8.

1.6 PHYSICAL LOCUS ANALYSIS

In recent years, as the human genome sequence has neared completion, practical approaches to physical characterization of a genetic locus have changed quite dramatically. The laborious laboratory-based process of contig construction using yeast and bacterial artificial chromosome (YAC and BAC) clones or cosmids, involving consecutive rounds of library screening, clone characterization and identifying overlaps between clones, has become largely redundant, as has clone screening for the identification of novel polymorphic markers and genes. Today this process, which took many months or even years, can be completed in an afternoon using web-based human genome browsers. This shifts the initial focus of a study from contig construction and characterization to very detailed locus characterization using a range of bioinformatics tools; it is now possible to characterize a locus *in silico* to a very high level of detail before any further laboratory work commences. When the wet work does start, good prior use of bioinformatics will have rendered many procedures superfluous and the study is far more efficient and focused as a result. Figure 1.1 illustrates some of the key stages in the genetic analysis of candidate genes and loci — the role of informatics at each stage of this process is explored in detail in this book and the relevant chapters addressing each issue are indicated.

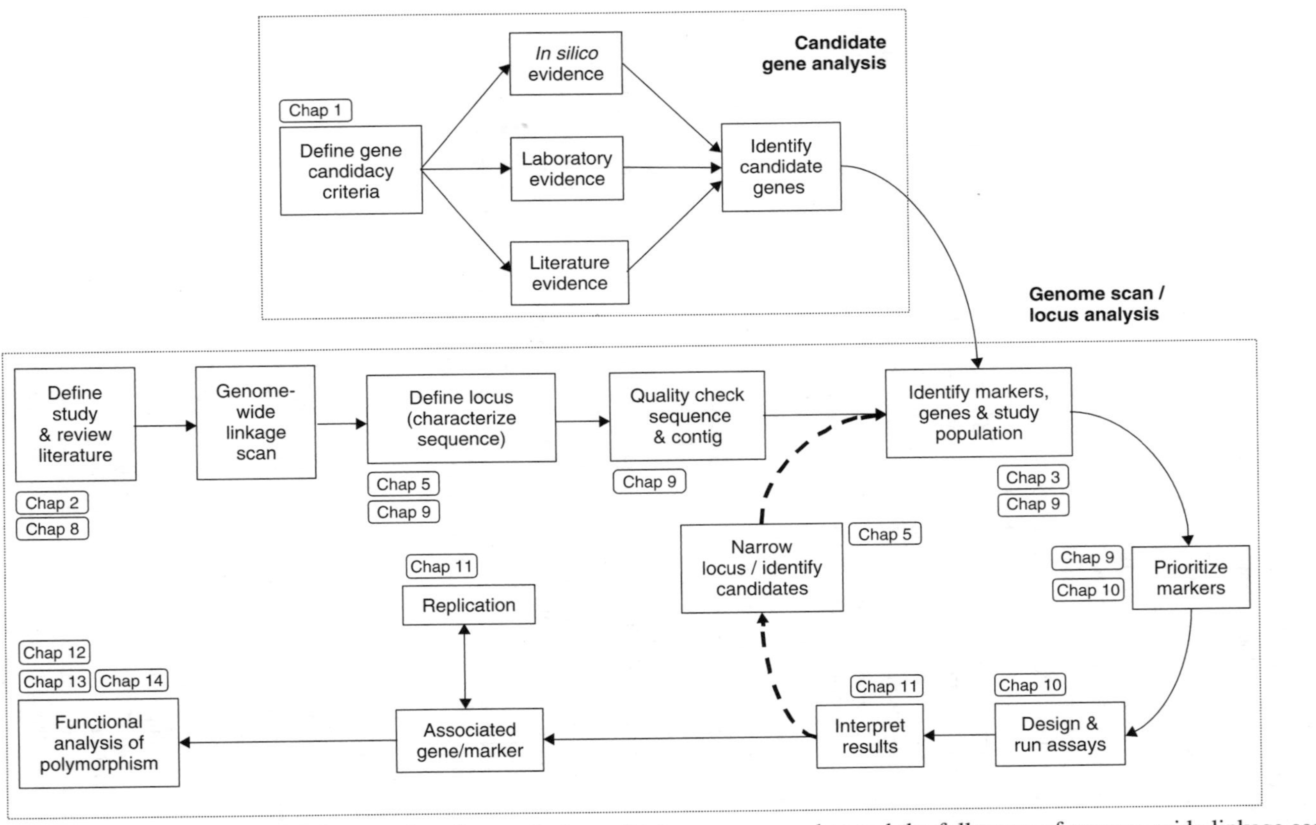

Figure 1.1 The genetic study process for complex disease, both candidate gene approaches and the follow-up of genome-wide linkage scans, highlighting chapters covering informatics aspects of each key step.

1.7 SELECTING CANDIDATE GENES FOR ANALYSIS

Candidate genes are typically selected for testing for association with a disease state on the basis of either (i) biological rationale; the gene encodes a product which the investigator has good reason to believe is involved in the disease process, (ii) the fact that the gene in question is located under a linkage peak, or (iii) both. The biggest problem with candidate gene analysis is that apparently excellent candidates are usually highly abundant and this surfeit of 'good' candidates is often difficult to rationalize.

Bioinformatics can be one of the most effective ways to help shorten, or more correctly prioritize, a candidate list without immediate and intensive laboratory follow-up. Firstly candidate criteria need to be determined based upon the phenotype in question. Detailed searches of the literature may help to flesh out knowledge of the disease and related pathways. Once a set of criteria is defined (for example which tissues are likely to be affected, which pathways are likely to be involved, and what types of genes are likely to mediate the observed phenotype), further literature review will help to 'round up the usual suspects', genes in known pathways with an established role in the phenotype under study. This is probably the most time-consuming step, but some tools can help to expedite this process, for example tools like OMIM can provide concise summaries of a disease area or gene family. Other databases encapsulate knowledge of pathways and regulatory networks, e.g. the *Kyoto Encyclopedia of Genes and Genomes* (KEGG; Kanehisa *et al.*, 2002). An alternative or parallel approach at this stage is to use a broader net to identify all genes which *could* be involved in the disease based on relaxed criteria such as tissue expression. Many *in silico* gene expression resources are available, including data derived from EST libraries, serial analysis of gene expression (SAGE; Velculescu *et al.*, 1995) data, microarray and RT-PCR data (see Chapter 15). For example, if the disease manifests in the lung, it is possible to distinguish genes that show lung expression from those that do not. This gives an opportunity to reduce emphasis on genes that show expression patterns which conflict with the disease hypothesis. However, it should be noted that electronic expression data is typically not comprehensive and care must be taken in using it to exclude the expression of a gene in a specific tissue. Low-level expression may not be detected by the method used; furthermore, gene expression may show temporal and spatial regulation — a gene may only be expressed during a specific phase of development or under particular conditions, e.g. cellular stress or differentiation.

1.8 PROGRESSING FROM CANDIDATE GENE TO DISEASE-SUSCEPTIBILITY GENE

In recent years, countless associations between genes and disease have been published, however many of these are likely to be spurious. Many reported associations show marginal p-values and subsequent studies often fail to replicate initial findings. Clearly p-values of around 0.05, generally accepted as the cut-off for a significant finding, will occur by chance for every 20 tests performed; this largely explains the general failure to reproduce promising primary results. However, real but very small effects giving marginal p-values are also difficult to replicate, leaving the investigator unsure as to the meaning of a failure to replicate. One approach for resolving the issue is to perform a rigorous meta-analysis using all available data, including both positive and negative associations. This type of analysis was recently used to demonstrate an association between the nuclear hormone receptor PPARγ and diabetes, using data (previously regarded as equivocal) drawn from a range of publications (Altshuler *et al.*, 2000b). Nonetheless, this approach relies on a lack of publication bias, i.e. the improbable assumption of an equal chance of publication for both positive and negative results.

Ultimately the biologist requires functional data to support an hypothetical genetic association; bioinformatics has a role to play here too. For example, DNA variants that alter subsequent amino-acid sequences can be checked for potential functional consequences using software tools (Chapters 12 and 14). Similarly, a thorough bioinformatic characterization of putative regulatory elements can give an indication of the possible impact of polymorphisms on *cis*-acting transcriptional motifs and the consequence on expression levels (Chapter 13). Bioinformatics can also assist in laboratory-based functional assessment of genes and polymorphisms; simple sequence manipulation tools coupled with genome sequence data can be used to design constructs for the *in vitro* and *in vivo* analysis of genes and polymorphisms using expression assays, transgenic mice and a host of other systems. However, perhaps the largest impact from bioinformatics on the field of functional characterization of genes will come from the development of powerful pattern recognition software for the identification of relationships between multitudes of transcripts analysed using microarrays. This approach has already proved useful in tumour classification by relating patterns of gene expression to response to chemotherapeutic agents (Butte *et al.*, 2000). An extension of this method should allow the elucidation of gene–gene interactions and the identification of common or converging biochemical pathways. Coupled with a knowledge of putative disease-related polymorphisms and comparable expression profiles in disease tissue, microarrays (together with the nascent field of proteomics; see Chapter 16) promise to be an extremely powerful future tool for the dissection of complex disease processes. Figure 1.2 illustrates approaches for gene characterization which are useful for both prioritizing candidate genes for analysis and establishing causality in a disease process. The chapter detailing each aspect is indicated.

1.9 COMPARATIVE GENETICS AND GENOMICS

We have already touched on the role of bioinformatics in relation to the identification of functionally important DNA motifs by cross-species comparison. This area is covered more fully in Chapters 9 and 12. Recently the sequencing of a number of genomes has been completed, including the yeasts *Saccharomyces cerevisiae* and *Schizosaccharomyces pombe*, the fruit fly *Drosophila melanogaster* and the nematode worm *Caenorhabditis elegans;* soon these will be joined by the puffer fish species *Fugu rubripes* and *Tetraodon nigroviridis*, the zebra fish *Danio rerio* and of course the mouse and rat. This has provided an unprecedented opportunity for large-scale genome comparisons, allowing researchers to make inferences not only with regard to the identification of conserved regulatory elements, but also about genome evolutionary dynamics. Whole genome availability also provides a complete platform for the design of *in vivo* paradigms of human disease, for example transgenic and gene knock-out animal models and more sophisticated spatially and temporally regulated conditional mutants.

Large-scale approaches to biochemical pathway dissection using expression microarrays in relatively simple organisms, particularly yeast, are also proving extremely promising. Whole genome expression profiles can be generated and correlated transcription profiles identified for related groups of genes. Coincident expression patterns are frequently indicative of subsequent protein–protein interactions and co-localization in protein complexes (Jansen *et al.*, 2002). Similar tissue-specific experiments can be performed for

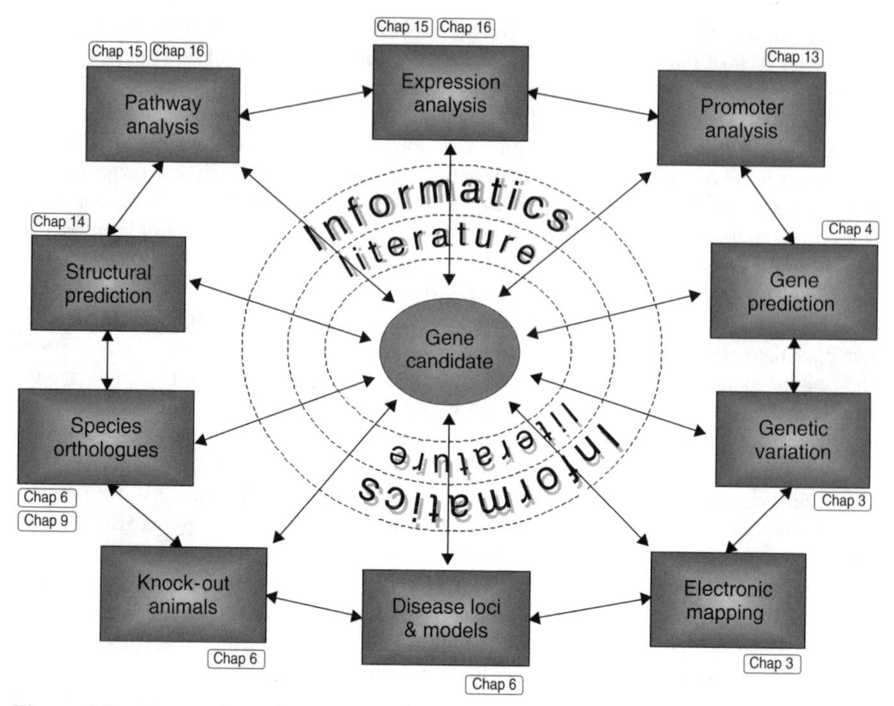

Figure 1.2 Approaches for gene characterization, indicating chapters detailing each aspect.

higher organisms, both for the purposes of identifying coincident transcription profiles for unravelling biochemical pathways and for comparison of diseased and normal tissues (see, for example Mody *et al.*, 2001; Saban *et al.*, 2001). Tissue derived from animal models such as mice can have advantages over using diseased human tissue: the disease model can be generated under a controlled environment, typically on an identical genetic background to the control tissue, and procurement of a significant number of high-quality tissue samples (essential for the extraction of good quality RNA) is more straightforward (see Chapter 15).

Thus far we have given a few examples of the impact of combining model organisms with high-throughput genomics technologies for improving our understanding of gene function and interaction, biochemical pathways and human disease (comparative genomics). Similar strides are being made in the field of comparative genetics (here we define genetics as phenotype-driven gene identification using genetic mapping procedures), particularly in the areas of mouse and rat genetics. The ability to perform controlled crosses such as inter-crosses and backcrosses (see Silver, 1995; Chapter 11) coupled with the development of fairly high density genetic maps over the last few years has rendered the mapping of monogenic traits in both mouse and rat a reasonably straightforward exercise. The impact of the completion of the mouse and rat genome sequences in the near future will be similar to the impact of the availability of the human genome on human genetics; indeed, the partially completed mouse and rat genomes are already giving significant improvements in speed of mapping and candidate gene identification. These developments together with recently implemented large-scale mutagenesis programmes for the generation of monogenic mutants (see Chapter 6) promise to provide a significant increase in the mutant mouse resource in terms of simple disease models.

Significant progress has also been made in mapping complex traits in both the mouse and rat in recent years, including the development of software packages for the identification of quantitative trait loci (QTL; see Chapter 11). However, although experimental crosses can be designed to maximize the chances of success (unlike human studies), complex trait analysis in model organisms is still plagued by the difficulties in identifying and precisely localizing genes of relatively small effect. QTL linkage peaks are typically broad due to lack of absolute correspondence between genotype and phenotype and a consequent inability to identify unequivocal recombinant animals. In an attempt to overcome this limitation, mapping methods using 'heterogenous stocks' have recently been developed (Mott *et al.*, 2000). The heterogenous stock comprises a mouse line resulting from inter-crossing several different inbred strains and maintaining the resulting mixed stock through several generations (typically 30–60). Each chromosome from a mouse derived from a heterogenous stock consists of a mosaic of DNA from the different founding strains, allowing a fine mapping approach based on a knowledge of the ancestral alleles in the original inbred lines. Mott *et al.* have developed publicly available software for the analysis of heterogenous stocks (see Chapter 11).

Perhaps one of the most exciting developments in model organism genetics is the fusion of classical genetics with high-throughput genomics techniques. Microarrays provide a means of checking all genes within a QTL linkage peak for subtle differences in expression levels, potentially pinpointing the culprit gene. This tactic was used successfully to reveal the role of *Cd36* in metabolic defects, following linkage analysis in the rat (Aitman *et al.*, 1999). As an extension of this method, a gene expression profile may be treated as a quantitative trait and used as a phenotypic measure in linkage analysis for the identification of genes influencing the expression level, as a route to biochemical pathway expansion. Jansen and Nap (2001) recently coined the phrase 'genetical genomics' for this type of approach.

1.10 CONCLUSIONS

We hope this book will help the geneticist to design and complete more effective genetic analyses. Bioinformatics can have far-reaching effects on the way that a laboratory scientist works but obviously it will never entirely replace the laboratory process and is simply another set of tools to expedite the research of the practising biologist. Misconceptions regarding the power of bioinformatics as a stand-alone science are perhaps among the biggest mistakes that computer-based bioinformatics specialists can make and may even explain a degree of prejudice against bioinformatics—perceived by some as an '*in silico* science' with little basis in reality. Taken to an extreme and without a balanced understanding of both the application of software tools and a good appreciation of basic biological principles, this is exactly what bioinformatics can be, but where bioinformatics proceeds as part of 'wet' and 'dry' cycles of investigation, both processes are stronger as a result. In this introduction we have briefly examined some of the experimental genetics processes which can be assisted by informatics; we now invite the reader to read on for more detailed coverage of each of these processes in the remaining chapters of this book.

REFERENCES

Aitman TJ, Glazier AM, Wallace CA, Cooper LD, Norsworthy PJ, Wahid FN, *et al.* (1999). Identification of Cd36 (Fat) as an insulin-resistance gene causing defective fatty acid and glucose metabolism in hypertensive rats. *Nature Genet* **21**: 76–83.

Aparicio S, Morrison A, Gould A, Gilthorpe J, Chaudhuri C, Rigby P, *et al.* (1995). Detecting conserved regulatory elements with the model genome of the Japanese puffer fish, *Fugu rubripes*. *Proc Natl Acad Sci USA* **92**: 1684–1688.

Altshuler D, Pollara VJ, Cowles CR, Van Etten WJ, Baldwin J, Linton L, *et al.* (2000a). An SNP map of the human genome generated by reduced representation shotgun sequencing. *Nature* **407**: 513–516.

Altshuler D, Hirschhorn JN, Klannemark M, Lindgren CM, Vohl MC, Nemesh J, *et al.* (2000b). The common PPARgamma Pro12Ala polymorphism is associated with decreased risk of type 2 diabetes. *Nature Genet* **26**: 76–80.

Antonellis A, Rogus JJ, Canani LH, Makita Y, Pezzolesi MG, Nam M, *et al.* (2002). A method for developing high-density SNP maps and its application at the Type 1 Angiotensin II Receptor (AGTR1) Locus. *Genomics* **79**: 326–332.

Butte AJ, Tamayo P, Slonim D, Golub TR, Kohane IS. (2000). Discovering functional relationships between RNA expression and chemotherapeutic susceptibility using relevance networks. *Proc Natl Acad Sci USA* **97**: 12182–12186.

Campuzano V, Montermini L, Molto MD, Pianese L, Cossee M, Cavalcanti F, *et al.* (1996). Friedreich's ataxia: autosomal recessive disease caused by an intronic GAA triplet repeat expansion. *Science* **271**: 1423–1427.

Collins A, Lonjou C, Morton NE. (1999). Genetic epidemiology of single-nucleotide polymorphisms. *Proc Natl Acad Sci USA* **96**: 15173–15177.

Debouck C, Metcalf B. (2000). The impact of genomics on drug discovery. *Ann Rev Pharmacol Toxicol* **40**: 193–207.

Fredman D, Siegfried M, Yuan YP, Bork P, Lehvaslaiho H, Brookes AJ. (2002). Hobbies: a human sequence variation database emphasizing data quality and a broad spectrum of data sources. *Nucleic Acids Res* **30**: 387–391.

Hugot JP, Chamaillard M, Zouali H, Lesage S, Cezard JP, Belaiche J, *et al.* (2001). Association of NOD2 leucine-rich repeat variants with susceptibility to Crohn's disease. *Nature* **411**: 599–603.

Huntington's Disease Collaborative Research Group. (1993). A novel gene containing a trinucleotide repeat that is expanded and unstable on Huntington's disease chromosomes. *Cell* **72**: 971–983.

Jansen RC, Nap JP. (2001). Genetical genomics: the added value from segregation. *Trends Genet* **17**: 388–391.

Jansen R, Greenbaum D, Gerstein M. (2002). Relating whole-genome expression data with protein–protein interactions. *Genome Res* **12**: 37–46.

Johnson GC, Esposito L, Barratt BJ, Smith AN, Heward J, Di Genova G, Ueda H, *et al.* (2001). Haplotype tagging for the identification of common disease genes. *Nature Genet* **29**: 233–237.

Kanehisa M, Goto S, Kawashima S, Nakaya A. (2002). The KEGG databases at GenomeNet. *Nucleic Acids Res* **30**: 42–46.

Kruglyak L. (1999). Prospects for whole-genome linkage disequilibrium mapping of common disease genes. *Nature Genet* **22**: 139–144.

Kruglyak L, Nickerson DA. (2001). Variation is the spice of life. *Nature Genet* **27**: 234–236.

Lai E, Riley J, Purvis I, Roses A. (1998). A 4-Mb high-density single nucleotide polymorphism-based map around human APOE. *Genomics* **54**: 31–38.

Miki Y, Swensen J, Shattuck-Eidens D, Futreal PA, Harshman K, Tavtigian S, *et al.* (1994). A strong candidate for the breast and ovarian cancer susceptibility gene BRCA1. *Science* **266**: 66–71.

Mody M, Cao Y, Cui Z, Tay KY, Shyong A, Shimizu E, *et al.* (2001). Genome-wide gene expression profiles of the developing mouse hippocampus. *Proc Natl Acad Sci USA* **98**: 8862–8867.

Mott R, Talbot CJ, Turri MG, Collins AC, Flint J. (2000). A method for fine mapping quantitative trait loci in outbred animal stocks. *Proc Natl Acad Sci USA* **97**: 12649–12654.

Pericak-Vance MA, Bass ML, Yamaoka LH, Gaskell PC, Scott WK, Terwedow HA, *et al.* (1998). Complete genomic screen in late-onset familial Alzheimer's disease. *Neurobiol Aging* **19** (1 Suppl): S39–S42.

Reich DE, Cargill M, Bolk S, Ireland J, Sabeti PC, Richter DJ, *et al.* (2001). Linkage disequilibrium in the human genome. *Nature* **411**: 199–204.

Risch NJ. (2000). Searching for genetic determinants in the new millennium. *Nature* **405**: 847–856.

Riordan JR, Rommens JM, Kerem B, Alon N, Rozmahel R, Grzelczak Z, *et al.* (1989). Identification of the cystic fibrosis gene: cloning and characterization of complementary DNA. *Science* **245**: 1066–1073.

Saban MR, Hellmich H, Nguyen NB, Winston J, Hammond TG, Saban R. (2001). Time course of LPS-induced gene expression in a mouse model of genitourinary inflammation. *Physiol Genomics* **5**: 147–160.

Saunders AM, Schmechel D, Pericak-Vance M, Enghild J, Salvesen GS, *et al.* (1993). Apolipoprotein E: high-avidity binding to beta-amyloid and increased frequency of type 4 allele in late-onset familial Alzheimer disease. *Proc Natl Acad Sci USA* **90**: 1977–1981.

Saunders AM, Strittmatter WJ, Schmechel D, St. George-Hyslop PH, Pericak-Vance MA, Joo SH, *et al.* (1993a). Association of apolipoprotein E allele E4 with late-onset familial and sporadic Alzheimer's disease. *Neurology* **43**: 1467–1472.

Sherry ST, Ward MH, Kholodov M, Baker J, Phan L, Smigielski EM, *et al.* (2001). dbSNP: the NCBI database of genetic variation. *Nucleic Acids Res* **29**: 308–311.

Silver LM. (1995). *Mouse Genetics: Concepts and Applications*. Oxford University Press: Oxford, UK.

Spielman RS, McGinnis RE, Ewen WJ (1993). Transmission test for linkage disequilibrium: the insulin gene region and insulin-dependent diabetes mellitus (IDDM). *Am J Hum Genet* **52**: 506–516.

Velculescu VE, Zhang L, Vogelstein B, Kinzler KW. (1995). Serial analysis of gene expression. *Science* **270**: 484–487.

CHAPTER 2

Internet Resources for the Geneticist

MICHAEL R. BARNES[1] and CHRISTOPHER SOUTHAN[2]

[1]*GlaxoSmithKline Pharmaceuticals*
Harlow, Essex, UK

[2]*Oxford GlycoSciences UK Ltd*
The Forum, 86 Milton Science Park
Abingdon OX14 4RY, UK

Bioinformatics for Geneticists. Edited by M.R. Barnes and I.C. Gray
 ISBNs: 0 470 84393 4; 0 470 84394 2 (PB)

2.1 INTRODUCTION

The World Wide Web ('the web') and our knowledge of human genetics and genomics are both expanding rapidly. By allowing swift, universal and largely free access to data, particularly the human genome sequence, the web has already played an important role in the study of human genetics and genomics. Increased data accessibility is dramatically changing the way the scientific community is communicating and carrying out research. The internet biology community is expanding daily with an organic development of websites, tools and databases, which could eventually replace the conventional scientific paper as the predominant form of communication. Already we are starting to see successful website/journal hybrids such as *Genome Biology* (http://genomebiology.com/) and biomednet (biomednet.com) which offer high quality peer-reviewed scientific articles and reviews alongside bioinformatics databases and tools. Many more established journals like *Nature* and *Science* are rapidly following suit with user-friendly websites, which offer much more than the full text of the journal.

The web is offering more than just information. Virtual research communities have been organized around databases and specialist research groups. These communities are even influencing the way bioinformatics tools are being developed, a good example being Ensembl the human genome browser developed at the EBI and Sanger Institute in Hinxton, Cambridgeshire (Hubbard *et al.*, 2002). In the spirit of open source community projects such as the free UNIX operating system Linux, the Ensembl development team has developed Ensembl on an 'open source' basis. This means all code is freely available to anyone who wishes to download it. But further still, Ensembl is developed by a 'virtual community' of developers from institutes, industry and academia around the world who are free to modify and add to the central software code (subject to a peer review). So the tools and interfaces, though primarily developed in Hinxton, may include contributions from developers in Singapore, North Carolina and New York or elsewhere.

2.1.1 Hypothesis Construction and Data Mining – essentials for Genetics

Genetics is a science which calls for analysis and interpretation across a wide range of biological research. Many chapters in this book deal with focused tools. Beyond these specialist applications however, geneticists need access to a wide range of databases and literature, both to update particular research areas and formulate new hypotheses. This requires expertise across the gamut of biological data on the internet. This ranges from the review literature to highly specific databases. This can illuminate biology from gene function to biological pathways. Effective data mining needs an understanding of the general principles by which it is organized, particularly the sequence-based data resources. This needs to be backed up by good scientific judgment concerning quality and significance.

An exhaustive description of biological data and databases on the internet would be beyond the scope of this book. Confucius might not have been thinking of internet searching when he said 'give a man a fish and he will live for a day, teach a man to fish and he will live forever', but the principle still applies. So, instead of reviewing the data resources themselves the most useful thing we can do here is to review search methods to help identify both current and future resources.

2.2 SUB-DIVISION OF BIOLOGICAL DATA ON THE INTERNET

Biological information on the internet can be roughly subdivided into two broad categories, which we will term 'the biological internet' and 'biological information on the internet'. This distinction may not be immediately apparent — we define 'the biological internet' as purpose-built biological tools and databases which index and contain detailed biological information, such as the human genome sequence, nucleotide and protein sequences, genetic markers, polymorphisms and the full range of biological literature. The majority of these tools and databases are maintained in a highly integrated form by major biological organizations such as NCBI and EMBL. We define 'biological information on the internet' as biological data which is less formally maintained on the web, this could include information on research laboratory homepages, conference abstracts, tools, boutique databases and any other data that scientists have seen fit to present on the web.

These distinctions are more clearly defined by the tools that are available to search the data. Firstly there are general purpose web search engines, such as Google, Lycos and Excite (see Table 2.1 for a full list), these tools index and search the full range of the internet and have the capability to identify webpages, tools and databases by simple keyword searching. A second category of tools are the specialist biological search tools, such as Entrez-PubMed and BLAST (see Chapter 4). The former uses keyword searching or accession number queries, the latter uses similarity searching to find related sequences.

The choice of search tool depends on the kind of information that needs to be retrieved. The scope of biological and genetic information on the internet is so broad that no single tool is available to index all data. The key point to understand is which tool is most suitable to identify a specific form of data. For example literature is most comprehensively indexed by PubMed or Scirus (see below), whereas nucleotide records can only be identified with any specificity by Entrez or BLAST. This is in contrast to a laboratory homepage or a boutique web resource. Unless a description is published in PubMed these resources may only be identified by a web search tool. If it is not clear what information needs to be retrieved then clearly both specific and general search tools should be used.

TABLE 2.1 Key Internet Search Engines with Reported Index Size (Equivalent to the Number of Documents Indexed)

Search Engine	URL	Reported Index Size
Google	http://www.google.com/	560 M
AltaVista	http://www.altavista.com/	350 M
FAST	http://www.alltheweb.com/	340 M
Northern Light	http://www.northernlight.com/	265 M
Excite	http://www.excite.com/	250 M
HotBot	http://www.hotbot.com/	110 M
Lycos	http://www.lycos.com/	110 M
MetaCrawler	http://www.metacrawler.com/	ND
Scirus	http://www.scirus.com/	69 M (science only)

2.3 SEARCHING THE INTERNET FOR GENETIC INFORMATION

The World Wide Web began as an information-sharing and retrieval project at the European particle physics laboratory CERN (Berners-Lee *et al.*, 1999). It has only recently evolved into the mass media beast that we all know. But just as the internet began, so it continues as an information-sharing resource for scientists in all fields. One cannot deny that commercial proliferation has not been an unmitigated success for the growth of the web but this has led many scientists to perceive the internet as a rising tide of irrelevant noise that has largely washed away any intrinsic value. This is a misconception. We will demonstrate that some web resources for biological sciences are both outstanding and indispensable. Internet biology suffers as much as any other field of scholarship from: data of dubious provenance, broken links, outdated sites and newsgroup spam. But it also contains valuable and novel data which can be crucial for scientific discovery. The skill is to recognize chaff and know how to sift the wheat from it. To do this we need tools that are capable of highlighting relevant information in an organized manner.

In the process of linking genotypes to phenotypes it is important to know about the function of a gene or gene family, for example to prioritize candidate disease-association genes. In such cases biological search tools and internet search tools may provide complementary results. To give an hypothetical example let us assume that a genetic locus associated with a familial form of basal cell carcinoma includes a novel gene with homology to WNT genes. With no knowledge of WNT genes it would be difficult to include or exclude this gene as a candidate. A search of PubMed would reveal a daunting range of over 1000 publications mentioning members of the WNT gene family. Some might contain specific information to link WNT genes to carcinoma but it would take a long time to read and digest all the available information. Using Google to search for 'WNT gene' would identify a range of conference abstracts and laboratory homepages. Towards the top of the hit-list this would include the 'World Wide WNT Window' (www.stanford.edu/~rnusse/wntwindow.html). This is an excellent summary of the whole WNT signalling pathway maintained by prominent researchers in the WNT signalling field. The page includes a detailed and regularly maintained summary of all genes in this highly complex pathway, which is currently unpublished. Examination of this pathway would identify the Patched receptor upstream, which has been shown to cause 80% of sporadic basal cell carcinomas. This is just one of many examples of how a thriving unpublished and unpublicized on-line research community can be identified by opportunistic internet searching.

2.4 WHICH WEB SEARCH ENGINE?

In a nutshell the availability of full-text search engines allows the web to be used as a searchable 15-billion-word encyclopedia. However, because the web is a distributed, dynamic, and rapidly growing information resource, it presents many difficulties for traditional information retrieval technologies. This why the choice of the search methodology used for searching can lead to very different results.

An important point to make is that all search engines are not the same. A common misconception is that most internet search engines index the same documents for a large proportion of the web. In fact the coverage of search engines may vary by an order of magnitude. An estimated lower boundary on the size of the indexable web is 0.8 billion pages (http://www.neci.nec.com/~lawrence/websize.html). Many engines index only a fraction of the total number of documents on the web and so the coverage of any one engine may be significantly limited. Combining the results of multiple engines has been shown to significantly increase coverage. This is done automatically with metasearch engines such as MetaCrawler (www.metacrawler.com), which search and combine the results of several search engines. Table 2.1 presents a selection of web search engines with direct applicability to biological searching. We also recommend the website, SearchEngineWatch.com, for reviews and reports on new search engines.

2.4.1 Google

It is apparent from Table 2.1 that Google offers the widest indexing capacity. This is an innovative search engine based on scientific literature citation indexes (Butler, 2000). Conventional search engines use algorithms and simple rules to rank pages based on the frequency of the keywords specified in a query. Google exploits the links between webpages to rank hits. Thus the highly cited pages of the web world with many links pointing to them are ranked highest in the results. This is an efficient searching mechanism which effectively captures the internet community 'word of mouth' on the best and most frequently used webpages.

2.4.2 Scirus

The greatest limitation for web search engines is unindexed databases. These include many of the databases that make up the biological internet, such as sequence databases and some subscription-based resources such as full-text journals, and commercial databases. Although limited material from these sites, such as front pages, documentation and abstracts are indexed by search engines, the underlying data is not available because of database firewalls and/or blocks on external indexing.

In an attempt to solve this problem, the publisher Elsevier has developed Scirus (http://www.scirus.com/). This is a joint venture with FAST, a Norwegian search engine company who have produced an excellent specialist scientific search engine. Scirus enhances its specificity and scope by only indexing resources with scientific content. These include webpages, full-text journals and Medline abstracts. This makes Scirus an effective tool for both web and literature searching tool. Both full text and PDF format journal content is indexed by performing a MetaSearch of the other major providers of full text—Elsevier's ScienceDirect and Academic Press's IDEAL collection. Scirus also searches the web for the same key words, including Medline, patents from the databases of the US Patent Office, science-related conferences and abstracts. The Medline database is provided on the BioMedNet platform, which requires a free BioMedNet login and password for access. Scirus offers the user several options to customize their searches to search only free sites, only membership sites or only specific sites. The advanced interface also allows boolean queries (see below). By March 2002 Scirus had indexed 69 million science-related pages, including PDF files and peer-reviewed articles, thereby covering the majority of the biologically relevant internet.

Although Scirus expands the scope of biological data searching beyond other search engines it falls short in some areas. For example the full-text journals are restricted to Elsevier and Academic Press. Coverage is also restricted by index pre-filtering that might miss some websites. Another disadvantage is that search results tend to be redundant. Although for literature searching there are alternative full text searching tools such as

HighWire (see below) Scirus is tantalizingly close to what a universal biological search engine should be.

2.5 SEARCH SYNTAX: THE MATHEMATICS OF SEARCH ENGINE USE

The best search engine in the world will not retrieve relevant results unless the query is correctly defined. This is easy to master and a few basic commands can turn a poor specificity keyword search into a highly targeted query. The key to successful sifting of the web is to select for the minimum number of irrelevant hits (maximize specificity) but avoiding the exclusion of relevant hits (minimize false negatives).

2.5.1 Using the '+ and –' Symbols to Filter Results

Sometimes it is necessary to ensure that a search engine finds pages that have all the words you enter, not just some of them. This can be achieved by using the '+' symbol. Similarly you may wish to exclude a specific word from your search by using the '–' symbol. These commands work with nearly all the major web search engines and are similar in function to the boolean operators 'AND' and 'NOT' respectively.

As an example let's say you wish to find information about human promoter prediction tools. You could search using [+ promoter + prediction + tool]. This will only retrieve pages that contain all three words. If the search returns excessive information by including tools for plant and bacterial promoter prediction, one could further refine the search by using the following search query [+ promoter + prediction + tool – plant – prokaryote]. This will subtract pages which mention plants and prokaryotes. Be aware though that this might filter out valid hits to tools which analyse *both* prokaryote and eukaryote sequences.

2.5.2 Using Quotation Marks to Find Specific Phrases

The most complex filtering syntax on our promoter prediction query still manages to retrieve over 1000 results, so we need to consider other methods of reducing the number of hits. One approach is to use a phrase search that will find only those pages where the terms appear in exactly the order specified. This is achieved by putting quotation marks around the phrase, so we might search with ['promoter prediction tool']. This retrieves six relevant hits but clearly many sources have been filtered out, so it is important to beware of over-specifying search terms.

2.5.3 Restricting the Searching Domain of a Query

A final measure that can be taken to fine tune your query is to restrict the internet domain. For example you can restrict your search to only identify hits in the .edu (educational) domain or to ignore hits from the .com (company) domain. This is achieved in Google and most other sites by using the [+ site:.edu] to include a domain or [– site:.com] to exclude a domain. This command can be extended further to search only a specific site, e.g. to search the NCBI website for SNP information try [+ SNP + site:ncbi.nlm.nih.gov].

Table 2.2 includes the search results obtained from the different variations on the search for promoter prediction tools, using both Google and Scirus. This shows the improvements

TABLE 2.2 Different Results Obtained from Different Query Targeting Methods. Results Compare the Number of Hits Returned by the General Search Engine Google and Specialist Science Search Engine Scirus

Query	Google Hits	Scirus Hits*
+ promoter + prediction + tool	4050	2379
'promoter prediction tool'	6	2
'promoter prediction tools'	14	8
+ promoter + prediction + tool − plant	2630	1312
+ promoter + prediction + tool − plant − bacterial	2080	936
+ promoter + prediction + tool − plant − bacterial − site:.com	1750	NA

*Queries to Scirus were designed using the equivalent boolean syntax in the advanced search form.

from filtering on the query. The final word on fine tuning web search queries is to be as flexible as possible. Try to use keywords which are likely to be specific to the kind of website or tool you are looking for. Sometimes it is useful to go to a page or tool similar to the one you are looking at to check for very specific words that might be shared by similar sites. For example, in the case of promoter prediction tools, a commonly occurring word was 'server'; exchanging this for 'tool' significantly improves the relevance of the hits.

2.6 BOOLEAN SEARCHING

Although the familiar boolean search commands (AND, OR, NOT) are widely used for many forms of database searching, including PubMed, they are not universally supported by all web search engines. Table 2.3 lists those supported by the most popular search engines. The functionality offered by AND and NOT mirrors the functionality of [+ and −]. Other commands have a distinct function, for example [SNP OR Analysis] will retrieve all webpages that contain the words SNP or analysis. The NEAR command is not

TABLE 2.3 Boolean Commands Supported by Popular Web Search Engines

Command	How	Supported by
Or	OR	AltaVista, Excite, Google, Lycos, Northern Light
	None	FAST, LookSmart,
And	AND	AltaVista, Excite, Lycos, Northern Light
	None	FAST, Google, LookSmart
Not	NOT	Excite, Lycos, Northern Light
	AND NOT	AltaVista
	None	FAST, Google, LookSmart,
Near	NEAR	AltaVista (10 words), Lycos (25 words)
	None	FAST, Google, LookSmart

widely supported but can be useful to help to identify two keywords in close proximity to each other.

2.7 SEARCHING SCIENTIFIC LITERATURE – GETTING TO 'STATE OF THE ART'

Effective mining of the literature is important at the stages of conception, design and construction of genetic studies. At the most basic level it is important to be aware of the 'state of the art' in a research area before embarking on new efforts. At the very least this avoids duplication of effort, but it can also provide previously unrecognized clues which need to be followed up. Unfortunately this important informatic process is still lacking truly innovative tools and databases. We are still struggling with tools that cover the fundamentals of literature searching, such as making the full text of *all* journals available for searching. Even with unlimited access to full text, the problems with effective literature mining are profound. Some of these problems stem from the limitation of language as a precise query tool—there is simply too much vocabulary to describe or specify the same target information. Some databases attempt to minimize the impact of this problem by the use of controlled vocabulary and gene nomenclature. But in the absence of such measures, flexible composition of queries becomes quite critical to obtain comprehensive coverage of a research area.

There are many commercial and publicly available tools and databases for mining scientific literature which vary in their data content. Some offer access to proprietary curated databases but they all employ essentially similar keyword-based interfaces with a facility for boolean operators to combine and subtract keywords.

2.7.1 PubMed

PubMed is the most widely used free literature searching tool for biologists. It forms part of the Entrez-integrated database retrieval system at the NCBI and is essentially a web interface to the Medline database which indexes >11 million journal abstracts. It also provides links to the full text of more than 1100 journals available on the web, although search queries are restricted to the text in abstracts. The interface allows the user to specify a search term (any alpha numeric string) and a search field (e.g. title, text word, journal or author). Queries retrieve abstracts from most of the major journals, although not all journals are indexed, particularly newer journals or journals with lower impact factors. There is a surprisingly stringent threshold applied before a journal will be considered for Medline indexing.

Many of the same guiding principles applied to searching the web also apply to PubMed, but there are some differences between this tool and other more general web search engines. Firstly the boolean operators are limited to the three main operators AND, OR and NOT. One major improvement over most web search engines is the availability of a wildcard function (*) to designate any character or combination of characters. The creative use of wildcards and boolean terms is important to widen the search without retrieving excessive and irrelevant results. For example, to find publications which present evidence of schizophrenia association on chromosome 8q21, an appropriate PubMed query might be [schizo* AND 8q*] searching the *text word* field. Using a wildcard search with 'schizo*' instead of 'schizophrenia' retrieves articles which mention schizoaffective, schizophrenia or schizophrenic, all of which may be relevant. By using a wildcard with

'8q' the search will retrieve nearby loci or larger loci which may encompass 8q21, e.g. 8q13–8q22. These are simple points but they are integral to a successful search strategy. Those using these facilities extensively will find additional searching guidelines on the NBCI website.

2.8 SEARCHING FULL-TEXT JOURNALS

Prospects for literature searching have improved recently with the greater availability of full-text articles. We have already described the advances offered by Scirus in searching full-text journals and the web simultaneously. Other highly recommended websites are HighWire which is approaching comprehensive coverage of available full-text journals and Medline (see below). However, searching scientific publications is still somewhat decentralized and there is still no completely comprehensive central tool to search all full-text journals, although it is possible to search the full text of most of the major genetics journals by visiting the top three or four major publishers. Table 2.4 lists the major sites which index the full text of a large range of science journals. As a benchmarking test we queried each tool, with a standard full-text query for the keyword [WNT], where searching Medline was also an option we identified the combined number of full text and Medline hits in parentheses. The highest number of results was retrieved with Scirus, however these results were very redundant. The HighWire tool seemed most effective in the benchmarking test, identifying a high number of hits with no redundancy.

2.8.1 HighWire

HighWire was set up as a non-profit making organization in 1995 by Stanford University to help universities and societies to publish on the web at low cost (Butler, 2000). Since its launch HighWire has expanded to become the world's second-largest scientific repository, after the US space agency NASA's Astrophysics Data System (which contains no biological information). Many journals available on the HighWire site make their content free immediately, or 1 or 2 years after print publication often coupled with an early view service for papers in press. In March 2002, HighWire had indexed 410,821 free full-text articles, derived from a list of 324 full-text journals. These are listed on the website along with Medline records from January 1948 through to April 2002. In our benchmark test

TABLE 2.4 Major Websites Providing Full-text Journal Access and Searching

Site/Publisher	Test Query Hits (with Medline)	URL
PubMed	(1615)	http://www.ncbi.nlm.nih.gov/entrez
Scirus	5061 (7015)*	http://www.scirus.com/
HighWire	2651 (3738)	http://highwire.stanford.edu/
Biomednet	1192 (2749)	http://www.bmn.com
ScienceDirect (Elsevier)	1264	http://www.sciencedirect.com
IDEAL	565	http://www.idealibrary.com/
Nature Publishing Group	255	http://www.nature.com/nature/
Wiley InterScience	196	http://www.interscience.wiley.com/

*Results from Scirus were redundant.

against other full-text search tools a comparative search of PubMed and HighWire with the keywords [Wnt16 OR Wnt-16] identified two papers with PubMed and eight papers with HighWire.

2.8.2 Literature Digests and Locus-specific Databases

The literature searching process can be simplified by searching locus-specific databases. The most widely used is On-line Mendelian Inheritance in Man (OMIM). As the name suggests, this focuses on Mendelian monogenic disorders, although it also offers some coverage of complex diseases. As a manually curated digest of the literature extracted from the full text of publications it can contain more information than PubMed. Although this has the disadvantage that not all entries are fully comprehensive or current, the database usually captures the most salient information and is therefore a good place to start. In addition OMIM is fully integrated with the NCBI database family. This facilitates rapid and direct linking between disease, gene sequence and chromosomal locus.

Other databases are available which provide curated information about genes and diseases which can also help to speed up the literature searching process. One of these is GeneReviews (www.geneclinics.org). This complements the molecular genetics emphasis of OMIM by offering a distinctly different focus. GeneReviews is a medical genetics information resource aimed at physicians and other healthcare providers. The site provides current, expert-authored, peer-reviewed, full-text articles describing the application of genetic testing to the diagnosis, management and genetic counselling of patients with specific inherited conditions. It also contains an international genetic testing Laboratory Directory and an international genetic and prenatal diagnosis Clinic Directory.

2.9 SEARCHING THE HEART OF THE BIOLOGICAL INTERNET – SEQUENCES AND GENOMIC DATA

So far we have reviewed a range of tools and approaches for searching the wider internet and the specialist scientific literature for biological information which may be useful for genetics. All of the tools reviewed so far may provide links, but will stop short of direct retrieval of actual biological database records, such as DNA or protein sequence records. This biological information is the heart of the biological internet. However, the flood of sequence data from genome sequencing has rapidly pushed biological sequence data beyond the reach of general internet searching tools. Instead sequence data can be searched and retrieved by using specialist bioinformatics tools based on sequence homology, map location, keyword, accession number and other features in the records. At a basic level this can be done by keyword searching using search tools such as, Entrez at the NCBI (Schuler *et al.*, 1996) or SRS at the EBI (Zdobnov *et al.*, 2002). Moving beyond simple searching methods the biological databases are constantly being updated and re-engineered to allow more powerful data query methods. These methods are covered in many other chapters throughout this book.

2.10 NUCLEOTIDE AND PROTEIN SEQUENCE DATABASES

There are three major organizations that collaborate to collect publicly available nucleotide and protein sequences. These organizations share data on a daily basis but they are distinguished by different international catchment areas for submissions, different formats and sometimes differences in the nature of their submitter annotations. Genbank is maintained by the NCBI in the United States (http://www.ncbi.nlm.nih.gov/Genbank/index.html). EMBL is maintained by the European Bioinformatics Institute in the United Kingdom (http://www.ebi.ac.uk/). The third member is the DNA Database of Japan (DDBJ) in Mishima, Japan (http://www.ddbj.nig.ac.jp/). All three organizations offer a wide range of tools for sequence searching and analysis but two integrated database query tools have become pre-eminent. These are Entrez from the NCBI and SRS from the EBI.

2.10.1 Entrez

Entrez (http://www.ncbi.nlm.nih.gov/Entrez/) is the backbone of the NCBI database infrastructure. It is an integrated database retrieval system that allows the user to search and browse all the NCBI databases through a single gateway. Entrez provides access to DNA and protein sequences derived from many sources, including genome maps, population sets and, as already described, the biomedical literature via PubMed and Online Mendelian Inheritance in Man (OMIM). New search features are being added to Entrez on a regular basis. Most recently facilities have been added to allow searches for DNA by 'ProbeSet' data from gene-expression experiments and for proteins by molecular weight range, by protein domain or by structure in the Molecular Modelling Database of 3D structures (MMDB).

2.10.2 Sequence Retrieval Server (SRS)

The sequence retrieval server (SRS) serves a similar role to Entrez, for the major European sequence databases. SRS is a flexible sequence query tool which allows the user to search a defined set of sequence databases and knowledge-bases by accession number, keyword or sequence similarity. SRS encompasses a very wide range of data, including all the major EMBL sequence divisions (Table 2.5). SRS goes one step further than Entrez by enabling the user to create analysis pipelines by selecting retrieved data for processing by a range of analysis tools, including ClustalW, BLAST and InterProScan.

2.11 BIOLOGICAL SEQUENCE DATABASES – PRIMARY AND SECONDARY

Anyone entering the heart of the biological internet encounters a bewildering number of accession numbers, identifiers and gene names. To get to grips with this flood of terminology it is important to understand the difference between primary and secondary databases and their associated accession numbers. This is not proposed as a rigorous definition but it does have a utility for understanding the information flow between sequence databases.

2.11.1 Primary Databases

Primary accession numbers have a number of key attributes; they refer to nucleic acid sequences derived directly from a sequencing experiment, the results are submitted by authors in a standardized format to GenBank, EMBL or DDBJ, the accession numbers are both unique and stable (if they are updated or amended by the submitting authors the accession number will signify a version change as .1, .2 etc.), the data records from every accession number can be retrieved, a contactable submitter is included in every record,

TABLE 2.5 Databases Indexed by the Sequence Retrieval Server at the EBI

Data Type	Database
Scientific literature	Medline, GO, GOA
Protein sequence libraries	European, Japanese and US protein patents, SWISS-PROT, SpTrEMBL
DNA sequence libraries	EMBL, Ensembl HUMAN, global DNA patents
Protein motifs	INTERPRO, PROSITE, PRINTS, PFAM, PRODOM, NICEDOM
DNA sequence related	UTR, UTRSITE, BLOCKS, TAXONOMY, GENETICCODE, REBASE, EPD, CPGISLAND, ENSEMBLCPG, UNIGENE
Transfac (Transcription factor analysis)	TFSITE, TFFACTOR, TFCELL, TFCLASS, TFMATRIX, TFGENE
Protein3DStruct	PDB, DSSP, HSSP, FSSP, RESID
Mutations	SWISSCHANGE, EMBLCHANGE, OMIM, HUMUT, HUMAN_MITBASE, P53LINK, Locus Specific Mutations (see Chapter 3)
SNPs	HGBASE, HGBASE_SUBMITTER
RH mapping	RHDB, RHEXP, RHMAP, RHPANEL
Metabolic pathways	LENZYME, LCOMPOUND, PATHWAY, ENZYME, EMP, MPW, UPATHWAY, UREACTION, UENZYME, UCOMPOUND
SRS pipelineapplications	FASTA, FASTX, FASTY, NFASTA, BLASTP, BLASTN, CLUSTALW, NCLUSTALW, PPSEARCH, RESTRICTIONMAP, HMMPfam, InterProScan, FingerPRINTScan, PfScan, BlastPRODOM, ScanRegExp

they are explicitly redundant in that all submissions are accepted regardless of partial or complete overlap with existing entries and lastly the growth rate remains close to exponential and now exceeds 16 million sequence records. The concept of authors' needs stretches to encompass consortia that run high-throughput sequencing projects. One of the most valuable and perhaps overlooked principals of these unique public repositories is that there is always (with the exception of patent data, see below) an identified individual or laboratory representative listed with the sequence record who can be contacted for any queries regarding experimental details, data quality and availability of source material. There is a large amount of information associated with primary sequence records. These include primary accession numbers, version numbers, protein ID numbers, gene identifier (GI) numbers, header records and feature identifiers. These cannot be covered in detail here but full descriptions are given in database guides (http://www.ebi.ac.uk/embl/index.html) and release notes (ftp://ftp.ncbi.nih.gov/genbank/gbrel.txt).

Geneticists should be encouraged to contact submitting authors in cases where anything seems non-obvious about primary data records for an mRNA or a finished genomic clone. They may have extra information that has a crucial bearing on the interpretation of genetic experiments. Authors may be difficult to track down if they have moved institutions but they are usually pleased to assist in the utilization of their data, because as with scientific publishing, this is the principle behind public sequence databases. Technical errors, anomalies, miss-annotation in submissions or artefacts are entirely the responsibility of submitting authors not the database administrators. Although we should be sanguine concerning anomalies in the high-throughput data divisions (EST, GSS, STS, HTG, HTC and SNP) if problems are pointed out authors can certainly amend or update their entries or in some cases may withdraw them. The primary data is deposited in good faith so authors should certainly not be harshly judged if an error has occurred in the rough and tumble of cloning, sequencing and submitter annotation. The exception to author responsibility for GenBank records is the patent division (gbPAT) where inventors are not equivalent to academic authors. These sequence records are processed by the US, European and Japanese patent offices and forwarded on to the databases. Although author contact may not be practical database users should be aware that patent applications are public documents and for an increasing number of gbPAT records the documentation can be accessed via the patent number on-line and free of charge (http://ec.espacenet.com/espacenet/ and http://www.uspto.gov/patft/). It is also possible to get to these patent full-text links directly from sequence entries via SRS.

2.11.2 Secondary Databases – Nucleic Acids and Proteins

By definition secondary databases are derived from the primary data. The word secondary should not be taken to imply lower value; indeed they include sources of the highest utility for genetic research. However they are defined, it is important to understand how they are linked back to the experimental data. The good news for geneticists is that there is now a comprehensive selection of high quality secondary databases that extract and collate subsets of mRNA, genomic or protein sequences from primary GenBank entries. The bad news is that the proliferation of features that make secondary databases so powerful also presents a bewildering range of options to the user. Testimony to both the good and bad news is given by the 2002 update of the Molecular Biology Database Collection (http://nar.oupjournals.org/cgi/content/full/30/1/1/DC1). This covers no less than 355 databases, up from 281 in 2001, of which the primary databases, GenBank, EMBL and DDJB, constitute only three entries. Although this compendium includes many non-human data sources almost all of these secondary databases contain information that could be pertinent to mammalian genetics. These review issues appear every January in *Nucleic Acids Research* and are definitely worth browsing. Are the genome portals secondary databases? This is where the definitions become blurred. Because NCBI generate their own genomic contig accessions (NT numbers) and Ensembl generate their own exon and gene identifiers they could be considered secondary databases. In so far as the UCSC genome portal marks up only external sequence record identifiers (primary and secondary) they are not strictly a secondary database. However, because they usefully give every type of gene prediction in the display a retrievable identity number, they could be considered as a secondary database.

The value of secondary databases includes the following:

- Distilling down a massive number of overlapping and/or redundant primary GenBank entries to a manageable range of genomic sections, unique transcripts and translated protein sequences
- Maintaining a running total of gene products, they partition human gene products and other vertebrates with extensive genomic data such as mouse, rat and zebra fish
- The inclusion of informative graphic displays for sequence features
- Providing access to a vast amount of pre-processed bioinformatic data

- Extensive interconnectivity through web hot-links
- Many of them are backed up by extensive institutional resources and expertise

However, users of these secondary databases also need to be aware of their shortcomings:

- They all suffer from the snapshot problem i.e. the time to re-build or update massive data sets means they are always out of date with respect to the new data cascading into the primary databases (given the complexity of the processes this is entirely expected but they often do not display the dates when the primary records were extracted)
- They all have different look-and-feel interfaces thereby necessitating regular practice to get the best out of them
- The web-based interoperativity can leave a lot to be desired; e.g. broken links, link-outs to databases that are not maintained to the same standards and overkill by linking out to too many similar sources
- Their automated annotation schema can be confounded by sequence artefacts (Southan *et al*., 2002)
- The overlap between utility and content between major databases is extensive but is never enough for any of them to be the mythical 'one-stop-shop'
- Non-redundant transcript and protein collections may seem conceptually similar but because they diverge in schema details and update frequency they all give different statistics
- Some secondary databases such as SwissProt keep sequence identifiers both unique and stable but for technical reasons others, such as UniGene EST clusters or Ensembl genes, may change identifiers between builds
- Many specialized 'boutique' databases are never updated when their originators move on or run out of resources
- Last but not least some secondary databases that initially had free access can become commercial and require a subscription fee

2.11.3 Nucleic Acid Secondary Databases

For the analysis of their results the geneticist must become acquainted with these feature-rich sources of gene product information. A key example, based around nucleic acid sequence but including protein of secondary databases is LocusLink/RefSeq (LLRS) for mRNAs. The LLRS system is built round a reference sequence (RefSeq) which is usually the longest available mRNA of those coding for the same protein. RefSeq includes splice variants and if only genomic sequence is available, such as for many of the 7TM receptors, the system defaults to the predicted coding sequence annotated as a 'CDS' in the database entry. For example there is no experimentally determined human rhodopsin mRNA in GenBank, only a model mRNA predicted from the genomic sequence U49742. This presents an immediate problem for the geneticist, as the untranslated region (UTR) of the rhodopsin locus, which defines the boundaries and functional regions of the gene may be extensive. Chapter 4 takes a detailed look at approaches to help define the true extent of gene loci.

The end-product of the RefSeq pipeline is a unique mRNA, coding sequence (CDS), or set of splice variants for those gene products where data or predictions are available.

The LocusLink side of things, as suggested by the title, is directed towards mapping the RefSeq gene products onto the genomic sequence and checking the consistency between the two. LocusLink has linked sections of key importance to the geneticist. These are: variation which assigns SNP data, OMIM which includes verified monogenic disease links, homologene which indicates close homologues in other species, UniGene which specifies ESTs clusters associated with the gene product, and PubMed that links to all publications that can be specifically linked to the primary GenBank accession numbers. There are also links to all three genome portals, NCBI, UCSC and Ensembl. There has been some confusion in the past where the portals could not synchronize their builds and track displays with GP version updates but this problem has been addressed and they should all be on version 28 (from December 2001) at the time of writing.

The RefSeq identifier is secondary in the sense that it is a supplementary identifier assigned to one particular mRNA chosen as the reference sequence. These accession numbers have the prefix NM_ for mRNA entries and NP_ for protein entries. The LocusLink/RefSeq system goes one step further in assigning a third identifier, XM_ for nucleic acid and XP_ for proteins, which are the genomic counterparts of the NM and NP numbers. A BLAST search against the NCBI protein database will show all three entries, the primary accession number, the NM_ and the XM_ entries. There is the added complication that the XP_ sequences have a variable evidence support level and include *ab-initio* genomic predictions both with and without EST support. Secondary accession numbers are also important for ESTs. ESTs can be considered as mRNA fragments that, with sufficient sampling (now just exceeding 4 million human entries in dbEST) can be clustered or assembled to form a contiguous extended transcription product and in some cases, the splice variants from the tissue types sampled for EST preparation. The main post-genomic utility of EST collections is as exon detectors. In addition to splice variants these can reveal possible gene transcription activity where no extended mRNA has been experimentally verified. The primary data source for ESTs is the dbEST division of GenBank.

The geneticist should be aware of two major secondary EST databases, UniGene (Wheeler *et al.*, 2002) and the TIGR human gene index (Liang *et al.*, 2000). The principles by which these different databases are constructed, are explained in the appropriate source references but in fact they both converge to a similar set of 'virtual' surrogate transcripts. In the TIGR case, the virtual transcripts assembled from overlapping ESTs can be retrieved; in the Unigene case, the individual EST reads can be batch downloaded. As with most secondary databases, built from the same source data, the two databases have both overlap and complementarity. The TIGR assemblies are particularly useful for extending the 3′ UTR of known mRNAs but the assemblies are re-compiled at long time intervals. UniGene is updated more frequently and is fully interlinked to the LocusLink/RefSeq system but the clusters are built on mRNAs from the preceding version of GeneBank.

2.11.4 STSs and SNPs

These are two of the most important data sources for the geneticist involved in disease mapping. The dbSTS database contains sequence and mapping data on short genomic landmark sequences. Although they have a primary sequence record and GB accession number they also have a number of alternative marker names. These have been cross-referenced into a secondary database called UniSTS that integrates all available marker and mapping data (http://www.ncbi.nlm.nih.gov/genome/sts/). The dbSNP database is an interesting exception in that it is not a division of GenBank so it is not strictly a primary database. The submissions (SS numbers) are equivalent to a primary record but overlapping sequences with the same polymorphism are collapsed into the Reference SNP Cluster Report with an RS number. This can be considered a secondary database where the RS numbers are non-redundant and stable. These RS numbers, currently at 2,640,509 for human, are integrated with other NCBI genomic data and primary GenBank records containing overlapping sequences deduced or stated to be from the same location. The HGVbase has a smaller set of 984,093 highly curated records (http://hgvbase.cgb.ki.se/). They have their own secondary accession/ID number and these can be queried and retrieved from the Ensembl genome annotation. Chapter 3 presents detailed examination of the major databases of genetic variation.

2.11.5 Protein Databases and Websites

A website of central importance in protein analysis is the Expert Protein Analysis System (ExPAS; http://www.expasy.ch/). In addition to protein analysis tools, such as PROSITE (http://www.expasy.ch/prosite/) and Swiss-3Image (http://www.expasy.ch/sw3d/) Swiss-Prot protein database contains high-quality annotation and web-linked cross-references to 60 other databases. It is accompanied by TrEMBL, a computer-annotated supplement that contains the translations of all coding sequences present in primary nucleotide sequence databases not yet in SwissProt. Sequence records are merged where possible to minimize the redundancy. Sequence conflicts and splice variants are indicated in the feature table of the corresponding entry. The combined database is referred to as SwissProt/TrEMBL (SPTR). Amongst the links in SPTR it is worth mentioning the InterPro system which is of very high utility for finding protein family-specific domain matches (Apweilwer *et al.*, 2000). Acquiring this information is one of the main goals of the bioinformatic analysis of proteins so it is useful to find that this piece of the work is already done and updated with new releases of InterPro. Other major sites provide PFAM, PROSITE, and other tools for protein sequence analysis. The Sanger Institute (http://www.sanger.ac.uk/) provides access and maintains PFAM and multiple other useful links and genomic tools, including three-dimensional protein structure prediction (http://genomic.sanger.ac.uk/123D/123D.shtml).

Any division between the universe of DNA and protein sequences is clearly artificial. Protein information can be accessed from within the LLRS system, just as it is also possible to link out to primary nucleic acid sequence record accession numbers from SPTR. However, the complementarity between LocusLink/RefSeq and SPTR is clear. The focus is on nucleic acid sequences in the former and protein sequences in the latter. The message for the user is that both sources will be essential for interpreting the results of genetic experiments.

2.12 CONCLUSIONS

In this chapter we have introduced the major data sources available on the internet that geneticists increasingly need to access for their research. The choice was based on our direct working experience of their utility. Rather than restrict ourselves to just cataloguing these, we have also included some discussion of the principles behind the organization of biological data, such as the concept of primary and secondary sequence databases. We have also demonstrated the power of web search engines, both of the specialist and common variety. Mastering these is essential for interrogating biological resources on the internet. They also allow the user to search for new developments, tools and databases. This is something we strongly recommend to future-proof your own research, even if we cannot future-proof this book!

REFERENCES

Apweiler R, Attwood TK, Bairoch A, Bateman A, Birney E, Biswas M, *et al.* (2000). InterPro — an integrated documentation resource for protein families, domains and functional sites. *Bioinformatics* **16**: 1145–1150.

Berners-Lee T, Fischetti M, Dertouzos M. (1999). *Weaving the Web: The Original Design and Ultimate Destiny of the World Wide Web by Its Inventor*. Harper: San Francisco.

Butler D. (2000). Biology back issues free as publishers walk HighWire. *Nature* **404**: 117.

Hubbard T, Barker D, Birney E, Cameron G, Chen Y, Clark L, *et al.* (2002). The Ensembl genome database project. *Nucleic Acids Res* **30**: 38–41.

Liang F, Holt I, Pertea G, Karamycheva S, Salzberg SL, Quackenbush J. (2000). Gene index analysis of the human genome estimates approximately 120,000 genes. *Nature Genet* **25**: 239–240.

Southan C, Cutler P, Birrell H, Connell J, Fantom KG, Sims M, *et al.* (2002). The characterization of novel secreted Ly-6 proteins from rat urine by the combined use of two-dimensional gel electrophoresis, microbore high performance liquid chromatography and expressed sequence tag data. *Proteomics* **2**: 187–196.

Schuler GD, Epstein JA, Ohkawa H, Kans JA. (1996). Entrez: molecular biology database and retrieval system. *Methods Enzymol* **266**: 141–162.

Wheeler DL, Church DM, Lash AE, Leipe DD, Madden TL, Pontius JU, *et al.* (2002). Database resources of the National Center for Biotechnology Information: 2002 update. *Nucleic Acids Res* **30**: 13–16.

Zdobnov EM, Lopez R, Apweiler R, Etzold T. (2002). The EBI SRS server — recent developments. *Bioinformatics* **18**: 368–373.

CHAPTER 3

Human Genetic Variation: Databases and Concepts

MICHAEL R. BARNES

GlaxoSmithKline Pharmaceuticals
Harlow, Essex, UK

Bioinformatics for Geneticists. Edited by M.R. Barnes and I.C. Gray
 ISBNs: 0 470 84393 4; 0 470 84394 2 (PB)

3.1 INTRODUCTION

Genetic variation is a key commodity for geneticists; not only as the much sought after basis of heritable phenotype, but also as a marker to aid in this search. For the wider biological research community, information on genetic variation can tell us many things about the functional parameters and critical regions of a gene, protein, regulatory element or genomic region. Study variation and a picture of the driving force of evolution begins to emerge. This knowledge can not only help us elucidate the function of genes and pathways by studying their function and dysfunction in normal and diseased states, it can also help us to understand the origins and diversity of mankind and other organisms. The availability of a complete human genome sequence finally puts this variation into context with all other biological data. In this chapter we will present an overview of the many forms of genetic variation, we will review current and past trends in the use of this data and highlight the key databases from which this data can be accessed and manipulated.

3.1.1 Human Genetic Variation

Human genetic variation and our environment are the two key factors that make each and every one of us different. Genetic variation takes many forms, although these variants arise from just two types of genetic mutation events. The simplest type of variant results from a single base mutation which substitutes one nucleotide for another. This mutation event accounts for the commonest form of variation, single nucleotide polymorphisms (SNPs). Many other types of variation result from the insertion or deletion of a section of DNA. At the simplest level this can result in the insertion or deletion of one or more nucleotides, so-called insertion/deletion (INDEL) polymorphisms. The most common insertion/deletion events occur in repetitive sequence elements, where repeated nucleotide patterns, so-called 'variable number tandem repeat polymorphisms' (VNTRs), expand or contract as a result of insertion or deletion events. VNTRs are further subdivided on the basis of the size of the repeating unit; minisatellites are composed of repeat units ranging from 10 to several hundred base pairs. Simple tandem repeats (STRs or microsatellites) are composed of 2–6-bp repeat units. The rarest insertion/deletion events involve deletions or duplications of regions ranging from a few kilobases to several megabases. These forms of variation were once thought to be restricted to rare genomic syndromes, however, sequencing of the human genome has presented a great deal of evidence to suggest that these events may be more common than previously expected.

The quantity of genetic variation in the human genome is something that until recently we have only been able to estimate by an educated guess. Empirical studies quite quickly identified that on average, comparison of chromosomes between any two individuals will generally reveal common SNPs (>20% minor allele frequency) at 0.3–1-kb average intervals, which scales up to 5–10 million SNPs across the genome (Altshuler *et al.*, 2000). The availability of a complete human genome has helped us considerably to estimate the number of potentially polymorphic STRs and minisatellites, as VNTRs over a certain number of repeats can be reliably predicted to be polymorphic. Viknaraja *et al.* (unpublished data) completed an *in silico* survey of potentially polymorphic VNTRs in the human genome and identified over 100,000 potentially polymorphic microsatellites. Other forms of variation such as small insertion/deletions are more difficult to quantify, although they are likely to fall somewhere between SNPs and VNTRs in numbers. Large deletions or duplications are the most unquantifiable form of variation in the genome. Quantification of these forms of variation is only possible by intensive cytogenetic methods (Gratacos *et al.*, 2001). They cannot be reliably identified from the genome sequence; in fact they are implicitly an obstacle to genome assembly, as large duplications are often incorrectly collapsed into a single assembly.

This huge quantity of genetic variation in the human genome led many to question the origin and maintenance of such a 'genetic load' in the human population. The traditional belief that most mutation was deleterious and subject to selection was quickly challenged by this data. In response to this observation Kimura (1983) and others formulated a 'neutral theory of evolution'. This theory proposed that most sequence variation does not directly impact phenotypic variation and so is not directly subjected to the forces of selection. Thus, the overwhelming majority of genetic variants are likely to be phenotypically neutral, while many will define the diverse phenotypes that define individual humans. However a certain undefined number of these alleles will have deleterious effects, either directly causing or increasing susceptibility to disease. Some of this variation, so-called mutations, will be rare in populations whilst others will be common, so-called polymorphisms that increase susceptibility to common diseases. It will not usually be possible to identify these deleterious alleles directly, instead genetics has developed around the concept of using markers to detect nearby deleterious alleles. Fortunately for geneticists, the huge quantity of common polymorphism across the human genome makes it very likely that one or more of these polymorphisms will be in close enough vicinity to a rarer disease allele to detect it by common co-inheritance (linkage disequilibrium) between the two alleles.

Thus, one of the primary objectives of genetics is to utilize polymorphisms across the genome as markers which show co-inheritance with the phenotype under study. SNPs are the most obvious choice for these studies as they are the commonest form of human variation. However this choice has not always been so clear. Despite the abundance of SNPs in the genome, without knowledge of the genome sequence, SNP identification is a laborious process. This has made SNP availability very limited until very recently. Instead geneticists have used microsatellites as markers. These highly polymorphic markers can be isolated by relatively simple molecular methods and can detect disease-causing mutations in family-based studies over a larger distance than SNPs, often over 20 MB. The extent of this linkage enables whole genome linkage studies with as few as 200–500 microsatellite markers. Such linkage studies have been very successful in mapping mutations causing single gene disorders or Mendelian traits, but have been largely unsuccessful in detecting the multiple genes responsible for the commoner complex diseases (Risch, 2000).

The primary approach proposed for mapping complex disease genes is to use markers to detect population-based allelic association or linkage disequilibrium (LD) between markers and disease alleles (see Chapter 8 for a detailed exploration of this area). These associations can be very strong even where the corresponding family linkage signal is weak or absent. This approach can localize disease alleles to very small regions, based on localized LD, which on average extends between 5–100 kilobases (kb) depending on a range of factors (Reich *et al.*, 2001). Detection of this association demands a massive increase in marker density with 200,000–500,000 markers estimated to be needed to cover the genome for an association scan compared to the 200–500 markers needed for a family-based linkage scan.

These population-based association studies call for ultra high-throughput genotyping methods. Technology developments to date suggest that SNPs are likely to be the most viable option for these studies for a number of reasons, but primarily because SNPs are more tractable to automated high-throughput analysis than microsatellite markers. Until very recently demand for SNPs completely outstripped SNP availability and so whole genome SNP association studies simply could not be attempted. This situation is now changing—completion of the genome has enabled several large-scale SNP discovery projects. Genetics is now entering a promising new era where marker resources and locus information are no longer the main factors limiting the success of complex disease gene hunting. The emphasis is now on good study design and carefully ascertained study populations. Effective informatics is critical to effectively exploit this data. More than ever, geneticists will need to be competent users of bioinformatics tools to construct sophisticated marker maps that can detect the full complexity of human genetic variation.

To find disease associations and ultimately disease alleles, it is necessary to study genetic variation at increasing levels of detail. At first, markers need to be identified at a sufficient density to build marker frameworks to detect linkage or association across the genome. Once this linkage or association is detected a denser framework of markers is needed to refine the signal. In the case of linkage analysis, marker density may not need to be increased beyond a few hundred kilobases as linkage is likely to remain intact over considerable distances in families. However in the case of association, marker density needs to be increased to a level at which all haplotype diversity in a population is captured (see Chapter 8). This may call for the construction of very dense marker maps down to a resolution of 5–10 kb. Ultimately, once LD is established between a marker and a phenotype it is necessary to identify all genetic variation across the narrowed locus, hopefully allowing the identification of the disease allele. This increasing resolution of analysis may involve a progression of bioinformatics tools and increasing ingenuity in the use of these tools as the requirements for detail increase. Variation can take many forms, any of which may have a bearing on the genetic mechanisms of disease. The very act of characterizing variation across a locus may help to cast light upon its genetic nature and the possible nature of the phenotype. For example, some genomic regions show hypermutability, while others show very low levels of mutation or polymorphism. The reasons for these differences are poorly understood, they may be based upon the physical properties of chromosomes, evolutionary selection or other unknown influences, all of which may have a bearing on disease.

3.1.2 The Genome as a Framework for Integration of Genetic Variation Data

Bioinformatics offers some powerful tools for detecting, organizing and analysing human genetic variation data. The value of these tools is totally dependent on the underlying quality and organization of the data. Ideally, variation data needs be available in an organized and centralized form that will allow complex queries and integration with other data sources. Without the benefit of a complete genome, such integration was little more than a pipe dream, but now we are presented with an opportunity to integrate data on the sequence framework. Generally it takes only two 20–30 base pairs of flanking sequence to unambiguously locate a sequence feature such as an SNP in the genome. This bioinformatics process is called electronic PCR (ePCR) and it is completely analogous to laboratory-based PCR. Two primers are used to map a sequence feature (e.g. a SNP). To validate the position both primers must map in the same vicinity spanning a defined distance, effectively producing an electronic PCR product. The possibilities for data integration are immense. For genetics, exact base pair localization of each variant allows the construction of absolutely precise physical maps, which can be accurately integrated with genetic maps. It is now possible to take a given region and place SNPs, mutations, microsatellites and insertion/deletions in exact order. Without a sequence map this simply would not have been possible as each marker may have been mapped by different laboratory methods — producing few directly comparable results (see Chapter 7 for a discussion of map integration issues).

3.2 FORMS AND MECHANISMS OF GENETIC VARIATION

In silico (bioinformatic) analysis of human sequence presents an opportunity to identify genetic variants by comparison of differences between two sequences. Most obviously *potential* SNPs can be identified by comparison of two sequences; these could be expressed sequence tags, cDNAs or genomic sequences. The same method can also be used to identify *potential* INDEL polymorphisms. *Potential* is a key word to apply to this *in silico* polymorphism discovery process which can be prone to false positives introduced by sequencing error and other issues (see Chapter 10).

Human genome sequence also gives us an opportunity to assess some of the less commonly studied forms of variation. Although under-represented in databases some potential forms of variation can be identified from a single DNA sequence, by sequence alone. Short tandem repeat sequences are the most obvious example of such variants, however, sequence analysis can also be used to identify minisatellites and segmental duplications which may also mediate large deletions or duplications. Our knowledge of these forms of variation is limited; this reflects studies to date which have focused on more technically tractable variants, such as SNPs, mutations and short tandem repeats. Databases have also as a matter of practicality tended to focus on these classes of variation, and in this chapter we will review these databases in detail. We will also attempt to draw the less studied forms of variation into context, reviewing the best tools to access this data. Where no database exists we will review the mechanisms which govern variation and which can assist detection by bioinformatics methods.

3.2.1 Single Nucleotide Variation: SNPs and Mutations

Terminology for variation at a single nucleotide position is defined by allele frequency. In the strictest sense, a single base change, occurring in a population at a frequency of $>1\%$ is termed a single nucleotide polymorphism (SNP). When a single base change occurs at $<1\%$ it is considered to be a mutation. However, this definition is often disregarded, instead 'mutations' occurring at $<1\%$ in general populations might more appropriately be termed low frequency variants. The term 'mutation' is often used to describe a variant identified in diseased individuals or tissues, with a proven role in the disease phenotype.

Mutation databases and polymorphism databases have generally been divided by this definition. Polymorphisms are generally considered widespread in populations and mutations are usually rare and are not generally thought to be spread widely in populations, but instead occur sporadically or are inherited in families in a Mendelian manner. A grey area exists, which argues against the rigidity of this division of data. Some autosomal recessive Mendelian mutations have been linked to complex disease susceptibility in a heterozygote form and indeed are relatively widely spread in populations. For example, homozygote mutations in the cystathione beta synthase gene cause homocystinuria, a rare disorder inducing multiple strokes at an early age. The heterozygotes do not share this severe disorder, but do have an increased lifetime risk of stroke (Kluijtmans *et al.*, 1996). In Caucasians the population frequency of homozygote homocystinuria mutations is only 1/126,000, but in the same population, heterozygote frequency is relatively high at 1/177. There are many other examples of 'Mendelian mutations' which actually exist at appreciable heterozygote levels in general populations, particularly isolated populations, e.g. mutations in the breast cancer susceptibility gene, BRCA1, have been found in 1–2% of Jewish populations (Bahar *et al.*, 2001) and mutations in the CFTR gene cause cystic fibrosis, the most common autosomal recessive disease in the Caucasian population, with a carrier frequency of around 2% (Roque *et al.*, 2001).

3.2.1.1 The Natural History of SNPs and Mutations

The presence of heterozygous 'Mendelian mutations' in general populations illustrates the point that it may not always be helpful to rigidly separate polymorphism and mutation data. Another factor which argues against division of these data is that both SNPs and mutations arise by the same mechanism, although selection may influence their spread in populations. Miller and Kwok (2001) presented a detailed review of the 'life cycle' of a single nucleotide variation, they defined SNP and mutation evolution in four phases (Figure 3.1):

(1) Appearance of a new variant allele by mutation
(2) Survival of the allele through early generations against the odds
(3) Increase of the allele to a substantial population frequency
(4) Fixation of the allele in populations

Each of these stages goes to the heart of the differences and similarities between SNPs and mutations. Both arise by the same mechanism; nucleotide substitution is DNA sequence context dependent — substitution rates are influenced by 5′ and 3′ nucleotides. This effect is most dramatic for CT and GA transitions; these CpG dinucleotides are methylated and tend to deaminate to either a TpG or CpA dinucleotide (Cooper and Youssoufian, 1988). This makes these dinucleotides the most likely locations for point mutation in the human genome, with $G > A$ or $C > T$ transitions accounting for 25% of all SNPs and mutations in the human genome (Miller and Kwok, 2001). In itself this molecular mechanism accounts for the deficiency of CG dinucleotides in the human genome. The creation of new CG dinucleotides is not an adequate counter balance against this effect, due to the lower frequency of tranversions back to CpG. While SNPs and mutations both arise in the same way, their survival in populations is likely to be quite different. Most newly arisen SNPs and mutations are likely to be lost in early generations by random sampling of the gene pool alone. For example if a heterozygous individual for a selectively neutral mutation has two offspring, there is a 0.75 probability that the mutation will be found in at least one child. If each generation has two children, the probability of loss of the new mutation is $1-(0.75)^g$, where g = generations. To give a worked example, this relates to a 94% probability of loss of a mutation or SNP in 10 generations (approximately

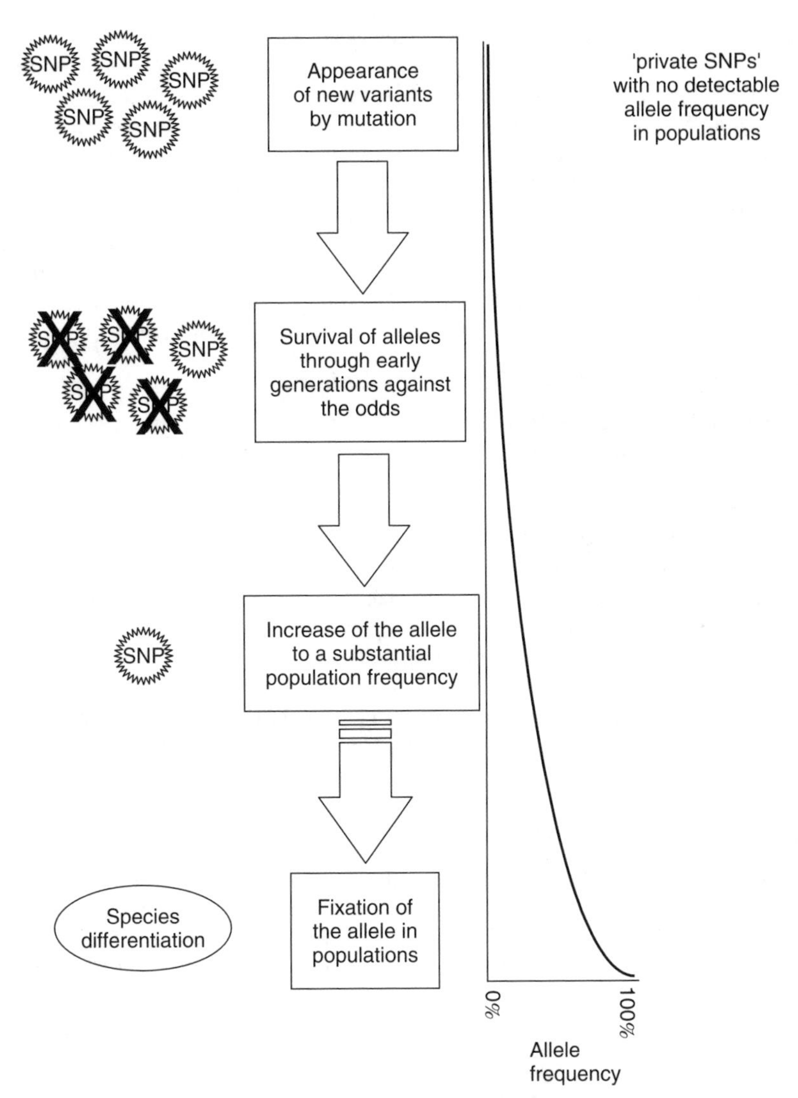

Figure 3.1 The life cycle of SNPs and mutations. SNP and mutation evolution occurs in four main phases: (1) appearance of a new variant allele by mutation; (2) survival of the allele through early generations against the odds; (3) increase of the allele to a substantial population frequency; (4) fixation of the allele in populations.

200 years). Where a heterozygous mutation has an early onset deleterious effect, natural selection is likely to further increase the rate of loss of the allele from populations. The same pressures do not apply to late onset diseases, perhaps explaining the proliferation of such diseases in humans.

If an SNP or mutation survives early generations and increases in frequency sufficiently to become homozygous in some individuals the risk of loss of the allele is reduced. At this stage the frequency of the allele in a population is likely to vary, with higher frequency alleles being consistently favoured, especially when populations are subject to severe bottlenecks in size. Reich *et al.* (2001) presented convincing evidence for such a bottleneck in recent Northern European population history. In the face of these fluctuations of allele frequency, an SNP or mutation will cease to exist in populations, either by disappearing or by reaching a 100% allele frequency, in which case the variant becomes an allele that helps to define a species. Interestingly there is no evidence of shared SNPs between species, a study of variation between the human and orang-utan X chromosome found that 2.9% of nucleotide sites differ, but no SNPs were shared (Miller *et al.*, 2001). This suggests that the lifetime of an SNP is considerably shorter than the divergence of these two species. Based on this data, Miller *et al.* (2001) estimated that the average period from original mutation to species fixation of an allele was 284,000 years.

3.2.1.2 SNP and Mutation Databases United?

The high level of interest in SNP data has led to the development of an excellent central SNP database — dbSNP at the NCBI (Sherry *et al.*, 2001). Mutation databases are still lagging behind SNPs in terms of data integration and visualization on the human genome. However the many commonalities between these two forms of data may have inspired the SNP database HGBase to re-align and rename itself HGVBASE — a central database of human genetic variation including SNP and mutation data (Fredman *et al.*, 2002). This is a valuable step which will make mutation data much more accessible to geneticists in a well-integrated form. Other highly specialized mutation databases exist, including HGMD, GDB and a large range of locus-specific databases. It is not yet clear to what extent mutation and SNP data will be integrated, but the availability of a complete human genome presents an unbeatable opportunity to bring these two sources of data together in a genomic context, without compromising the necessary integrity of either form of data.

3.2.2 Tandem Repeat Polymorphisms

Tandem repeats or variable number repeat polymorphisms (VNTRs) are a very common class of polymorphism, consisting of variable length sequence motifs that are repeated in tandem in a variable copy number (Figure 3.2). VNTRs are only surpassed in quantity by SNPs in the human genome. They have been found in all organisms studied, although they

Repeat type	Example
Mononucleotide	**A**A**A**A**A**A**A**A**A**A**A**A**A**A**A**A**A**A
Dinucleotide	**CA**CA**CA**CA**CA**CA**CA**CA**CA**
Triplet/trinucleotide	**CAG**CAG**CAG**CAG**CAG**CAG**CAG**CAG**CAG**
Tetranucleotide	**TAAG**TAAG**TAAG**TAAG**TAAG**TAAG**TAAG**
Pentanucleotide etc.	**GAATT**GAATT**GAATT**GAATT**GAATT**GAATT
Repeat terminology	**Example**
Perfect STR	CACACACACACACACACACACACACACA
Imperfect STR	CACA**T**ACACACACACACAC**G**CACACACA
Interrupted STR	CACACACACAC**GGG**CACACACACACACA
Compound STR	CACACACACACACA**TGTGTGTGTGTGTG**

Figure 3.2 Tandem repeat types and terminology.

tend to occur at higher frequencies in organisms with large genomes. Viknaraja *et al.* (unpublished data) analysed the draft human genome sequence (December 2001 freeze) and identified several hundred thousand potentially polymorphic VNTRs. However there is little or no information on the heterozygosity and polymorphic nature of the vast majority of these polymorphisms. VNTRs have traditionally been subdivided into subgroups based on the size of the tandem repeat unit. Repeated sequences of one to six bases are termed microsatellites or short tandem repeats (STR), larger tandem repeats in units of 14–100 bp are termed minisatellites. Microsatellites and minisatellites are generally thought to show different mutational mechanisms which are influenced by sequence properties and lengths. In microsatellites the predominant mutational mechanism is thought to be DNA slippage during replication. In minisatellites the predominant mechanism appears to be gene conversion and unequal crossing over (Goldstein and Schlotterer, 1999). The distinction between microsatellites and minisatellites is somewhat arbitrary for repeat units between 7 and 13 bp and it has been suggested that highly repeated sequences or sequences which are more likely to form loops in these size categories should be called minisatellites. This somewhat vague definition may be academic, in effect microsatellites and minisatellites have quite different properties, dictated by their repeat size, copy number and the perfection of the repeat. For the specific needs of a genetic study it may be necessary to pick the tandem repeat which conforms most closely to the heterozygosity requirements for the marker (see Chapter 8). The polymorphic nature of a VNTR is thought to depend upon a range of factors: the number of repeats, their sequence content, their chromosomal location, the mismatch repair capability of the cell, the developmental stage of the cell (mitotic or meiotic) and/or the sex of the transmitting parent. (Debrauwere *et al.*, 1997).

Aside from their utility as highly polymorphic genetic markers, much evidence exists to demonstrate that tandem repeats exert a functional effect when located in or near gene coding or regulatory regions. Thus VNTRs in themselves can be candidates for disease-causing genetic variants. The best characterized of these are the triplet repeat expansion diseases. Triplet repeat expansion is an insertion process that occurs during meiosis. Insertion of new repeats is strongly favoured over loss of repeats — pathological triplet repeat expansions manifest through successive generations with worsening symptoms known as ‘anticipation’, as the repeat expands with increasingly pathological results. Most triplet repeat expansions have been identified in monogenic diseases and may occur in almost any genic region. Over five triplet repeat classes have been described so far, causing a range of diseases including, Fragile X, myotonic dystrophy, Friedreich’s ataxia, several spinocerebellar ataxias and Huntington’s disease (Usdin and Grabczyk, 2000). Spinocerebellar ataxia 10 (SCA10) is notably caused by the largest tandem repeat seen in the human genome (Matsuura *et al.*, 2000). In general populations the SCA10 locus is a 10–22mer ATTCT repeat in intron 9 of the SCA10 gene; in SCA10 patients, the repeat expands to >4500 repeat units, which makes the disease allele up to 22.5 kb larger than the normal allele.

Tandem repeats have also been associated with complex diseases, for example different alleles of a 14mer VNTR in the insulin gene promoter region, have been associated with different levels of insulin secretion. Different alleles of this VNTR have been robustly linked with type I diabetes (Lucassen *et al.*, 1993) and in obese individuals they have also been associated with the development of type II diabetes (Le Stunff *et al.*, 2000). Kubota (2001) took the concept of triplet repeat anticipation to an extreme by suggesting that every human chromosome suffers from a burden of accumulating trinucleotide repeats. Thus, he predicted the ‘mortality’ of human chromosomes with the passage of generations, eventually leading to a deficiency of replication and to the mortality of *Homo sapiens* as a species! This is certainly a controversial theory, but the basic concept is interesting and illustrates that the burden of VNTR-mediated genetic disease is only likely to increase.

The value of tandem repeats as markers and functional elements is clear, although for practical reasons the focus of genetics is shifting to SNPs. However, VNTR markers will probably continue to be a fundamental tool and to overlook them could be unwise, as often a highly polymorphic VNTR may be more informative than several SNPs. In comparison to the relatively low heterozygosity of SNPs, much less dense VNTR maps are needed to match the equivalent detection power of a high density SNP map (see Chapter 7). A single polymorphic VNTR may even be as informative as a complex SNP haplotype. The drawback of tandem repeats are mainly technological—detection methods cannot currently match the highly automated microtitre plate-based or DNA chip-based assays that have characterized modern SNP assay development, although technology developments may eventually alter this situation (Krebs *et al.*, 2001).

In comparison to the hundreds of thousands of VNTR polymorphisms in the genome, only 18,000 VNTRs have been genetically characterized. Several highly characterized subsets of these markers have been arranged into well-defined linkage marker panels by the Marshfield Institute and Genethon (see Chapter 7 for details). These panels vary in marker spacing to allow different density genome scans. Almost all genetically characterized VNTRs are stored centrally in several sources, including GDB, CEPH and dbSTS (see below). Potentially polymorphic novel VNTRs can be identified from genomic sequence using the tandem repeat finder tool (Benson, 1999; http://c3.biomath.mssm.edu/trf.html). A complete analysis of the human genome sequence using tandem repeat finder is presented in the UCSC human genome browser in the 'simple repeats' track (see Chapter 9).

3.2.3 Insertion/Deletion Polymorphisms and Chromosomal Abnormalities

While tandem repeat polymorphisms are in themselves a major form of variation in genomes, they may also mediate other forms of variation by predisposing DNA to localized rearrangements between homologous repeats. Such rearrangements give rise to Insertion/Deletion (INDEL) polymorphisms. Indels appear to be quite common in most genomes studied so far, this probably reflects their association with common VNTRs. Indels have been associated with an increasing range of genetic diseases, for example, Cambien *et al.* (1992) found association between coronary heart disease and a 287-bp Indel polymorphism situated in intron 16 of the angiotensin converting enzyme (ACE). This Indel, known as the ACE/ID polymorphism, accounts for 50% of the inter-individual variability of plasma ACE concentration. The molecular mechanism of insertion/deletion polymorphism is still poorly understood, many different molecular mechanisms may account for an Indel event, although most are likely to be DNA sequence dependent. As discussed earlier, localized sequence repetitiveness in the form of direct tandem repeats or inverted repeats or 'symmetric elements', have been shown to predispose DNA to insertion/deletion events (Schmucker and Krawczak, 1997). Darvasi and Kerem (1995) found evidence to suggest that slipped-strand mispairing (SSM) was a common mechanism for insertion/deletion events. Analysis of sequences surrounding 134 disease-causing Indel mutations in the coding regions of three genes, the cystic fibrosis transmembrane conductance regulator, beta globin and factor IX, found that 47% of Indel mutations occurred within a unit repeated tandemly two- to seven-fold. The proportion of SSM mutations was significantly higher than expected by chance. The estimated net proportion of deletion and insertion mutations attributed to SSM was 27%. Further mechanisms have been proposed; Deininger and Batzer (1999) suggested that many INDELs may be caused by the insertion of Alu elements, which number in excess of 500,000 copies in the human genome providing abundant opportunities for unequal homologous recombination events.

Although Indel polymorphisms are likely to be very widely distributed throughout the genome, relatively few have been characterized and there is no central database collating this form of polymorphism. The Marshfield website maintains the most comprehensive single source of short insertion/deletion polymorphisms (SIDPs), over 2000 are maintained in a form which can be searched by chromosome location. Other databases such as dbSNP and HGVBASE also capture SIDPs to some extent. Larger Indels are generally overlooked in databases unless associated with a specific gene or study, in which case they appear in GDB, OMIM and other similar sources.

3.2.4 Gross Chromosomal Aberrations

While minor Indel polymorphisms are thought to be relatively common in human populations, gross chromosomal abnormalities such as deletions, inversions or translocations were thought to be rare. Nevertheless as our knowledge of the genome develops an increasing number of clinically characterized genomic syndromes are being identified. Some of these affect multiple genes and cause pronounced phenotypes including velocardiofacial syndrome (VCFS) a deletion syndrome on 22q11.2 (Gong *et al.*, 1996) and Charcot-Marie-Tooth disease type 1A (CMT1A) a duplication syndrome on 17p11.2 (Thomas, 1999). Other much more subtle genomic syndromes are emerging which suggest that these syndromes may in fact be more common than previously believed. DUP25 is an interstitial duplication of 17 Mb at 15q24–26, which is associated with joint laxity and panic disorder (Gratacos *et al.*, 2001). Changes in dosage of one or more of the 59+ genes in the DUP25 region are likely to contribute to the subtle clinical phenotype. Detection of DUP25 was not easy as it shows non-Mendelian transmission precluding straightforward linkage analysis. Instead researchers used laborious cytogenetic methods to detect the duplication. This analysis identified DUP25 in 90% of patients with one or more anxiety disorders, and in 80% of subjects with joint laxity and remarkably in 7% of French population-based controls.

These genomic disorders are generally thought to be caused by aberrant recombination at region- or chromosome-specific low-copy repeats, known as segmental duplications (Emanuel and Shaikh, 2001). This new class of repetitive DNA element has only been identified very recently, largely as a result of human genome sequencing. Segmental duplications result from the duplication of large segments of genomic DNA that range in size from 1 to 400 kb. These duplications can mediate interchromosomal or intrachromosomal recombination events. Knowledge that relatively common diseases can be caused by recurrent chromosomal duplications and deletions has demonstrated that potential for genomic instability could be directly related to the structure of the regions involved. The sequence of the human genome offers to add insight and understanding to the molecular basis of such recombination 'hot spots'. This insight is already being gained, in the case of VCFS on 22q11.2 complete genomic sequence across the region has revealed four segmental duplications flanking the VCFS deletion region (Shaikh *et al.*, 2001).

Availability of information on known deleted or duplicated regions varies greatly; some have been narrowed to fairly well-defined critical regions, others are very poorly defined. Details of some of the more extensively characterized deletion/rearrangement syndromes are captured in GDB and OMIM, although in most cases information is spread throughout the literature and basically needs to be hunted down on a case by case basis. The UCSC human genome browser is a particularly useful ally in this hunt (see Chapter 5), as it annotates large duplicated regions in the human genome. The objective of this annotation is primarily to identify duplication errors in human genome contig assembly, but this also effectively identifies segmental duplications, such as the duplications flanking the VCFS region on 22q11.2.

3.2.5 Somatic Mutations

A completely distinct category of human mutations arises somatically during the process of tumourgenesis. These mutations may take many forms, the most commonly characterized are somatic point mutations identified during the screening of candidate genes in tumour tissues. Cytogenetic studies of human neoplasias have also identified a number of chromosomal aberrations involving large deletions and duplications (Shapira, 1998). As somatic mutations are not inherited it is obviously important to avoid mixing somatic point mutation data with human polymorphism and mutation data.

3.2.5.1 Somatic Point Mutations

Screening of candidate genes for point mutations in tumour material has identified a number of key genes with a role in cancer. There is no central database containing point mutation data identified during these screens, although some locus-specific databases do exist, it is not possible to list all these specialist resources. In some cases is may be possible to identify locus-specific databases by a gene-specific websearch (e.g. using SCIRUS, see Chapter 2). In most cases mutation data needs to be identified directly from the literature.

3.2.5.2 Genomic Aberrations in Cancer

Almost 100,000 neoplasia-associated chromosomal abnormalities have been characterized at the molecular level, revealing previously unknown genes that are closely associated with tumourigenesis. It is not clear if somatic chromosomal aberrations and genomic syndromes share any common mechanisms, such as mediation by segmental duplications, although this is a possibility. Prospects for informatic and laboratory study of chromosomal aberrations in cancer are assisted by the availability of a centralized database to capture this data, the Mitelman map of chromosome aberrations in cancer. This resource has been integrated into the NCBI MapViewer tool and the Cancer Genome Anatomy Project (CGAP) (see Table 3.1).

3.3 DATABASES OF HUMAN GENETIC VARIATION

The vast range of human genetic variation is still largely uncharted and what information exists cannot be derived from a single database. At best the data needs to be gathered from several databases or worse still the data may not be readily available in a database at all, in which case detailed literature and internet searching or bioinformatic analysis approaches may be necessary. Having described the main forms of human variation, we will now introduce the key databases for mining this information. We will also examine how these genetic databases integrate with other databases and the human genome sequence to add a full genomic context to variation, to help in the characterization of a potential genetic lesion. Table 3.1 presents a selection of the best tools and databases for this purpose.

TABLE 3.1 Genetic Variation-Focused Databases and Tools on the Web

Mutation databases	
OMIM	http://www.ncbi.nlm.nih.gov/Omim/
HGMD	http://www.hgmd.org
GDB Mutation Waystation	http://www.centralmutations.org/.
HUGO mutation database initiative	http://www.genomic.unimelb.edu.au/mdi/
Central databases (SNPs and mutations)	
HGVbase	http://hgvbase.cgb.ki.se/
Sequence variation database (SRS)	http://srs.ebi.ac.uk/
dbSNP	http://www.ncbi.nlm.nih.gov/SNP/
The SNP consortium (TSC)	http://snp.cshl.org/
Genetic marker maps (microsatellites, STSs other markers)	
Marshfield maps	http://research.marshfieldclinic.org/genetics/
Genome Database (GDB)	http://www.gdb.org
dbSTS	http://www.ncbi.nlm.nih.gov/STS/
UniSTS	http://www.ncbi.nlm.nih.gov/genome/sts/
Somatic and non-nuclear mutation databases	
MitoMap	http://www.gen.emory.edu/mitomap.html
Mitelman Map	http://cgap.nci.nih.gov/Chromosomes/Mitelman
Gene-orientated SNP and mutation visualization	
LocusLink	http://www.ncbi.nlm.nih.gov/LocusLink/
PicSNP	http://picsnp.org
Protein Mutation Database	http://www.genome.ad.jp/htbin/www_bfind?pmd
Go!Poly	http://61.139.84.5/gopoly/
GeneLynx	http://www.genelynx.org
SNPper	http://bio.chip.org:8080/bio/snpper-enter
GeneSNPs	http://www.genome.utah.edu/genesnps/
CGAP SNP database	htpp://lpgws.nci.nih.gov/
Genome-orientated for SNP and mutation visualization	
Ensembl	http://www.ensembl.org
Human Genome Browser (UCSC-HGB)	http://genome.ucsc.edu/index.html
Map Viewer	http://www.ncbi.nlm.nih.gov/cgi-bin/Entrez/hum_srch

3.4 SNP DATABASES

The deluge of SNP data generated over the past 2 years can primarily be traced to two major overlapping sources: The SNP Consortium (TSC) (Altshuler *et al.*, 2000) and members of the Human Genome Sequencing Consortium, particularly the Sanger Institute and Washington University. The predominance of SNP data from this small number of closely related sources has facilitated the development of a central SNP database — dbSNP at the NCBI (Sherry *et al.*, 2001). Other valuable databases have developed using dbSNP data as a reference, these tools and databases bring focus to specific subsets of SNP data, e.g. gene-orientated SNPs, while enabling further data integration around dbSNP.

3.4.1 The dbSNP Database

The National Center for Biotechnology Information (NCBI) established the dbSNP database in September 1998 as a central repository for both SNPs and short INDEL polymorphisms. In May 2002 (Build 104) dbSNP contained 4.2 million SNPs. These SNPs collapse into a non-redundant set of 2.7 million SNPs, known as Reference SNPs (RefSNPs). Approximately 10% of these RefSNPs do not currently map to the draft human genome, which leaves 2.43 million SNPs with potential utility for genetic mapping. These quantities of SNPs give a high level of coverage across the genome. One study estimated that 85% of all known exons are within 5 kb of an SNP in the dbSNP database (International SNP Map Working Group, 2001). These figures will have undoubtedly improved considerably by the time this book comes to press.

3.4.1.1 The Reference SNP Dataset (RefSNPs)

The non-redundant RefSNP dataset is produced by clustering SNPs at identical genomic positions and creating a single representative SNP (designated by an 'rs' ID). The sequence used in the RefSNP record is derived from the SNP cluster member with the longest flanking sequence; this sequence is derived from one individual and is not a composite sequence assembled from the cluster. The RefSNP record collates all information from each member of the cluster, e.g. frequency information. The availability of the RefSNP dataset considerably streamlines the process of integrating SNPs with other data sources. External resources generally use the RefSNP dataset which makes the RefSNP ID the universal SNP ID in the SNP research community. RefSNPs have also become an integral part of the NCBI data infrastructure, so that the user can effortlessly browse to dbSNP from diverse NCBI resources, including LocusLink, Map View and Genbank.

3.4.1.2 Searching dbSNP

There are a bewildering range of approaches for searching dbSNP. The database can be searched directly by SNP accession number, submitter, detection method, population studied, publication or a sequence-based BLAST search. The database also has a complex search form which allows more flexible freeform queries (http://www.ncbi.nlm.nih.gov/SNP/easyform.html). This allows the user to select SNPs which meet several criteria, for example it is possible to search for all validated non-synonymous SNPs in gene coding regions on chromosome 1 (Figures 3.3 and 3.4). The advanced form also includes a separate interface for retrieving all SNPs between two STS markers or two golden path locations.

There are many other tools which use the dbSNP dataset, e.g. LocusLink, SNPper and the human genome browsers (Table 3.1). These tools can offer powerful alternative interfaces for searching dbSNP, but be aware that third party tools and software may use filtering or repeat masking protocols, which can lead to the exclusion of SNPs with poor quality or short flanking sequence, or SNPs in repeat regions. If it is important to identify *all* SNPs in a given gene or locus then it is worth consulting several different tools and comparing the results. Some of the best SNP visualization tools are discussed later in this chapter.

3.4.1.3 Submitting Data to dbSNP

The dbSNP database accepts direct data submissions from researchers by e-mail or FTP. The submission process is generally intended for large batch submissions involving hundreds or thousands of SNPs, using a text flatfile submission format. Each SNP submission

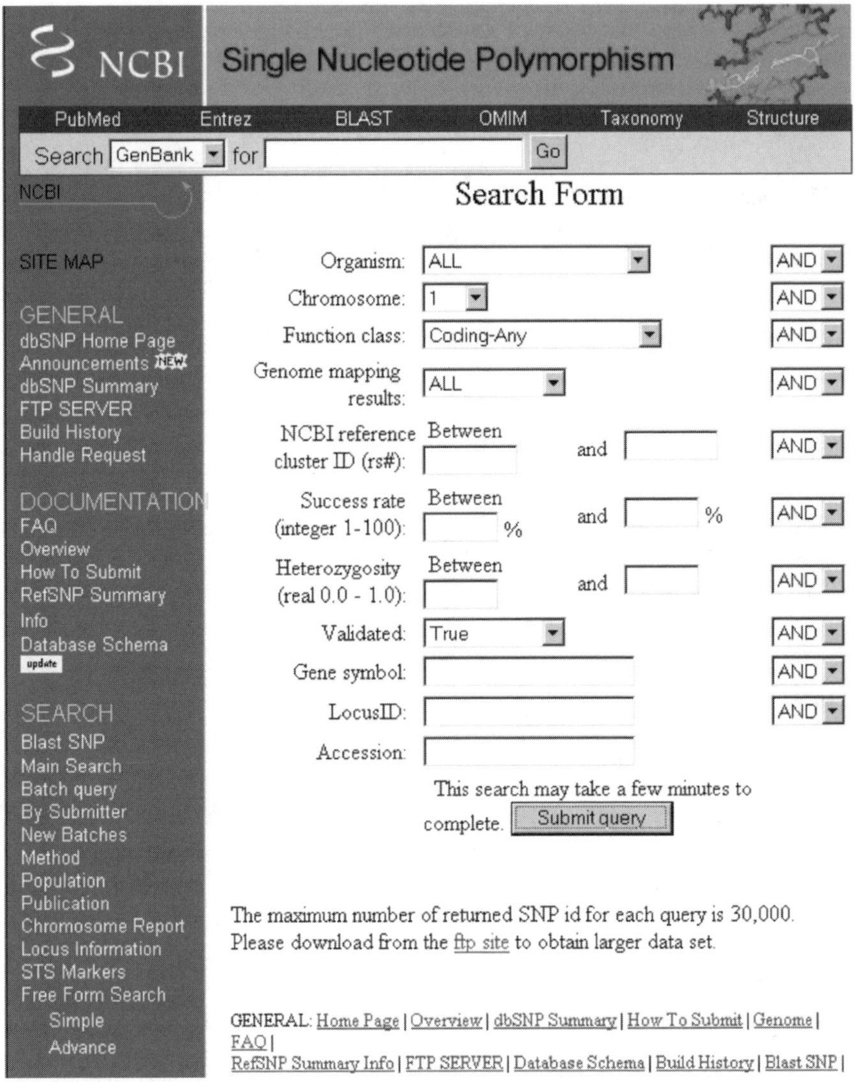

NCBI Single Nucleotide Polymorphism
PubMed Entrez BLAST OMIM Taxonomy Structure
Search GenBank for Go
NCBI
SITE MAP
GENERAL
dbSNP Home Page
Announcements NEW
dbSNP Summary
FTP SERVER
Build History
Handle Request
DOCUMENTATION
FAQ
Overview
How To Submit
RefSNP Summary Info
Database Schema update
SEARCH
Blast SNP
Main Search
Batch query
By Submitter
New Batches
Method
Population
Publication
Chromosome Report
Locus Information
STS Markers
Free Form Search
Simple
Advance

Search Form
Organism: ALL AND
Chromosome: 1 AND
Function class: Coding-Any AND
Genome mapping results: ALL AND
NCBI reference cluster ID (rs#): Between and AND
Success rate (integer 1-100): Between % and % AND
Heterozygosity (real 0.0 - 1.0): Between and AND
Validated: True AND
Gene symbol: AND
LocusID: AND
Accession:
This search may take a few minutes to complete. Submit query
The maximum number of returned SNP id for each query is 30,000. Please download from the ftp site to obtain larger data set.
GENERAL: Home Page | Overview | dbSNP Summary | How To Submit | Genome | FAQ |
RefSNP Summary Info | FTP SERVER | Database Schema | Build History | Blast SNP |

Figure 3.3 The dbSNP freeform search interface.

contains many elements to describe the SNP, but primarily it should contain a report describing how to assay the SNP, the SNP sequence information and if available the SNP allele frequency. While the submission format is suitable for bulk submissions it may present the occasional submitter some problems. Preparation of any more than a handful of SNPs in this format really requires some grasp of a text manipulation language such as perl (Stein, 2001). In this case it may be a good idea to find a friendly perl programmer or contact dbSNP directly for guidance and assistance in the preparation of the submission. A web interface for form-based submission is currently in development, which should alleviate this problem.

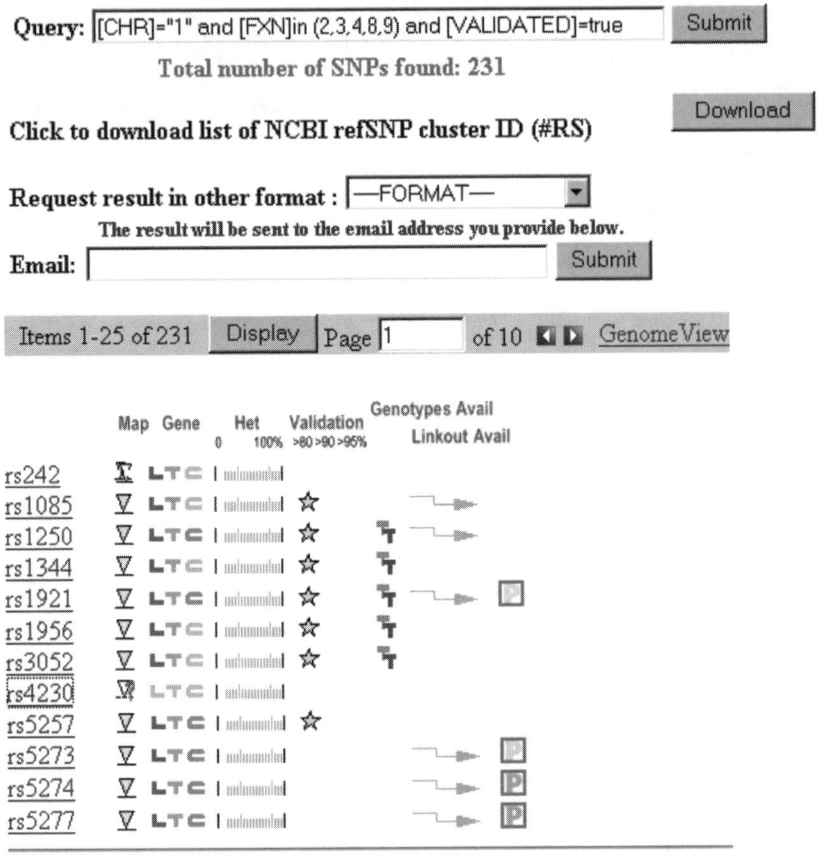

Query: [CHR]="1" and [FXN]in (2,3,4,8,9) and [VALIDATED]=true Submit

Total number of SNPs found: 231

Click to download list of NCBI refSNP cluster ID (#RS) Download

Request result in other format : —FORMAT—

The result will be sent to the email address you provide below.

Email: Submit

Items 1-25 of 231 Display Page 1 of 10 GenomeView

Map Gene Het Validation Genotypes Avail Linkout Avail

rs242
rs1085
rs1250
rs1344
rs1921
rs1956
rs3052
rs4230
rs5257
rs5273
rs5274
rs5277

Figure 3.4 Search results from a dbSNP freeform search.

3.4.1.4 Key SNP Data Issues

The sequencing of the human genome has provided a massive boost to human polymorphism discovery efforts. Table 3.2 presents a breakdown of dbSNP submission sources. From this table it is clear that 94% of SNPs in dbSNP originate from three main sources: the TSC, the Sanger Institute and the Kwok Laboratory (informatic analysis of data from the Whitehead Institute and Washington University). SNPs sourced from the TSC were identified by the major genome sequencing centres by detection of high-confidence base differences in aligned sequences primarily from reduced representation shotgun (RRS) sequencing (Altshuler *et al*., 2000) and also by alignment of genomic clones (Mullikin *et al*., 2000). RRS sequencing involves sequencing of random clones from the genomes of many individuals. This method has several advantages over other SNP identification methods, in that it does not require previous knowledge of genomic sequence or PCR, and it provides haploid genotypes, the alleles of which are easier to call (see Chapter 10 for on overview of these methods). The later two sources, SANGER and KWOK account for 64% of dbSNP SNPs. These represent SNPs generated by the major human genome sequencing centres. These SNPs were identified by overlapping genomic sequence reads.

TABLE 3.2 Main SNP Submission Sources in the dbSNP Database (BUILD 104)

Source	Total submissions	RefSNP clusters	Primary SNP ID method
TSC	1,279,099	1,275,272	Shotgun and Genomic
Kwok (WASHU)	1,182,884	493,536	Genomic overlap
Sanger	1,529,560	1,348,534	Genomic overlap
Lee	99,505	46,942	EST trace mining
Yusuke	73,720	73,720	SNP disc (Japanese)
Perlegen	25,326	25,315	Microarray (Chr21 only)
HGBASE	13,100	13,081	Various
CGAP	12,881	12,733	EST trace mining
Other	13,367	ND	Various
Total	4,229,442	2,673,925	

In the wake of the TSC and the genomic overlap SNP discovery projects, further SNP submissions to dbSNP will continue from the genome centres in the final stages of genome finishing, but further growth of dbSNP will depend on the next steps after completion of the human genome. The human genome is likely to be repeatedly re-sequenced in the next few years, either entirely or across defined regions. This will in turn generate further SNPs by comparison of genomic overlaps. The Sanger Institute has already announced a 5-year plan to re-sequence all known human exons in 96 individuals. This should detect 95% of SNPs with a frequency of >1%. Inevitably novel SNPs will become increasingly rare, based on a law of diminishing returns. Based on the observed SNP density in the genome, estimates suggest that the dbSNP dataset may currently represent 20–30% of common SNPs in the human genome. Different SNP discovery projects have sampled variation at very different levels. The TSC SNPs were discovered using a publicly available panel of 24 ethnically diverse individuals (Collins *et al.*, 1998). This panel would have a 95% chance of detecting SNPs down to a frequency of 5%. SNPs identified by genomic sequence overlap (which comprise 64% of dbSNP data), offer the shallowest sampling of human variation. Genomic overlap SNPs are candidate SNPs identified by comparison of two individuals, this approach has some major drawbacks, the SNP discovery method is more error prone (heterozygotic SNPs are often missed) and many SNPs discovered by this method are likely to be 'private' SNPs which are restricted to the individual and not generally represented in populations (see below for more details on candidate SNP issues).

Aside from the major SNP data submissions from the genome centres, dbSNP also accepts direct SNP submissions from researchers and most journals now require SNP submission to dbSNP before publication (a practice which needs to be encouraged). These have been estimated to add to dbSNP at a rate of about 100 primarily gene-orientated SNPs per month.

3.4.1.5 Candidate SNPs – SNP to Assay

As we have already demonstrated, the dbSNP dataset has one overwhelming caveat — most of the SNPs are 'candidate' SNPs of unknown frequency and are unconfirmed in a laboratory assay. This translates to the simple fact that many SNPs do not exist at a detectable frequency in any population. Over 60% of the SNPs in dbSNP were detected by statistical methods for identification of 'candidate' SNPs by comparison of DNA sequence traces from overlapping clones. Marth *et al.* (2001) investigated the reliability of these candidate

SNPs in some depth completing two pilot studies to determine how well candidate SNPs would progress to working assays in three common populations. In both studies, they found that between 52–54% of the characterized SNPs turn out to be common SNPs (above >10%) for each population. Significantly, between 30 and 34% of the characterized SNPs were not detected in each population. These results suggest that if a candidate SNP is selected for study in a common population, there is a 66–70% chance that the SNPs will have detectable minor allele frequency (1–5%) and a 50% chance that the SNPs are common in that population (>10%). Put another way, ~17% of candidate SNPs will have no detectable variation in common populations, these 'monomorphic' SNP candidates, are likely to represent 'private SNPs', which exist in the individual screened but not appreciably in populations. This probably reflects the massive increase in population size and admixture over the past 500 years (Miller and Kwok, 2001). Beyond validation of the SNP, the last hurdle is assay design — many SNPs are located in repetitive or AT rich regions, which makes assay design difficult, this can account for a further 10–30% fallout, depending on the assay technology.

Any genetic study needs to take these levels of attrition between SNP and working assay into account (Table 3.3). There is only one solution to this problem — to determine the frequency of the 2 million or so public SNPs in common ethnic groups. This is now widely recognized in the SNP research community and several public groups including the TSC are already undertaking or seeking to undertake large-scale SNP frequency determination projects.

3.4.2 Human Genome Variation Database (HGVbase)

The Human Genome Variation database, HGVbase, previously known as HGbase (Brookes *et al.* 2000; http://hgvbase.cgb.ki.se/), was initially created in 1998 with a remit to capture all intra-genic (promoter to end of transcription) sequence polymorphism. One year later, the remit of the database expanded to a whole genome polymorphism (and nominally mutation) database, this ambitious expansion in remit was supported by the establishment of a European consortium comprising teams at the Karolinska Institute, Sweden, the European Bioinformatics Institute, UK and at the European Molecular Biology Laboratory, Germany. At this point, HGbase encompassed the same classes of variants as dbSNP. Both HGVbase and dbSNP make regular data exchanges to allow data synchronization. In November 2001, the HGbase project adopted the new name HGVbase (Human Genome Variation database; Fredman *et al.*, 2002). This change reflected another change in the scope of the database as it took on a HUGO endorsed role as a central repository for mutation collection efforts undertaken in collaboration with the Human Genome Variation Society (HGVS).

TABLE 3.3 Pitfalls from Candidate SNP to Assay (From Marth *et al.*, 2001)

SNP to assay convertion steps	Remaining RefSNPs (% attrition)
Reference SNP identified	2.4 M
Not mapped to human genome	2.16 M (10%)
Assay design not possible or assay fails	1.84 M (15%)
Not polymorphic in study population	1.52 M (17%)
Frequency <20% in chosen population	1.26 M (50%)
SNPs (>20% frequency) with assay available	0.63 M

There is no doubt that dbSNP has assumed the *de facto* position of the primary central SNP database. To accommodate this, HGVbase has assumed a complementary position, with a broader remit covering all single nucleotide variation — both SNPs and mutations. HGVBASE is also taking a distinct approach to dbSNP by seeking to summarize all known SNPs as a semi-validated, non-redundant set of records. HGVbase is seeking to address some of the problems associated with candidate SNPs and so in contrast to the automated approach of dbSNP, HGVbase is highly curated. The curators are aiming to provide a more-extensively validated SNP data set, by filtering out SNPs in repeat and low complexity regions and by identifying SNPs for which a genotyping assay can successfully be designed. The HGVbase curators have also identified SNPs and mutation data from the literature, particularly older publications before database submission was the norm. HGVbase currently contains 1.45 M non-redundant human polymorphisms and mutations (release 13–March 2002).

HGVbase is a highly applied database, which also provides some useful tools for experimental design, including a tool for defining haplotype tags — 'Tag 'n Tell'. This tool will find a minimum set of markers that uniquely characterize (or 'tag') chosen haplotypes. According to user preferences, not all entered haplotypes have to be considered in the tag-selection process, this is useful for determining optimal haplotype tag sets to capture common haplotypes (see Chapter 9 for an example of haplotype tagging using this tool).

The HGVbase search interface is relatively simple, tools are available to facilitate BLAST searching and keyword queries of the database. As these options are relatively limited, other tools which access HGVbase data, are a better option — most are from the EMBL and EBI organizations, including Ensembl and SRS (Table 3.3; described below). The *in silico* quality control approach adopted by HGVbase is valuable, particularly for the broader biological community of SNP data consumers. For the geneticist, HGVbase serves to identify SNPs with a higher chance of converting from 'candidate SNPs' to informative SNP assays. If you take the cost of failed assays into account, this is a valuable objective, although if all available SNPs need to be identified it may still be important to search dbSNP and other resources.

3.5 MUTATION DATABASES

The polymorphism data stored in dbSNP is valuable biological information that helps to define the natural range of variation in genes and the genome, however most of the polymorphisms can be assumed to be functionally neutral. By contrast human mutation data is functionally defined and has obvious implications for the nature and prevalence of disease and the pathways underlying disease. This makes the study of naturally-occurring mutations important for the understanding of human disease pathology, particularly the relationships between genotype and phenotype and between DNA and protein structure and function. A large number of Mendelian disease mutations have been identified over the past 20 years. These have helped to define many key biological mechanisms, including gene regulatory motifs and protein–protein interactions (see Chapter 13). Many highly specialized locus-specific databases (LSDBs) have been established to collate this data. This chapter could not hope to cover all these databases, but there are now several centralized resources which index and provide links to some of the larger resources. Other 'boutique' databases can sometimes be identified by general web searching (see Chapter 2).

3.5.1 The Human Gene Mutation Database (HGMD)

The HGMD was established in April 1996 to collate published germline mutations responsible for human inherited disease. In October 2001, HGMD contained 26,637 mutations in 1153 genes. The scope of HGMD is limited to mutations leading to a defined inherited phenotype, including a broad range of mechanisms, such as point mutations, insertion/deletions, duplications and repeat expansions within the coding regions of genes. Somatic mutations and mutations in the mitochondrial genome are not included. HGMD invites submissions from researchers but most records are curated directly from mutation reports in more than 250 journals and directly from the LSDBs which are comprehensively linked. To be included, there must be a convincing association between the mutation and the phenotype. All mutations in HGMD are represented in a non-redundant form which unfortunately does not conserve all the redundant mutations constituting the cluster, so it is not possible to determine if mutations are identical by descent, also data is lost on the frequency of mutations. The HGMD search interface is primarily text based and targeted searching tends to rely on knowledge of the correct HUGO nomenclature for a gene.

3.5.2 Sequence Variation Database (SRS)

The sequence variation database forms part of the Sequence Retrieval Server (SRS) at the EBI, Hinxton UK. SRS is a flexible sequence query tool which allows the user to search a defined set of sequence databases by accession number, keyword or sequence similarity. Several categories of sequence variation are encompassed by SRS, including HGVbase and a large number of locus specific databases which are listed in Table 3.4.

3.5.3 The Protein Mutation Database (PMD)

The Protein Mutation Database (PMD) is unique among genetic variation databases as it contains both natural and artificial mutation data derived from human proteins (Kawabata *et al.*, 1999). The artificial mutation data is derived from the literature and mainly consists of site-directed and random mutagenesis data. It is important to clearly delineate artificial data and so each record is clearly defined as either natural or artificial. The database gives detailed description of the functional or structural effects of the mutations if known and provides links to the original publications. Relative differences in activity and/or stability, in comparison with the wild-type protein, are also indicated. PMD contains 119,190 natural and artificial mutations (January 2002) and these can be searched by keyword or sequence similarity (BLAST), a complete report on the mutated protein sequence is displayed which allows the user to see the position of altered amino acids. Where 3D structures have been experimentally determined, PMD displays mutated residues in a different colour on the 3D structure.

The Protein Mutation Database is very valuable for the functional analysis of proteins. The detailed functional characterization of mutations gives the user an opportunity to compare known mutations with variations in orthologous residues in related proteins. The data is also useful to aid in the delineation of the functional domains of proteins in the database and other homologous proteins (see Chapter 14 for further examination of such approaches for mutation analysis).

3.5.4 On-line Mendelian Inheritance in Man (OMIM)

OMIM is an on-line catalogue of human genes and their associated mutations, based on the long running catalogue Mendelian Inheritance in Man (MIM), started in 1967 by Victor McKusick at Johns Hopkins (Hamosh *et al.*, 2000). OMIM is an excellent resource for providing a brief background-biology on genes and diseases, it includes information on the most common and clinically significant mutations and polymorphisms in genes. Despite the name, OMIM also covers complex diseases in varying degrees of detail.

TABLE 3.4 Locus-Specific Databases Indexed by the Sequence Variation Database

Name	Description	Entries
General mutation databases		**74,117**
EMBLCHANGE	Sequence change features from EMBL	32,863
SWISSCHANGE	Sequence change features from SWISS-PROT	17,294
OMIMALLELE	Alleles from OMIM	9344
HUMUT	Protein Mutation Databank	14,616
Mitochondrial genome		**9401**
HUMAN_MITBASE	Human mitochondrial DNA variants	9401
Locus-specific mutation databases		**240,73**
P53LINK	p53 mutations database	14,834
APC	APC mutation database	825
BTKBASE	Bruton's tyrosine kinase mutations	454
VWF	von Willebrand factor gene variations	144
CFTR	Cystic fibrosis mutation database	809
PAH	Phenylalanine hydroxylase mutations	289
HAEMA	Haemophilia A, Factor VIII mutations	604
HAEMB	Haemophilia B	1722
LDLR	Low-density lipoprotein receptor	283
PAX6	PAX6 mutation database	118
EMD	Emery–Dreifuss muscular dystrophy	87
L1CAM	Neuronal cell adhesion molecule gene mutations	91
CD40LBASE	CD40 ligand defects	60
G6PD	Glucose-6-phosphate dehydrogenase variants	122
ANDROGENR	Androgen receptor mutations	514
RDS	Retinal degeneration slow gene mutations	33
RHODOPSIN	Rhodopsin gene mutations	133
FANCONI	Fanconi anaemia mutation database	32
HEXA	Hexosaminidase A mutations	89
XCGDBASE	X-linked chronic granulomatous disease	303
DMD	Duchenne/Becker muscular dystrophy	184

(continued overleaf)

TABLE 3.4 *(continued)*

Name	Description	Entries
FVII	Factor VII mutation database	176
ATM	Ataxia–telangiectasia mutation database	200
P16	CDKN2A/P16NK4A mutation database	146
GAA	Acid alpha-glucosidase mutation database	83
OTC	Ornithine transcarbamylase (OTCase) mutations	105
IL2RGBASE	Interleukin-2 receptor gamma mutations	161
BIOMDB	Database of tetrahydrobiopterin deficiency mutations	78
Central databases		**984,093**
HGVbase	Human Genome Variation database (SNPs and mutations)	984,093

In January 2002, the database contained over 13,285 entries (including entries on 9837 gene loci and 982 phenotypes). OMIM is curated by a dedicated but small group of curators, but the limits of a manual curation process mean that entries may not be current or comprehensive. With this caveat aside OMIM is a very valuable database, which usually presents a very accurate digest of the literature (it would be difficult to do this automatically). A major added bonus of OMIM is that it is very well integrated with the NCBI database family, this makes movement from a disease to a gene to a locus and vice versa fairly effortless.

3.6 GENETIC MARKER AND MICROSATELLITE DATABASES

3.6.1 dbSTS and UniSTS

dbSTS is an NCBI database containing sequence and mapping data for Sequence Tagged Sites (STSs) (Olson *et al.*, 1989). These STSs can include polymorphic sequences such as short tandem repeats (STRs), or non-polymorphic sequences. In fact any unique genomic landmark which can be amplified by PCR can be used as an STS marker. Both polymorphic and non-polymorphic STS markers have been used to construct extensive high resolution radiation hybrid maps of the human gene, while polymorphic markers have been used to construct genetic maps (see Chapter 7). The dbSTS database maintains complete records for over 133,202 STS markers, including ~18,000 STR markers and gives key information for each record such as primer sequences, map location and marker aliases. Searching dbSTS can be achieved in many ways. The UniSTS interface allows direct searches by keyword, the NCBI Map View application allows searching by genomic location or locus, while dbSTS is also available for BLAST searching by NCBI BLAST. This array of search options makes the dbSTS database a very reliable source for retrieval of both genetic and physical STS map markers.

3.6.2 The Genome Database (GDB)

The genome database (GDB) was established ahead of most other genetics databases in 1990 as a central repository for mapping information from the human genome project. Throughout the early 1990s GDB was the dominant genome database and served as the primary repository for genetic map-related information. In January 1998, after several years of uncertain US government funding, GDB funding was officially terminated. By December 1998 funding from another source was found, but at a significantly lower level. By this time other databases had inevitably overtaken GDB as 'central genome databases' (Cuticchia, 2000). Today GDB is still one of the most comprehensive sources for some forms of genetic data, including tandem repeat polymorphisms (it contains over 18,000), it also contains an eclectic range of information on fragile sites, deletions, disease genes and mutations, collected by a mixture of curation and direct submission. GDB development is ongoing and the historical focus of the database on genetic maps is broadening to a more integrated view of the genome ultimately down to the sequence level (which unfortunately is currently lacking). Plans to finally integrate a sequence map might well make GDB a prominent genetic resource again, although political issues still threaten to halt these aspirations (Bonetta, 2001).

The GDB graphical search interface was a truly pioneering tool of the field and was the first to introduce the kind of graphical map viewing applications that Ensembl and UCSC now excel at. Unfortunately the originals are not always the best and the graphical GDB interface is now starting to look very tired indeed. However, GDB also has a more productive text/table based search interface. This allows complex queries, for example it is possible to retrieve all known polymorphic or non-polymorphic markers between two markers. Advanced filters can also be used, for example markers above a defined level of heterozygosity can be retrieved. Results are retrieved and ordered based on the genetic distances of the markers, along with a very roughly estimated Mb location. As the markers are ordered by genetic distance, many markers cannot be resolved beyond a certain level, therefore markers with identical genetic distances are presented in an arbitrary order. However, high level order is quite reliable and supported by LOD scores. Clarification of genetic marker order and distance is a complex process, which involves integrating multiple maps ultimately down to the level of the human genome to build up a consensus order and distance. These issues of map and marker integration will be examined in detail in Chapter 7.

3.7 NON-NUCLEAR AND SOMATIC MUTATION DATABASES

3.7.1 MITOMAP

The sequencing of the human mitochondrial genome (mtDNA) was a landmark in genomics, being the first component of the human genome to be completely sequenced (Anderson *et al.*, 1981). The mitochondrial genome consists of a 16,569-bp closed circular molecule in the mitochondrion — each of the several thousand mtDNAs per cell encodes a control region encompassing a replication origin and the promoters, a large (16 S) and small (12 S) rRNA, 22 tRNAs, and 13 polypeptides. All of the mtDNA polypeptides are components of the mitochondrial energy generating pathway, oxidative phosphorylation, which is functionally essential and evolutionarily constrained (Wallace *et al.*, 1995). Despite this selection pressure, maternally inherited mtDNA has a very high mutation rate — mtDNA mutates 10–20 times faster than nuclear DNA as a result of inadequate proofreading by mitochondrial DNA polymerases and limited mtDNA repair capability. As a result mtDNA mutations might be expected to be relatively common — this is supported by the relative abundance of mitochondrial disorders described to so far — although it is also important to note that such mutations, being comparatively easy to identify by sequencing, are likely to have been among the first to be characterized.

More than 100 mitochondrial diseases have now been described, including a broad spectrum of degenerative diseases involving the central nervous system, heart, muscle, endocrine system, kidney and liver. Information on the phenotypes and causative mutations of these diseases are covered briefly in OMIM and in detail in the mitochondrial mutation database, MITOMAP (Kogelnik *et al.*, 1998). The MITOMAP database (Table 3.1) integrates information on all known mtDNA mutations and polymorphisms with the broad spectrum of available molecular, genetic, functional and clinical data, into an integrated resource which can be queried from a variety of different perspectives.

MITOMAP places the clinical mutation dataset of over 150 disease-associated mutations into their genomic context. It also encompasses information on over 100 mtDNA rearrangements, including nucleotide positions of breakpoint junctions and sequences of associated repeat elements. Clinical characteristics are associated with the mutations and are accessible both through associated datasets in MITOMAP as well as through linkage to OMIM. MITOMAP also provides information on nuclear genes which impinge on mtDNA structure and function. Finally, a population variation dataset provides access to known mtDNA haplotypes and their continental distributions and population frequencies.

3.7.1.1 Searching MITOMAP

MITOMAP is searchable by gene, disease and enzyme — users can refine their search by function, polymorphism, or references (author, title, journal, year or keyword). MITOMAP data has been collated from published literature on the mitochondrial genome and regular searches are made to capture new publications. The database also accepts direct submissions, including over 199 unpublished polymorphisms and mutations.

3.7.2 The Mitelman Chromosome Abberations Map

Cytogenetic studies over the past few decades have revealed clonal chromosomal aberrations in over 100,000 human neoplasms. Many of these have been characterized at the molecular level, revealing previously unknown genes that may be closely associated with tumourigenesis. Information on chromosome changes in neoplasia has grown rapidly, making it difficult to identify all recurrent chromosomal aberrations. The Mitelman Map of Chromosome Aberrations in Cancer (Mitelman *et al.*, 1997) was first published over 15 years ago to compile this information; the database now contains over 7100 references encompassing some 100,000 aberrations in 97 different histological types of cancers. The catalogue has evolved from a book to a CD-ROM published by John Wiley and now it is also available as a web-based database (http://cgap.nci.nih.gov/Chromosomes/ Mitelman; Mitelman *et al.*, 2002).

The Mitelman database actually consists of three databases. A generalized search form, allows one to search by abnormality, breakpoint, number of clones, number of chromosomes, sex, age, race, country, series, hereditary disorder, topography, immuno-phenotype, morphology, tissue, previous tumour, treatment, reference and/or cytogenetic characteristics to determine frequencies of balanced and unbalanced translocations. The results of a search provide a variety of information. For example, if you select a breakpoint and a gene, the search retrieves relevant PubMed references, diagnoses, the specific chromosome aberration and all genes involved. The Mitelman map is an extremely complex and detailed database so it is well worth consulting the 'Help' section for specific instructions before commencing a search. A more immediately accessible breakdown of the recurrent neoplasia-associated aberrations described by Mitelman are presented by the NCBI MapView tool. This data is an updated version of the survey appearing in the April 1997 Special Issue of *Nature Genetics* (Mitelman *et al*., 1997). To view the Mitelman abberations across chromosome 22, for example, try the following URL: http://www.ncbi.nlm.nih.gov/cgi-bin/Entrez/maps.cgi?ORG=hum&MAPS=ideogr,mit &CHR=22

For cancer geneticists, the Mitelman database benefits greatly from inclusion in the Cancer Genome Anatomy Project (CGAP). CGAP and NCBI are also collaborating closely which has allowed information on chromosomal aberrations to be closely linked with the other CGAP and NCBI resources including mapped SNPs, FISH mapped BACs, and GeneMAP99. The CGAP catalogue is of particular value, serving as a comprehensive index to breakpoints, clones (BACs, cDNA), genes (expression, sequence, tissue), libraries and SNPs (primer pairs, linkage and physical maps). The Mitelman database is undoubtedly the most comprehensive listing of clinical cytogenetic studies in existence, integration of this data with MapViewer and soon hopefully with other viewers such as Ensembl, creates a great opportunity to study the genetics and the biological process of chromosomal aberration right down to the sequence level; this should in turn help to provide insight into the molecular mechanisms of tumourigenesis.

3.8 TOOLS FOR SNP AND MUTATION VISUALIZATION – THE GENOMIC CONTEXT

The human genome is the ultimate framework for organization of SNP and mutation data and so genome viewers are also one of the best tools for searching and visualizing polymorphisms. The three main human genome viewers, Ensembl, the UCSC Human Genome Browser (UCSC-HGB) and the NCBI Map Viewer (Table 3.1), all maintain variable levels of SNP annotation on the human genome, although none maintain annotation of mutation data. Most of the information in these viewers overlap, but each contains some different information and interpretation and so it usually pays to consult at least two viewers, if only for a second opinion. Consultation between viewers is easy as all three now use the same whole genome contig, known as 'the golden path' and so they link directly between viewers to the same golden path coordinates.

User defined queries with these tools can be based on many variables, STS, markers, DNA accessions, gene symbol, cytoband or golden path coordinate. This places SNPs and mutations into their full genomic context, giving very detailed information on nearby genes, transcripts and promoters. Ensembl and UCSC-HGB both show conservation between human and mouse genomes, UCSC-HGB also includes tetradon and fugu (fish) genome conservation. This may be particularly useful for identification of SNPs in potential functional regions, as genome conservation is generally restricted to genes (including undetected genes) and regulatory regions (Aparicio *et al*., 1995). We examine the use of these tools in detail in Chapters 5, 9 and 12.

3.9 TOOLS FOR SNP AND MUTATION VISUALIZATION – THE GENE CONTEXT

For the biologist or candidate gene hunting geneticist, SNP information may be of most interest when located in genes or gene regions, where implicitly each SNP can be evaluated for potential impact on gene function or regulation. Many tools are available to identify and analyse such SNPs and almost all are based on the dbSNP dataset, but most have somewhat different approaches to the presentation of data (see Table 3.1 for a list of these tools). Choice of tool may be a matter of personal preference so it is probably worth taking a look at a few. The drawback of using some of these tools is that some are maintained by very small groups so sometimes tools may not be comprehensive or current. New tools are constantly appearing in this area so it is often worth running a web search to look for new and novel contributions to this research area — for example 'SNP AND gene AND database' is all you need to enter as a search term in a general web search engine.

3.9.1 LocusLink

The NCBI LocusLink database is a reliable tool for gene-orientated searching of dbSNP. It can be queried by gene name or symbol, query results will show a purple 'V' link if SNP records have been mapped to a gene. Clicking on this link will take you to a report detailing all RefSNP records mapped across the gene. Almost all NCBI tools integrate directly with dbSNP; LocusLink is the central NCBI 'gene view' which links out to a wide range of resources, it also includes a RefSNP gene summary (a purple V or VAR link). This summary details all SNPs across the entire gene locus including upstream regions, exons, introns and downstream regions. Non-synonymous SNPs are identified and the amino acid change is recorded, analysis even accommodates splice variants. LocusLink has the advantage of the NCBI support so it is probably one of the most comprehensive and reliable data sources for gene-orientated SNP information.

Although LocusLink benefits from the reliability bestowed by the infrastructure and resources available at the NCBI, several other tools present gene-focused data with a subtly different approach. Some of these are worth trying, again the tool for you may be a matter of personal preference so try a few. There are many tools which fit into this category, some of these are listed in Table 3.1, but for the purposes of this chapter we will only review two of the more outstanding tools: SNPper and CGAP-GAI.

3.9.2 SNPper

SNPper is a web-based tool developed by the Children's Hospital Informatics Program (CHIP), Boston (Riva and Kohane, 2001). The SNPper tool maps dbSNP RefSNPs to known genes, allowing SNP searching by name (e.g. using the dbSNP 'rs' name), or by the golden path position on the chromosome. Alternatively, you can first find one or more genes you are interested in and find all the SNPs that map across the gene locus, including flanking regions, exons and introns. SNPper produces a very effective gene report (Figure 3.5) which displays SNP positions, alleles and the genomic sequence surrounding the SNP. It also presents very useful text reports which mark up SNPs across the entire genomic sequence of the gene and another report which marks up all the amino acid-altering SNPs on the protein.

The great strength of SNPper lies in its data export and manipulation features. At the SNP report level, SNPs can be sent directly to automatic primer design through a Primer3 interface. At a whole gene level or even at a locus level, SNP sets can be defined and refined and e-mailed to the user in an excel spreadsheet with SNP names in the first column and flanking sequences in the second, ready for primer design.

SNPper currently contains information on around 1,900,000 SNPs and 12,479 genes (January 2002). These correspond to all the unambiguously mapped known SNPs and

SNPper

Gene: IFNAR2

Name:	interferon (alpha, beta and omega) receptor 2	XmlXport	
Sequence:	Fasta - Annotated - Protein	**Strand:**	+
Transcript Position:	chr21:31460142-31492817	**Length:**	32676
Coding Sequence Position:	chr21:31471990-31492784	**Length:**	20795

Look up this gene in:

Genbank (mRNA):	NM_000874	**Genbank (prot):**	NP_000865	**Entrez:**	IFNAR2	**LocusLink:**	3455
PubMed:	IFNAR2	**OMIM:**	602376	**Unigene:**	IFNAR2	**Ensembl:**	IFNAR2

Exons:

#	Start		End		Length
1	31460142	(0)	31460284	(142)	143
2	31471907	(11765)	31472045	(11903)	139
3	31473740	(13598)	31473782	(13640)	43
4	31475018	(14876)	31475142	(15000)	125
5	31476785	(16643)	31476958	(16816)	174
6	31478776	(18634)	31478922	(18780)	147
7	31482729	(22587)	31482898	(22756)	170
8	31490664	(30522)	31490795	(30653)	132
9	31492628	(32486)	31492817	(32675)	190
XmlXport				**Total:**	1263

Refine SNPset

SNPset: SS397 **Total number of SNPs:** 29
Size: 33932 **Average distance:** 1170
Resolution: 0 **Visible SNPs:** 29
Restrict to: TSC SNPs; Validated SNPs; Promoter; 3' UTR; Exons; Coding sequence; Introns; Exon/intron boundary
New resolution: Set

Known SNPs:

SNPset: SS397

Source:	IFNAR2
Created on:	**01/17/2002 08:38:53**
SNPs:	**29** (avg dist: **1170**)
Spacing:	**0**
Commands:	Save this SNPset Refine this SNPset Email this SNPset to yourself XmlXport SNP graph Get flanking sequences

Name	Position	Genepos	Role
rs1476415	chr21:31456148	-15842 A/C	Promoter
rs2843981	chr21:31458136	-13854 A/T	Promoter
rs2248202	chr21:31461643	-10347 A/C	Intron
rs2300370	chr21:31462320	-9670 A/G	Intron
rs2248412	chr21:31463294	-8696 A/G	Intron
rs2248420	chr21:31463541	-8449 C/T	Intron
rs1051393	chr21:31472018	28 G/T	Exon, Coding sequence
rs2834156	chr21:31473920	1930 C/T	Intron
rs2834157	chr21:31474308	2318 A/G	Intron
rs2236756	chr21:31474686	2696 A/C	Intron
rs2834158	chr21:31474976	2986 C/T	Intron, Exon/intron boundary

Figure 3.5 The SNPper gene report. The report displays SNP positions, alleles and the genomic sequence surrounding the SNP. It also presents text reports which mark up SNPs across the entire genomic sequence of the gene and amino acid-altering SNPs on the protein.

genes in the human genome. By restricting the database to known genes, they have considerably simplified their task as all the gene annotation is well defined. SNPper uses this advantage to maximum effect by presenting the data very clearly and informatively. SNPper is a highly recommended tool for the laboratory-based geneticist.

3.9.3 CGAP-GAI (htpp://lpgws.nci.nih.gov/)

The Cancer Genome Annotation Project (CGAP)/Genetic Annotation Initiative (GAI) database is a valuable resource which identifies SNPs by *in silico* prediction from alignments of expressed sequence tags (ESTs) (Riggins and Strausberg, 2001). The database was established specifically to mine SNPs from ESTs generated by CGAP's Tumour Gene Index project (Strausberg *et al.*, 2000), which is generating more than 10,000 ESTs per week from over 200 tumour cDNA libraries. The analysis also encompasses other public EST sources.

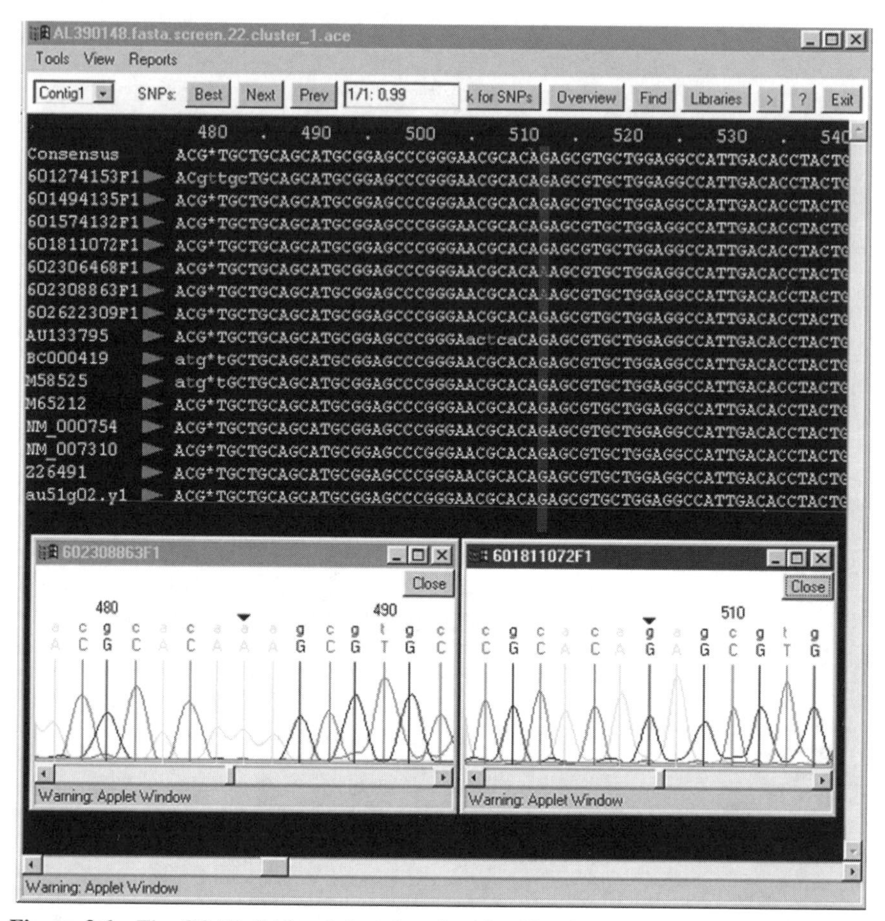

Figure 3.6 The CGAP-GAI web interface for identification of candidate SNPs in ESTs. The JAVA view of trace data helps to support the base call of a potential SNP in an EST, although laboratory investigation is the only reliable SNP confirmation.

Candidate SNPs in ESTs can easily be viewed with the CGAP-GAI web interface in a graphical JAVA assembly (Figure 3.6). SNPs in ESTs are identified by an automated SNP-calling algorithm, mining EST data with greater than 10 reads from the same transcribed region yielded predicted SNPs with an 82% confirmation rate (Riggins and Strausberg, 2001). All SNPs which meet the stringent calling criteria are submitted to dbSNP. It is also worthwhile searching CGAP directly if you are interested in a specific gene. The threshold for automated SNP detection is set very high, so many potential SNPs evade automatic detection, but these candidate SNPs can be identified quite easily by eye, simply by looking for single base conflicts where sequence is otherwise high quality. The JAVA view of trace data helps to support the base call of a potential SNP in an EST (Figure 3.6), although laboratory investigation is the only completely reliable SNP confirmation. Intriguingly this resource could potentially contain some somatic mutations from tumour ESTs which would probably be discarded by the automatic detection algorithm which requires some degree of redundancy to call the SNP.

3.10 CONCLUSIONS

The last few years have revolutionized our knowledge of polymorphism and mutation in the human genome. SNP discovery efforts and processing of genome sequencing data have yielded several million base positions and several hundred thousand VNTRs that might be polymorphic in the genome. This information is complemented by a more select collection of mutation data painstakingly accumulated over many years of disease-gene hunting and mutation analysis. The sheer scale of this data offers tremendous opportunities for genetics and biology. We are now entering a new phase in genetics where we can begin to design experiments to capture the full genetic diversity of populations. This may herald a revolution in genetics allowing rapid association of genes with diseases, alternatively it may simply identify further downstream bottlenecks in the progression to validated disease genes. The literature is already replete with reports of genetic associations and still more failures to replicate associations, but progressions from associated marker to validated disease gene are rare indeed. This may be the real challenge for genetics — to cast new insight into the structure and function of genes, proteins and regulatory regions. To achieve this we will need to integrate diverse sources of data to build up complete pictures of biological systems and their interactions with disease. Again an understanding of mutation and polymorphism may be an important aid in this process — with mutations representing the extreme boundaries beyond which genes begin to dysfunction and polymorphisms perhaps representing the functional range within which genes can operate. Our knowledge of the breadth and variety of human genetic variation can only increase our understanding of the mechanisms of disease and more importantly it may help us to define targets for intervention.

REFERENCES

Altshuler D, Pollara VJ, Cowles CR, Van Etten WJ, Baldwin J, Linton L, *et al.* (2000). A SNP map of the human genome generated by reduced representation shotgun sequencing. *Nature* **407**: 513–516.

Anderson S, Bankier AT, Barrell BG, de Bruijn MHL, Coulson AR, Drouin J, *et al.* (1981). Sequence and organization of the human mitochondrial genome. *Nature* **290**, 457–465.

Aparicio S, Morrison A, Gould A, Gilthorpe J, Chaudhuri C, Rigby P, *et al.* (1995). Detecting conserved regulatory elements with the model genome of the Japanese puffer fish, *Fugu rubripes. Proc Natl Acad Sci USA* **92**: 1684–1688.

Bahar AY, Taylor PJ, Andrews L, Proos A, Burnett L, Tucker K, *et al.* (2001). The frequency of founder mutations in the BRCA1, BRCA2, and APC genes in Australian Ashkenazi Jews: implications for the generality of U.S. population data. *Cancer* **92**: 440–445.

Benson G. (1999). Tandem repeats finder: a program to analyze DNA sequences. *Nucleic Acids Res* **27**: 573–580.

Bonetta L. (2001). Sackings leave gene database floundering. *Nature* **414**: 384.

Brookes AJ, Lehväslaiho H, Siegfried M, Boehm JG, Yuan YP, Sarkar CM, *et al.* (2000). HGBASE: A database of SNPs and other variations in and around human genes. *Nucleic Acids Res* **28**: 356–360.

Cambien F, Poirier O, Lecerf L, Evans A, Cambou J-P, Arveiler D, *et al.* (1992). Deletion polymorphism in the gene for angiotensin-converting enzyme is a potent risk factor for myocardial infarction. *Nature* **359**: 641–644.

Collins FS, Brooks LD, Chakravarti A. (1998). A DNA polymorphism discovery resource for research on human genetic variation. *Genome Res* **8**: 1229–1231.

Cooper DN, Youssoufian H. (1988). The CpG dinucleotide and human genetic disease. *Hum Genet* **78**: 151–155.

Cuticchia AJ. (2000). Future vision of the GDB Human Genome Database. *Hum Mut* **15**: 62–67.

Darvasi A, Kerem B. (1995). Deletion and insertion mutations in short tandem repeats in the coding regions of human genes. *Eur J Hum Genet* **3**: 14–20.

Debrauwere H, Gendrel CG, Lechat S, Dutreix M. (1997). Differences and similarities between various tandem repeat sequences: minisatellites and microsatellites. *Biochimie* **79**: 577–586.

Deininger PL, Batzer MA. (1999). Alu repeats and human disease. *Mol Genet Metab* **67**: 183–193.

Emanuel BS, Shaikh TH. (2001). Segmental duplications: an 'expanding' role in genomic instability and disease. *Nature Rev Genet* **2**: 791–800.

Fredman D, Siegfried M, Yuan YP, Bork P, Lehvaslaiho H, Brookes AJ. (2002). HGVbase: a human sequence variation database emphasizing data quality and a broad spectrum of data sources. *Nucleic Acids Res* **30**: 387–391.

Goldstein DB, Schlotterer C. (Eds) (1999). *Microsatellites—Evolution and Applications*. Oxford University Press: Oxford, UK.

Gong W, Emanuel BS, Collins J, Kim DH, Wang Z, Chen F, *et al.* (1996). A transcription map of the DiGeorge and velo-cardio-facial syndrome minimal critical region on 22q11. *Hum Mol Genet* **5**: 789–800.

Gratacos M, Nadal M, Martin-Santos R, Pujana MA, Gago J, Peral B, *et al.* (2001). A polymorphic genomic duplication on human chromosome 15 is a susceptibility factor for panic and phobic disorders. *Cell* **106**: 367–379.

Hamosh A, Scott AF, Amberger J, Valle D, McKusick VA. (2000). Online Mendelian Inheritance in Man (OMIM). *Hum Mut* **15**: 57–61.

International SNP Map Working Group (2001). A map of human genome sequence variation containing 1.42 million single nucleotide polymorphisms. *Nature* **409**: 928–933.

Kawabata T, Ota M, Nishikawa K. (1999). The protein mutant database. *Nucleic Acids Res* **27**: 355–357.

Kimura M. (1983). *The Neutral Theory of Molecular Evolution*. Cambridge University Press: Cambridge, UK.

Kluijtmans LA, van den Heuvel LP, Boers GH, Frosst P, Stevens EM, van Oost BA, *et al.* (1996). Molecular genetic analysis in mild hyperhomocysteinemia: a common mutation in the methylenetetrahydrofolate reductase gene is a genetic risk factor for cardiovascular disease. *Am J Hum Genet* **58**: 35–41.

Kogelnik AM, Lott MT, Brown MD, Navathe SB, Wallace DC. (1998). MITOMAP: a human mitochondrial genome database—1998 update. *Nucleic Acids Res* **26**: 112–115.

Krebs S, Seichter D, Forster M. (2001). Genotyping of dinucleotide tandem repeats by MALDI mass spectrometry of ribozyme-cleaved RNA transcripts. *Nature Biotechnol* **19**: 877–880.

Kubota S. (2001). The extinction program for *Homo sapiens* and cloning humans: trinucleotide expansion as a one-way track to extinction. *Med Hypotheses* **56**: 296–301.

Le Stunff C, Fallin D, Schork NJ, Bougneres P. (2000). The insulin gene VNTR is associated with fasting insulin levels and development of juvenile obesity. *Nature Genet* **26**: 444–446.

Lucassen AM, Julier C, Beressi JP, Boitard C, Froguel P, Lathrop M, *et al.* (1993). Susceptibility to insulin dependent diabetes mellitus maps to a 4.1-kb segment of DNA spanning the insulin gene and associated VNTR. *Nature Genet* **4**: 305–310.

Marth GT, Korf I, Yandell MD, Yeh RT, Gu Z, Zakeri H, *et al.* (1999). A general approach to single-nucleotide polymorphism discovery. *Nature Genet* **23**: 452–456.

Marth G, Yeh R, Minton M, Donaldson R, Li Q, Duan S, *et al.* (2001). Single-nucleotide polymorphisms in the public domain: how useful are they? *Nature Genet* **27**: 371–372.

Matsuura T, Yamagata T, Burgess DL, Rasmussen A, Grewal RP, Watase K, *et al.* (2000). Large expansion of the ATTCT pentanucleotide repeat in spinocerebellar ataxia type 10. *Nature Genet* **26**: 191–194.

Miller RD, Kwok PY. (2001). The birth and death of human single-nucleotide polymorphisms: new experimental evidence and implications for human history and medicine. *Hum Mol Genet* **10**: 2195–2198.

Miller RD, Taillon-Miller P, Kwok PY. (2001). Regions of low single-nucleotide polymorphism incidence in human and orang-utan xq: deserts and recent coalescences. *Genomics* **71**: 78–88.

Mitelman F, Mertens F, Johansson B. (1997). A breakpoint map of recurrent chromosomal rearrangements in human neoplasia. *Nature Genet* **15**: 417–474.

Mitelman F, Johansson B, Mertens F (Eds) (2002). Mitelman Database of Chromosome Aberrations in Cancer http://cgap.nci.nih.gov/Chromosomes/Mitelman.

Mullikin JC, Hunt SE, Cole CG, Mortimore BJ, Rice CM, Burton J, *et al.* (2000). An SNP map of human chromosome 22. *Nature* **407**: 516–520.

Olson M, Hood L, Cantor C, Botstein D. (1989). A common language for physical mapping of the human genome. *Science* **245**: 1434–1435.

Reich D, Cargill M, Bolk S, Ireland J, Sabeti P, Richter D, *et al.* (2001). Linkage disequilibrium in the human genome. *Nature* **411**: 199–204.

Riggins GJ, Strausberg RL. (2001). Genome and genetic resources from the Cancer Genome Anatomy Project. *Hum Mol Genet* **10**: 663–667.

Risch N. (2000). Searching for genetic determinants in the new millennium. *Nature* **405**: 847–856.

Riva AA, Kohane IS. (2001). A web-based tool to retrieve human genome polymorphisms from public databases. *Proc AMIA Symp* 558–562.

Roque M, Godoy CP, Castellanos M, Pusiol E, Mayorga LS. (2001). Population screening of F508del (DeltaF508), the most frequent mutation in the CFTR gene associated with cystic fibrosis in Argentina. *Hum Mut* **18**: 167.

Schmucker B, Krawczak M. (1997). Meiotic microdeletion breakpoints in the BRCA1 gene are significantly associated with symmetric DNA sequence elements. *Am J Hum Genet* **61**: 1454–1456.

Shaikh TH, Kurahashi H, Emanuel BS. (2001). Evolutionarily conserved duplications in 22q11 mediate deletions, duplications, translocations and genomic instability. *Genet Med* **3**: 6–13.

Shapira SK. (1998). An update on chromosome deletion and microdeletion syndromes. *Curr Opin Pediatr* **10**: 622–627.

Sherry ST, Ward MH, Kholodov M, Baker J, Phan L, Smigielski EM, *et al.* (2001). dbSNP: the NCBI database of genetic variation. *Nucleic Acids Res* **29**: 308–311.

Stein LD. (2001). Using Perl to facilitate biological analysis. *Methods Biochem Anal* **43**: 413–449.

Strausberg RL, Buetow KH, Emmert-Buck MR, Klausner RD. (2000). The cancer genome anatomy project: building an annotated gene index. *Trends Genet* **16**: 103–106.

Thomas PK. (1999). Overview of Charcot-Marie-Tooth disease type 1A. *Ann NY Acad Sci* **883**: 1–5.

Usdin K, Grabczyk E. (2000). DNA repeat expansions and human disease. *Cell Mol Life Sci* **57**: 914–931.

Wallace DC, Shoffner JM, Trounce I, Brown MD, Ballinger SW, Corral-Debrinski M, *et al.* (1995). Mitochondrial DNA mutations in human degenerative diseases and aging. *Biochim Biophys Acta* **1272**: 141–151.

CHAPTER 4

Finding, Delineating and Analysing Genes

CHRISTOPHER SOUTHAN

Oxford GlycoSciences UK Ltd
The Forum, 86 Milton Science Park
Abingdon OX14 4RY, UK

4.1 INTRODUCTION

This chapter will describe ways to interrogate human genome (HG) data with the results of genetic experiments in order to locate known genes on the current Golden Path (GP) chromosomal assemblies. It will also describe the assessment of evidence for genes that do not yet have experimental support and some analytical choices that may reveal more about them. In addition to some general aspects of gene detection some specific examples will be worked through in some detail. This illustrates technical subtleties that are not easy to capture at the overview level. As an introduction to the HG, GP and gene annotation the following chapter by Semple is recommended. Chapter 2 also provides some useful background on the organization of sequence databases. A caveat needs to be added here that many roads lead to Rome. Some particular ways of hacking through the genome jungle

Bioinformatics for Geneticists. Edited by M.R. Barnes and I.C. Gray
 ISBNs: 0 470 84393 4; 0 470 84394 2 (PB)

are implicitly recommended by being used as the examples in this chapter. They will also be restricted to public databases and web tools. These are the personal choices of the author based on an assessment of their availability and utility. Other experts may propose alternative routes to the same information, either using different public resources, locally downloaded datasets, Unix-based tools, commercial software or subscription databases.

Genetic investigations are concerned with discerning the complex relationships between genotype and phenotype. The statement that phenotype is determined by the biochemical consequences of gene expression is equally obvious. However, the reason for making this explicit is to recommend that those performing and interpreting genetic experiments may find it more useful to conceptualize the gene as a cascade of evidence that connects DNA to a protein product rather than abstract ideas about what might constitute a gene locus. The idea of focusing on gene products also makes it easier to design experiments to verify predicted transcripts and proteins. It must also be remembered that many gene products are non-message RNA molecules but they will not be covered in this chapter. Before describing the evidence used to classify gene products it is necessary to define some of the terminology encountered in the literature and database descriptions. These are variously classified as known, unknown, hypothetical, model, predicted, virtual or novel. There are no widely accepted definitions of these terms but their usage in this chapter will be as follows. A known gene product is experimentally supported and would be expected to give close to a 100% identity match to a unique GP location. The term 'unknown' is typically applied to gene products that are supported experimentally but that lack any detectable homology or experimentally determined function. The term 'predicted', also referred to as 'model' or 'hypothetical' by the NCBI, will be reserved for an mRNA or protein ORF predicted from genomic DNA. Virtual mRNAs will refer to constructs assembled from overlapping ESTs that exceed the length of any single component. The term 'novel' has diminishing utility and will simply refer to a protein with no extended identity hits in the major protein databases.

4.2 THE EVIDENCE CASCADE FOR GENE PRODUCTS

So what kinds of evidence need to be considered before we assess the likelihood of a stretch of genomic DNA giving rise to a gene product and what kind of numbers can be assigned to these evidence levels? The most solid evidence of a gene is the experimental verification of the protein product by mass spectrometry and/or Edman sequencing. Although these techniques are commonly used to analyse proteins produced by heterologous expression *in-vitro* surprisingly few genes from *in-vivo* or cell line sources have been verified at this level. From the entire SP/TR collection of human proteins only 311 are cross-referenced as having at least a fragment of their primary structure identified directly from a 2D-PAGE experiment (http://ca.expasy.org/ch2d/) (Hoogland *et al.*, 2000). Numerous mass spectrometry-based identifications and peptide sequences from human proteins are reported in the literature but little of this data has been formally submitted to the public databases and therefore has not been captured by SwissProt or other secondary databases. However, even this most direct of gene product verifications is rarely sufficient to confirm the entire open reading frame (ORF). For example secreted proteins are characterized by the removal of signal peptides and frequent C-terminal processing. This precludes defining the N and C translation termini by protein chemical means.

The next level down in the evidence cascade is of course an extended mRNA. There are currently 48,681 human mRNAs in GenBank. However transcript coverage is by no means complete as they collapse down by shared identity to a set of 13,429 human transcripts (excluding splice variants) in the NCBI RefSeq collection (http://www.ncbi.nlm.nih.gov/LocusLink/RSstatistics.html) (Pruitt and Maglott, 2001). Although this collection attempts to provide a non-redundant snapshot of gene transcription it must be remembered that they are not all full-length transcripts. If the databases do not contain an extended mRNA the assembly of overlapping and/or clone-end clustered ESTs can be considered as a virtual mRNA (Schuler, 1997). The ESTs have the additional utility that many of them can be ordered as clones. Alternatively, the virtual consensus sequence, backed up by comparisons to the genomic DNA, can be used for PCR cloning. The fact that 94% of known mRNAs are covered by at least one EST makes them strong supporting evidence for a transcript, especially if they include a plausible splice junction and are derived from multiple clones from different tissue cDNA libraries (http://www.ncbi.nlm.nih.gov/UniGene/). The TIGR gene indexes are a useful source of pre-assembled virtual sequences that they term tentative human consensus sequences or THCs (Quackenbush *et al.*, 2001). These can also be selected in the UCSC genome display. The use of unspliced ESTs as evidence for a transcribed gene is unreliable as they can arise from genomic contamination. However human EST-to-genome matches for exon detection can be further supported where orthologous ESTs from other vertebrates, such as mouse or rat, match uniquely in the same section of GP. If an assembly of mouse ESTs is consistent with a human gene model then the existence of an orthologous human transcript is strongly implicated.

The protein databases occupy the centre of the evidence cascade for gene products. Those mRNAs that translate to an open reading frame (ORF) are experimentally supported even if they are not full-length and/or there can be ambiguity about the choice of potential initiating methionines. However, the fact that the protein databases have now expanded to include human ORFs derived solely from genomic predictions (described in the next section) means that the evidence supporting them as gene products becomes circular. The highest curation level is provided by SwissProt sequences from the Human Proteomics Initiative set (HPI) (http://ca.expasy.org/sprot/hpi/hpi_stat.html). The March 2002 number comprised 7895 unique gene products and 2039 splice variants (O'Donovan *et al.*, 2001). The SwissProt/TrEMBL (SP/TR) total for human proteins in February 2002 was 24,147, including splice variants (http://www.ebi.ac.uk/proteome/HUMAN/interpro/stat.html). The current Ensembl release, 4.28.1, contains 21,619 proteins classified as 'knowns' by an identity above 95% to a human SP/TR entry (Hubbard *et al.*, 2002). The International Protein Index (IPI) maintains a database of cross references between the data sources SwissProt, TrEMBL, RefSeq and Ensembl. This provides a minimally redundant yet maximally complete set of human proteins with one sequence per transcript (http://www.ebi.ac.uk/IPI/IPIhelp.html). The March 2002 release contains 65,082 protein sequences but this includes 28,350 XP RefSeq ORFs predicted by the NCBI which are not supported by mRNAs.

The next level of evidence can be classified as genomic prediction i.e. where a cDNA, a translated ORF and a plausible gene splice pattern can be predicted from a stretch of genomic DNA (Burge and Karlin, 1997). This proceeds more effectively on finished sequence or at least where unfinished sequence contains the exons in the correct order. This is done after filtration of repeats which can be considered as another link in the evidence chain. A very high local repeat density certainly suggests where exons are unlikely but the converse is not true i.e. the absence of repeats does not prove the presence of genes. The shortcomings of *ab initio* gene prediction have been pointed out but the geneticist should at least be aware of possible false positives and false negatives (Guigo *et al.*, 2000). The Ensembl statistics of the ratio of genes predicted by Genscan over genes with a high evidence-supported threshold is currently 7.5 : 1 (http://www.ensembl.org/Homo_sapiens/stats/). Although this clearly represents over-prediction some may be 'genes-in-waiting' which more accumulated evidence may verify, for example by the cloning of an extended mRNA. Looking for a consensus or at least common exons from a number of gene prediction programs with different underlying gene model assumptions can strengthen this type of evidence but this can become a circular argument where the programs are both trained and benchmarked with known genes. For unfinished genomic sequence the presence of gaps and local miss-ordering of contigs within the clone degrades the performance of *ab initio* methods. The most effective way of filtering down genomic predictions without experimental evidence is homology support i.e. the predicted protein shows extended similarity with other proteins. This is described in detail in the Ensembl documentation but in essence all possible protein similarity sections from translated DNA are identified and used to build homology-supported gene predictions using GeneWise (Birney and Durbin, 2000). The advantage of gene detection by homology is that the entirety of protein sequence space can be used. The caveat is that predicted gene products with low similarity to extant proteins would be discarded in this filter, although the entire set of Genscan predictions are preserved for searching in Ensembl and can also be displayed at UCSC.

The next link in the evidence chain is a special case of the similarity principle but in this case utilizing comparisons between the genomes of other vertebrates such as mouse and fish for which extended data are now available (Wiehe *et al.*, 2001). Mouse genome assemblies have recently appeared on the Ensembl and UCSC sites. Although the initial assembly is only ~20% the total depth in the trace archives and HTGS divisions is approaching complete coverage. Cross-species data can be assessed at three levels. The first is a simple DNA similarity on pieces of mouse DNA known to be syntenic from the location of known mouse genes and/or the extended similarity score which, with appropriate masking, locates it uniquely to a human locus. This approach is termed phylogenetic footprinting (Susens and Borgmeyer, 2001). The problem for gene product detection is that this is too sensitive i.e. mouse/human syntenic regions have many conserved similarity 'patches' outside the boundaries of known exons. They are likely to be important for functions not yet understood but are difficult to discriminate from potential coding regions. The second level is mouse BLAT as used on the UCSC site. This goes a step back by doing a translation similarity comparison rather than direct DNA-to-DNA. This makes it more likely to pick up reading frame similarities across exons. The third level is the so-called exofish. By the detection of translation similarities at the amino acid level this is capable of detecting those exons that are conserved between human and fish. This will be more useful when exofish updates to a complete fish genome rather than a partially assembled one.

The last link in the evidence chain, the *in silico* recognition of transcriptional control regions, is circumstantial but is likely to increase in utility (Kel-Margoulis *et al.*, 2002). These could include potential start sites in proximity to CpG islands, promoter elements, transcription factor binding sites, and potential polyadenylation acceptor sites in 3′ UTR. When considered in isolation these signals have poor specificity but taken in combination with a consensus gene prediction and conservation of these putative control regions between human and mouse, they can become a useful part of the evidence chain.

In summary there is currently direct experimental evidence for ~15,000 genes and strong evidence to support a lower gene limit of around 25,000. The confirmation rates for the types of evidence listed above has not been calibrated experimentally so we cannot come up with any kind of scoring function to rank gene likelihood. Going to the extremities of the evidence cascade, for example with the 65,082 ORFs from the IPI or the 62,271 UniGene clusters containing at least two ESTs, would result in a higher upper limit. This uncertainty becomes a key issue for genetic experiments. Let us suppose, for example, that a linkage study has defined a trait within the genomic region bounded by two microsatellite markers. If the lower limit gene number is true then the investigator merely needs to check the annotations from any of the three gene portals to produce a list of gene products between the positioned markers from which to choose candidates for further work. If the upper limit is true this approach has a major limitation because many of the genes between the markers will not be annotated. However, the different levels of gene evidence described above can be visualized in the display tracks of the genome viewers. Consideration of the evidence will enable the geneticist to decide what experiments need to be designed to confirm potential novel gene products. An example of working through this evidence is given in the examples below.

4.3 SHORTCOMINGS OF THE STANDARD GENE MODEL

One of the conclusions that could be drawn from the draft human genome sequence was that the standard gene model of a defined gene locus → a single mRNA species → a single protein, is no longer adequate to describe the increasingly complex relationship between the genome and its products. Attempts to fit transcript data into the standard gene model highlight a number of 'grey' areas. The first of these is delineating the extreme 5′ and 3′ ends of the mRNA transcripts (Pesole *et al.*, 2002; Suzuki *et al.*, 2002). The fact that many mRNAs are labelled as partial is testimony to the difficulty of finding library inserts that are complete at the 5′ end. In many cases the mRNAs are considered finished when a plausible ORF has been delineated. However, very few cDNAs are full-length in that they have been 'walked out' to determine the true 5′-most initiation of transcription in the 5′ UTR. The same problem applies to the UTR at the 3′ end. There may be substantial stretches of 3′ UTR extending downstream of the first polyadenylation position at which further cloning attempts have ceased. The problem is compounded by the poor performance of gene prediction programs for 5′ and 3′ ends. The first step towards resolving uncertainties about transcript extremities, is to survey the coverage of all available cDNA sequences, whether nominally full-length or partial, ESTs and patent sequences. These can often extend the UTR sections. The second grey area concerns pseudogenes. In some cases genomic sequence is so severely degraded that transcription is unlikely. However, from the current pseudogene listing in RefSeq of 1598 loci, at least 30 are recorded as having detectable transcripts (http://www.ncbi.nlm.nih.gov/LocusLink/statistics.html). The third grey area is gene product heterogeneity. In some cases there may be alternative upstream initiation methionines or alternatively spliced exons in the 5′ UTR. The causes for 3′ heterogeneity include variations in the pattern of intron splicing from a pre-mRNA, as well as alternative poladenylation positions inside the 3 UTR. The fourth grey area concerns overlapping genes. As genomic annotation proceeds we can find more examples of this both from gene products reading from opposite strands and same-strand genes in close proximity.

Considering these grey areas as a whole, they can all be seen as deviations from the simple gene model. Many individual examples of such complexities had been documented before the genome draft of May 2001. However, it is only since then that assessments of their overall incidence could be made, most recently for completed chromosomes such as 20 (Deloukas *et al.*, 2001). It is therefore essential for the geneticist to keep an open mind about the extremities and plurality of gene products.

4.4 LOCATING KNOWN GENES ON THE GOLDEN PATH

Genes can be located by one of the following: a section of raw sequence data, a primary accession number, a secondary accession number, a similarity search, a gene product name, or a set of Golden Path (GP) coordinates. Each of these has advantages and disadvantages and, although the three gene portals are generally consistent, they may not give the same answers in every case. Bearing in mind that only the first two of these are stable and (almost) free of potential ambiguity it is better to use at least two ways to define and store the results, for example a section of raw sequence and a gene name, or a primary accession number and a set of GP coordinates. The BACE gene will be used as an example of a known gene to locate. The potential complexity of this task is illustrated by the example of the Ensembl gene report for BACE that includes no less than 46 separate terms (Figure 4.1).

4.4.1 Raw Sequence Data

The availability of GP means that most features can now be unambiguously located in the genome with as little as 100 bp. This means that storing a sequence string, preferably with a longer sequence context of 200–1000 bp, is a useful method of locking-on to a genomic location. It is also immune to the vagaries of shifting secondary accession numbers, naming ambiguities or GP sequence finishing that can change the genomic coordinates. Performing nucleotide searches against GP using tools such as BLAT (UCSC) or SAHA (Ensembl) or BLAST (NCBI), means that sequence matches can be quickly located. The disadvantage for raw sequence is that it has to be stored in its entirety, it may contain errors, it needs the operation of a similarity search to be located and similarity matches across repeat containing sections or duplicated regions of the genome need close inspection to sort out. This can be a particular problem for STSs and SNPs

Ensembl gene ID	ENSG00000160610
Genomic Location	**View gene in genomic location:** 120549397 - 120575715 bp (120.5 Mb) on chromosome 11 **This gene is located in sequence:** AP001822.4.1.134278
Description	BETA-SECRETASE PRECURSOR (EC 3.4.23.-) (BETA-SITE APP CLEAVING ENZYME)(BETA-SITE AMYLOID PRECURSOR PROTEIN CLEAVING ENZYME) (ASPARTYLPROTEASE 2) (ASP 2) (ASP2) (MEMBRANE-ASSOCIATED ASPARTIC PROTEASE 2) (MEMAPSIN-2). [Source:SWISSPROT;Acc:P56817]
Prediction Method	This gene was predicted by the Ensembl analysis pipeline from either a GeneWise or Genscan prediction followed by confirmation of the exons by comparisons to protein, cDNA and EST databases
Predicted Transcripts	1: ENST00000292095 [View supporting evidence] [View protein information]
Links	**This Ensembl gene corresponds to the following other database identifiers** **EMBL:** AB032975 [align] AB050438 [align] AF190725 [align] AF200193 [align] AF200343 [align] AF201468 [align] AF204943 [align] AF338816 [align] **GO:** GO:0004194 GO:0005624 GO:0005887 GO:0006508 GO:0008798 GO:0009405 GO mapping is inherited from swissprot/sptrembl **HUGO:** Search GeneCards for BACE **LocusLink:** 23621 [align] **MIM:** 604252 **RefSeq:** NM_012104 [Target %id: 93; Query %id: 87] [align] **SWISSPROT:** BACE_HUMAN [Target %id: 93; Query %id: 87] [align] [Search GO] **SpTrEMBL:** Q9BYB9 [Target %id: 91; Query %id: 74] [align] [Search GO] Q9ULS1 [Target %id: 87; Query %id: 87] [align] [Search GO] **protein_id:** AAF04142 [align] AAF13715 [align] AAF17079 [align] AAF18982 [align] AAF26367 [align] AAK38374 [align] BAA86463 [align] BAB40933 [align]
InterPro	IPR001461 Pepsin (A1) aspartic protease [View other Ensembl genes with this domain] IPR001969 Eukaryotic and viral aspartic protease active site [View other Ensembl genes with this domain]
Protein Family	ENSF00000001079 : BETA SECRETASE PRECURSOR EC 3.4.23.- BET This cluster contains 2 Ensembl gene member(s)
Export Data	Export gene data in EMBL, GenBank or FASTA

Figure 4.1 The Ensembl gene report page for BACE (release 4.28.1).

if the GP match is in the region of 98 to 95% identity. Within this range it is difficult to discriminate technical sequencing errors from multiple genomic locations or assembly duplication errors. It can also be useful to search the primary genomic data, especially if GP is not complete in that section. For example although BACE is linked by Ensembl to AP001822 as the finished GP sequence, a database search reveals another four matching primary genomic accession numbers from chromosome 11, AP000892 (finished at version 4) with AC020997, AP000685 and AP000761 still unfinished. One less obvious advantage of these five overlapping genomic contigs is that if they proceed to finishing more SNP positions may be revealed. As described below the genome portals capture mRNA entries for most gene products unless they are very recent. However, because of the thin annotation they do not capture sequences from the patent divisions. A BLAST search of gbPAT with any BACE mRNA gives 18 high-identity DNA matches. These are clearly mRNAs that could be usefully compared with all other mRNA sequences for polymorphisms, splice variants or UTR differences. However users should be aware that not only are some of these 18 entries identical versions of the same sequence derived from multiple claims in the patent documents but they may also be identical to a public accession number if the authors and inventors are from the same institution. Another reason for using raw sequence data for gene product checking is because all secondary databases suffer from the snapshot effect where updates lag behind the content of the primary databases. For example the SNP or EST assignments made for BACE in the secondary databases (see below) could be checked by BLAST searches against the updates of dbSNP or dbEST (remember the latest EST data needs to be searched in 'month' as well as dbEST).

4.4.2 Primary Accession Numbers

Because these uniquely define stretches of sequence they are stable except where genomic and occasionally mRNAs, undergo version changes. They can be used in any of the major genome query portals to go directly to a genomic location. The disadvantage is redundancy for mRNAs, short sequence context for some STSs, both redundancy and large multi-gene sequence tracts for genomic mRNA, and very recent accessions may not be indexed in genome builds. If the query fails to connect to a genome feature the sequences can be searched as raw sequence. Taking the BACE example there are eight mRNA accession numbers listed in Figure 4.1 that can be used as a genome portal query. Interrogating UCSC with BACE retrieves nine mRNA entries, LocusLink connects directly to only three but the UniGene cluster Hs.49349 connects to 12. Users need to be aware that although an mRNA accession number can provide a specific route into GP the variable number of links to the genome portals is related to their update frequency.

4.4.3 Secondary Accession Numbers

From Figure 4.1 we can read eight secondary accession numbers that designate protein translations for each of the BACE mRNAs. It also has three RefSeq numbers NM_012104 for the mRNA, NP_036236 for the protein and NT_009151 for the genomic contig. There is one SwissProt accession BACE_HUMAN (P56817) and one TREMBL splice variant Q9BYB9. The LocusID, 23621, in turn links out to many other accession numbers which point to the BACE genome sequence. These include the Hs.49349 UniGene cluster that includes 336 ESTs with primary accession numbers. Via the LocusLink Variation link the RefSNP numbers can be located. In this case they consist of 43 intronic SNPs, three within the mRNA, including one (rs539765) which causes an Arg > Cys exchange, and seven SNPs in the 3′ UTR. It is possible to use a RefSNP (rs) number to go directly to the SNP location in Ensembl or UCSC. However because of multiple GP matches in Ensembl it is necessary to know the genomic location beforehand.

4.4.4 Gene Names

Including abbreviations Figure 4.1 there are nine synonyms or aliases for this enzyme. This illustrates the problem where gene products are given different names by different authors. The best way to cross-check names, spelling variations and frequency of use, is to search PubMed. Checking title lines only is more specific but does not capture all occurrences. In this case a title search found a new name extension, BACE1, with five citations compared with 22 for BACE. This seems logical since the discovery of the BACE2 paralogue on chromosome 21. However, the Human Gene Nomenclature Committee have not been consistent because they have only listed BACE and BACE2 as official symbols even though they have listed ACE1 as an alias for ACE since the recent discovery of ACE2 (http://www.gene.ucl.ac.uk/nomenclature/). The most frequent specific term was 'beta-secretase precursor' at 30 citations. The alternative 'membrane-associated aspartic protease 2′ gave eight citations and 'beta-site app cleaving enzyme' was the least frequent with only two. Paradoxically this has been chosen for the LocusLink name. The least specific name was aspartylprotease 2 with two false positives and ASP 2 with 143 title matches, also mostly false positives. The imprecision of name searching was reinforced by checking ASP-2 with three matches and ASP2 with five. Only one was a true positive and two of the citations referred to ASP2 as an odorant-binding protein from the honeybee. The complexity of the aliases for just one gene product makes it clear that any gene name lists, for example as candidate genes to be screened for mutations, must be backed-up by accession numbers and/or raw sequence. It also illustrates the need to cross-check aliases and their spellings when attempting a comprehensive literature search on a particular gene product. The formal sequence-literature links that can be followed in Entrez, LocusLink or SwissProt are not comprehensive because they are dependent on the journal–author–database system that usually only makes these links explicit for a new accession number. Much important literature remains outside this system. Review articles, for example, do not typically include primary accession numbers when describing genes so the specificity of literature searches remains dependent on the name links. Information trawling with gene names can also be done with the standard internet search portal. Putting the term 'beta-site app cleaving enzyme' into the Google search engine gave 249 hits (http://www.google.com/). The listing included duplicates but very few false positives.

4.4.5 Genome Coordinates

Since the adoption of a unified GP assembly this method of genomic location has become more reliable but users are advised to check the synchronization of new GP versions between the three portals. Users should refer to the individual portals for the details of using these coordinates but for the BACE example the NCBI showed a region described as 120,533K–120,594K, the Ensembl viewer specified the coordinates as 120549397–120575715 bp (with a zoom setting 120.5 Mb) on chromosome 11 and the UCSC viewer designated the position in the form chr11 : 120545299–120599798.

4.5 GENE PORTAL INSPECTION

From the descriptions above it should be possible to locate any known gene or genetic marker such as an STS or a SNP. Descriptions of the genome viewer features for Ensembl, UCSC and NCBI are included in the chapter by Semple. However two examples are included below (Figures 4.2 and 4.3) because they illustrate technical differences and highlight the deviations from the standard gene model. The UCSC display (Figure 4.2) includes 12 mRNA sequences for BACE where Ensembl (Figure 4.1) has included accession number links for only eight. The display in Figure 4.2 also shows there are significant differences in the lengths of the 5′ and 3′ ends. Clearly AF201468 (5878 bp) and AB032975 (5814 bp) are the longest reads but in fact AB032975 is labelled as a partial CDS because of what may be a sequencing error at the 5′ end. The matches to the spliced ESTs together with the rat and mouse mRNAs suggest the 5′ UTR may be full-length for these entries i.e. they extend to the start of transcription. This is in contrast to the shorter 5′ ends for the majority of mRNAs. A detailed analysis of the 3′ ends by EST distribution profiles indicates that the different UTR lengths in this case arise not from incomplete cloning but from three alternative polyadenylation positions (Southan, 2001). Further heterogeneity is illustrated by three splice variants affecting exons 3 and 4. The representative mRNAs are AB050436, AB050437 and AB050438. There is also an alternative protein reading frame from AF161367, a partial mRNA cloned from CD34+ stem cells. Opening up the spliced EST tracks in the viewer shows individual ESTs corresponding to these splice forms. Approximately midway between exons 1 and 2 (from the 5′ end) is a spliced EST, AL544727, derived from spleen. This suggests the possibility of another splice form but this would need analysis for canonical splice sites and experimental verification. Similarly an EST from spinal cord AL589586 suggests an alternative exon just on the 5′ side of exon 3. Although the rat and mouse mRNAs displayed in Figure 4.2 show the same exon positions as most human sequences there are suggestions of splice variants in non-human ESTs but these tracks were not expandable in the version tested.

The NCBI display for BACE mRNAs and ESTs (Figure 4.3) shows concordance and discrepancies with the UCSC display (Figure 4.2). The exon positions are identical. They include the same RefSeq mRNA and genomic secondary accession numbers. The EST matches are in broad agreement towards the 3′ end but two additional potential exon matches are indicated at the 5′ end. Although these may be unspliced matches that would need further investigation, one of these coincides with the XM_084660 reference sequence

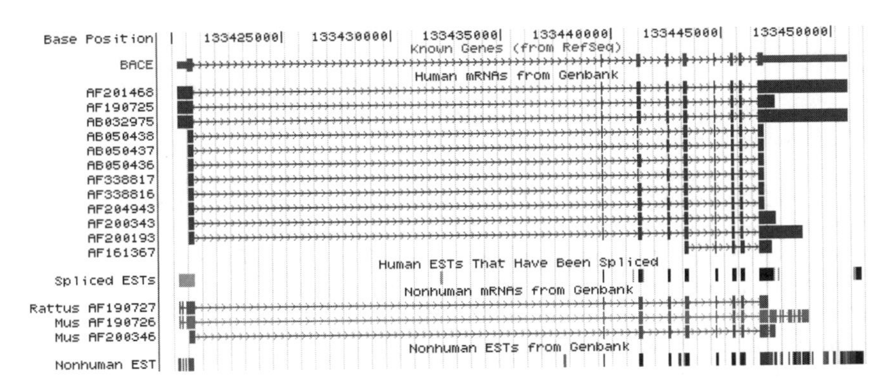

Figure 4.2 The UCSC display for BACE mRNAs and ESTs.

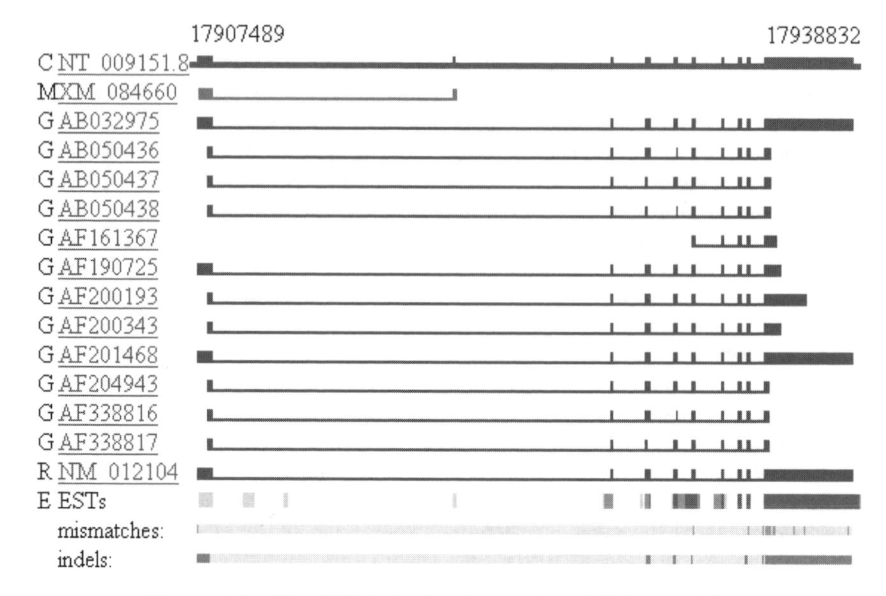

Figure 4.3 The NCBI display for BACE mRNAs and ESTs.

predicted by NCBI from the contig NT_009151. There is no mRNA verification for this prediction so it will be of interest to see if additional EST data will appear and, if not, how long this prediction will be maintained as genome annotation. The mismatches and INDEL tracks are a useful feature unique to NCBI. The mismatches within the set of 12 mRNA sequences could represent SNPs or technical sequence errors. The INDELs also show major length discrepancies. In Figure 4.2 these highlight the three splice positions in agreement with UCSC but the INDEL in exon 8 could not be interpreted from the link provided.

4.6 LOCATING GENES WHICH ARE NOT PRESENT IN THE GOLDEN PATH

Estimates suggest the GP is still missing ~2.5% of the genome, there are still small gaps in the unfinished sections and the latest Ensembl release locates only 92% of known proteins (http://www.ensembl.org/Dev/Lists/announce/msg00070.html). This means that some genetic markers in close proximity to genes are either not covered by GP or are not fully annotated in unfinished sequence. Two human proteins that have no matches on the current GP version 28 from December 2001 illustrate this problem. The first of these, spP83110 serine protease HTRA3, has an mRNA entry AY040094. The second protein spP83105 serine protease HTRA4 has an mRNA accession but the entire ORF is covered by two long EST reads AL545759 and AL576444. Because it has a full length mRNA HTRA3 has a LocusLink ID of 94031 but no mapping links. Searching HTRA3 by BLASTN against the NCBI nr nucleotide database, containing 1,184,532 sequences, hits only the probable mouse orthologous mRNA, AY037300, at 86% identity within the reading frame. However checking monthly updates at 811,100 sequences reveals a 99% identity to a new genomic entry AC113611 of 190,038 bp from chromosome 4. This sequence was also in the unfinished High Throughput Genomic Sequences (HTGS) division, with 47,855 sequences, along with the probable rat orthologous genomic section, AC110369, at 87% identity. There were no mouse genome matches from this search. A check on the nucleotide patent databases, with 582,838 sequences, showed a new mRNA match, AX338509 from patent WO0183775. The HTRA3 mRNA has EST matches to UniGene cluster Hs.60440 with four STSs from chromosome 4. Presumably these STSs will be located on GP when the AC113611 genomic sequence is assembled into chromosome 4. Checking the chromosome 4 SNPs at 105,568 sequences by BLAST search, recorded no hits within the 2552 bp mRNA of AY040094 but found over 100 matches within the repeat-masked sections of AC113611. Using the same sequence to BLAST against the 115,608 sequences in the STS division gives eight hits above 95% identity, although only three looked like unique matches. Interestingly the HTRA3 mRNA AY040094 has no STS matches although four chromosome 4 STSs were picked up in the UniGene entry. A possible explanation is that the cluster included clone links to ESTs that extend past the 3′ end of the mRNA.

Performing the same database checks for the HTRA4 ESTs, AL545759 and AL576444, produces a different pattern of findings. There were no hits in nr or gbPAT. However, the HTGS search located extended identity hits to no less than four genomic entries. These comprised of three recently sequenced sections of chromosome 8 AC108863, AC105089, AC105088, and a short match to an entry without a chromosomal assignment, AC107926. Checking for HTRA4 in LocusLink could find no IDs because of the absence of a full-length mRNA. It was picked up as the UniGene cluster Hs.322452 with nine ESTs but no mapping information was included even though our search update had located it to chromosome 8. No reading frame SNPs could be detected from the 92,110 chromosome 8 entries. By using the genomic contig, AC108863, (198,743 bp) as a BLAST query only three SNP identity matches were detected, rs1467190, rs2010445 and rs2056170, but three STS markers G60989, G23343, and G04735, were located.

In summary; although these two gene products cannot be located on the latest GP a series of manual database checks have established a mixture of patent mRNAs, unfinished genomic matches, ESTs, STSs and SNPs. It will be interesting to track how soon these features find their way into the GP annotation pipelines. If genetic studies should need this location data in the interim, the searches have established that HTRA3 probably has enough SNPs in the genomic vicinity for association studies, but that there is a very low SNP density in proximity to HTRA4. If the overlapping genomic coverage for HTRA4 could detect all the exons it might be possible to assemble a 'mini golden path' across this particular section. However if it became necessary to re-order and re-assemble the contigs within the unfinished entries this would be a challenging task to perform with web-based tools.

4.7 ANALYSING A NOVEL GENE

Sooner or later experimental results will locate a piece of GP where there are no fully annotated known genes. Figures 4.4, 4.5 and 4.6 show selected tracks from the Ensembl, UCSC and NCBI displays between the 3′ side of the BACE gene and the 5′ end of the next known gene PCSK7. The known genes are marked in brown in Ensembl and blue in UCSC. The latter are mRNA mappings and therefore include the UTR sections. Let us assume a genetic linkage study had found significant associations in this area, either from the two STS markers or the 50 or so SNPs that lie in this interval but are outside the boundaries of the two neighbouring genes. The question immediately arises as to what

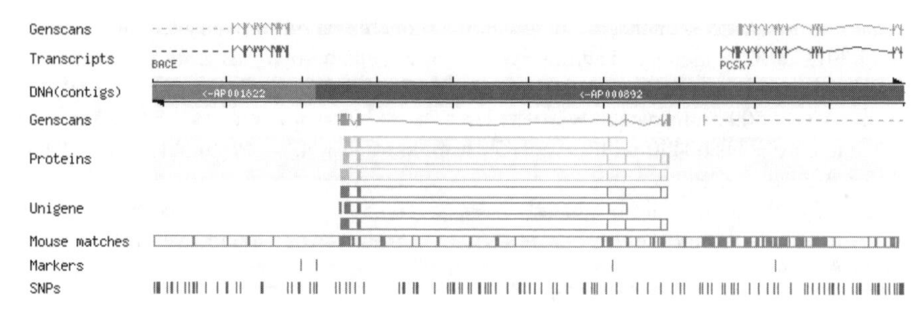

Figure 4.4 The Ensembl display for the unknown gene between BACE (left) and PCSK7 (right).

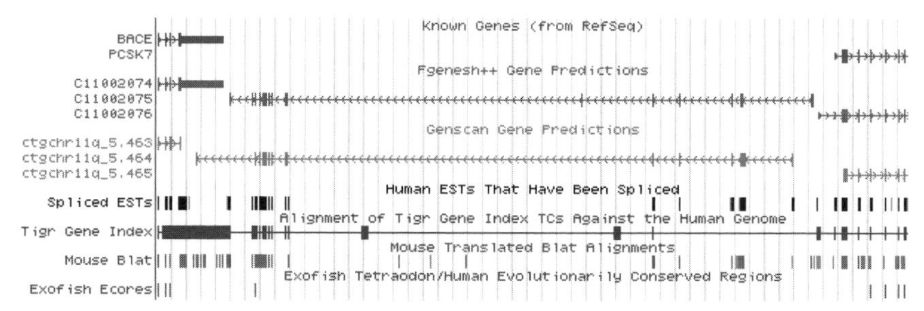

Figure 4.5 The UCSC display for the unknown gene between BACE (left) and PCSK7 (right).

other gene product(s) might be located between the two knowns. The first step is to check the continuity of this section of GP. This can be done in any of the viewers and in this case there is complete clone overlap across this section.

Inspection of all three displays indicates a possible novel gene product with a variety of supporting evidence. They include gene predictions which include both common and different exon positions. The UCSC Genscan prediction number 464 overlaps with the 3′ UTR of BACE making this a less plausible (but still possible) exon. Reading vertically down the Ensembl tracks first we see evidence for three protein homologies (yellow) as judged by the matches in register with the Genscan exon predictions. These are Q96RS9, a novel DZIP3, Q02455 a myosin-like peptide from yeast and P53804 a tetratricopeptide repeat protein. There is the same pattern of exon matches to three UniGene cluster entries (red) Mm.3679 *Mus musculus* for the tetratricopeptide repeat domain protein, Hs.165662 for *Homo sapiens* KIAA0675 unknown protein and Hs.118174 for *Homo sapiens* TTC3 tetratricopeptide repeat domain 3. There is a denser pattern of matches to mouse DNA (pink) that includes many sections outside the Genscan predicted exons.

Moving down the UCSC tracks in Figure 4.5 we see the spliced ESTs (black) in register with Genscan exons. However these identity EST matches are not equivalent to the homology-based UniGene matches in Ensembl. Interestingly the internal exon predicted only by Fgenes has no spliced EST support. Exploring the EST coverage further we see that the (brown) THC tracks include an assembly that matches the predicted exons at the BACE end of the Fgenesh++ prediction. The NCBI tracks go into more detail by not only mapping UniGene cluster components directly back to putative genomic exons by

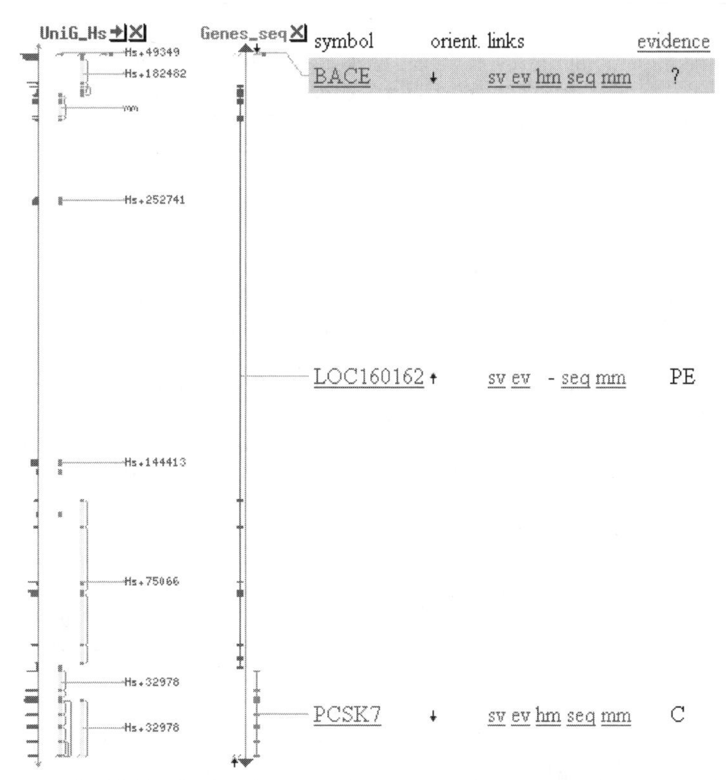

Figure 4.6 The NCBI display for the unknown gene between BACE (top) and PCSK7 (bottom). The leftmost track shows the EST distribution. The next track to the right marks the UniGene clusters. The central track is the gene prediction for LOC160162 and the gene structure for the N-terminal section of PCSK7 (bottom).

identity matches but also, on the left hand edge, showing an identity block proportional to the number of EST matches. Surprisingly there are five EST clusters which raises the possibility of more than one gene. The mouse BLAT track (brown) is equivalent to the Ensembl (pink) mouse track but the translation mode filters down to fewer features. The exofish track in UCSC (blue) supports just one single exon at the 5′ end of the putative novel gene compared with many conserved exons in both gene neighbours. In isolation this would be considered as weak evidence for the gene product. However it could simply mean that this predicted protein is not conserved between fish and human or the puffer fish ORFs are not complete across this section.

Up to this point our analysis of the genomic region between the 3′ end of BACE and the 5′ end of PKSC7 points strongly to the presence of a gene product on the basis of gene prediction and EST coverage. So where do we go from here? One option is to do some searches with the available mRNA and protein sequence from the Fgenesh++ prediction (numbered C11002075 in Figure 4.5) that can be downloaded from the UCSC site. The result brings us a long way forward in the evidence cascade because we record an 81% protein identity to what is likely to be the recently deposited mouse orthologue mRNA, BC023073. Interestingly this level of similarity should result in this gene passing the Genwise threshold for marking a novel gene position (black) in the next release of Ensembl. At these similarity levels we can back-check this mouse sequence against human GP by the very fast BLAT search (http://genome.ucsc.edu/cgi-bin/hgBlat?command=start). The result (Figure 4.7) clearly supports both the orientation (3′-to-3′ relative to BACE) and seven of the exons from C11002075. However the mouse sequence is clearly missing the 5′ end.

The next step involved searching the entire genomic DNA section of 54 kb from which C11002075 was predicted against human ESTs. This was performed using MEGABLAST with a 90% match stringency and masking of the repeat sections in the genomic query section. The result (Figure 4.8) is equivalent in principal to the UniGene clusters in the NCBI viewer but it is easier to pick out the ESTs that bridge several exons. Another reason for doing this analysis is that over 1 million human ESTs have been added to dbEST since the UniGene clusters were built. We can identify three ESTs that cover 35 kb of genomic sequence across three exons and performing the analogous search against mouse ESTs, with an 80% identity cut-off, finds a long EST spanning the four central exons. This gives us more confidence of a single rather than multiple gene products. The next step was to search ESTs against the TIGR THCS to establish if any virtual mRNAs could be found. In fact two of these, THC856832 and THC796698, represented the 5′ and 3′ ends respectively and to join these assemblies a bridging EST was found, BM055167. By using a web version of the CAP3 assembler (http://bio.ifom-firc.it/ASSEMBLY/assemble.html) it was possible to construct an extended virtual mRNA of 2720 bp. This was translated into a protein of 474 amino acids using the translation tool (http://ca.expasy.org/tools/dna.html) (Figure 4.9).

So far so good, but what else can we do to verify this putative novel protein *in silico*? The first step is a cross-check for reading frame consistency and species orthologues by performing TBLASTN against all ESTs (Figure 4.10). The results show the complete coverage of the entire ORF by human ESTs but also suggests potential splice variants

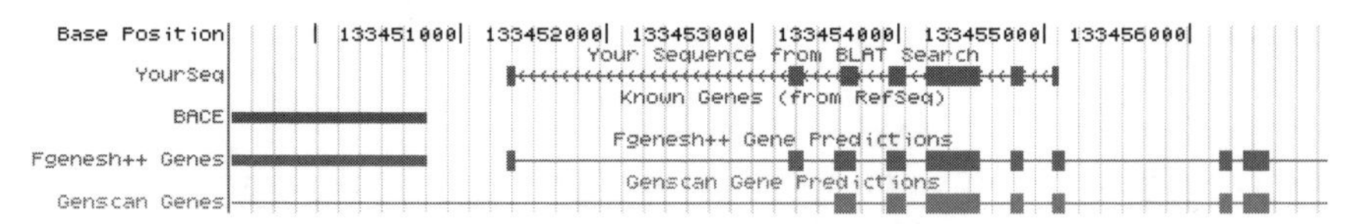

Figure 4.7 The alignment of the mouse protein from BC023073 after a BLAT search against the UCSC GP. The BACE gene is on the left hand side.

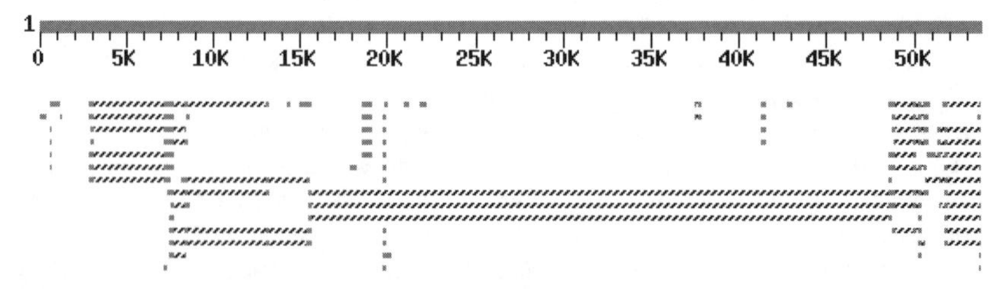

Figure 4.8 Result of a MEGABLAST search of the genomic sequence between BACE and PCSK7 against human ESTs. The solid lines indicate gaps in the same ESTs. The solid sections are putative exon matches.

```
gcgggtcctgtccctcccccactttcctcccggggcgcggcgcgggagagcataatggc
agcgtctgaggttgctggtgttgtggccaatgcccccagtcctccggaatcttctagttt
atgtgcttccaaatcagacgaaggtctcccagatggtctaagcaccaaagactctgcaca
gaagcagaagaactcgcctctgttgagtgtaagtagccaaacaataaccaaggagaataa
cagaaatgtccatttggagcactcagagcagaatcctggttcatcagcaggtgacacctc
agcagcgcaccaggtggtttaggagaaaacttgatagccacagcccttgtctttctggc
agtgggtctcagtctgatttgaaggatgtggccagcacagcaggagaggagggggacaca
agccttcgggagagcctccatccagtcactcggtctcttaaggcagggtgccatactaag
cagcttgcctccaggaattgctctgaagagaaatccccacaaacctccatcctaaaggaa
ggtaacagggacacaagcttggatttccgacctgtagtgtctccagcaaatggggttgaa
ggagtccgagtggatcaggatgatgatcaagatagctcttccctgaagctttctcagaac
attgctgtacagactgactttaagacagctgattcagaggtaaacacagatcaagatatt
gaaaagaatttggataaaataatgacagagagaaccctgttgaaagagcgttaccaggag
                     M  T  E  R  T  L  L  K  E  R  Y  Q  E
gtcctggacaaacagaggcaagtggagaatcagctccaagtgcaattaaagcagcttcag
V  L  D  K  Q  R  Q  V  E  N  Q  L  Q  V  Q  L  K  Q  L  Q
caaaggagagaagaggaaatgaagaatcaccaggagatattaaaggctattcaggatgtg
Q  R  R  E  E  E  M  K  N  H  Q  E  I  L  K  A  I  Q  D  V
acaataaagcgggaagaaacaaagaagaagatagagaaagagaagaaggagtttttgcag
T  I  K  R  E  E  T  K  K  K  I  E  K  E  K  K  E  F  L  Q
aaggagcaggatctgaaagctgaaattgagaagctttgtgagaagggcagaagagaggtg
K  E  Q  D  L  K  A  E  I  E  K  L  C  E  K  G  R  R  E  V
tgggaaatggaactggatagactcaagaatcaggatggcgaaataaataggaacattatg
W  E  M  E  L  D  R  L  K  N  Q  D  G  E  I  N  R  N  I  M
gaagagactgaacgggcctggaaggcagagatcttatcactagagagccggaaagagtta
E  E  T  E  R  A  W  K  A  E  I  L  S  L  E  S  R  K  E  L
ctggtactgaaactagaagaagcagaaaaagaggcagaattgcaccttacttacctcaag
L  V  L  K  L  E  E  A  E  K  E  A  E  L  H  L  T  Y  L  K
tcaactcccccaacactggagacagttcgttccaaacaggagtgggagacgagactgaat
S  T  P  P  T  L  E  T  V  R  S  K  Q  E  W  E  T  R  L  N
ggagttcggataatgaaaaagaatgttcgtgaccaatttaatagtcatatccagttagtg
G  V  R  I  M  K  K  N  V  R  D  Q  F  N  S  H  I  Q  L  V
aggaacggagccaagctgagcagccttcctcaaatccctactcccactttacctccaccc
R  N  G  A  K  L  S  S  L  P  Q  I  P  T  P  T  L  P  P  P
ccatcagagacagacttcatgcttcaggtgtttcaacccagtccctctctggctcctcgg
P  S  E  T  D  F  M  L  Q  V  F  Q  P  S  P  S  L  A  P  R
atgcccttctccattgggcaggtcacaatgcccatggttatgcccagtgcagatccccgc
M  P  F  S  I  G  Q  V  T  M  P  M  V  M  P  S  A  D  P  R
tccttgtctttcccaatcctgaaccctgccctttcccagcccagccagccttcctcaccc
S  L  S  F  P  I  L  N  P  A  L  S  Q  P  S  Q  P  S  S  P
cttcctggctcccatggcagaaatagccctggcttgggttcccttgtcagcccccacggt
L  P  G  S  H  G  R  N  S  P  G  L  G  S  L  V  S  P  H  G
ccacacatgccccctgccgcctccatcccacctcccccaggcttgggcggtgttaaggct
P  H  M  P  P  A  A  S  I  P  P  P  P  G  L  G  G  V  K  A
tctgctgaaactccccggccccaaccagtagacaaactggagaagatcctggagaagctg
S  A  E  T  P  R  P  Q  P  V  D  K  L  E  K  I  L  E  K  L
ctgacccggttcccacagtgcaataaggcccagatgaccaacattcttcagcagatcaag
L  T  R  F  P  Q  C  N  K  A  Q  M  T  N  I  L  Q  Q  I  K
acagcacgtaccaccatggcaggcctgaccatggaggaacttatccagttggttgctgca
T  A  R  T  T  M  A  G  L  T  M  E  E  L  I  Q  L  V  A  A
cgactggcagaacatgagcgggtggcagcaagtactcagccacttggtcgcatccgggcc
R  L  A  E  H  E  R  V  A  A  S  T  Q  P  L  G  R  I  R  A
ttgttccctgctccactggcccaaatcagtaccccaatgttcttgccttctgcccaagtt
L  F  P  A  P  L  A  Q  I  S  T  P  M  F  L  P  S  A  Q  V
tcatatcctggaaggtcttcacatgctccagccacctgtaagctatgtctaatgtgccag
S  Y  P  G  R  S  S  H  A  P  A  T  C  K  L  C  L  M  C  Q
aaactcgtccagcccagtgagctgcatccaatggcgtgtacccatgtattgcacaaggag
K  L  V  Q  P  S  E  L  H  P  M  A  C  T  H  V  L  H  K  E
tgtatcaaattctggcccagaccaacacaaatgacacttgtccctttgtccaactctt
C  I  K  F  W  A  Q  T  N  T  N  D  T  C  P  F  C  P  T  L
aaatgacggacctgactggggaggaagaagaagagaaactgatgtgaacaggaagcgcgg
K
gttcaagatttctaaaactctatatttatacagtgacatatactcatgccatgtacattt
ttattatataggtaatgtgtgtatagaaagtctgtattccaatgttcgtaaatgaaacta
tgtatattatgcagaaacagtctgttcccccctcatcttgcaattcctttgggggatgcag
attgtagggaagatgatgtttagtttggccttgaaattatgatatccctgcccagggctg
ttttcaaatacaatataaaaaccacctaggaacctgctgttgctctaaggccattctgct
ttggtttggctcagcctctagtccatttccttaaggctcatgtatgcagatttaaagcct
ggtgctcacccactgtccaaccagatgccttgcttaccgaaagcctccagaagcctcagt
attgttttagccactctactccaaatggataaaatgagactctgattgaggaaaaaaaag
taaccctagtagtttgaaa
```

Figure 4.9 Predicted ORF for a novel protein. This was produced by assembling the appropriate assemblies and ESTs into a virtual mRNA. This was then translated to give the putative full-length protein sequence.

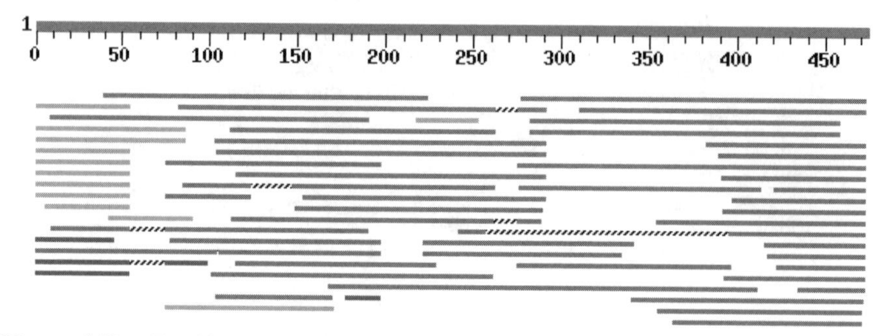

Figure 4.10 Checking for continuity of reading frame by translation searching (TBLASN) of the unknown ORF against all ESTs. The hatched lines represent deletions in ESTs that could represent splice variants.

in these matches, for example AI351632, represented as hatched lines in Figure 4.10. In addition to a bovine sequence BE75593 we also see a likely orthologous match to AL640079 from a toad. The support for the ORF now seems unassailable. The next step using BLAT again, is to map it back to GP (Figure 4.11). This reveals the matching of 15 exons from putative 5′ UTR to 3′ UTR. This is consistent with the Fgenes++ prediction at the 5′ end but this included two extra exons at the 3′ end. The fact that the virtual mRNA butts up very close to both neighbouring genes suggests that this could be a full-length transcript.

Clearly the analysis of what, for example, might be a candidate disease-associated gene, has to move on from the identification of an ORF to the assignment of function that is both mechanistically plausible and experimentally testable. The subject of assigning functions to new proteins is outside the scope of this chapter. However the two basic steps are a protein database search and motif analysis. The protein search (Figure 4.12) only shows significant similarity scores over the C-terminal section of the protein but the hits include the same proteins assigned as UniGene homologies by Ensembl. A comprehensive domain analysis using InterPro recognizes two domains (Kriventseva *et al.*, 2001; Southan, 2000). One of the domains identified, IPR000694, is a proline-rich domain that may be involved in protein–protein interactions (Figure 4.13). However, the motif recognition specificity is low and therefore this could be a spurious match arising from a general high proline composition. An SRS query shows 1152 of these domains have been recorded in Ensembl (Zdobnov *et al.*, 2002). The second domain, IPR001841, is more specific because it only occurs 187 times in the Ensembl gene set. The RING-finger is a specialized type of Zn-finger of 40 to 60 residues that binds two atoms of zinc, and is probably also involved

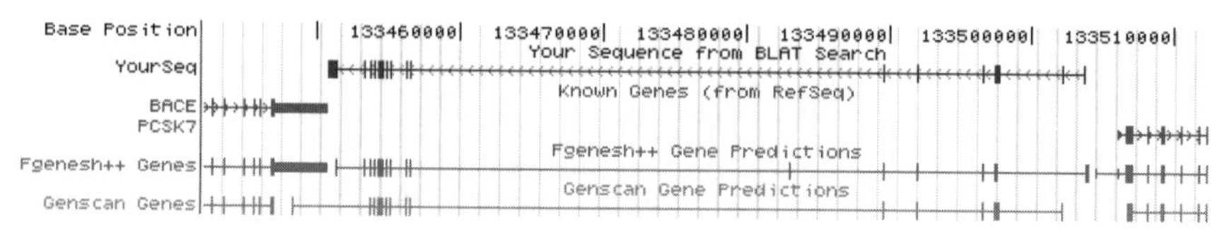

Figure 4.11 Matching the virtual mRNA back against GP using the BLAT search at UCSC. This delineates 15 exons with the gene reading in the opposite orientation to its neighbouring genes, i.e. 3′ end to the left, on the same strand.

1
0
50
100
150
200
250
300
350
400
450

```
gi|18606012|gb|AAH23073.1|   (BC023073) Unknown (protein for ...   453   e-126
gi|18595990|ref|XP_088700.1|   (XM_088700) similar to tetratr...    50   5e-05
gi|7662244|ref|NP_055463.1|   (NM_014648) KIAA0675 gene produ...    50   6e-05
gi|14765961|ref|XP_032867.1|   (XM_032867) similar to tetratr...    49   9e-05
gi|1730008|sp|P53804|TTC3_HUMAN  TETRATRICOPEPTIDE REPEAT PR...     44   0.004
gi|2662364|dbj|BAA23666.1|   (D83327) DCRR1 [Homo sapiens]          44   0.005
gi|10835037|ref|NP_003307.1|   (NM_003316) tetratricopeptide ...    43   0.006
```

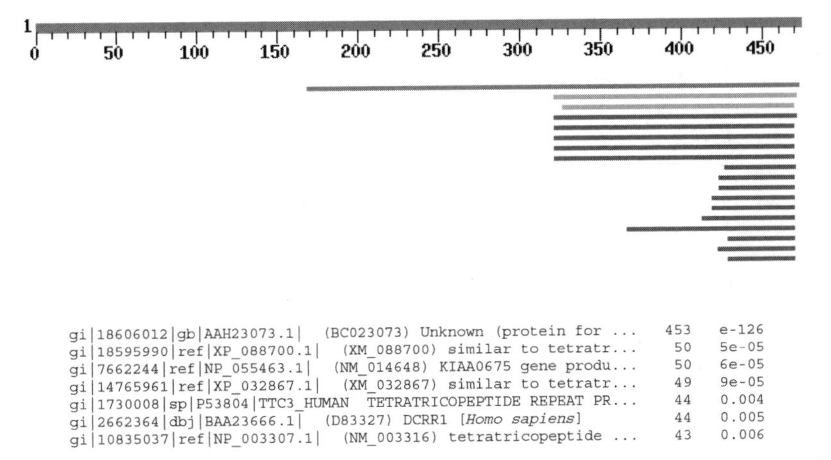

Figure 4.12 The sequence similarity scores of the novel ORF against the NCBI non-redundant protein database.

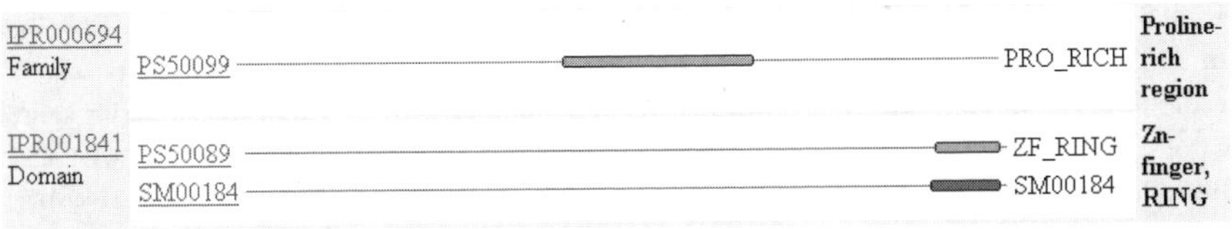

IPR000694
Family
PS50099
PRO_RICH
Proline-rich region
IPR001841
Domain
PS50089
ZF_RING
SM00184
SM00184
Zn-finger, RING

Figure 4.13 The InterPro domain/protein family analysis result for the novel ORF. The praline-rich domain is defined from a Prosite profile. The zinc finger is defined by both a Prosite profile and a SMART domain.

in mediating protein–protein interactions. They can also bind DNA however, since they contain many Lys, Ser and Thr residues. In fact combining the two domain searches finds intersecting hits (i.e. containing both domains) for only 17 Ensembl proteins. Inspecting the graphical displays shows one of these gene products, ESP0000020915, to be similar in domain orientation and spacing to the novel protein. Unfortunately the trail went cold here because this identifier has been changed in the latest Ensembl release and the SRS link to the protein sequence was dead.

So how did the three major gene portals do? Quite well considering they all included the potential novel gene product as a gene prediction although they disagreed on exon number. They also displayed key supporting evidence in different forms of track annotation. Only a small subset of the display options has been presented here. Was the use of all three portals essential? Strictly speaking we could have accessed sufficient supporting evidence from each one. However to collect all the available data it was necessary to use all three. The other aspect is that each portal has particular facilities that even if not unique at the technical level is easier to use at one portal compared to the other three. Consequently this kind of detailed analysis becomes a *de facto* three-stop-shop. For example the UniGene homology assignments, available from Ensembl, were all correct as judged by the agreement with the protein similarities (red tracks in Figure 4.4). Having said that, one of the direct protein homology assignments (yellow tracks in Figure 4.4), the myosin-like peptide from yeast, was probably erroneous because of the low complexity of the query protein amino acid composition. In terms of markers, SNPs and genes Ensembl does particularly well for combined export options. The UniGene identity matches on the NCBI display together with the graphical stacks proportional to the number of EST matches are useful but in this case what is likely to be a single transcript was split into four clusters. One of these was illegible on the graphic and two others are dubious because of being unspliced. The UCSC displays were useful to see the two alternative gene models as well as being the only source of the TIGR EST assemblies. Another useful facility on this site is the ability of BLAT to display the hits of any externally constructed model or new database sequence. This can then be compared directly with the other display options (e.g. in Figures 4.6 and 4.11). The NCBI have recently introduced a gene model builder that can reproduce some of the steps above (http://www.ncbi.nlm.nih.gov/PMGifs/Genomes/ModelMakerHelp.html).

4.8 COMPREHENSIVE DATABASE SEARCHING

The protein matches and the InterPro analysis have already given functional clues about our novel protein. However if this particular gene product was located in close proximity to an SNP with a disease association we would need to find out as much as possible, not only to provide more supporting evidence for the gene product but also testable predictions about function that can be followed up. Performing a comprehensive search is not a trivial exercise since it involves 17 divisions of GenBank and sources of trace data that have not yet been submitted to GenBank. So where do we start? The two large repositories labelled nr protein or nucleotide on the NCBI BLAST server are a useful first choice (http://www.ncbi.nlm.nih.gov/blast/Blast.cgi). We have already checked nr protein at 891,607 sequences but we need to compliment this with month, which in this case yields another 61,254 protein sequences but no additional high-scoring hits. The search against nr nucleotide with 1,192,858 sequences records three extended matches. This includes the mouse sequence already described, BC023073, and the primary accession number of the finished genomic section AP000892. The third match, XM_100696, is a secondary accession number for a reference mRNA sequence predicted by the NCBI Annotation Project from a genomic contig NT_009151. This is the same prediction labelled LOC160162 in Figure 4.5. There is an accompanying 56-residue predicted ORF that is in the NCBI protein database but has no supporting evidence. Inspection of the genomic location suggests it may be a spurious prediction.

Checking public patented proteins at 88,019 sequences gave no hits. However the patent nucleotide division, gbPAT, at 581,001 sequences, gives three solid hits, AX321627, AX192589 and AX072029. The first of these is a 2114-bp DNA from patent WO0172295. The document indicates this protein was isolated from a lung cancer sample (http://ep.espacenet.com/). These hits constitute a partial mRNA level of confirmation for the novel protein but a reciprocal check (i.e. a BLASTN of AX321627 against the nr nucleotide database) indicates this clone may be a chimera from two separate gene products. A search against a commercial patent database, containing 673,453 protein sequences, reveals identity matches for the N-terminal section from patent WO200060077 and a C-terminal identity match from WO200055350, both of which are reported as cancer-associated transcripts (http://www.derwent.com/geneseq/index.html). Checking the GSS division by TBLASTN gives four genomic hits; AZ847251 from mouse, AG114530 from chimpanzee, BH306228 from rat and BH406519 from chicken. Using BLAST against the

TABLE 4.1 Useful Resources for Gene Finding and Analysis

Site description	URL
Ensembl at EBI/Sanger Centre	http://www.ensembl.org/
Human Genome Browser at UCSC	http://genome.ucsc.edu/
Map Viewer at NCBI	http://www.ncbi.nlm.nih.gov/cgi-bin/Entrez/map_ search
Protein Atlas of the genome	http://www.confirmant.com/
SWISS-2DPAGE database	http://ca.expasy.org/ch2d/
Ensembl 4.28.1 announcement	http://www.ensembl.org/Dev/Lists/announce/msg00070.html
NCBI gene model builder	http://www.ncbi.nlm.nih.gov/PMGifs/Genomes/ModelMakerHelp.html
UniGene EST clusters	http://www.ncbi.nlm.nih.gov/UniGene/
InterPro at EBI	http://www.ebi.ac.uk/interpro/
Proteome analysis at EBI	http://www.ebi.ac.uk/proteome/
Google general search portal	http://www.google.com/
RefSeq at NCBI	http://www.ncbi.nlm.nih.gov/LocusLink/refseq.html
International Protein Index	http://www.ebi.ac.uk/IPI/IPIhelp.html
Derwent sequence patent databases	http://www.derwent.com/geneseq/index.html
BLAST at NCBI	http://www.ncbi.nlm.nih.gov/BLAST/
BLAT at UCSC	http://genome.ucsc.edu/cgi-bin/hgBlat?command = start)
DAS — distributed annotation	http://biodas.org/
Exofish at Genoscope	http://www.genoscope.cns.fr/externe/tetraodon/
Fgenesh at Sanger Institute	http://genomic.sanger.ac.uk/gf/Help/fgenesh.html
Expasy translation tool	http://ca.expasy.org/tools/dna.html
CAP3 nucleotide assembly tool	http://bio.ifom-firc.it/ASSEMBLY/assemble.html
GeneWise at Sanger Institute	http://www.sanger.ac.uk/Software/Wise2/
Genscan at MIT	http://genes.mit.edu/GENSCAN.html
SSAHA at Sanger Institute	http://www.sanger.ac.uk/Software/analysis/SSAHA/

Ensembl mouse peptides detected a C-terminal similarity that is a zinc finger domain match. However both the human and mouse mRNA have unique and solid hits against mouse chromosome 9.40 Mb. This suggests the gene product is derived from this locus although it has not been annotated yet by Ensembl. Interestingly the gene lies between two odour receptors, unlike the human positioning between BACE and PCSK7, showing the position is non-syntenic.

Drawing detailed conclusions from these results is outside the scope of this chapter but the example makes clear how much extra information a comprehensive database search can yield. Was the protein unknown and/or novel? The difficulty of answering this question illustrates the diminishing utility of these terms. The protein has at least one function-related motif that can be recognized at high specificity so it can no longer be classified as an unknown. It remains novel only in the strict sense of not being represented in the current protein databases. It is not novel in the wider sense because both the mRNA and ORF were substantially covered as predicted by sequence data entries in the public and patent databases respectively.

4.9 CONCLUSIONS AND PROSPECTS

The geneticist is in the fortunate position of having access to secondary databases and GP genomic viewers of increasing quality, content and utility. This is making the process of finding and analysing gene products easier. However the examples used in this chapter also show that there are many subtle details in genomic annotation and the implications of these will take some time to unravel. This requires comprehensive inspection and may ultimately need experimental verification. The expansion of web-linked interoperativity and interrogation tools means that new options will already be available by the time this is in print. One consequence of these advances could be the perception of a diminished necessity to perform bioinformatic analysis. Although this is true in the sense that secondary databases include an increasing amount of 'pre cooked' bioinformatic data, there is a paradox in that the more sophisticated the public annotation becomes the more important it is to understand the underlying principles. For example, it is important to be able to discriminate between gene products defined by *in-vitro* data or only by *in-silico* prediction.

So what of the future? There are four developments worth highlighting. The first is that the combination of increasing transcript coverage, finished golden path and extensive mouse synteny data will diminish the uncertainty limits of gene numbers. The ability to pick out SNP haplotype blocks in relationship to gene products, already available as tracks on the UCSC display options for chromosome 21 will be a big step forward for association studies (Patil *et al.*, 2001). The proliferation of DAS servers will enable more groups to share their own specialized annotation tracks with the wider community (http://biodas.org/). Last but not least defining gene products at the protein level is likely to have a major impact on annotation quality, and efforts are already underway to do this on a genome-wide scale (http://www.confirmant.com/).

REFERENCES

Birney E, Durbin R. (2000). Using GeneWise in the Drosophila annotation experiment. *Genome Res* **10**: 547–548.

Burge C, Karlin, S. (1997). Prediction of complete gene structures in human genomic DNA. *J Mol Biol* **268**: 78–94.

Deloukas P, Matthews LH, Ashurst J, Burton J, Gilbert JG, Jones M, *et al.* (2001). The DNA sequence and comparative analysis of human chromosome 20. *Nature* **414**: 865–871.

Guigo R, Agarwal P, Abril JF, Burset M, Fickett JW. (2000). An assessment of gene prediction accuracy in large DNA sequences. *Genome Res* **10**: 1631–1642.

Hoogland C, Sanchez JC, Tonella L, Binz PA, Bairoch A, Hochstrasser DF, *et al.* (2000). The 1999 SWISS-2DPAGE database update. *Nucleic Acids Res* **28**: 286–288.

Hubbard T, Barker D, Birney E, Cameron G, Chen Y, Clark L, *et al.* (2002). The Ensembl genome database project. *Nucleic Acids Res* **30**: 38–41.

Kel-Margoulis OV, Kel AE, Reuter I, Deineko IV, Wingender E. (2002). TRANSCompel: a database on composite regulatory elements in eukaryotic genes. *Nucleic Acids Res* **30**: 332–334.

Kriventseva EV, Biswas M, Apweiler, R. (2001). Clustering and analysis of protein families. *Curr Opin Struct Biol* **11**: 334–339.

O'Donovan C, Apweiler R, Bairoch A. (2001). The human proteomics initiative (HPI). *Trends Biotechnol* **19**: 178–181.

Patil N, Berno AJ, Hinds DA, Barrett WA, Doshi JM, Hacker CR, *et al.* (2001). Blocks of limited haplotype diversity revealed by high-resolution scanning of human chromosome 21. *Science* **294**: 1719–1723.

Pesole G, Liuni S, Grillo G, Licciulli F, Mignone F, Gissi C, *et al.* (2002). UTRdb and UTRsite: specialized databases of sequences and functional elements of 5′ and 3′ untranslated regions of eukaryotic mRNAs. Update 2002. *Nucleic Acids Res* **30**: 335–340.

Pruitt KD, Maglott DR. (2001). RefSeq and LocusLink: NCBI gene-centered resources. *Nucleic Acids Res* **29**: 137–140.

Quackenbush J, Cho J, Lee D, Liang F, Holt I, Karamycheva S, *et al.* (2001). The TIGR Gene Indices: analysis of gene transcript sequences in highly sampled eukaryotic species. *Nucleic Acids Res* **29**: 159–164.

Schuler GD. (1997). Pieces of the puzzle: expressed sequence tags and the catalog of human genes. *J Mol Med* **75**: 694–698.

Southan C. (2000). Website review: Interpro (the integrated resource of protein domains and functional sites). *Yeast* **17**: 327–334.

Southan C. (2001). A genomic perspective on human proteases as drug targets. *Drug Discov Today* **6**: 681–688.

Susens U, Borgmeyer U. (2001). Genomic structure of the gene for mouse germ-cell nuclear factor (GCNF). II. Comparison with the genomic structure of the human GCNF gene. *Genome Biol* **2**: research No. 0017.

Suzuki Y, Yamashita R, Nakai K, Sugano S. (2002). DBTSS: DataBase of human Transcriptional Start Sites and full-length cDNAs. *Nucleic Acids Res* **30**: 328–331.

Wiehe T, Gebauer-Jung S, Mitchell-Olds T, Guigo R. (2001). SGP-1: prediction and validation of homologous genes based on sequence alignments. *Genome Res* **11**: 1574–1583.

Zdobnov EM, Lopez R, Apweiler R, Etzold T. (2002). The EBI SRS server — recent developments. *Bioinformatics* **18**: 368–373.

SECTION 2

THE IMPACT OF COMPLETE GENOME SEQUENCES ON GENETICS

CHAPTER 5

Assembling a View of the Human Genome

COLIN A. M. SEMPLE

Bioinformatics
MRC Human Genetics Unit
Edinburgh EH4 2XU
UK

5.1 INTRODUCTION

The miraculous birth of the draft human genome sequence took place against the odds. It was only made possible by parallel revolutions in the technologies used to produce, store and analyse the sequence data and by the development of new large-scale consortia to organize and obtain funding for the work (Watson, 1990). The initial flood of sequence has subsided as the sequencing centres begin the task of converting the fragmented draft sequences into a finished, complete sequence for each chromosome. The steady progress of the cloned fragments of the human genome towards a finished state can be observed in the Genome Monitoring Table (Beck and Sterk, 1998; http://www.ebi.ac.uk/genomes/mot/),

Bioinformatics for Geneticists. Edited by M.R. Barnes and I.C. Gray
 ISBNs: 0 470 84393 4; 0 470 84394 2 (PB)

but although we can examine the sequences in public databases we have yet to comprehensively interpret them. There is a need to relate the raw sequence data to what we already know about human genetics and biology in general, this is the process of genome annotation. Preliminary annotation of a genome is a semi-automated process, with human curators interpreting the results of various computer programs. In practical terms, preliminary annotation currently consists of determining the position of known markers, known genes and repetitive sequence in combination with efforts to delineate the structure of novel genes. Eventually we would like to know much more, including the multifarious interactions of the genome's contents with one another and the environment, their expression in the biology of the cell and role in human physiology. These additional layers of annotation will come from the patient laboratory work of the next several decades but a prerequisite for this work is a complete (or nearly complete) genome sequence and an accurate preliminary annotation which is available to the total scientific community. This chapter will aim to describe the sources of freely available annotation, their strengths, their shortcomings and some likely future developments. All websites referred to in the text are listed in Table 5.1.

TABLE 5.1 The Websites Referred to in the Text

Site Description	URL
Genomic sequence assemblies	
CG Human Genome Assembly	http://public.celera.com
NCBI Human Genome Assembly	http://www.ncbi.nlm.nih.gov/genome/guide/human/
UCSC Human Genome Assembly	http://genome.ucsc.edu/
Annotation browsers	
Ensembl at EBI/Sanger Institute	http://www.ensembl.org/
Genome Channel at ORNL	http://compbio.ornl.gov/channel/
Human Genome Browser at UCSC	http://genome.ucsc.edu/
Map Viewer at NCBI	http://www.ncbi.nlm.nih.gov/cgi-bin/Entrez/map_search
Data sources	
ArrayExpress at EBI	http://www.ebi.ac.uk/arrayexpress/
COGs database at NCBI	http://www.ncbi.nlm.nih.gov/COG/
dbSNP at NCBI	http://www.ncbi.nlm.nih.gov/SNP/index.html
DOTS at University of Pennsylvania	http://www.allgenes.org/
FlyBase at EBI	http://fly.ebi.ac.uk:7081/
Genome Monitoring Table at EBI	http://www.ebi.ac.uk/genomes/mot/
GEO at NCBI	http://www.ncbi.nlm.nih.gov/geo/
IHGMC FPC map at Washington University in St Louis	http://genome.wustl.edu/cgi-bin/ace/GSCMAPS.cgi?

TABLE 5.1 *(continued)*

Site Description	URL
InterPro at EBI	http://www.ebi.ac.uk/interpro/
Mouse Genome Database at Jackson Laboratory	http://www.informatics.jax.org/
Mouse Atlas Database at MRC Human Genetics Unit	http://genex.hgu.mrc.ac.uk/
OMIM at NCBI	http://www.ncbi.nlm.nih.gov/Omim/
Pfam at Sanger Institute	http://www.sanger.ac.uk/Software/Pfam/
Proteome Analysis at EBI	http://www.ebi.ac.uk/proteome/
RefSeq at NCBI	http://www.ncbi.nlm.nih.gov/LocusLink/refseq.html
Saccharomyces Genome Database at Stanford University	http://genome-www.stanford.edu/Saccharomyces/
UniGene at NCBI	http://www.ncbi.nlm.nih.gov/UniGene/
Software	
ACEDB (Sanger Institute)	http://www.acedb.org/
Acembly (NCBI)	http://www.ncbi.nih.gov/IEB/Research/Acembly/help/AceViewHelp.html
Apollo (EBI)	http://www.ensembl.org/apollo/
BLAST (NCBI)	http://www.ncbi.nlm.nih.gov/BLAST/
BLAT (UCSC)	http://genome.ucsc.edu/cgi-bin/hgBlat?command=start)
DAS (Cold Spring Harbor Laboratory)	http://biodas.org/
EMBOSS (EMBnet)	http://www.uk.embnet.org/Software/EMBOSS/
Exofish (Genoscope)	http://www.genoscope.cns.fr/externe/tetraodon/
ePCR (NCBI)	http://www.ncbi.nlm.nih.gov/genome/sts/epcr.cgi
Fgenesh (Sanger Institute)	http://genomic.sanger.ac.uk/gf/Help/fgenesh.html
Gene Ontology Consortium	http://www.geneontology.org/
GENEWISE (Sanger Institute)	http://www.sanger.ac.uk/Software/Wise2/
GENSCAN (MIT)	http://genes.mit.edu/GENSCAN.html
GrailEXP (ORNL)	http://compbio.ornl.gov/grailexp/
HMMER (WUSTL)	http://hmmer.wustl.edu/
NIX at (HGMPRC)	http://www.hgmp.mrc.ac.uk/Registered/Webapp/nix/
Phrap (University of Washington)	http://bozeman.genome.washington.edu/index.html
RepeatMasker (Uni. of Washington)	http://ftp.genome.washington.edu/RM/RepeatMasker.html
SIM4 (Penn State University)	http://bio.cse.psu.edu/
Spidey (NCBI)	http://www.ncbi.nlm.nih.gov/IEB/Research/Ostell/Spidey/
SSAHA (Sanger Institute)	http://www.sanger.ac.uk/Software/analysis/SSAHA/
Twinscan (WUSTL)	http://genes.cs.wustl.edu/

5.2 GENOMIC SEQUENCE ASSEMBLY

Any discussion of computational sequence annotation should begin with a consideration of the sequence data itself. Genomic sequence data has traditionally come from many sources: studies of transcribed sequences, individual genes and genetic/physical markers from mapping studies. Over the past decade we have entered the era of large-scale efforts to sequence entire genomes and the most abundant sources of sequence have become the sequencing vectors from these efforts. In practical terms this has meant that we acquire many fragments, from a few hundred bases to a few hundred kilobases in length, of a genome which must then be assembled computationally to produce a continuous sequence. In the case of the human genome, two unfinished 'draft' sequences have been produced using different methods, one by the International Human Genome Sequencing Consortium (IHGSC) and one by Celera Genomics (CG).

The IHGSC began with a BAC (bacterial artificial chromosome) clone-based physical map of the genome (IHGSC, 2001). This map was constructed by digesting each clone with restriction enzymes and deriving a characteristic pattern or fingerprint. All of the fingerprints are then processed by a program called FPC (Soderlund *et al.*, 2000) which produces BAC clone contigs on the basis of the shared fragments in their fingerprints (International Human Genome Mapping Consortium (IHGMC), 2001; http://genome.wustl.edu/cgi-bin/ace/GSCMAPS.cgi?). A selection of clones from this map covering the vast majority of the genome, were then 'shotgun sequenced' (Sanger *et al*, 1982). The fragments of each clone were then assembled into initial sequence contigs based upon overlaps between shotgun sequencing reads. The collection of initial sequence contigs from a single clone, make up the sequence data for a BAC clone in GenBank. As more shotgun sequencing of the clone is carried out, the initial sequence contigs are re-assembled with the new sequences and the database sequence entry for the clone is updated accordingly. Gradually the initial sequence contigs increase in length and decrease in number, until the sequence of the clone is finished and is represented by a single contig 100–200 kb in length. The program used to assemble the initial sequence contigs is called Phrap (Green, unpublished data; http://bozeman.genome.washington.edu/index.html) and takes sequencing quality estimates for each base into account. CG used the whole-genome shotgun method where the entire genome is randomly fragmented and each of the cloned fragments is sequenced (Venter *et al.*, 2001). Sequences from these cloned fragments are produced as mate-pairs: 150–800 bp sequencing reads from either end of the clone with known relative orientation and approximate spacing. A mixture of clones of different sizes was used: 2, 10, 50 and 100 kb. CG assembled their sequence data with that produced by the IHGSC and published an analysis of this early CG draft genome assembly (Venter *et al.*, 2001). Sequences from this assembly are available, under a variety of restrictions, from the CG draft genome publication site (http://public.celera.com), however the CG raw sequencing data and subsequent versions of the CG draft genome assembly are not publicly available. In spite of the differences between the two efforts to sequence the human genome, both groups had to address the fundamental problem of assembling incomplete data. In both cases the strategy was broadly to merge overlapping sequences into contigs and then to order contigs relative to one another using various types of mapping data.

The published IHGSC assembly was produced using a program called 'GigAssembler' devised at the University of California at Santa Cruz (UCSC) (Kent and Haussler, 2001). GigAssembler began with initial sequence contigs from GenBank at a given point (a 'freeze' dataset). All sequences were repeat masked using the RepeatMasker program (Smit and Green, unpublished data; http://ftp.genome.washington.edu/RM/RepeatMasker.html) to highlight known repetitive sequence. Within each IHGMC physical map contig (IHGMC, 2001) the initial sequence contigs from BAC clones belonging to it were assembled into consensus 'raft' sequences using sequence overlaps between fragments. The first joins were made between the best matching fragments. These rafts were ordered and orientated relative to one another using bridging sequences from other sources (mRNA, EST, plasmid and BAC end pairs) and FPC contig data. For instance the 5′ end of a single mRNA may be found within one raft while the 3′ end matches another raft. Repeated tracts of the letter 'N' were inserted between rafts to give a sequence for each IHGMC map contig. The published version of the UCSC assembly and all subsequent versions are freely available online (http://genome.ucsc.edu/).

The CG draft genome assembly was carried out by a program described as a 'compartmentalized shotgun assembler' (CSA) (Huson *et al*, 2001) using both CG sequence data and IHGSC initial sequence contigs from GenBank (as of 1 September 2000 for the published CG assembly) fragmented into smaller sequences a few hundred base pairs long. The CSA began by comparing all CG mate-pair fragments with all the initial sequence contig fragments and avoiding matches based upon repetitive sequence. Repetitive sequence was identified using comparisons to a library of known repeats (analogously to RepeatMasker) but also by additional procedures to detect sequence likely to represent unknown repeat sequences. The mate-pair fragment pairs matching more than one initial sequence contigs were then used as bridging sequences to order and orientate the initial sequence contig fragments within and between BAC clones. Essentially the paired CG fragments are used as high resolution mapping data to re-assemble both IHGSC BAC sequences and the broader genomic regions they originate from. The result was a set of 'scaffolds' consisting of ordered, oriented sequence contigs separated by gaps of estimated sizes. CG fragments not matching IHGSC initial sequence contigs were also assembled using a different algorithm (Myers *et al.*, 2000) to give additional scaffolds containing sequence not represented in IHGSC data. Scaffolds were then positioned relative to one another based upon sequence overlaps and bridging mate-pair fragments. The derived order of scaffolds was then manually curated to identify mistakes by examining sequence alignments by eye and confirming or rejecting orders based on external physical mapping data such as those from the IHGMC.

A third assembly method, using repeat masked data from the IHGSC, was produced by the National Centre for Biotechnology Information (NCBI) using a computational protocol (NCBI, unpublished data; http://www.ncbi.nlm.nih.gov/genome/guide/build.html) based upon the BLAST algorithm (Altschul *et al.*, 1997). The NCBI approach also began by finding an order for adjacent BACs but in this case it was derived from BAC sequence overlaps (detected using a variant of BLAST), fluorescence *in situ* hybridization (FISH) chromosome assignment and STS content. The sequence fragments from these overlapping BACs were then merged into consensus 'meld' sequences. As with the UCSC method, these melds were then ordered and orientated based on ESTs, mRNAs and paired plasmid reads before being combined into a single NCBI genomic sequence contig with melds separated by runs of the letter 'N'. NCBI contigs were ordered and oriented relative to one another according to matches to mapped STS markers and paired BAC end sequences.

The assembly protocols used by UCSC, CG and NCBI differ in terms of the amount and variety of input data and the algorithms used; it would therefore be surprising if they gave identical assemblies as output. Of particular interest are the relative rates of misassembly (sequences assembled in the wrong order and/or orientation) and the relative coverage achieved by the three protocols. Unfortunately the UCSC group are alone in having published assessments of the rate of misassembly in the contigs they produce. Using artificial datasets they found that on average ~10% of assembled fragments were assigned the wrong orientation and ~15% of fragments were placed in the wrong order by their protocol (Kent and Haussler, 2001). Two independent assessments of UCSC assemblies have come to similar conclusions. Katsanis *et al.* (2001) examined various UCSC consecutive draft genome assembly releases and reported that 10–15% of EST sequences identified within them appeared to be on wrongly assembled genomic sequences. In agreement with this, Semple *et al.* (2002) observed 19 and 11% of erroneously ordered marker sequences in two consecutive UCSC assemblies for a ~5.8 Mb region of chromosome 4. The latter study also found wide variation in coverage (23–59% of the available IHGSC sequence data included) and rates of misassembly (2.08–4.74 misassemblies per Mb) between consecutive UCSC and NCBI assemblies and the published CG assembly for the same region. These analyses indicate that the lowest rate of misassembly is produced by the CG protocol, followed by the UCSC and lastly the NCBI protocols. However, the CG protocol also produced the lowest coverage, including only around half the sequence data recruited into the UCSC and NCBI assemblies. Olivier *et al.* (2001) compared orders of TNG radiation hybrid map STSs produced by UCSC and CG protocols. They found widespread differences, such that 36% of TNG STS pairs were present in orders that differed between UCSC and CG assemblies. The TNG order was consistent with the CG assembly order slightly more often than with the UCSC assembly order. The UCSC website provides a variety of comparisons of its assemblies to genetic, physical and cytogenetic mapping data and these comparisons represent a useful resource for users to assess the likely degree of misassembly in a region of interest.

Unsurprisingly, it has been shown that differences between assemblies do indeed result in differences in annotation. Semple *et al.* (2002) found variable amounts of tandemly duplicated and interspersed repeat sequence between UCSC, NCBI and CG derived assemblies and more striking differences in annotation were also identified by Hogenesch *et al.* (2001) between CG and UCSC assemblies. Hogenesch *et al.* (2001) found large differences between the genes found in CG and UCSC assemblies, such that more than one-third of the genes identified in one assembly were not found in the other. Thus, genomic sequence annotation can only be as good as the underlying genomic sequence assembly and, as we have seen, accurate assembly of draft sequence fragments is far from error free.

The human genome is widely reported to be due for completion in 2003 but at the moment around one-quarter of publicly available human genome sequence is still categorized as 'draft' or unfinished. Relatively small, problematic regions of gapped draft sequence may well persist beyond 2003, since certain regions of the genome are simply not present within existing clone libraries and are also recalcitrant to subcloning (Hattori *et al.*, 2000). Specialized technologies are required to close such gaps in the clone map. It therefore seems likely that draft assemblies of some small regions of the human genome will be with us for some time to come. Also a fraction of the genome (perhaps 5%) consists of large (>10 kb) duplicated segments which share 90–98% sequence identity. Regions containing such duplicated segments are notoriously difficult to assemble accurately and are not only found in pericentromeric and subtelomeric regions but also across the rest of the genome, including the gene-rich regions that sequence annotators are primarily interested in (Eichler, 2001). A comparison of the completed sequence of chromosome 20 with the preceding public CG and UCSC draft assemblies of the same chromosome identified 'major discrepancies' (Hattori and Taylor, 2001). These authors concluded that the draft assemblies were probably confounded by large duplicated regions.

5.3 ANNOTATION FROM A DISTANCE: THE GENERALITIES

If some troublesome regions of the genome are set to continue as problems for cloning, sequencing and assembling, this is a minor concern in comparison to the comprehensive annotation of genomic sequence. At almost every level, computational annotation of genomic sequence is error prone and incomplete. Of course, the aim of computational annotation in common with much of bioinformatics, is to provide a preliminary set of predictions that must then be tested by 'wet' laboratory work. The aim is a rapid first pass or 'base line' annotation as the most comprehensive genomic annotation resource Ensembl (Hubbard *et al.*, 2002) puts it. From the computational point of view this enterprise is hugely successful: merely by considering the statistical qualities of the raw sequence data we can detect the presence of most protein-coding human genes. We can then identify the presence of known, structural domains within the conceptually translated products of these predicted genes and make informed guesses about functional roles and subcellular localization. When one looks at a raw BAC sequence entry from GenBank it is easy to appreciate the scale of these achievements but the view from the wet laboratory bench can be different. The broad success of computational gene prediction is little consolation to the bench geneticist who has to sift through numerous artifactual exon predictions only to find later that his gene of interest was not detected by any of the algorithms used. What is broadly impressive to the bioinformaticist can be just plain wrong to those dealing with specifics. In a recent excellent introduction to genomic sequence annotation Lincoln Stein has defined three, hierarchical levels of annotation: the most fundamental nucleotide level, followed by protein level and then process level (Stein, 2001).

5.3.1 Nucleotide Level

Nucleotide level is the point at which the raw genomic sequence is analysed and forms the basis for subsequent levels of interpretation. The first step is to identify as many known genomic landmarks as possible; these are generally markers from previous mapping studies, repeats and known genes already in public databases. This can be done quickly and accurately by a variety of programs. Markers from previous genetic, physical and cytogenetic maps are placed upon the genomic sequence by algorithms designed to find short, almost exact sequence matches such as the ePCR program (Schuler, 1997; http://www.ncbi.nlm.nih.gov/genome/sts/epcr.cgi), BLASTN (Altschul *et al.*, 1990), SSAHA (Ning *et al.*, 2001; http://www.sanger.ac.uk/Software/analysis/SSAHA/) and BLAT (Kent, unpublished data; http://genome.ucsc.edu/cgi-bin/hgBlat?command=start). Identifying these markers is essential to allow the genomic sequence to be seen in relation to the previous, pre-genome sequence literature, for example on human disease genetics. The newest type of markers, single nucleotide polymorphisms or SNPs, are also identified in the sequence to facilitate the next generation of disease gene mapping studies. Similar algorithms, extended to incorporate information on gene structure, are used to identify the positions of known mRNAs within the genomic sequence, examples include Spidey (Wheelan *et al.*, 2001; http://www.ncbi.nlm.nih.gov/IEB/Research/Ostell/Spidey/), SIM4 (Florea *et al.*, 1998; http://bio.cse.psu.edu/) and est2genome which is available from the EMBOSS package (Rice *et al.*, 2000; http://www.uk.embnet.org/Software/EMBOSS/). Just as the efforts to assemble genomic sequence take measures to identify and exclude repetitive sequence, an important part of annotation is to identify interspersed and simple repeats. The most widely used program for this task is RepeatMasker.

The central problem of nucleotide-level annotation is the prediction of gene structure. Ideally we would like to correctly delineate every exon of every gene but in large, repeat-rich eukaryotic genomes, liberally scattered with long genes with many exons, this task has turned out to be more difficult than expected. *Ab initio* gene prediction algorithms (that rely only on the statistical qualities of genomic sequence data) identify most protein coding genes reliably in prokaryotic genomes but the task is more complex in eukaryotic genomes (Burge and Karlin, 1998). Fundamentally the problem is gene density, whereas in prokaryotic genomes and yeast more than two-thirds of the genome is protein coding sequence only a few percent of the human genome fits this description. Additional problems are added by overlapping genes, alternatively spliced exons and the paucity of differences between intergenic sequence and introns. The gene prediction literature is full of metaphors involving needles and haystacks, and with good cause. The 13-Mb *S. cerevisiae* yeast genome provides a sobering example, completed in 1996 and initially thought to contain 6274 genes, the sequence has provided a steady trickle of additional genes that had been overlooked. Since publication of the yeast genome a further 202 genes have been discovered, most appear to have been missed because they are relatively short or overlap a previously annotated gene on the opposite strand (Kumar *et al.*, 2002). At the same time, new analyses of these yeast sequences using a variety of statistical analyses and comparative genomics approaches have suggested that several hundred of the originally annotated genes may be spurious (Malpertuy *et al.*, 2000; Zhang and Wang, 2000).

This brings us to the use of sequence similarity in gene prediction. In practice genome annotators use a combination of information to make predictions of gene structures: *ab initio* exon predictions (predictions of coding sequence made by a program on the basis of statistical measures of features such as codon usage, initiation signals, polyA signals and splice sites), repetitive sequence content and similarity to expressed sequences and proteins. These different strands of evidence are usually combined and evaluated by human annotators who use graphical interfaces, such as those provided by NIX (unpublished data; http://www.hgmp.mrc.ac.uk/Registered/Webapp/nix/) and ACEDB (Eeckman and Durbin, 1995; http://www.acedb.org/), to view all the evidence simultaneously. A recent trend in gene prediction is the design of programs that automatically incorporate evidence based on sequence similarity into their predictions. Among the best and most widely used *ab initio* algorithm is Genscan (Burge and Karlin, 1997; http://genes.mit.edu/GENSCAN.html). Guigo *et al.* (2000) tested its success in artificially produced sequence data designed to mimic human BAC sequences. At the same time they tested algorithms that use sequence similarity to make their predictions, such as GeneWise (Birney and Durbin, 2000; http://www.sanger.ac.uk/Software/Wise2/). The results showed a clear advantage to including evidence from sequence similarity where the similarity was strong. In such cases GENEWISE could correctly identify 98% of coding bases present while generating a comparatively low level of artifactual exons (2%) and missing 6% of real exons. Where levels of similarity were more modest however the performance of algori GENEWISE declined to below that of GENSCAN. GENSCAN was foun 89% of coding bases at the cost of a rather high level of artifactual exons (41 14% of real exons missed. Guigo *et al.* (2000) suggest that the success of all the programs tested is expected to be lower in real genomic sequence. Another comparison of gene prediction programs using *D. melanogaster* genomic sequence identified similar levels of performance for the programs tested and also indicated an advantage to algorithms including similarity-based evidence in predictions (Reese *et al.*, 2000). Shortcuts to the structures of many genes may come from a large collection of full-length mouse cDNAs (Kawai *e al.*, 2001) and large human cDNA collections (Kikuno *et al.*, 2002), which are expected to grow rapidly over the next few years.

As we amass genomic sequence data from many organisms the reach of computational annotation based upon sequence similarity is increasing. New methods aimed at the prediction of non-coding features in the genome, such as regulatory regions and non-coding RNAs (ncRNAs) are evolving rapidly. Whereas protein coding exons have a distinctive statistical fingerprint ncRNAs do not, or at least they do not appear to from our present, limited knowledge of them (Eddy, 2001). For better understood classes of ncRNAs, such as tRNAs, prediction methods involving secondary structure prediction have been successful (Lowe and Eddy, 1997) but for novel ncRNAs the only effective methods are based on comparative genomics (Rivas *et al.*, 2001). The same is true for novel regulatory sequences, where only a fraction of transcription factor binding sites have been identified to date (Wingender *et al.*, 2001). Even incomplete, fragmentary sequence data from other organisms has been used with some success to predict putative regulatory regions (Chen *et al.*, 2001). This approach is examined in some detail in Chapter 7.

5.3.2 Protein Level

Once we have a gene prediction that we believe, the next step is to assign a possible function to the encoded protein; this is the central task of protein-level annotation. Most computationally assigned functions are derived from sequence similarity. A pair of proteins that align along 60% or more of their lengths with significant similarity (e.g. E <0.01 in a BLASTP search of a large public database) are very likely to be homologous — that is derived from a common ancestor. Such a pair of sister proteins may be paralogues, derived from a duplication event, or orthologues, that exist as a result of a speciation event. For every homologous pair identified in this way additional searches may verify that each member of the pair identifies the other member as the best match within the organism of interest. This makes it likely that the pairs identified are likely to be orthologues (Huynen and Bork, 1998), which is desirable since orthologues are likely to share the same function (Jordan *et al.*, 2001) whereas functional diversification between paralogues is thought to be common (Li, 1997). In most cases this strategy of reciprocal sequence similarity searches to identify orthologues is successful (Chervitz *et al.*, 1998) and is the rationale that underlies the construction of the Clusters of Orthologous Groups of proteins (COGs) database (Tatusov *et al.*, 2000; http://www.ncbi.nlm.nih.gov/COG/). However, caution is necessary when dealing with the results of such analyses. For example, a novel human gene may be directly descended from a common ancestor of a yeast gene (in which case the two genes are orthologues and are likely to share the same function), or it may be descended from a duplicated sister yeast gene (and the two genes are really paralogues) with a different function. Without a complete picture of the related family of proteins we are dealing with, it can be difficult to decide. Definitive evidence for orthology versus paralogy can come from comprehensive phylogenetic analysis but even then, when dealing with larger families and/or incomplete data, it can be difficult. As a result, it is not uncommon to find mistaken computational predictions of function that are not supported by further experiment (Iyer *et al.*, 2001).

In the absence of any detailed knowledge about the evolutionary pedigree of the protein under study, similarity may sometimes still imply functional similarity. For example two proteins only 30% identical may share much of their biochemistry but have different substrates (Todd *et al.*, 2001). In spite of their divergence they may share a common functional domain. There are a variety of protein domain databases and they are widely used in genome annotation. For example, version 7 of the Pfam database contains 3360 domains that match 69% of proteins in public sequence databases, with domains represented by alignments between regions of proteins containing them (Bateman *et al.*, 2002; http://www.sanger.ac.uk/Software/Pfam/). Statistical models of these alignments are constructed and searched using the elegant HMMER software package (Eddy, 1998; http://hmmer.wustl.edu/). The Interpro database (Apweiler *et al.*, 2000; http://www.ebi.ac.uk/interpro/), which amalgamates several databases (including Pfam) covering protein domains, families and functional sites, was used by the IHGSC to provide the publicly available annotation for the draft human genome. Interpro entries provide links to additional information including functional descriptions, references to the literature and structural data. Since the IHGSC draft genome publication, the EBI (European Bioinformatics Institute; http://www.ebi.ac.uk/proteome/) has maintained and updated annotation for the set of known and predicted human proteins using Interpro but their most recent analyses match only around 60% of the set. Thus even our most strenuous efforts to gain clues to protein function, often based upon rather distant homology, tell us nothing about 40% of human proteins.

5.3.3 Process Level

Ultimately the goal of genetics is to understand the relationship between genotype and phenotype. There is a large gap between annotation at the nucleotide or protein level and an understanding of how a given protein influences phenotype. Even in the best case, with a known gene coding for a protein containing well-studied domains, there are always questions that remain to be asked. How does the protein interact or complex with other proteins? Where does it localize within the cell? Which cellular processes and organelles is it involved with? In which tissues and at which developmental stages does it act? The answers to these questions provide process-level annotation. The most important applications of our knowledge about the human genome are in medicine, to discover the variations and aberrations that underlie disease. Process level annotation provides a rational way to select the best candidate genes for involvement in disease. For example, when it was first submitted to GenBank in 1997 a certain gene (accession number U80741) was annotated as '*Homo sapiens* CAGH44 mRNA' and 'polyglutamine rich'. Due to the painstaking work of Lai *et al.* (2001) on a region associated with speech disorders we now know this gene as FOXP2, the first gene found to be involved in human language acquisition disorders. Before their work FOXP2 appeared to be one of many transcription factors, expressed in many tissues and best studied in *D. melanogaster*. With better process level annotation FOXP2 may have been identified earlier as a good candidate for involvement in disease.

The main source of process-level annotation is the scientific literature but, even with modern access through the web, the literature is a 20th century resource unsuited to 21st century needs. What we have is a dizzying array of terms for a single gene, function or process and no accepted way of organizing this information, added to this are all the vagaries and idiosyncrasies of human language. What is needed is a structured resource with a limited number of terms for genes and descriptions of their functions, organized so that it is easily processed automatically by computer programs. A recent initiative, called the Gene Ontology (GO) project has provided a framework to achieve this (Gene Ontology Consortium, 2001; http://www.geneontology.org/). GO consists of an hierarchical set of structured vocabularies to describe the molecular functions, biological processes, and cellular components associated with gene products. With the known and predicted genes in a genome annotated using GO it is possible to quickly retrieve, for example, all genes encoding transmembrane receptors, all genes involved in apoptosis, or all genes encoding products localized to the cytoskeleton. The hierarchical nature of GO means that subsets of these categories can also be retrieved, for example all G-protein coupled receptors within the transmembrane receptor category. GO annotation has already been adopted by databases for several model organism genomes, including the *Saccharomyces* Genome Database (Dwight *et al.*, 2002; http://genome-www.stanford.edu/Saccharomyces/), FlyBase (FlyBase Consortium, 2002; http://fly.ebi.ac.uk:7081/) and the Mouse Genome Database (Blake *et al.*, 2002; http://www.informatics.jax.org/). At the moment GO annotations are added to genes in these databases manually by trained biologist curators browsing the scientific literature. In the longer term, with the rapidly increasing number of completed genomes, this process will become increasingly automated. Efforts are already underway to develop software that will automatically extract information from the literature to be incorporated into the GO annotation of a gene (Raychaudhuri *et al.*, 2002).

The scale of the problem of providing process-level annotation for every human gene is prompting the development of large-scale technologies to generate data on many genes at once. Large-scale parallel measurement of gene expression for entire genomes is now possible and should give good data on the developmental timing and tissue specificity of many human genes, from which it is possible to infer process-level annotation (Noordewier and Warren, 2001). An important step on the way to designating the processes a protein is involved in, is to define the proteins with which it interacts, and work is well underway to elaborate the web of interacting proteins and complexes that define the *S. cerevisiae* proteome (Gavin *et al.*, 2002; Ho *et al.*, 2002). However these high-throughput methods are known to generate false positives and negatives; that is they identify some artifactual interactions and miss some genuine interactions. Thus, high-throughput technologies may eventually provide useful process-level annotation for many, if not most, human genes but there will always be an indispensable role for conventional, detailed laboratory studies of smaller scale. New databases and analyses will be necessary to make sense of the network of genetic interactions that underlie the phenotype. A good example is the Mouse Atlas and Gene Expression Database Project (Baldock *et al.*, 2001; http://genex.hgu.mrc.ac.uk/) which aims to describe the patterns of gene expression responsible for the emergence of anatomical structure during mouse development. It will enable gene expression data to be viewed in the context of three-dimensional embryo sections.

5.4 ANNOTATION UP CLOSE AND PERSONAL: THE SPECIFICS

Even given the difficulties and shortcomings in computational annotation discussed above, several well-resourced groups have undertaken the task of compiling, maintaining and updating freely accessible annotation for the entire human genome. There are now four well-designed websites offering users the chance to browse annotation of the draft human genome. All four sites offer a graphical interface to display the results of various analyses, such as gene predictions and similarity searches, for draft and finished genomic sequence. These interfaces are indispensable for allowing rapid, intuitive comparisons between the features predicted by different programs. For instance, one can see at once where an exon prediction overlaps with interspersed repeats or an SNP. But the four sites are not equivalent and there are important distinctions between them in terms of the data analysed, the analyses carried out and the way the results are displayed.

5.4.1 Ensembl

Ensembl is a joint project between the EBI and the Sanger Institute (http://www.sanger.ac.uk/). The Ensembl database (Hubbard *et al.*, 2002; http://www.ensembl.org/), launched in 1999, was the first to provide a window on the draft genome, curating the results of a series of computational analyses. Until January 2002 (release 3.26.1) Ensembl used the UCSC draft sequence assemblies as its starting point but it is now based upon NCBI assemblies. The Ensembl analysis pipeline consists of a rule-based system designed to mimic decisions made by a human annotator. The idea is to identify 'confirmed' genes that are computationally predicted (by the GENSCAN gene prediction program) and also supported by a significant BLAST match to one or more expressed sequences or proteins. Ensembl also identifies the positions of known human genes from public sequence database entries, using GENEWISE to predict their exon structures. The total set of Ensembl genes should therefore be a much more accurate reflection of reality than *ab initio* predictions alone but it is clear that many novel genes are missed (Hogenesch *et al.*, 2001). Of the novel genes that are detected many, if not most are expected to be incomplete for two main reasons. Firstly, as we have seen, while GENSCAN can detect the presence of most genes in a genomic sequence it is substantially less successful in predicting their correct exonic structures (as with other *ab initio* gene predictions). Secondly, any prediction is entirely dependent upon the quality of the genomic sequence and where the sequence is gapped or wrongly assembled the missing exons may not be present for the software to find.

Many other genomic features have been included as Ensembl has developed: different repeat classes, cytological bands, CpG island predictions, tRNA gene predictions, expressed sequence clusters from the UniGene database (Wheeler *et al*, 2002; http://www.ncbi.nlm.nih.gov/UniGene/), SNPs from the dbSNP database (Sherry *et al.*, 2001; http://www.ncbi.nlm.nih.gov/SNP/index.html), disease genes found in the draft genome from the OMIM database (On-line Mendelian Inheritance in Man database; Wheeler *et al.*, 2002; http://www.ncbi.nlm.nih.gov/Omim/) and regions of homology to mouse draft genomic sequences. GENSCAN-predicted exons that have not been incorporated into Ensembl-confirmed genes may also be viewed. This means that the display can be used as a workbench for the user to develop personalized annotation. For example, one may discover novel exons by finding GENSCAN exon predictions which coincide with good matches to a fragment of the draft mouse genome, or novel promoters by finding matches to the draft mouse genome that occur upstream of the 5′ end of a gene. Once you have identified a gene of interest you can link to a wealth of information at external sites such as the Interpro protein domains it encodes and its expression profile according to the SAGEmap repository (Lash *et al.*, 2000). Eventually Ensembl aims to become a platform for studies in comparative genomics and already it is possible while browsing the human genome to jump to an homologous region of the mouse genome via a match to a mouse genomic sequence fragment. Substantial thought and effort has evidently gone into the Ensembl site design. The result is certainly a user-friendly experience, and not just by the standards of computational biology. The web interface to the database achieves the laudable aim of providing seamless access to the human genome. The user can sink down through cytogenetic ideograms of whole chromosomes, to large unfinished sequence contigs several Mb long and then to smaller fragments of individual BAC clones only kb long. Along the way a graphical display shows the relative positions of genes and the other features.

Figure 5.1 shows the Ensembl display for the genomic region around the FOXP2 gene mentioned earlier. The region is shown at three levels of resolution. The upper panel shows the position of the region as a small red box on a cytogenetic ideogram of chromosome 7. The middle panel shows an exploded view of this box, including the structure of the draft genome assembly, the relative positions of various markers and a simple overview of the gene content. The bottom panel gives a detailed view of a subsection (indicated again by a red box) of the middle panel. This detailed view is the business end of the browser

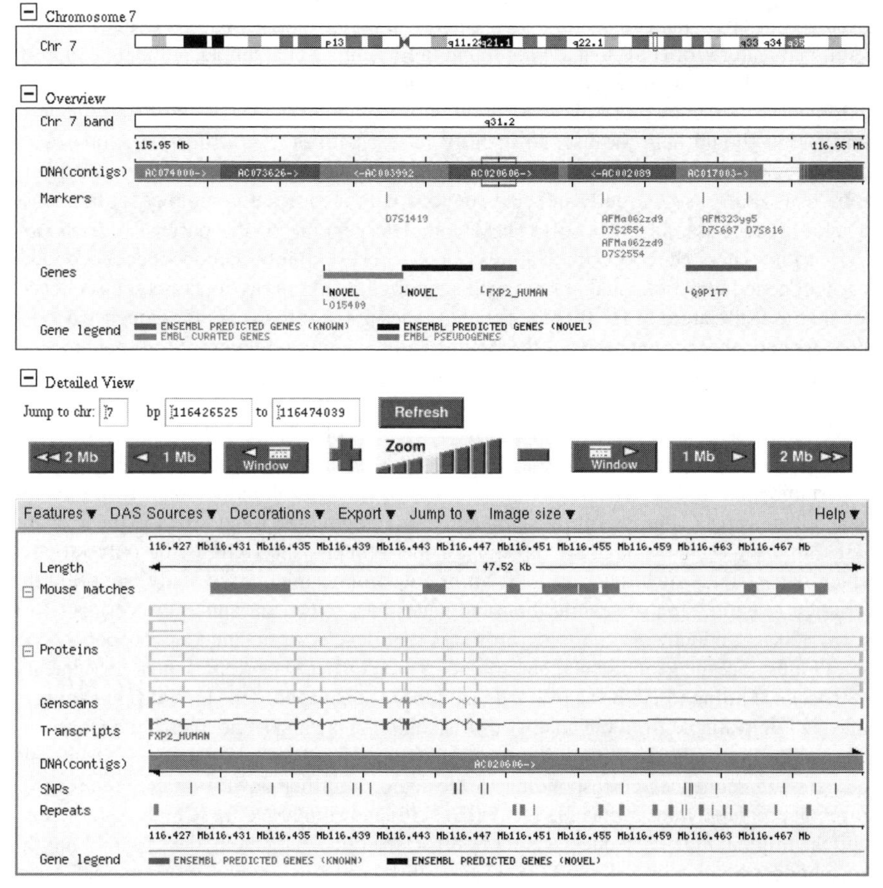

Figure 5.1 The genomic region around the FOXP2 gene according to Ensembl (See Colour Plates).

and is easily customized, via pull-down menus, to display any desired combination of the available features. In Figure 5.1 the combination chosen shows the positions of matches to the mouse genome in relation to GENSCAN-predicted exons and similarities to protein sequences, which allows a user to define non-coding conserved regions that may be of regulatory importance. Using this display one could also select SNPs that lie outside repetitive sequences; an important consideration for PCR-based SNP assays.

Data retrieval is extremely well catered for in Ensembl, with text searches of all database entries, BLAST searches of all sequences archived and the availability of bulk downloads of all Ensembl data and even software source code. Ensembl annotation can also be viewed and added to interactively on your local machine using the Apollo viewer (http://www.ensembl.org/apollo/).

5.4.2 UCSC Human Genome Browser (HGB)

The UCSC Human Genome Browser (HGB) bears many similarities to Ensembl, it too provides annotation of the NCBI assemblies (as well as UCSC assemblies) and it displays a similar array of features, including confirmed genes from Ensembl. The range of features displayed in HGB (and Ensembl) often change between releases but generally there are additional features of HGB that are not found in Ensembl. For example, at the time of writing HGB includes predictions from two *ab initio* gene prediction programs: GENSCAN and Fgenesh (Salamov and Solovyev, 2000; http://genomic.sanger.ac.uk/gf/Help/fgenesh.html). This should help the user to identify false positives (i.e. artifactual exons) from either program and concentrate on exons predicted by both programs that are most likely to be real. HGB also currently indicates regions with significant homology to the mouse genome as in Ensembl but also to the incomplete genome of the pufferfish *Tetraodon nigroviridis*. These HGB-specific features can provide useful information when one is dealing with gene predictions that are not well supported by similarity to expressed sequence. Another useful feature of HGB is the detailed description of the genomic sequence assemblies. Graphical representations of the fragments making up a region of draft genome can be displayed, showing the relative size and overlaps of each fragment and also whether any gaps between fragments are bridged by mRNAs or paired BAC end sequences. This means that one can get an idea of the likely degree of misassembly in a draft region. There is an increasing amount of data becoming available from large-scale gene expression studies and public repositories have emerged for their curation, such as the NCBI Gene Expression Omnibus (http://www.ncbi.nlm.nih.gov/geo/) and ArrayExpress at the EBI (http://www.ebi.ac.uk/arrayexpress/). At the moment, the HGB is the only browser which incorporates such data, in the form of data from a microarray study exploring the variation in expression of several thousand genes in a screen for anti-cancer drugs (Ross *et al*., 2000). Undoubtedly the other browsers will develop to include similar data.

In Figure 5.2 the genomic neighbourhood of the FOXP2 gene (represented by sequence U80741) according to HGB (as of 6 August 2001) is displayed. This provides the kinds of information available from the analogous Ensembl display and some interesting additional data. At the top of the display there are indications of the size, cytogenetic band and the genomic sequences corresponding to the region. Further down one can compare an Ensembl predicted transcript (ENST00000265436) and similar NCBI Acembly predictions with the original FOXP2 sequence entry (U80741). Notice that neither the Ensembl nor the Acembly predictions find all the FOXP2 exons that we know are present from U80741, at the same time both *ab initio* prediction algorithms (GENSCAN and Fgenesh) have split the gene into more than one prediction. These are all familiar problems in genomic sequence

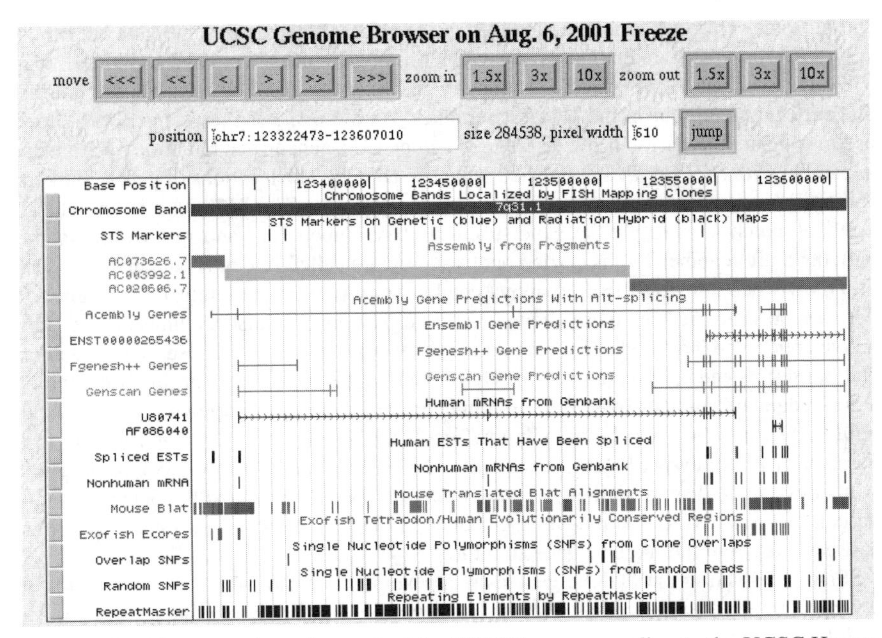

Figure 5.2 The genomic region around the FOXP2 gene according to the UCSC Human Genome Browser.

annotation. Notice also that the Ensembl prediction has a number of additional exons 3′ of the last U80741 exon. This is because U80741 does not contain the full coding sequence of FOXP2 and the Ensembl prediction is based upon a later sequence entry (AF337817) which does. This illustrates another common problem: different annotation sources may be based upon different sequence data, depending on what is available at the time. As with Ensembl, the HGB display of the region shows regions of homology to the mouse genome but also to the pufferfish genome (identified by a program called Exofish, see http://www.genoscope.cns.fr/externe/tetraodon/). It is apparent that the evolutionary distance between humans and fish means that the Exofish results are more helpful in defining exons rather than regulatory regions. However there are still regions upstream of the first U80741 exon that appear to be well conserved across the human, mouse and pufferfish genomes. Such regions may define the promoter of the FOXP2 gene.

Data retrieval is facilitated by text, BLAT (a faster, less sensitive algorithm than BLAST) searches and bulk downloads of annotation or sequence data. As with Ensembl, the HGB website has been well designed and is sympathetic to the naive user, but the HGB graphical interface is more Spartan. If Ensembl is Disney then HGB is Southpark. The positive side of this is that HGB will usually display a region on your local web browser more quickly than Ensembl can. Both the Ensembl and HGB interfaces offer users the ability to jump between their respective views of a region and so, when they are both annotating the same version of the same NCBI assembly, they can easily be used as complementary resources.

5.4.3 NCBI Map Viewer (NMV)

As the human genome nears completion the problems of dealing with draft sequence data will recede and the main task will be to curate the finished sequences representing each chromosome. This task will be undertaken at the NCBI. Whereas Ensembl and HGB both previously provided annotation of the UCSC draft genome assemblies the NCBI Map Viewer (NMV; http://www.ncbi.nlm.nih.gov/cgi-bin/Entrez/map_search) has always displayed features present in the NCBI assemblies. As the name suggests, the NMV shows useful comparisons between a wide range of cytogenetic, genetic and radiation hybrid maps in parallel with NCBI draft and finished sequence contigs. The locations of genes, markers, and SNPs are indicated on the contig sequences. As with Ensembl, there is an analysis protocol which aims to predict gene structures based upon EST and mRNA alignments with the draft genome. This is carried out by a program called Acembly (unpublished data; http://www.ncbi.nih.gov/IEB/Research/Acembly/help/AceViewHelp.html) which aims to derive gene structure from these alignments alone. The program also attempts to give alternative splice variants of genes where its alignments suggest them. These gene structures and transcripts end up as records in the NCBI RefSeq database, which is slowly compiling a non-redundant curated dataset representing current knowledge of known genes (Wheeler *et al.*, 2002; http://www.ncbi.nlm.nih.gov/LocusLink/refseq.html). Like the Ensembl protocol many Acembly-predicted structures (the NCBI estimate 42%) are incomplete. These structures can be displayed alongside *ab initio* gene models predicted by GenomeScan (a variant of GENSCAN) and matching UniGene clusters to allow users to make their own assessments about the likeliest gene structure.

Figure 5.3 shows the FOXP2 gene as it appears in the NMV which shows features on a vertical rather than horizontal display. The genomic sequence contig the gene occurs on (NT_023632) is shown in the leftmost column, followed by BLAST matches to three UniGene expressed sequence clusters. This gene is typical in having more than one UniGene cluster representing it, particularly at the 3′ end as ESTs are more commonly sequenced from the 3′ ends of mRNAs. In the next columns are a GenomeScan prediction which misses some exons and a depiction of XM_059813: the model of FOXP2 that Acembly has constructed by aligning expressed sequences with this region of the genome. SNPs from the NCBI dbSNP database are also displayed with those occurring within the gene highlighted, however there is no indication of repetitive sequence. In the rightmost column the FOXP2 gene structure is displayed according to the XM_059813 model.

The NMV offers tabulated downloads of data and it is possible to BLAST search the NCBI assembly (via the NCBI BLAST site: http://www.ncbi.nlm.nih.gov/BLAST/) and view the matching regions using the NMV. All annotated genes are connected to NCBI LocusLink which provides links to associated information such as related sequence accession numbers, expression data, known phenotypes and SNPs.

5.4.4 ORNL Genome Channel (GC)

The ORNL (Oak Ridge National Laboratory) Genome Channel (GC; http://compbio.ornl.gov/channel/) consists of a series of tools for visualizing and querying the NCBI human genome sequences and those of other organisms assembled and annotated by ORNL and collaborators. The GC browser provides the usual categories of nucleotide-level annotation: repetitive sequences, CpG islands, polyA sites and marker positions. The GC gene prediction protocol is pitched somewhere between Ensembl and HGB: GrailEXP (Uberbacher *et al.*, 1996; http://compbio.ornl.gov/grailexp/) and GENSCAN predictions are given where they are supported by BLAST matches to expressed sequence along with known genes from RefSeq or GenBank entries. Sequence similarity results are not viewable as independent features (as in the other browsers), only as evidence associated with predicted exons. This is rash considering the number of coding sequences missed by *ab initio* algorithms and unhelpful where one is interested in non-coding regions such as UTRs.

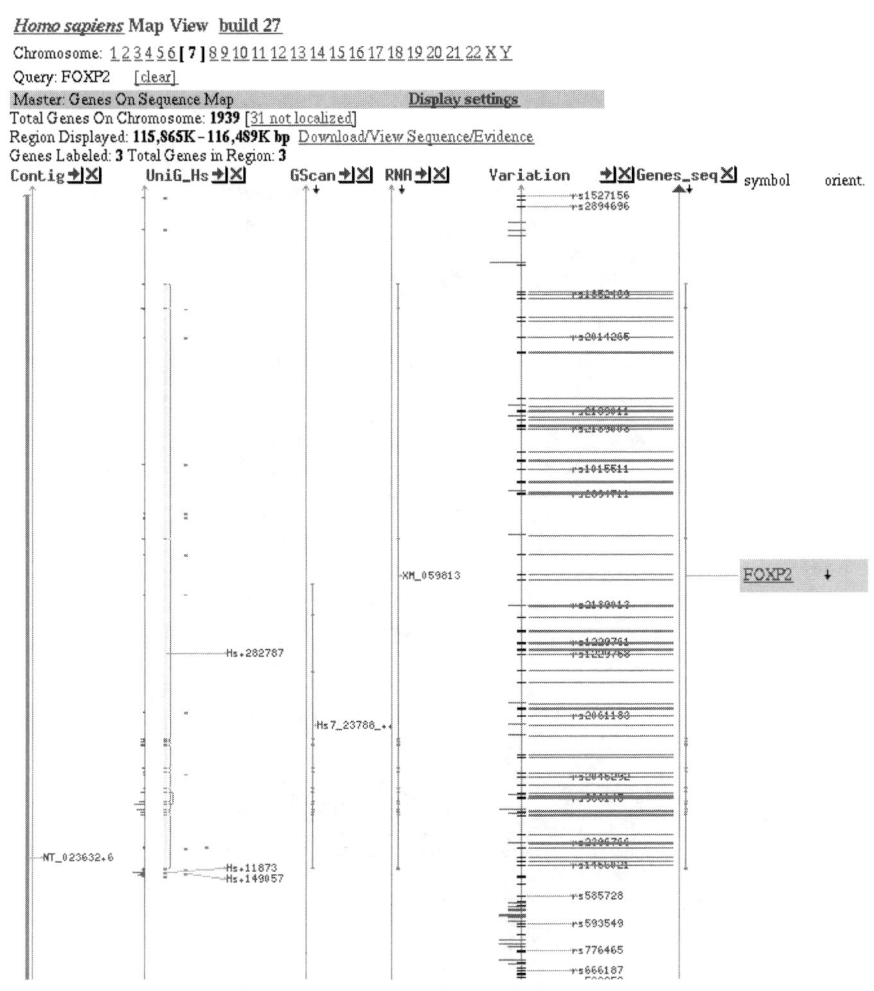

Figure 5.3 The genomic region around the FOXP2 gene according to the NCBI Map Viewer (See Colour Plates).

The only kind of sequence similarity results displayed independently are gene predictions derived from transcripts from the Database of Transcribed Sequences (DoTS; unpublished data; http://www.allgenes.org/) which clusters and assembles expressed sequences. On the platforms I tested (Netscape running in UNIX and Microsoft Internet Explorer in Windows NT), the graphical display itself also has a problem: several features (different classes of repeats, CpG islands and polya sites) appear on top of one another, which makes it difficult to see what is going on. On the positive side GC does allow users to submit their own sequences to the suite of BLAST searches and gene prediction programs underlying the GC analysis pipeline. None of the other sites allow this. Downloads of genomic DNA and the mRNA and peptide sequences for the predicted genes in GC are available. The GC browser's view of the FOXP2 gene and flanking regions is provided in Figure 5.4. The central horizontal band displays the clones making

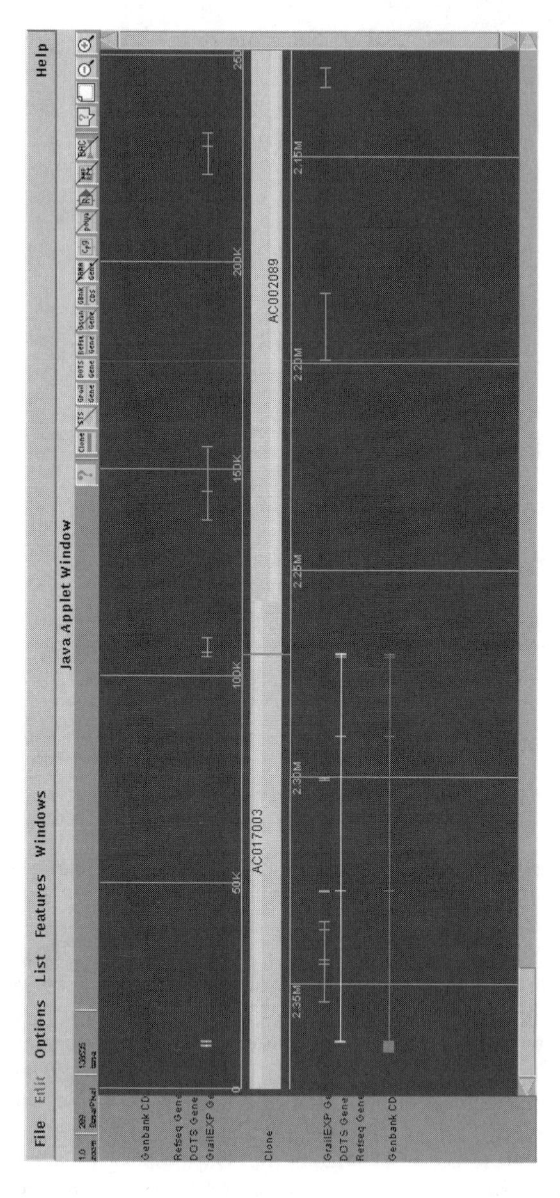

Figure 5.4 The genomic region around the FOXP2 gene according to ORNL Genome Channel (See Colour Plates).

up this NCBI genomic sequence contig, the vertical line intersecting one of the clones represents a CpG island. Repeats and polya sites also appear as lines within this band and gene predictions on either strand are displayed in the panels above and below it. At the time of writing it is not possible to view homologies to other genome sequences or the positions of SNPs. More information on the features that are displayed is available from other windows.

5.5 ANNOTATION: THE NEXT GENERATION

In spite of difficulties with the quality of genomic sequence assemblies and the errors and omissions of computational annotation the browsers discussed above remain extremely useful tools for the cautious biologist. They undoubtedly indicate the presence of most coding sequence in a given fragment of genomic sequence and indicate their location in the genome based on the best genomic sequence available. In addition they have a stab at predicting gene structures for novel genes that should be accurate if the gene in question is known or has a close homologue which is known. Most aspects of the analysis carried out are the subjects of active research, and improvements in performance due to the inclusion of new sequence data and annotation software will be ongoing. The downside of these developments is that all annotation of genomic sequence is potentially in flux and one should not assume that the representation of a region will remain the same between different software or data releases.

At some time in 2003 discussions of draft sequence assembly should be academic for more than 90% of the human genome and large finished contigs, tens of megabases long will be curated at NCBI. The main tasks with regard to the primary sequence data will then relate to data curation rather than assembly. Annotation of these sequences, on the other hand, should still be at a relatively early stage. Even at nucleotide level there is much to be done, particularly in exploiting the data available from model organism genome sequencing projects. There have already been notable successes in using comparative genomics to predict gene structures using the Twinscan program (Korf *et al.*, 2001; http://genes.cs.wustl.edu/). The cutting edge of nucleotide-level annotation is in defining regulatory regions: transcription start sites (TSSs), transcription factor binding sites and promoter modules (Werner, 2001). Here again, comparative genomics is already a rich source of information simply using existing sequence search algorithms such as BLAST (Levy *et al.*, 2001). At a higher level, gene expression is also regulated by the large-scale topology of chromosomes, and annotation may eventually indicate features such as chromosome domains (genomic regions that bind histone-modifying proteins) and matrix attachment sites (regions that facilitate the organization of DNA within a chromosome into loops). However, defining the genes whose transcription is regulated from such features may be an insoluble problem computationally, since they may regulate transcription from a given TSS, from several different TSSs of the same gene or multiple genes in a region.

At the protein and process levels of annotation there is also progress, for instance in our ability to detect more remote homologies and gain clues about function. Homologous proteins, sharing a common three-dimensional structure and function, need not share detectable sequence similarity. There is therefore increasing interest in annotation by similarity at the level of protein structure (Gough and Chothia, 2002). The genome sequence is already changing the way we study biology as we start to fill in the gaps between genetics, cellular function and development. Rather than studying a particular gene or protein we are increasingly able to study all elements in a system of interest, a group of proteins that participate in a complex for example. We might start with a single protein and identify others in the proteome that potentially interact with it, on the basis of the presence of domains known to interact. In the process we may discover previously unknown connections with other complexes or biochemical pathways that can be included in the annotation of the relevant sequences. Studies on this scale are prompting the development of multidisciplinary groups that study the behaviour and perturbation of entire biological systems (Ideker *et al.*, 2001). In the end this should provide a genome sequence with contents which can be browsed at the level of their genomic neighbourhood but also at the level of the interactions, complexes and processes that they participate in and the phenotypes they influence.

This review has only provided a brief introduction to the fields of computational draft genome assembly and annotation but it should be evident that what has already been achieved has involved innovations as great as those in the biotechnology that led to the production of the sequence data itself. At the same time, problems remain at every level and are the subjects of active research. As a result many different groups around the world are working on interpreting the data avalanche that is modern genetics and communication and comparison of results becomes difficult. The Distributed Annotation System (DAS; Dowell *et al.*, 2001; http://biodas.org/) aims to provide a framework for people to exchange data easily using the web. It promises a future without the current confusion of incompatible interfaces and data formats, and an increase in the open exchange of data and ideas.

ACKNOWLEDGEMENTS

Colin Semple enjoys the financial support of the UK Medical Research Council. Martin S. Taylor provided comments on an earlier version of this manuscript.

REFERENCES

Altschul SF, Gish W, Miller W, Myers EW, Lipman DJ. (1990). Basic local alignment search tool. *J Mol Biol* **215**: 403–410.

Altschul SF, Madden TL, Schaffer AA, Zhang J, Zhang Z, Miller W, *et al.* (1997). Gapped BLAST and PSI-BLAST: a new generation of protein database search programs. *Nucleic Acids Res* **25**: 3389–3402.

Apweiler R, Attwood TK, Bairoch A, Bateman A, Birney E, Biswas M, *et al.* (2000). InterPro — an integrated documentation resource for protein families, domains and functional sites. *Bioinformatics* **16**: 1145–1150.

Baldock R, Bard J, Brune R, Hill B, Kaufman M, Opstad K, *et al.* (2001). The Edinburgh Mouse Atlas: using the CD. *Brief Bioinform* **2**: 159–169.

Bateman A, Birney E, Cerruti L, Durbin R, Etwiller L, Eddy SR, *et al.* (2002). The Pfam protein families database. *Nucleic Acids Res* **30**: 276–280.

Beck S, Sterk P. (1998). Genome-scale DNA sequencing: where are we? *Curr Opin Biotechnol* **9**: 116–120.

Birney E, Durbin R. (2000). Using GeneWise in the Drosophila annotation experiment. *Genome Res* **10**: 547–548.

Blake JA, Richardson JE, Bult CJ, Kadin JA, Eppig JT. (2002). The Mouse Genome Database (MGD): the model organism database for the laboratory mouse. *Nucleic Acids Res* **30**: 113–115.

Burge CB, Karlin S. (1997). Prediction of complete gene structure in human genomic DNA. *J Mol Biol* **268**: 78–94.

Burge CB, Karlin S. (1998). Finding the genes in genomic DNA. *Curr Opin Struct Biol* **8**: 346–354.

Chen R, Bouck JB, Weinstock GM, Gibbs RA. (2001). Comparing vertebrate whole-genome shotgun reads to the human genome. *Genome Res* **11**: 1807–1816.

Chervitz SA, Aravind L, Sherlock G, Ball CA, Koonin EV, Dwight SS, *et al.* (1998). Comparison of the complete protein sets of worm and yeast: orthology and divergence. *Science* **282**: 2022–2028.

Dowell RD, Jokerst RM, Day A, Eddy SR, Stein L. (2001). The Distributed Annotation System. *BMC Bioinformatics* **2**: 7.

Dwight SS, Harris MA, Dolinski K, Ball CA, Binkley G, Christie KR, *et al.* (2002). Saccharomyces Genome Database (SGD) provides secondary gene annotation using the Gene Ontology (GO). *Nucleic Acids Res* **30**: 69–72.

Eddy SR. (1998). Profile hidden Markov models. *Bioinformatics* **14**: 755–763.

Eddy SR. (2001). Non-coding RNA genes and the modern RNA world. *Nature Rev Genet* **2**: 919–929.

Eeckman FH, Durbin R. (1995). ACeDB and macace. *Methods Cell Biol* **48**: 583–605.

Eichler EE. (2001). Segmental duplications: what's missing, misassigned, and misassembled and should we care? *Genome Res* **11**: 653–656.

Florea L, Hartzell G, Zhang Z, Rubin GM, Miller W. (1998). A computer program for aligning a cDNA sequence with a genomic DNA sequence. *Genome Res* **8**: 967–974.

FlyBase Consortium. (2002). The FlyBase database of the Drosophila genome projects and community literature. *Nucleic Acids Res* **30**: 106–108.

Gavin A-C, Bosche M, Krause R, Grandi P, Marzioch, Bauer A, *et al.* (2002). Functional organization of the yeast proteome by systematic analysis of protein complexes. *Nature* **415**: 141–147.

Gene Ontology Consortium. (2001). Creating the gene ontology resource: design and implementation. *Genome Res* **11**: 1425–1433.

Gough J, Chothia C. (2002). SUPERFAMILY: HMMs representing all proteins of known structure. SCOP sequence searches, alignments and genome assignments. *Nucleic Acids Res* **30**: 268–272.

Guigo R, Agarwal P, Abril JF, Burset M, Fickett JW. (2000). An assessment of gene prediction accuracy in large DNA sequences. *Genome Res* **10**: 1631–1642.

Hattori M, Taylor TD. (2001). Part three in the book of genes. *Nature* **414**: 854–855.

Hattori M, Fujiyama A, Taylor TD, Watanabe H, Yada T, Park HS, *et al.* (2000). The DNA sequence of human chromosome 21. *Nature* **405**: 311–319.

Ho Y, Gruhler A, Heilbut A, Bader GD, Moore L, Adams SL, *et al.* (2002). Systematic identification of protein complexes in *Saccharomyces cerevisiae* by mass spectrometry. *Nature* **415**: 180–183.

Hogenesch JB, Ching KA, Batalov S, Su AI, Walker JR, Zhou Y, *et al.* (2001). A comparison of the Celera and Ensembl predicted gene sets reveals little overlap in novel genes. *Cell* **106**: 413–415.

Hubbard T, Barker D, Birney E, Cameron G, Chen Y, Clark L, *et al.* (2002). The Ensembl genome database project. *Nucleic Acids Res* **30**: 38–41.

Huynen MA, Bork P. (1998). Measuring genome evolution. *Proc Natl Acad Sci USA* **95**: 5849–5856.

Ideker T, Galitski T, Hood L. (2001). A new approach to decoding life: systems biology. *Annu Rev Genomics Hum Genet* **2**: 343–372.

International Human Genome Sequencing Consortium. (2001). Initial sequencing and analysis of the human genome. *Nature* **409**: 860–892.

Iyer LM, Aravind L, Bork P, Hofmann K, Mushegian AR, Zhulin IB, *et al.* (2001). *Quod erat demonstrandum*? The mystery of experimental validation of apparently erroneous computational analyses of protein sequences. *Genome Biol* **2**.

Jordan IK, Kondrashov FA, Rogozin IB, Tatusov RL, Wolf YI, Koonin EV. (2001). Constant relative rate of protein evolution and detection of functional diversification among bacterial, archaeal and eukaryotic proteins. *Genome Biol* **2**.

Katsanis N, Worley KC, Lupski JR. (2001). An evaluation of the draft human genome sequence. *Nature Genet* **29**: 88–91.

Kawai J, Shinagawa A, Shibata K, Yoshino M, Itoh M, Ishii Y, *et al.* (2001). Functional annotation of a full-length mouse cDNA collection. *Nature* **409**: 685–690.

Kent WJ, Haussler D. (2001). Assembly of the working draft of the human genome with GigAssembler. *Genome Res* **11**: 1541–1548.

Kikuno R, Nagase T, Waki M, Ohara O. (2002). HUGE: a database for human large proteins identified in the Kazusa cDNA sequencing project. *Nucleic Acids Res* **30**: 166–168.

Korf I, Flicek P, Duan D, Brent MR. (2001). Integrating genomic homology into gene structure prediction. *Bioinformatics* **17** (Suppl. 1): S140–S148.

Kumar A, Harrison PM, Cheung KH, Lan N, Echols N, Bertone P, *et al.* (2002). An integrated approach for finding overlooked genes in yeast. *Nature Biotechnol* **20**: 58–63.

Lai CS, Fisher SE, Hurst JA, Vargha-Khadem F, Monaco AP. (2001). A forkhead-domain gene is mutated in a severe speech and language disorder. *Nature* **413**: 519–523.

Lash AE, Tolstoshev CM, Wagner L, Schuler GD, Strausberg RL, Riggins GJ, *et al.* (2000). SAGEmap: a public gene expression resource. *Genome Res* **10**: 1051–1060.

Levy S, Hannenhalli S, Workman C. (2001). Enrichment of regulatory signals in conserved non-coding genomic sequence. *Bioinformatics* **17**: 871–877.

Li WH. (1997). *Molecular Evolution*. Sinauer Associates: Sunderland, MA, USA.

Lowe TM, Eddy SR. (1997). tRNAscan-SE: a program for improved detection of transfer RNA genes in genomic sequence. *Nucleic Acids Res* **25**: 955–964.

Malpertuy A, Tekaia F, Casaregola S, Aigle M, Artiguenave F, Blandin G, *et al.* (2000). Genomic exploration of the hemiascomycetous yeasts: 19. Ascomycetes-specific genes. *FEBS Lett* **487**: 113–121.

Myers EW, Sutton GG, Delcher AL, Dew IM, Fasulo DP, Flanigan MJ, *et al.* (2000). A whole-genome assembly of Drosophila. *Science* **287**: 2196–2204.

Ning Z, Cox AJ, Mullikin JC. (2001). SSAHA: a fast search method for large DNA databases. *Genome Res* **11**: 1725–1729.

Noordewier MO, Warren PV. (2001). Gene expression microarrays and the integration of biological knowledge. *Trends Biotechnol* **19**: 412–415.

Olivier M, Agarwal A, Allen J, Almendras AA, Bajorek ES, Beasley EM, *et al.* (2001). A high-resolution radiation hybrid map of the human genome draft sequence. *Science* **291**: 1298–1302.

Raychaudhuri S, Chang JT, Sutphin PD, Altman RB. (2002). Associating genes with gene ontology codes using a maximum entropy analysis of biomedical literature. *Genome Res* **12**: 203–214.

Reese MG, Hartzell G, Harris NL, Ohler U, Abril JF, Lewis SE. (2000). Genome annotation assessment in *Drosophila melanogaster*. *Genome Res* **10**: 483–501.

Rice P, Longden I, Bleasby A. (2000). EMBOSS: the European Molecular Biology Open Software Suite. *Trends Genet* **16**: 276–277.

Rivas E, Klein RJ, Jones TA, Eddy SR. (2001). Computational identification of noncoding RNAs in *E. coli* by comparative genomics. *Curr Biol* **11**: 1369–1373.

Ross DT, Scherf U, Eisen MB, Perou CM, Rees C, Spellman P, *et al.* (2000). Systematic variation in gene expression patterns in human cancer cell lines. *Nature Genet* **24**: 227–235.

Salamov AA, Solovyev VV. (2000). *Ab initio* gene finding in Drosophila genomic DNA. *Genome Res* **10**: 516–522.

Sanger F, Coulson AR, Hong GF, Hill DF, Petersen GB. (1982). Nucleotide sequence of bacteriophage lambda DNA. *J Mol Biol* **162**: 729–773.

Schuler GD. (1997). Sequence mapping by electronic PCR. *Genome Res* **7**: 541–550.

Semple CAM, Morris SW, Porteous DJ, Evans KL. (2002). Computational comparison of human genomic sequence assemblies for a region of chromosome 4. *Genome Res* (in press).

Sherry ST, Ward MH, Kholodov M, Baker J, Phan L, Smigielski EM, *et al.* (2001). dbSNP: the NCBI database of genetic variation. *Nucleic Acids Res* **29**: 308–311.

Soderlund C, Humphray S, Dunham A, French L. (2000). Contigs built with fingerprints, markers, and FPC V4.7. *Genome Res* **10**: 1772–1787.

Stein L. (2001). Genome annotation: from sequence to biology. *Nature Rev Genet* **2**: 493–503.

Tatusov RL, Galperin MY, Natale DA, Koonin EV. (2000). The COG database: a tool for genome-scale analysis of protein functions and evolution. *Nucleic Acids Res* **28**: 33–36.

Todd AE, Orengo CA, Thornton JM. (2001). Evolution of function in protein superfamilies, from a structural perspective. *J Mol Biol* **307**: 1113–1143.

Uberbacher EC, Xu Y, Mural RJ. (1996). Discovering and understanding genes in human DNA sequence using GRAIL. *Methods Enzymol* **266**: 259–281.

Venter JC, Adams MD, Myers EW, Li PW, Mural RJ, Sutton GG, *et al.* (2001). The sequence of the human genome. *Science* **291**: 1304–1351.

Watson JD. (1990). The human genome project: past present, and future. *Science* **248**: 44–49.

Werner T. (2001). Cluster analysis and promoter modelling as bioinformatics tools for the identification of target genes from expression array data. *Pharmacogenomics* **2**: 25–36.

Wheelan SJ, Church DM, Ostell JM. (2001). Spidey: a tool for mRNA-to-genomic alignments. *Genome Res* **11**: 1952–1957.

Wheeler DL, Church DM, Lash AE, Leipe DD, Madden TL, Pontius JU, *et al.* (2002). Database resources of the National Center for Biotechnology Information: 2002 update. *Nucleic Acids Res* **30**: 13–16.

Wingender E, Chen X, Fricke E, Geffers R, Hehl R, Liebich I, *et al.* (2001). The TRANSFAC system on gene expression regulation. *Nucleic Acids Res* **29**: 281–283.

Zhang CT, Wang J. (2000). Recognition of protein coding genes in the yeast genome at better than 95% accuracy based on the Z curve. *Nucleic Acids Res* **28**: 2804–2814.

CHAPTER 6

Mouse and Rat Genome Informatics

JUDITH A. BLAKE*, JANAN EPPIG and CAROL J. BULT

The Jackson Laboratory
600 Main Street
Bar Harbor, ME 04609, USA

* corresponding author

Bioinformatics for Geneticists. Edited by M.R. Barnes and I.C. Gray
 ISBNs: 0 470 84393 4; 0 470 84394 2 (PB)

6.1 INTRODUCTION

Mouse and rat genome informatics is grounded in work on mouse and rat genetics and physiology that has been on-going since the early 20th century. The mouse, with its short generation time, small size, and plethora of phenotypic variants excelled as a tool for genetic investigations, especially after the conceptualization and creation of inbred strains, work begun by C. C. Little (Little and Tyzzer, 1916). Genetic crosses between inbred strains led to detailed mapping of genes and phenotypes, the construction of linkage groups, the development of chromosomal mapping techniques and the investigation of genetic components of phenotypes including diseases. Of particular significance was the development of specialized strains for genetic testing and technologies for manipulating the mouse genome. Standard inbred strains, their various derivatives, and 'boutique' mice developed through mutagenesis and genetic engineering have become essential tools. Coupled with advances in micro-technologies that are enabling detailed physiological studies in mice, the rich understanding of mouse genetics is accelerating the studies of genotype–phenotype relationships.

The rat, in contrast, was valued especially for its larger size relative to the mouse, and thus better suitability for physiological studies and experimental interventions. For rat, much is known about diseases, component factors in resistance/susceptibility, and specific networks of disease processes. Areas of research have been broad, including immunology, cancer, diseases of specific organ systems (cardiovascular, urogenital, skeletal, behaviour, growth and metabolism), neurological diseases, haematologic disorders, toxicology, histology, endocrinology, pathophysiology, and pharmacology (Gill *et al*., 1989; James and Lindpaintner, 1997). The genetics of the rat lagged behind until recently, when genomic tools (expressed sequence tags or ESTs, radiation hybrid and physical maps) for rat have rapidly been created and developed.

Today, rat and mouse are both strong animal models for the investigation of biology particularly with regard to human biology and disease. The availability of two rodent animal models is also fortuitous because it permits the examination of genetic and phenotypic variation between two closely related organisms and the ability, then, to contrast that information with knowledge about the biology of humans.

6.1.1 Bioinformatics for Mouse and Rat Geneticists

The term 'bioinformatics' is used to refer to many aspects of the intersection of computer science, biology, and information science. The term is often equated with the informatics challenges of the genome projects. There are several reasons for this. First, the genome sequencing efforts generate enormous volumes of electronic data that must be organized, stored, and analysed using powerful computers and sophisticated algorithms from the inception of the project. Second, substantial fiscal resources are being devoted to these projects, so the advancement of the informatics component is both absolutely necessary and well funded. Finally, there is the high visibility of the genome projects, with frequent newsflashes about the discovery of new and interesting genes. As a result of these forces, many scientists think of bioinformatics as an endeavour focused solely on the management and analysis of sequence data.

However, all aspects of biological investigation benefit from the ordered assembly of the information and from the use of computer technologies to store, query, sort and manage biological data. Prior to a database implementation, many structured datasets about mouse genetics and heritable mutants were maintained manually. The first gene description catalogue for mouse was published in 1941 by Dr George Snell (Snell, 1941). As early as the 1950s Dr Margaret Green began compiling mouse linkage and mapping data on index cards. Linkage maps were drawn by hand and published annually in the *Mouse Newsletter* from 1965–1994. Compilations of mutant genes and polymorphic loci, chromosome atlases, and lists of synteny homologies between mouse and man were irregularly published in journals (cf. Eppig, 1992; Nadeau *et al.*, 1991; Staats, 1985) in addition to books such as *Genetic Variants and Strains of the Laboratory Mouse* (Green, 1981; Lyon and Searles, 1989; Lyon *et al.*, 1996). During the 1980s many of these resources began to be maintained electronically and resulted in an early publicly accessible mouse database GBASE (Doolittle *et al.*, 1991) and the Encyclopedia of the Mouse Genome software tools (Eppig *et al.*, 1994). During the 1990s, this sweep of information about the genetics and biology of the laboratory mouse was integrated and brought fully into electronic form with the construction of the Mouse Genome Database (http://www.informatics.jax.org/; Richardson *et al.*, 1995) and the development of computer programs to manipulate and query the data such as MapManager (Manly, 1993). In addition, large-scale mapping projects redefined the management of genetic data (Dietrich *et al.*, 1992) and led to the construction of additional bioinformatics resources for mouse geneticists.

Compilations for rat information developed in a different way. Billingham and Silvers published the first compilation of rat strain information in 1959 (Billingham and Silvers, 1959). A standard nomenclature for rat strains emerged in 1973 (Festing and Staats, 1973). Rat strain descriptions were catalogued (Greenhouse *et al.*, 1990), and later maintained electronically by M. F. W. Festing and made publicly available in the model organism databases. Gene data was published sporadically and accumulated slowly due to the emphasis of rat researchers on physiology rather than genetics. The pressure for databases and computational tools for rat has been a recent occurrence. Although RatMap, which exclusively curates mapped genes was started in 1993, the need for resources to manage genomic data (simple sequence length polymorphisms or SSLPs, ESTs, comprehensive gene data, genomic sequence, etc.) was not recognized as critical until the joint US–German Rat Genome Project began generating large volumes of data in the mid/late 1990s. This recognition led to the development of the Rat Genome Database (http://rgd.mcw.edu) described more fully below.

6.1.2 Data Integration: The Challenge and the Conundrum

The advent of the Internet and the development of the www permitted the development of multiple sites committed to the presentation of biological data relative to the mouse and rat. Some, such as the sequence repositories GenBank (http://www.ncbi.nlm.nih.gov/; Wheeler *et al.*, 2001) and EMBL (http://www.ebi.ac.uk/embl/; Stroesser *et al.*, 2001), include mouse and rat sequences along with sequences from all other species. Others, such as the Whitehead Institute for Biomedical Research/MIT Center for Genome Research site (http://www-genome.wi.mit.edu/cgi-bin/mouse/index) provide specialized mouse datasets such as the pages for the 'Genetic and Physical Maps of the Mouse Genome'. For investigators, the reality is that information about the genetics and genomics of the laboratory mouse and the rat are found throughout cyberspace. Standards for nomenclature or descriptions of experimental data are not uniformly implemented, and it is often difficult to equate information at one site with information at another. Consequently, the investigator spends much time looking for data, collecting the data, and then manipulating the data before being able to explore and mine the data for knowledge. This has not gone unnoticed by data providers, but efforts to standardize and integrate information are often stymied by the variety of data types, the variability in data annotation, and the diversity of needs of the users. This presents a conundrum for bioinformatics professionals. Scientists do not want to be forced to use standard nomenclature or terminologies in the publication of their own work, but they do want to find a suite of information about a set of genes or sequences without having to do the data integration themselves.

The solution is easy to define, but hard to implement. It is dependent more on the sociology of doing science rather than the need for a technological solution. Data integration requires the implementation of standards and structures across multiple information resources (Bult *et al.*, 2000). Key strategies for data integration are the use of accessioned data entities, the application of nomenclature standards for key objects such as genes and strains, and the use of controlled, structured vocabularies and ontologies for functional annotation of biological information. Most of the larger data providers of interest to mouse and rat geneticists are now working to implement shared standards and to provide curated links between the different resources. Much harder is the integration of the scientific literature. As yet, most authors are unaware of and/or are not required to use standard nomenclature for genes, proteins, anatomy or biochemical reactions in the publication of laboratory research results. The result is that it is more difficult than it needs to be to bring experimental data into electronic form and to integrate it with other information. Hopefully, the use of data and nomenclature standards will become more common as scientists of all types recognize the value of bioinformatics resources and consequently appreciate the necessity and the power of data integration.

6.2 THE MODEL ORGANISM DATABASES FOR MOUSE AND RAT

One approach to integration of information about mouse and rat has been the construction of model organism databases. Several issues swirl around informatics sites devoted to model organisms. On the one hand, better interoperability among large data providers might obviate the need for an organisms-specific site. On the other hand, for model organisms such as *Saccharomyces*, *Caenorhabditis elegans*, *Drosophila* and others, including mouse and rat, there is a need for a central site that integrates all kinds of information about these well-studied species. Various approaches to shared data structures and standards are continually under discussion and have resulted in the increased similarities and links between the model organisms databases. Will there ultimately be one information system for all biology? Or will there continue to be specialized model organism sites loosely connected with other bioinformatics servers? The interconnectivity and transparency between bioinformatics resources continues to evolve, and it is imprudent to envisage bioinformatics systems just a few years hence. Today there exist model organism databases for the mouse and the rat, the Mouse Genome Database and the Rat Genome Database. Both work to provide comprehensive access to experimental and consensus data about these model organisms.

6.2.1 The Mouse Genome Database

The Mouse Genome Database (MGD) (http://www.informatics.jax.org) is the original model organism database for the laboratory mouse (Blake *et al.* 2001). Derived from the merger of several small specialized databases in 1994, MGD now focuses on the integrated representation of genotype (sequence) to phenotype data for the mouse with a particular emphasis on information about genes and gene products. MGD provides official gene nomenclature for the research community and works closely with human and rat genome curators to implement common standards for annotation of genes and other genome features. As part of the Mouse Genome Informatics (MGI)) system (see below), MGD focuses on data integration are through representations of relationships between genes, sequences and phenotypes, the representation of mouse mapping data, the association of genes to the Gene Ontology (GO), the description of targeted mutations and other alleles, and the curation of mammalian orthologies.

6.2.2 Mouse Genome Informatics

MGD is one component of the Mouse Genome Informatics (MGI) consortium based at The Jackson Laboratory. Other components of the MGI consortium include the Gene Expression Database (GXD; Ringwald *et al.*, 2001), the Mouse Tumor Biology Database (MTB; Bult *et al.*, 2001) and the Mouse Genome Sequencing Project (MGS). GXD focuses on the presentation of detailed experimental data about time and place of gene expression during development. MTB provides web-based access to mouse models of human cancers including experimental data and genotype-specific information. MGS works with the public mouse genome sequencing coalition to link the emerging genome with the mouse biological information. Overall, then, the MGI project provides the research community with a canonical set of mouse genes, their official names and genome locations, sequences, mammalian homologies, expression and functional information, phenotypic alleles and variants, associated literature and extensive links to other bioinformatics resources. This highly-integrated system is complemented with many cross-links to genetic and genomic resources for other organisms.

6.2.3 RatMap

RatMap (http://ratpmap.gen.gu.se) focuses on presenting the subset of rat genes, DNA markers, and quantitative trait loci (QTL) that are localized to chromosomes. RatMap maintains a highly-curated set of data, including nomenclature, chromosomal assignment and localization, mapping method statements, human and mouse homologues, references, and links to nucleotide sequences, UniGene and Rat Genome Database (RGD). In addition, RatMap maintains the rat idiograms and current cytogenetic maps. RatMap also provides a 'gene and position predictor' (GAPP) report that presents predicted positions for over 6000 rat genes based on conserved syntenic chromosomal segments between mouse and rat (Helou *et al.*, 2001).

6.2.4 The Rat Genome Database (RGD)

The Rat Genome Database (RGD, http://rgd.mcw.edu; Twigger *et al.*, 2002) is a collaborative effort between the Bioinformatics Research Center at the Medical College of

Wisconsin, The Jackson Laboratory and the National Center for Biotechnology Information (NCBI) to gather, integrate and make available data generated from ongoing rat genetic and genomic research efforts. Initially released in 2000, RGD includes curated data on rat genes, QTL, ESTs, sequence tagged sites (STSs) and microsatellite markers as well as details of inbred rat strains. RGD also contains detailed information on nomenclature, genetic and RH maps, mouse and human homologies, Gene Ontology data, and includes key literature citations. Research tools that are provided include 'VCMap', a sequence-based homology tool and gene prediction and RH mapping tools. RGD is introducing disease-based curation for disease processes frequently studied in the rat. Integration of the emerging rat genomic sequence is also planned.

6.3 MOUSE GENETIC AND PHYSICAL MAPS

The genetic map of the mouse has been built over time through the contributions of many research groups, using a variety of methods, including, but not limited to, backcross, intercross and complex cross analyses, congenic strain analysis and recombinant inbred and recombinant congenic strain analyses. Chromosomal rearrangements, somatic cell hybrids and *in situ* hybridization are used to supplement these methods. These diverse methods, utilizing a wide variety of laboratory and wild-derived mouse strains, have been used to develop the consensus linkage map for mouse (MGD, http://www.informatics.jax.org/searches/linkmap_form.shtml). For many purposes, this map is a standard for understanding the overall genomic organization of the mouse and for identifying potential candidate genes for diseases in particular regions.

6.3.1 Mouse DNA Mapping Panels

The development of large interspecific and intersubspecific crosses, for which progeny DNA are stored for cumulative genotyping, provides single-source high-resolution linkage maps containing thousands of markers and with well-defined crossover points (cf. Avner *et al.*, 1988; Copeland and Jenkins, 1991; Dietrich *et al.*, 1992; European Backcross Collaborative Group, 1994; Rowe *et al.*, 1994). Any newly discovered gene for which DNA polymorphism is detectable between the original parental strains can be mapped immediately without setting up a *de novo* cross and the cumulative data can be used to explore questions of recombination distribution across the genome and crossover interference. These DNA backcross panels are, however, not suitable for mapping new genes that are only defined by phenotype.

Genotyping data for individual progeny from many of these DNA mapping panels are available through the Mouse Genome Database (http://www.informatics.jax.org/searches/crossdata_form.shtml). In addition, maps can be generated using these data via the MGD Map Building tool at http://www.informatics.jax.org/searches/linkmap_form.shtml. Two of these DNA mapping panels are also maintained at specific websites: The Jackson Laboratory DNA Mapping Panels (http://www.jax.org/resources/documents/cmdata/bkmap) and the Whitehead Institute for Biomedical Research/MIT DNA Mapping Panels (http://www-genome.wi.mit.edu/cgi-bin/mouse/index#genetic).

6.3.2 Mouse Radiation Hybrid Maps

Recombination maps from DNA mapping panels provide unambiguous placement of gene order. However, for very closely linked genes, these maps may not be able to resolve locus order. For mouse, a radiation hybrid (RH) panel (T31) of 100 cell lines developed from a 3000-rad irradiated primary cell line from mouse embryo fused with hamster fibroblast has been developed (McCarthy *et al*., 1997). Radiation hybrids can be used for high throughput mapping and high resolution of locus order because each hybrid cell line contains a highly fragmented subset of the mouse genome. The co-retention of mouse genes across the 100-cell panel is indicative of their relative distance apart, assuming random chromosomal breakage and leads to the construction of RH maps (cf. Van Etten *et al*., 1999). Two complementary databases serve as community resources for gathering, distributing and analysing the T31 RH data.

6.3.3 The Jackson Lab Radiation Hybrid Map

The JAX RHmap provides web-based access to a comprehensive, integrated database that includes all typing data, retention frequency and log of the odds (LOD) scores for markers typed on the T31 panel, as well as RH framework maps for many of the chromosomes (http://www.jax.org/resources/documents/cmdata/rhmap/). All publicly available T31 data from large genome centres at the Whitehead Mouse RH Database (http://www-genome.wi.mit.edu/mouse_rh/index.html), the UK Mouse Genome Centre (http://www.mgc.har.mrc.ac.uk/physical/est_mapping/est.html) and Genoscope–CNS (http://www.genoscope.cns.fr/externe/English/Projets/Projet_ZZZ/rhmap.html), as well as from many individual laboratories are included.

The website includes an electronic submission interface for depositing RH typing data from users, data error checking and quality control, technical support, data analysis and the development of RH maps. All data, with references and experimental notes can be viewed or downloaded. Data are shared with the Mouse Genome Database (MGD) and the EBI data repository (RHdb, below).

6.3.4 The EBI Radiation Hybrid Database

The European Bioinformatics Institute (EBI) Radiation Hybrid Database (RHdb) is a repository for the raw data for constructing radiation hybrid maps, STS data, scores and experimental conditions (Rodriguez-Tomé and Lijnzaad, 2001; http://www.ebi.ac.uk/RHdb/index.html). The EBI RHdb is designed to be a species-neutral database, and currently contains human, mouse, and rat RH data. Data content relies entirely on submissions from data providers and research groups. Maps are not assembled from the accumulating data, but maps may be submitted by data developers.

6.3.5 Mouse Physical Maps

Two genome centres have produced physical maps for mouse that are accessible via the Internet: the Whitehead Institute/MIT (http://www-genome.wi.mit.edu/cgi-bin/mouse/index#phys) and the UK Mouse Genome Centre at Harwell (http://www.mgc.har.mrc.ac.uk/physical/phys.html). Whitehead Institute/MIT data include contigs and STS content mapping across the entire mouse genome and utilizes existing SSLP markers that characterize the MIT genetic map of the mouse to tie the physical and recombination maps together. The UK Mouse Genome Centre data consists of physical maps of selected regions of the genome that are being developed in association with individual research interests, notably regions of chromosomes 13 and X. Data from these sites are integrated into MGD, as well as being available from the originator's site.

A physical map of the genome of the C57BL/6J strain of laboratory mouse has been constructed using Bacterial Artificial Chromosome (BAC) clones (Gregory *et al.*, 2002). This map serves as the framework for the Mouse Genome Sequencing initiative (described below). The current BAC map for the mouse was derived from 305,768 BAC clones from two libraries: RPCI23 (female) and RPCI24 (male) (Osoegawa *et al.*, 2000). These libraries are available for distribution to the scientific community through the BACPAC Resource at the Children's Hospital Oakland Research Institute (http://www.chori.org/bacpac). The RPCI23 library is also available through Research Genetics (http://www.resgen.com/products/RPCI23MBAC.php3).

The clones from the RPCI BAC libraries were fingerprint mapped at the Genome Sequencing Centre in Vancouver, British Columbia (Marra *et al.*, 1997; http://www.bcgsc.bc.ca/projects/mouse_mapping/). The fingerprint data were combined with BAC end sequence data (Zhao *et al.*, 2001; http://www.tigr.org/tdb/bac_ends/mouse/bac_end_intro.html) to produce a mouse physical map that contains 296 contigs and covers an estimated 2,739 Mb (Gregory *et al.*, 2002). The average length of the contigs is 9.3 Mb. Of the 296 contigs, 228 can be localized to a chromosome. Approximately 97% of the total clone coverage for the mouse genome (2,658 Mb in 211 contigs) can be aligned to the human genome sequence.

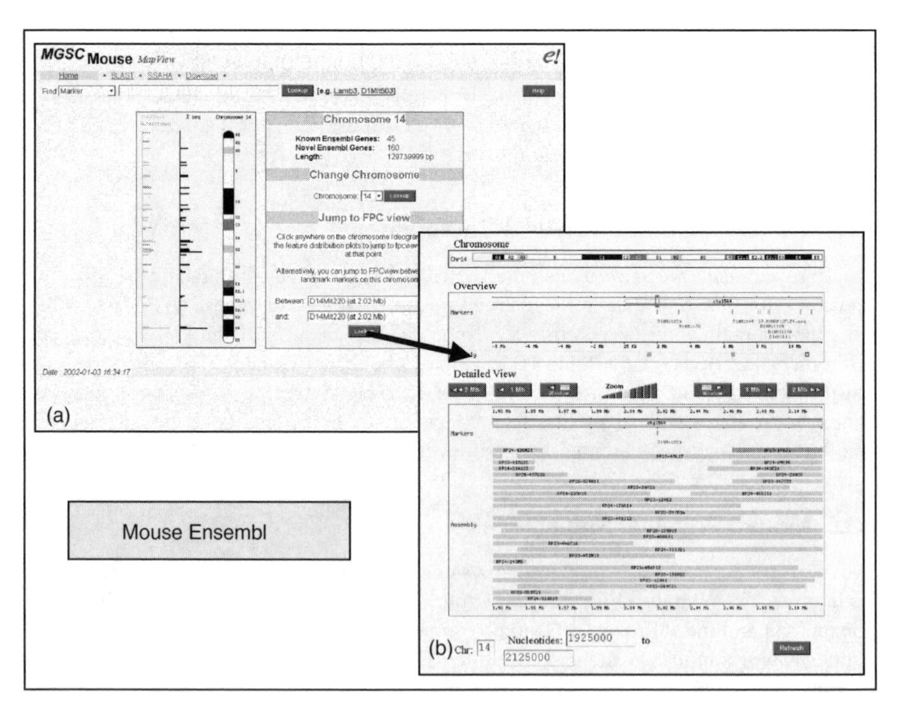

Figure 6.1 Mouse Ensembl. A graphical representation of the clone-based physical map for the proximal end of mouse chromosome 14 from Ensembl. This browser allows users to search for regions of a chromosome between two STS markers and to view the current clone coverage in the selected area. Because the browser is web-based, users do not have to download and install special software to view the BAC map (See Colour Plates).

There are three ways to view the current status of the mouse BAC physical map. Researchers can download and install a software product called FPC from the Sanger Institute (http://www.sanger.ac.uk/Software/fpc/) (Soderlund *et al.*, 2000) and use this software to graphically display the BAC clone fingerprint data generated by the Genome Sequence Centre in Vancouver. A similar display tool called the internet Contig Explorer (iCE) is available from the Genome Sequence Centre in Vancouver (http://ice.bcgsc.bc.ca/). An option for viewing the map that does not require the installation of software is to view the physical map using the Ensembl mouse browser at the Sanger Institute (http://mouse.ensembl.org/; or the mouse MapViewer of NCBI (http://www.ncbi.nlm.nih.gov).

The ultimate physical map, of course, is the genome sequence itself. Despite the expectation that the mouse genome will be available soon, the need for genetic maps and other physical maps will not disappear. The mouse sequence will continue to be built, reassembled and re-annotated for many years to come, making the physical contig map an important resource for anchoring this new information as it develops. Genetic maps will be needed indefinitely, for the mapping of QTLs, spontaneous mutations and other phenotypes with undetermined molecular defects. In addition, genetic maps are essential for studying chromosome structure and function, and recombination itself.

6.4 RAT GENETIC AND PHYSICAL MAPS

6.4.1 Rat Genetic Maps

The early development of rat genetics paralleled that of the mouse, with the establishment of genetic linkage between albino and pink-eyed dilution in both mouse and rat (Castle and Wachter, 1924; Dunn, 1920). Haldane (Haldane, 1927) recognized that, if these genes were homologues, they represented conserved synteny over evolutionary time. Subsequently, research geneticists focused on mouse and the rat became the major tool for physiologists. Thus, the development of the rat genetic map began to lag behind that of the mouse. As of 1991 there were 214 genes mapped in rat (Levan *et al.*, 1991) in contrast to nearly 3000 genes mapped in mouse (Hillyard *et al.*, 1991). This disparity in the number of genes mapped has continued to this day, with 1576 genes currently mapped in rat (RatMap, 2002) versus 18,983 in mouse (MGD, 2002). Maps of rat genes are largely cytogenetic rather than recombination maps and are maintained by RatMap (http://ratmap.gen.gu.se). After a century of concentrated use of rat by physiologists, rat genetics is now undergoing a revival as genomic tools are developed and its genome is finally being sequenced.

The resurgence of interest in the rat map has paralleled the development of genomic resources for rat. In the 1990s the first rat genome projects were begun to generate ESTs, YAC and BAC libraries, and SSLP maps. There were a number of backcrosses and intercrosses made among rat strains that were used to develop SSLP maps with several hundred to a few thousands markers (cf. Bihoreau *et al.*, 1997; Brown *et al.*, 1998; Dracheva *et al.*, 2000; Watanabe *et al.*, 2000; Wei *et al.*, 1998). Most of these SSLP maps are not yet integrated, although SSLP data and maps for some of the crosses are available through RGD. Data from two F2 intercrosses have been integrated and the resulting map containing 4786 SSLP markers can be found at the Whitehead Institute (http://www-genome.wi.mit.edu/rat/public/). In parallel, a large collaborative Allele Characterization Project was begun to establish allele sizes of 8000 SSLPs among 48 genetically and physiologically important inbred rat strains (http://www.brc.mcw.edu/LGR/research/lgr_acp.

html). Data generated from this project will provide investigators with a means of quickly selecting informative markers for new and existing mapping crosses.

6.4.2 Rat Radiation Hybrid Maps

A rat whole genome radiation hybrid panel (T55) generated by Linda McCarthy in Peter Goodfellow's laboratory has been used to construct high-resolution maps of the rat genome (http://www.well.ox.ac.uk/rat_mapping_resources/rat_radiation_hybrid_maps.html). The first radiation hybrid map was based on 5255 markers and included both microsatellites and known genes (Watanabe *et al.*, 1999). Another map using the same panel was constructed as a framework map using 2000 evenly spaced markers (http://rgd.mcw.edu/RHMAPSERVER/; Steen *et al.*, 1999). Both sites provide RH map web servers for users to map their markers — users submit data to the Rat RH Map Server and a map placement with a summary report is returned.

6.4.3 Rat Physical Maps

In contrast to the mouse, the rat has no genome-wide clone-based physical maps, only a few for specific regions such as the MHC locus (Gunther and Walter, 2001; Ioannidu *et al.*, 2001). Most of the 'physical map' for the rat genome consists of the cytogenetic maps that are maintained in RatMap and include a fair amount of FISH data. A physical BAC map of the rat is in preparation, as part of the NHGRI-sponsored rat genome sequencing initiative. A BAC library (CHORI-230) from the BN/SsNHsd/MCW (Brown Norway) strain of laboratory rat has been prepared using the same methods as were used for the mouse BAC libraries (http://www.chori.org/bacpac/) (Osoegawa *et al.*, 2000). The BAC clones from this library are being fingerprint mapped by the Genome Sequencing Centre in Vancouver, Canada (http://www.bcgsc.bc.ca/projects/rat_mapping/). There are currently (late 2001) 136,195 clones in their database. The BAC ends for this library are being sequenced at The Institute for Genomic Research (TIGR; http://www.tigr.org/tdb/bac_ends/rat/bac_end_intro.html).

6.5 GENOME SEQUENCE RESOURCES

6.5.1 Mouse Genome Sequencing Initiative

The initiative to sequence the genome of the laboratory mouse was announced by the National Human Genome Research Institute (NHGRI) of NIH in September 1999 as part of an overall 'action plan' for mouse genomics (Battey *et al.*, 1999). The goals of the initiative were to have a working draft of the genome of the C57BL/6J strain of mouse completed by 2003 and the finished genome sequence by 2005. The initial strategy for obtaining the mouse genome sequence was to build a physical BAC map of the genome as the BAC clones were sequenced (http://www.nhgri.nih.gov/NEWS/MouseRelease.htm).

In October of 2001 the strategy for obtaining the mouse sequence changed to include a whole genome shotgun approach. Part of the rationale for this change in sequencing strategy was that the shotgun sequences for the mouse genome could be used to assist in the identification of genes in the working draft of the human genome. The sequencing centres of the Sanger Institute, Washington University Medical Centre and the Whitehead Institute for Biomedical Research were funded to generate whole genome shotgun data for the mouse (http://www.nih.gov/science/models/mouse/).

Simultaneously with this shift in sequencing strategy, NIH launched a program to sequence mouse BAC clones that covered genomic regions of high biological interest. Individual investigators were invited to submit applications requesting specific BACs to be sequenced. Several sequencing centres, including the Cold Spring Harbor Laboratory, Harvard University Medical School and the University of Oklahoma were funded to sequence these BACs (http://www.nih.gov/science/models/bacsequencing/). The NIH BAC sequencing program was initially restricted to clones from specific BAC libraries for the mouse. However the program now accepts applications for the sequencing of clones from any BAC library and also from organisms other than mouse.

Several other sequencing centres around the world are using their sequencing capacity for regional and/or comparative sequencing of the mouse genome. For example, the DOE-funded Joint Genome Institute focused on sequencing segments of the mouse genome that are homologous to human chromosome 19 (http://bahama.jgi-psf.org/pub/ch19/; Dehal *et al.*, 2001). The Medical Research Council (MRC) is focusing on sequencing of mouse chromosomes 2, 4, 13 and mouse–human comparative sequencing for chromosome X (http://mrcseq.har.mrc.ac.uk/). Although the primary focus of the Baylor College of Medicine genome centre is now on sequencing the rat genome, it originally focused on sequencing BACs across mouse chromosome 11.

The NCBI maintains a status report of the progress of the mouse genome sequence project (http://www.ncbi.nlm.nih.gov/genome/seq/MmHome.html) as well as a registry of BAC clones that are being sequenced under the auspices of the Trans-NIH BAC Sequencing Program (http://www.ncbi.nlm.nih.gov/genome/clone/cstatus.html).

6.5.2 Mouse Genome Sequence Resources

There are several ways to access mouse genome sequence (here we focus on freely-accessible public resources). The whole genome shotgun data for the mouse can be found in a 'Trace Archive' maintained by the NCBI and can be searched via BLAST (http://www.ncbi.nlm.nih.gov/blast/mmtrace.html). A similar resource is maintained at the European Bioinformatics Institute (EBI; http://www.ebi.ac.uk/blast2). As of December 2001, there were over 31 million sequencing reads available in these archives; greater than six times the coverage of the mouse genome.

The Mouse Genome Sequencing Consortium has released an annotated draft assembly of the mouse genome to the research community (Mouse Genome Sequencing Consortium, 2002). The current draft assembly covers over 96% of the genome; a complete genome sequence for the laboratory mouse is anticipated by 2005. The draft genome and the associated annotations can be accessed using the Ensembl genome browser (http://www.ensembl.org), NCBI's Map Viewer (http://www.ncbi.nlm.nih.gov), and the University of Santa Cruz's genome browser (http://genome.ucsc.edu).

Other genome resources include MouseBLAST (Figure 6.2), a server maintained by the MGS group at The Jackson Laboratory that allows researchers to connect mouse sequence data with the wealth of biological knowledge about the mouse available in the MGI. Finally, the Mouse Genome Resources pages at NCBI (http://www.ncbi.nlm.nih.gov/

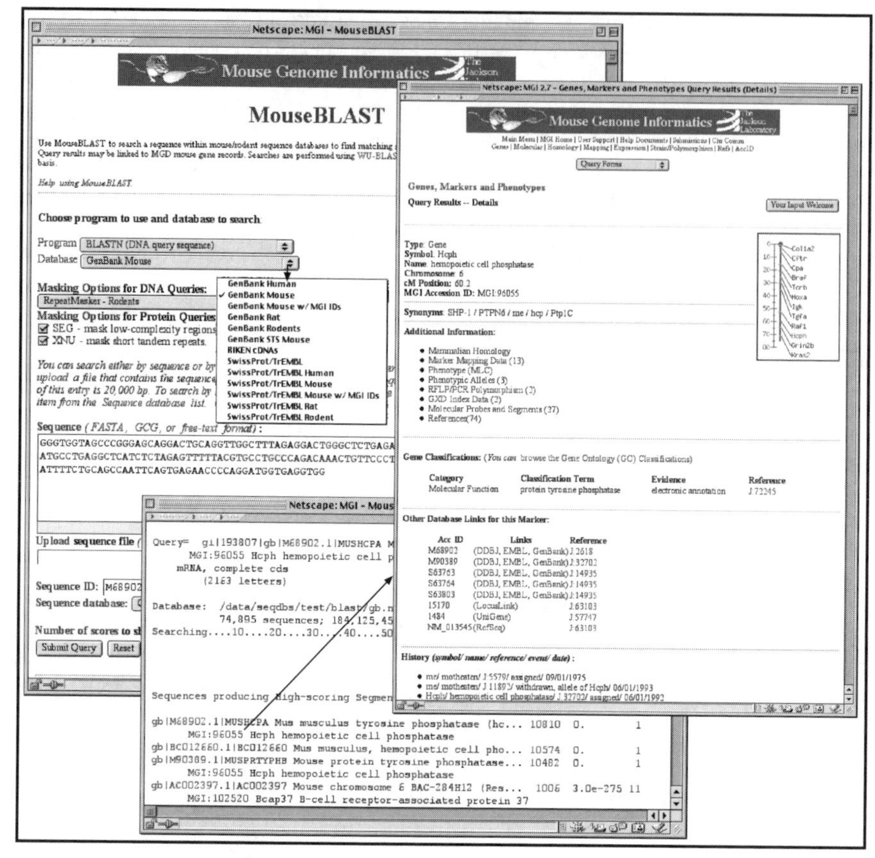

Figure 6.2 MouseBLAST. The MouseBLAST resource available from the Mouse Genome Informatics database website. MouseBLAST returns links to MGI Gene Detail pages as part of a standard BLAST report. The Gene Detail pages in MGI provide a wealth of information about homology, map location, phenotype associations, gene expression data, references and gene function annotation for each gene.

genome/guide/M_musculus.html) provide a compendium of links to various mouse genome resources.

6.5.3 Mouse cDNA Clone Resources

Several groups have undertaken initiatives to obtain full-length cDNA clones and sequences for every mouse gene. The RIKEN Institute from Japan has collected and sequenced over 60,000 cDNA clones for the mouse (http://genome.gsc.riken.go.jp/; The RIKEN Exploration Research Group Phase II Team and the FANTOM Consortium, 2001). The sequences for these clones are publicly available. The Mammalian Gene Collection, an NIH initiative, has a goal to provide a complete set of full-length (open reading frame) sequences and cDNA clones of expressed genes for human and mouse (http://mgc.nci.nih.gov/; Strausberg *et al.*, 2000).

6.5.4 Rat Genome Sequencing Initiative

In February 2001, the National Heart, Lung, and Blood Institute (NHLBI) announced funding support for the sequencing of the rat genome (http://www.nhgri.nih.gov/NEWS/nih_expands_programs.html). Three sequencing centres have been funded to produce enough genome sequence data to have a working draft of the rat genome by 2004: Celera Genomics, Baylor College of Medicine Genome Sequencing Centre and Genome Therapeutics, Inc.

6.6 COMPARATIVE GENOMICS

The sequencing of both the mouse and rat genomes promises to stimulate research based on comparative genome organization and comparative analysis between the human, mouse

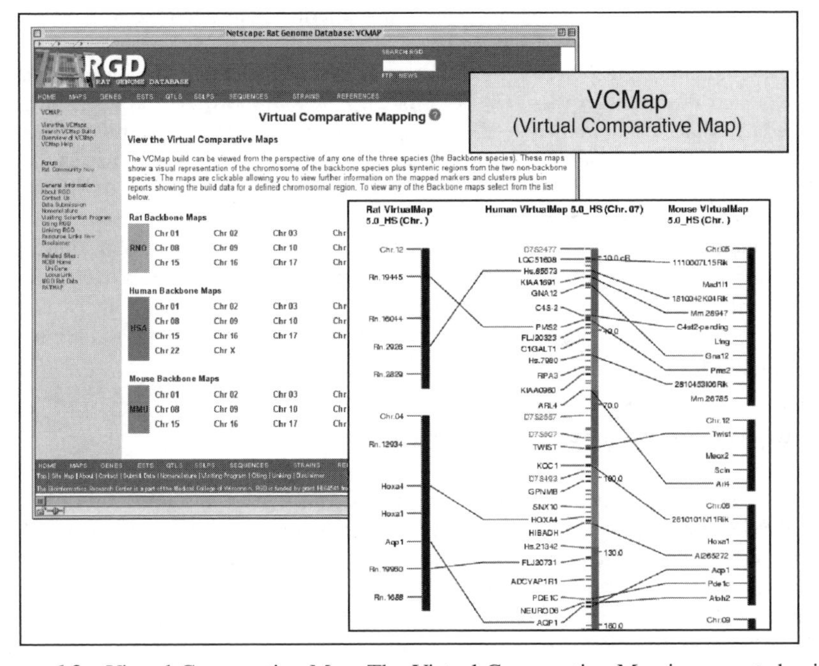

Figure 6.3 Virtual Comparative Map. The Virtual Comparative Map is generated using sequence-based algorithms that predict syntenic regions inferred from homology among mapped sequences. Sequence comparisons between ESTs and cDNAs from human, mouse and rat are combined with Radiation Hybrid map locations to define regions of synteny. Locations for unmapped markers in a species are then predicted based on the map location of the orthologous marker in a syntenic region of another species. The forepanel shows a virtual comparative map using human as the backbone map (centre) and syntenic regions of rat (left) and mouse (right). Mapped genes, UniGenes and STSs are shown, with lines connecting predicted homologues among the species. Data sources for the virtual maps are RGD, NCBI and MGD. The virtual comparative maps are available at http://rgd.mcw.edu/VCMAP/ (See Colour Plates).

and rat. Research papers based on comparison of large conserved segments between mouse and human are being published (Dehal *et al.*, 2001; Glusman *et al.*, 2001). Another approach is to use genome comparisons for elucidation of a suite of comparable genome features such as transcription factors (Wasserman *et al.*, 2000). Computational approaches to uncovering conserved regions such as exons or regulatory sites facilitate the discovery of new important genome features (Oeltjen *et al.*, 1997).

The direct comparison of genomic sequence from conserved linkage groups between mouse and human (and other organisms) has proven to be an effective strategy for identifying biologically relevant regions (coding and non-coding) in genomes. Two of the most commonly used tools for this effort are VISTA (http://www-gsd.lbl.gov/vista/; Mayor *et al.*, 2001) and PIPMAKER (http://bio.cse.psu.edu/pipmaker/; Schwartz *et al.*, 2000). These resources allow researchers to submit large genomic sequence regions to be aligned and analysed for the presence of conserved sequence elements. The VISTA group provides a set of pre-aligned sequences of mouse and human from finished genomic data in GenBank (http://pipeline.lbl.gov/). Applications include determining all of the protein-coding segments in both species, locating regulatory signals, understanding the mechanisms and history of genome evolution and deducing the similarities and differences in gene organization between the species of interest.

Other comparative map viewers incorporate information about the rat. One resource is the Gene and Position Predictor (GAPP) produced by RatMap which provides predicted comparative maps using known gene orthologues and zoo-FISH data (http://gapp.gen.gu.se/Description.html; Nilsson *et al.*, 2001). A different type of predictive map is the Virtual Comparative Map (VCMap) (http://rgd.mcw.edu/VCMAP/; Figure 6.3). These maps are generated using sequence-based algorithms that predict syntenic regions inferred from homology among mapped sequences. The Otsuka GEN Research Institute posts a genome-wide comparative map of the rat based primarily on extensive RH mapping data (http://ratmap.ims.u-tokyo.ac.jp/cgi-bin/comparative_home.pl). Finally, maps of curated orthologues for mouse/rat/human are available from MGD (http://www.informatics.jax.org/menus/homology_menu.shtml).

6.7 FROM GENOTYPE TO PHENOTYPE

Beyond a generalized representation of the mouse and rat are the intricacies of differences due to differing genetic backgrounds that can be revealed by comparisons between strains, among the rodent species, between rodents and other mammals and even between more distantly related organisms. The publication of the mouse genome sequence and the promise of the rat genome sequence in the near future will facilitate systematic genome-wide approaches to investigate normal and disordered cellular and physiological states. Genome-wide surveys of gene expression or genotype variation will enhance the gene-by-gene approach to the assessment of gene function. Scientists have long known of the importance of genetic background in the analysis of gene function or dysfunction due to the phenotypic variability resulting from epistatic interactions. Now, it may be possible to precisely assess the effect of genotype variability on the expression, function and interaction of gene products. As ever, the challenge for bioinformaticians will be to integrate the data from various experimental approaches into a coherent representation of the model organism. Ideally, one would like to query for a set of gene products expressed at the same time/state, evaluate the effect of genotype on the function and phenotypic presentation of variant gene products or compare 'snapshots' of cellular component sets between tissues or strains of rodents.

6.7.1 Genetic Variants

Genetically-engineered strains of mice including mice altered by gene transfer (transgenics), homologous recombination (gene targeting) and chemical mutagenesis provide powerful new tools for biomedical research. The use of these strains has become critical for basic research and for investigating causes of and potential treatments for human disease. The number of genes in mice that have enough characterization to be given descriptive names now exceeds 12,000, perhaps one-third or one-quarter of the estimated total number of genes. Genome manipulation techniques that target specific genes (e.g. knock-outs, knock-ins, and conditional mutations) or that identify sequence variants (e.g. microsatellites and single nucleotide polymorphisms or SNPs), are providing new alleles for biological analysis. Although many factors can contribute to a phenotype, a widely used research approach focuses on the isolated effects of single genes and their mutant alleles on biological systems. An alternative approach is to study quantitative traits where multiple genes contribute to the observed phenotypes. Here a one-to-one relationship between gene and phenotype does not exist and, as in humans, the discovery of the genes underlying complex traits such as obesity and hypertension continues to be challenging, but should become more tractable as new mapping resources are developed.

6.7.2 Mouse Single Nucleotide Polymorphism (SNP) Databases

SNP technologies are being exploited for the investigation of human syndromes and diseases (Schork *et al.*, 2000). Human SNP resources such as dbSNP (see Chapter 3) provide access to high-density SNP maps for humans. Large-scale discovery and genotyping of SNPs in mice is underway (Lindblad-Toh *et al.*, 2000) and a limited quantity of mouse SNP data is already available in the Roche mouse SNP database (http://mousesnp.roche.com/) and the Whitehead/MIT SNP database (http://www-genome.wi.mit.edu/snp/mouse/). With the sequencing of large genomic regions of multiple mouse inbred strains, further SNP sets for mouse will be defined and could facilitate computer-based identification of QTL loci between inbred strains; one group has already reported some success using this method, but the approach is controversial at present (Chesler *et al.*, 2001; Darvasi, 2001; Grupe *et al.*, 2001).

6.7.3 Induced Mutant Resources

The rapid generation of many induced mutants of the mouse through the use of technologies such as homologous recombination and targeted knock-outs has created the need for a central facility to collect and distribute them to the scientific community. The Induced Mutant Resource (IMR) (http://www.jax.org/resources/documents/imr/) at The Jackson Laboratory is an example of a national clearing-house for the collection and distribution of a subset of genetically-engineered mice. The IMR maintains an on-line database to provide information about these strains. This information includes a description of the mutant phenotype, husbandry requirements and links to related resources. Another resource providing mouse mutants to the community is the Mutant Mouse Regional Resource Centres (http://www.mmrrc.org/). The MMRRC strive to enhance the availability of genetically-engineered mice for the study of human biology and disease. The European Mouse Mutant Resource (EMMA) (http://emma.rm.cnr.it/) is another repository for mouse mutant stocks.

6.7.4 Resources for Mouse Strain Characterization

Inbred strains in mouse have been specifically generated to facilitate the study of the genetic component of phenotypes including disease phenotypes by being able to isolate the impact of the mutant gene on a standard genetic background. With the advent of many new technologies, molecular information about the whole genome is becoming available for different inbred strains, and the need for standard evaluation of differences between inbred strains is apparent. New initiatives to study strain characteristics in mice and rats are underway with the attendant development of bioinformatics resources.

The Mouse Phenome Database (MPD; http://www.jax.org/phenome/) was established to provide a collection of baseline phenotypic data on commonly used and genetically diverse inbred mouse strains. Many institutions and investigators are involved in this effort to provide standard sets of strain characteristics for the most commonly used strains of mice. The MPD will enable investigators to identify appropriate strains for physiological testing and disease onset and susceptibility.

6.7.5 Phenotypic Variants

In contrast with the reliance on the gene-by-gene approach to discovery of functions and roles for genes and for the investigation of diseases and disorders, a recent development has been the use of systematic large-scale phenotype-driven mutagenesis studies in the mouse. This approach uses chemical or physical disruption of the genome followed by identification of putative mutants using a series of phenotypic screens for particular traits. This phenotype-driven approach to genome characterization has an important role to play in linking gene identification with gene function. This approach will allow researchers to better understand the molecular basis of diseases through the identification of mutants that develop the same or similar phenotypes but that have mutations in different genes. Furthermore, a full appreciation of the genetic basis of a disease requires that the phenotypes associated with multiple alleles of the same gene be studied to identify hypomorphs, alleles that confer gain of function, etc. Although it is unclear how much of the genome can be saturated with this approach, these projects will provide the community with a vast array of new phenotypes for biological analysis.

6.7.6 ENU Mutagenesis Centres

Several public large-scale ENU mutagenesis projects are already underway and are providing new models for the study of disease and gene function to the community (Brown and Nolan, 1998; De Angelis *et al.*, 2000; Justice *et al.*, 1999; Nolan *et al.*, 2000) (Table 6.1). Some of the mutagenesis centres are working in several disease areas to identify new mutants including the ENU Mutagenesis Programme at Harwell (http://www.mgu.har.mrc.ac.uk/mutabase/), the RIKEN Mouse Functional Genomics Group (http://www.gsc.riken.go.jp/Mouse/), and the GSF ENU Mouse Mutagenesis Screen Project (http://www.gsf.de/isg/groups/enu-mouse.html). Several of these mutagenesis centres are focusing on the identification of new mutant mice to serve as models for neurological disorders (Moldin *et al.*, 2001) including the Neuroscience Mutagenesis Facility at The Jackson Laboratory (http://www.jax.org/nmf/), the Neurogenomics Centre at Northwestern University (http://genome.northwestern.edu/), the Tennessee Mouse Genome Consortium (http://www.tnmouse.org/) and the McLaughlin Research Institute (http://www.montana.edu/wwwmri/enump.html). The mutagenesis facility at the Baylor College of Medicine (http://www.mouse-genome.bcm.tmc.edu/ENU/MutagenesisProj.asp) is focusing on developmental defects. The Medical Genome Centre in Australia focuses on cancer-related phenotypes (http://jcsmr.anu.edu.au/group_pages/mgc/CancerGenLab.html). The Mouse Heart, Lung,

TABLE 6.1 Mouse Mutagenesis Centres and Databases

Mutagenesis Centre	Disease Focus	URL
ENU Mutagenesis Programme (Harwell)	General	http://www.mgu.har.mrc.ac.uk/mutabase/
RIKEN Mouse Functional Genomics Group	General	http://www.gsc.riken.go.jp/Mouse/
GSF ENU Mouse Mutagenesis Screen Project	General	http://www.gsf.de/isg/groups/enu-mouse.html
Neuroscience Mutagenesis Facility at The Jackson Laboratory	Neurological	http://www.jax.org/nmf/
Neurogenomics Centre at Northwestern University	Neurological	http://genome.northwestern.edu/
Tennessee Mouse Genome Consortium	Neurological	http://www.tnmouse.org/
McLaughlin Research Institute	Neurological	http://www.montana.edu/wwwmri/enump.html
Baylor College of Medicine	Developmental disorders	http://www.mouse-genome.bcm.tmc.edu/ENU/MutagenesisProj.asp
Medical Genome Centre (Australia)	Cancer	http://jcsmr.anu.edu.au/group_pages/mgc/CancerGenLab.html
The Mouse Heart, Lung, Blood, and Sleep Disorders Centre (JAX)	Cardiovascular	http://www.jax.org/hlbs/index.html

Blood, and Sleep Disorders Centre at The Jackson Laboratory is focusing on the identification of new mutants for cardiovascular diseases (http://www.jax.org/hlbs/index.html).

6.8 FUNCTIONAL GENOMICS

In the post-genome world, mouse and rat models will be heavily used for investigation of gene function and disease pathogenesis (Schimenti and Bucan, 1998; Temple *et al.*, 2001; Zheng *et al.*, 1999). With the completion of the mouse genome, attention can move to genome-wide screens for gene expression and systematic investigation of gene function. The inclusion of functional information with gene annotations first appeared in the sequence data repositories. From the start, issues of quality control for data associations were evident. Evaluation of sequence similarities often led to the transfer of function information from one gene annotation report to another without experimental verification or any statement about the basis for the function assertion. The first detailed functional classification was developed to catalogue the genes of *Escherichia coli* (Riley, 1993). Since then, functional annotation schemes have been developed for single organisms, multi-organism databases, and for pathway-related systems (see Rison *et al.*, 2000 for review).

6.8.1 Gene Ontology

A recent effort initiated by several of the model organism databases has been the development of ontologies describing aspects of biology common to all organisms (GO Consortium, 2000, 2001). These 'controlled structured vocabularies' include defined terms and relationships for the domains of 'molecular function', 'biological process' and 'cellular component'. The Gene Ontology (GO) project (http://www.geneontology.org) is now a consortium of model organism databases, sequence information centres and other genome data providers. In addition to the development of the ontologies, the genome annotation groups contribute gene–GO association files to a central GO repository. MGI and RGD provide detailed GO annotations for mouse and rat genes respectively. A GO database (http://www.godatabase.org/) holds the ontologies, their definitions and relationships, and the contributed sets of gene–GO association files. The AMIGO browser (Figure 6.4) provides access to the data.

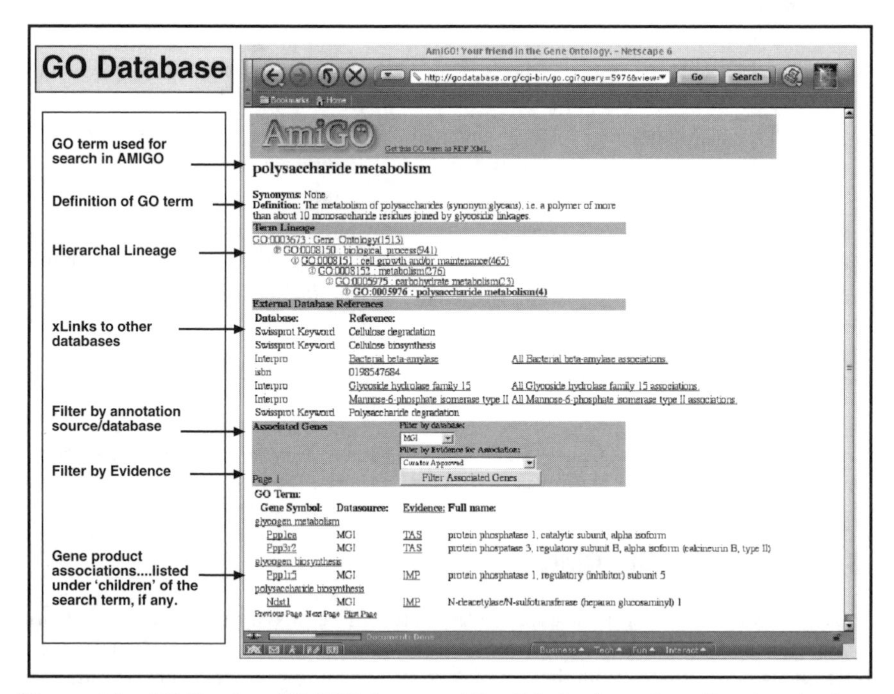

Figure 6.4 GO Database/AMIGO browser. The GO database (http://www.godatabase.org/) and AMIGO browser are recent additions to the tools and resources of the Gene Ontology project (http://www.geneontology.org/). Here a detail page from a query on the controlled GO term 'Polysaccharide metabolism' is displayed. The definition of the term and its relationship to other terms is shown. There are cross-links to other external keyword sets. The detail page has been expanded to show all mouse gene products annotated to this term. The gene product associations detail can be filtered by source of annotation (MGI, RGD, other contributing model organism or genome annotation groups, etc.) and by type of evidence (cf. sequence similarity, mutant phenotype, direct assay, etc).

6.9 RODENT DISEASE MODELS

The experimental manipulation of mice and rats for the purpose of creating animal models for human disease is implicit in the scientific endeavours detailed here. Mouse and rat models will continue to be the best models for experimental manipulation of the mammalian genome for the foreseeable future (Bedell, *et al.*, 1997). The fact that inbred strains exist, providing consistent homogenous genetic backgrounds for experimentation, allows the genesis of diseases characteristic of particular inbred strains to be studied, as well as the development and testing of therapeutic interventions. The occurrence of spontaneous or induced single gene mutations in these strains allows precise detailed studies of the multiple effects of that particular mutation. Targeted mutations that produce knock-out or conditional mutations permit researchers to mimic the molecular defect of human diseases. Comparative studies have uncovered many rodent mutations that reflect their counterpart human disease. Multigenic diseases and quantitative trait loci can be dissected in mice and rats using controlled crosses and through creation of specialized strains, such as congenics and consomics, which place particular parts of the genome from one strain onto the background of another strain (cf. Kwitek-Black and Jacob, 2001; Sugiyama *et al.*, 2001).

In addition to the discovery or creation of models that reflect the underlying genetics of particular disease states, researchers may also find it useful to study animal models that reflect phenotypic similarity alone. That is to say, there is a phenotypic similarity in the animal model to a human disease condition, and the animal model is useful for studying that phenotype, even though we do not know that the underlying genetic dysfunction is exactly the same. For example, many cancers have unknown genetic aetiology, but particular strains or mutants prone to the development of particular cancers can serve as effective animal models (Hann and Balmain, 2001).

The strength of using rat and mouse models clearly lies in the physiological research that has gone before and that is accelerating with micro-technology developments. A fuller understanding of the genetics of these organisms, coupled with the imminent availability of their genome sequences, will enhance our ability to analyse the functions of gene products and to dissect the molecular basis of phenotypes.

6.10 SUMMARY

With new technologies and methods, the pace of data acquisition only quickens. Simultaneously, there are now intense efforts underway to improve data integration and to support rapid access to and interactive use of molecular and related biological information. Biological databases and information resources existed long before the advent of computers and the internet. We are, however, yet developing and realizing the capacity that computers give us to use the databases not just as archives, but also as research tools. The future of computerized scientific databases and information resources will be in their ability to rapidly retrieve and manipulate data in response to complex queries. The full value of the information they contain can then be exploited to address outstanding scientific inquiries.

ACKNOWLEDGEMENTS

This work is supported in part by NIH/NHGRI grant HG00330, DOE grant FG99ER 62850, and HG02273. We thank Greg Cox, Cathy Lutz and Tim O'Brien for their critical reading of this paper; W. John Boddy for preparation of figures and Janice Ormsby for technical review.

REFERENCES

Avner P, Amar L, Dandolo L, Guenet JL. (1988). Genetic analysis of the mouse using interspecific crosses. *Trends Genet* **4**: 18–23.

Battey J, Jordan E, Cox D, Dove W. (1999). An action plan for mouse genomics. *Nature Genet* **21**: 73–75.

Bedell MA, Largaespada DA, Jenkins NA, Copeland NG. (1997). Mouse models of human disease. Part 11: Recent progress and future directions. *Genes Develop* **11**: 11–43.

Billingham RE, Silvers WK. (1959). Inbred animals and tissue transplantation immunity. *Transplant Bull* **6**: 399–406.

Bihoreau M-T, Gaugier D, Kato N, Hyne G, Lindpainter K, Rapp JP, *et al.* (1997). A linkage map of the rat genome derived from three F2 crosses. *Genome Res* **7**: 434–440.

Blake JA, Richardson JE, Bult CJ, Kadin JA, Eppig JT and the Mouse Genome Database Group (2001). The Mouse Genome Database (MGD): the model organism database for the laboratory mouse. *Nucleic Acids Res* **30**: 1–3.

Brown DM, Matise TC, Koike G, Simon JS, Winer ES, Zangen S, *et al.* (1998). An integrated genetic linkage map of the laboratory rat. *Mammal Genome* **9**: 521–530.

Brown SD, Nolan PM. (1998). Mouse mutagenesis — systematic studies of mammalian gene function. *Hum Mol Genet* **7**: 1627–1633.

Bult CJ, Krupke DM, Näf D, Sundberg JP, Eppig JT. (2001). Web-based access to mouse models of human cancers: the Mouse Tumor Biology (MTB) Database. *Nucleic Acids Res* **29**: 95–97.

Bult CJ, Richardson JE, Blake JA, Kadin JA, Ringwald M, Eppig JT and the Mouse Genome Informatics Group (2000). Mouse genome informatics in a new age of biological inquiry. In *Proceedings of the IEEE International Symposium on Bio-Informatics and Biomedical Engineering*, pp. 29–32.

Castle WE, Wachter WL. (1924). Variations of linkage in rats and mice. *Genetics* **9**: 1–12.

Chesler EJ, Rodriguez-Zas SL, Mogil JS. (2001). *In silico* mapping of mouse quantitative trait loci. *Science* **294**: 2423.

Copeland NG, Jenkins NA. (1991). Development and applications of a molecular genetic linkage map of the mouse genome. *Trends Genet* **7**: 113–118.

Darvasi A. (2001). *In silico* mapping of mouse quantitative trait loci. *Science* **294**: 2423.

De Angelis MH, Flaswinkel H, Fuchs H, Rathkolb B, Soewarto B, Marschall S, *et al.* (2000). Genome-wide, large-scale production of mutant mice by ENU Mutagenesis. *Nature Gene* **25**: 444–447.

Dehal P, Predki P, Olsen AS, Kobayashi A, Folta P, Lucas S, *et al.* (2001). Human chromosome 19 and related regions in mouse: conservative and lineage-specific evolution. *Science* **293**: 104–111.

Dietrich W, Katz H, Lincoln SE, Shin H-S, Friedman J, Dracopoli NC, *et al.* (1992). A genetic map of the mouse suitable for typing intraspecific crosses. *Genetics* **131**: 423–447.

Doolittle DP, Hillyard AL, Davisson MT, Roderick TH, Guidi JN. (1991). GBASE — The genomic database of the mouse, In *Fifth International Workshop on Mouse Genome Mapping, Lunteren, Netherlands*, p. 27.

Dracheva SV, Remmers EF, Chen S, Chang L, Gulko PS, Kawahito Y, *et al.* (2000). An integrated genetic linkage map with 1,137 markers constructed from five F2 crosses of autoimmune disease-prone and -resistant inbred rat strains. *Genomics* **63**: 202–226.

Dunn LC. (1920). Linkage in mice and rats. *Genetics* **5**: 325–343.

Eppig JT. (1992). Mouse DNA clones and probes. *Mammal Genome* **3**: 300–330.

Eppig JT, Blackburn RE, Bradt DW, Corbani LE, Davisson MT, Doolittle DP, *et al.* (1994). *The Encyclopedia of the Mouse Genome, an update. Third International Conference on Bioinformatics and Genome Research, Tallahassee*, p. 73.

European Backcross Collaborative Group (1994). Towards high resolution maps of the mouse and human genomes — a facility for ordering markers to 0.1 cM resolution. *Hum Mol Genet* **3**: 621–627.

Festing MFW, Staats J. (1973). Standardized nomenclature for inbred strains of rats, fourth listing. *Transplantation* **16**: 221–245.

Gill TJ, Smith GJ, Wissler RW, Kunz HW. (1989). The rat as an experimental animal. *Science* **245**: 269–276.

Glusman G, Rowen L, Lee I, Boysen C, Roach JC, Smit AF, *et al.* (2001). Comparative genomics of the human and mouse T Cell receptor loci. *Immunity* **15**: 337–349.

GO Consortium. (2000). Gene Ontology: Tool for the unification of biology. *Nature Genet* **25**: 25–29.

GO Consortium. (2001). Creating the gene ontology resources: design and implementation. *Genome Res* **11**: 1425–1433.

Green MC. (Ed.) (1981). *Genetic Variants and Strains of the Laboratory Mouse*, 1st edn. Fischer Verlag: Stuttgart.

Greenhouse DD, Festing MFW, Hasan S, Cohen AL. (1990). Catalogue of inbred strains of rats. In *Genetic Monitoring of Inbred Strains of Rats*, Hedrich HJ (Ed.), Gustav Fischer: Stuttgart, pp. 120–132.

Gregory SG, *et al.* (2002). A Physical map of the mouse genome. *Nature* **418**: 743–50.

Grupe A, Germer S, Usuka J, Aud D, Belknap JK, Klein RF, *et al.* (2001). *In silico* mapping of complex disease-related traits in mice. *Science* **292**: 1915–1918.

Gunther E, Walter L. (2001). The major histocompatibility complex of the rat (*Rattus norvegicus*). *Immunogenetics* **53**: 520–542.

Haldane JBS. (1927). The comparative genetics of colour in rodents and carnivora. *Biol Rev Cambridge Phil Soc (London)* **2**: 199–212.

Hann B, Balmain A. (2001). Building ‘validated’ mouse models of human cancer. *Curr Opin Cell Biol* **13**: 778–784.

Helou K, Walentinsson A, Levan G, Stahl F. (2001). Between rat and mouse zoo-FISH reveals 49 chromosomal segments that have been conserved in evolution. *Mammal Genome* **12**: 765–771.

Hillyard AL, Doolittle DP, Davisson MT, Roderick TH. (1991). Locus map of mouse. *Mouse Genome* **89**: 16–30.

Ioannidu S, Walter L, Dressel R, Gunther E. (2001). Physical map and expression profile of genes of the telomeric class I gene region of the rat MHC. *J Immunol* **166**: 3957–3965.

James MR, Lindpainter K. (1997). Why map the rat? *Trends Genet* **13**: 171–173.

Justice MJ, Noveroske JK, Weber JS, Zheng B, Bradley A. (1999). Mouse ENU mutagenesis. *Hum Mol Genet* **8**: 1955–1963.

Kwitek-Black AE, Jacob HJ. (2001). The use of designer rats in the genetic dissection of hypertension. *Curr Hyperten Rep* **3**: 12–18.

Levan G, Szpirer J, Szpirer C, Klinga K, Hanson C, Islam MQ. (1991). The gene map of the Norway rat (*Rattus norvegicus*) and comparative mapping with mouse and man. *Genomics* **10**: 699–718.

Lindblad-Toh K, Winchester E, Daly MJ, Wang DG, Hirschhorn JN, Laviolette JP, *et al.* (2000). Large-scale discovery and genotyping of single-nucleotide polymorphisms in the mouse. *Nature Genet* **24**: 381–386.

Little CC, Tyzzer EE. (1916). Further studies on inheritance of susceptibility to a transplantable tumor of Japanese waltzing mice. *J Med Res* **33**: 393–398.

Lyon MF, Searle AG. (Eds) (1989). *Genetic Variants and Strains of the Laboratory Mouse*, 2nd edn. Oxford University Press: Oxford.

Lyon MF, Rastan S, Brown SDM. (Eds) (1996). *Genetic Variants and Strains of the Laboratory Mouse*, 3rd edn. Oxford University Press: New York.

Manly KF. (1993). A Macintosh program for storage and analysis of experimental genetic mapping data. *Mammal Genome* **4**: 303–313.

Marra MA, Kucaba TA, Dietrich NL, Green ED, Brownstein B, Wilson RK, *et al.* (1997). High throughput fingerprint analysis of large-insert clones. *Genome Res* **7**: 1072–1084.

Mayor C, Brudno M, Schwartz JR, Poliakov A, Rubin EM, Frazer KA, *et al.* (2001). VISTA: visualizing global DNA sequence alignments of arbitrary length. *Bioinformatics* **16**: 1046–1047.

McCarthy LC, Terrett J, Davis ME, Knights CJ, Smith AL, Critcher R, *et al.* (1997). A first-generation whole genome-radiation hybrid map spanning the mouse genome. *Genome Res* **7**: 1153–1161.

MGD. (2002). Statistics for number of localized genes. ftp://ftp.informatics.jax.org/pub/informatics/reports/MGD_Stats.sql.rpt [1 January 2002].

Moldin SO, Farmer ME, Chin HR, Battey JF Jr. (2001). Trans-NIH neuroscience initiatives on mouse phenotyping and mutagenesis. *Mammal Genome* **12**: 575–581.

Mouse Genome Sequencing Consortium (2002). Initial Sequencing and Comparative analysis of the Mouse genome. Nature, in press.

Nadeau JH, Grant P, Kosowsky M. (1991). Mouse on human homology map. *Mouse Genome* **89**: 31–36.

Nilsson S, Helou K, Walentinsson A, Szpirer C, Nerman I, Stahl F. (2001). Rat–mouse and rat–human comparative maps based on gene homology and high-resolution zoo-FISH. *Genomics* **74**: 287–298.

Nolan PM, Peters J, Strivens M, Rogers D, Hagan J, Spurr N, *et al.* (2000). A systematic, genome-wide, phenotype-driven mutagenesis programme for gene function studies in the mouse. *Nature Genet* **25**: 440–443.

Oeltjen JC, Malley TM, Muzny DM, Miller W, Gibbs RA, Belmont JW. (1997). Large-scale comparative sequence analysis of the human and murine Bruton's tyrosine kinase loci reveals conserved regulatory domains. *Genome Res* **7**: 315–329.

Osoegawa K, Tateno M, Woon PY, Frengen E, Mammoser AG, Catanese JJ, *et al.* (2000). Bacterial artificial chromosome libraries for mouse sequencing and functional analysis. *Genome Res* **10**: 116–128.

RatMap (2002). Statistics for number of localized genes. http://ratmap.gen.gu.se [1 January 2002].

Richardson JE, Eppig JT, Nadeau JH. (1995). Building an integrated mouse genome database. *IEEE Eng Med Biol* **14**: 718–724.

Riley M. (1993). Functions of the gene products of *Escherichia coli*. *Microbiol Rev* **57**: 862–952.

Ringwald M, Eppig JT, Begley DA, Corradi JP, McCright IJ, Hayamizu TF, *et al.* (2001). The Mouse Gene Expression Database (GXD). *Nucleic Acids Res* **29**: 98–101.

Rison SCG, Hodgman TC, Thornton JM. (2000). Comparison of functional annotation schemes for genomes. *Funct Integ Genomics* **1**: 56–69.

Rodriguez-Tomé P, Lijnzaad P. (2001). RHdb: the Radiation Hybrid database. *Nucleic Acids Res* **29**: 165–166.

Rowe LB, Nadeau JH, Turner R, Frankel WN, Letts VA, Eppig JT, *et al.* (1994). Maps from two interspecific backcross DNA panels available as a community genetic mapping resource. *Mammal Genome* **5**: 253–274.

Schimenti J, Bucan M. (1998). Functional genomics in the mouse: phenotype based on mutagenesis screens. *Genome Res* **8**: 698–710.

Schork NJ, Fallin D, Lanchbury JS. (2000). Single nucleotide polymorphisms and the future of genetic epidemiology. *Clin Genet* **58**: 250–264.

Schwartz S, Zhang Z, Frazer KA, Smit A, Riemer C, Bouck J, *et al.* (2000). PipMaker: A web server for aligning two genomic DNA sequences. *Genome Res* **10**: 577–586.

Snell GD. (1941). Genes and chromosome mutation. In *Biology of the Laboratory Mouse*, 1st edn, Snell GD. (Ed.). McGraw-Hill: New York, pp. 234–247.

Soderlund C, Humphrey S, Dunhum A, French L. (2000). Contigs built with fingerprints, markers and FPC V4.7. *Genome Res* **10**: 1772–1787.

Staats J. (1985). Standardized nomenclature for inbred strains of mice: eighth listing. *Cancer Res* **45**: 945–977.

Steen RG, Kwitek-Black AE, Glenn C, Gullings-Handley J, Van Etten W, Atkinson OS, *et al.* (1999). A high-density integrated genetic linkage and radiation hybrid map of the laboratory rat. *Genome Res* **9**, AP1–8, insert.

Strausberg RL, Feingold EA, Klausner RD, Collins FC. (2000). The mammalian gene collection. *Science* **286**: 455–457.

Stroesser G, Baker W, van den Broek A, Camon E, Garcia-Pastor M, Kanz C, *et al.* (2001). The EMBL nucleotide sequence database. *Nucleic Acids Res* **29**: 17–21.

Sugiyama F, Yagami K, Paigen B. (2001). Mouse models of blood pressure regulation and hypertension. *Curr Hyperten Rep* **3**: 41–48.

Temple LKF, McLeod RS, Gallinger S, Wright JG. (2001). Defining disease in the genomics era. *Science* **293**: 807–808.

The RIKEN Exploration Research Group Phase II Team and the FANTOM Consortium. (2001). Functional annotation of a full-length mouse cDNA collection. *Nature* **409**: 685–690.

Twigger S, Lu J, Shimoyama M, Chen D, Pasko D, Long H, *et al.* (2002). Rat Genome Database (RGD): mapping disease onto the genome. *Nucleic Acids Res* **30**: 125–128.

Van Etten WJ, Steen RG, Nguyen H, Castle AB, Slonim DK, Ge B, *et al.* (1999). Radiation hybrid map of the mouse genome. *Nature Genet* **22**: 384–387.

Wasserman WW, Palumbo M, Thompson W, Fickett JW, Lawrence CE. (2000). Human–mouse genome comparisons to locate regulatory sites. *Nature Genet* **26**: 225–228.

Watanabe TK, Bihoreau MT, McCarthy LC, Kiguwa SL, Hishigaki H, Tsuji A, *et al.* (1999). A radiation hybrid map of the rat genome containing 5,255 markers. *Nature Genet* **22**: 27–36.

Watanabe TK, Ono T, Okuno S, Mizoguchi-Miyakita A, Yamasaki Y, Kanemoto N, *et al.* (2000). Characterization of newly developed SSLP markers for the rat. *Mammal Genome* **11**: 300–305.

Wei S, Wei K, Moralejo DH, Yamada T, Izumi K, Matsumoto K. (1998). An integrated genetic map of the rat with 562 markers from different sources. *Mammal Genome* **9**: 1002–1007.

Wheeler DL, Church DM, Lash AE, Leipe DD, Madden TL, Pontius JU, *et al.* (2001). Database resources of the National Center for Biotechnology Information. *Nucleic Acids Res* **29**: 11–16.

Zhao S, Shatsman S, Ayodeji B, Geer K, Tsegaye G, Krol M, *et al.* (2001). Mouse BAC ends quality assessment and sequence analyses. *Genome Res* **11**: 1736–1745.

Zheng BJ, Mills AA, Bradley A. (1999). A system for rapid generation of coat color-tagged knockouts and defined chromosomal rearrangements in mice. *Nucleic Acids Res* **27**: 2354–2360.

CHAPTER 7

Genetic and Physical Map Resources — an Integrated View

MICHAEL R. BARNES

GlaxoSmithKline Pharmaceuticals, Harlow, Essex, UK

Bioinformatics for Geneticists. Edited by M.R. Barnes and I.C. Gray
 ISBNs: 0 470 84393 4; 0 470 84394 2 (PB)

7.1 INTRODUCTION

Not so many years ago, maps of the human genome were restricted to a handful of very low resolution diallelic RFLP marker maps of specific loci. Physical mapping following linkage analysis required a laborious laboratory-based process of contig construction using yeast and bacterial artificial chromosome (YAC and BAC) clones or cosmids. This involved consecutive rounds of library screening and clone characterization to identify overlaps between clones and build contigs. In recent years, as the human genome sequence nears completion, practical approaches to the characterization of genomic loci have changed quite dramatically. Today the process which took many months or even years can be completed in an afternoon using web-based resources. These tools might lead us to believe that the human genome sequence is the only map we need to know, but it actually represents just one dimension of a multifaceted map. Other maps including genetic, cytogenetic and radiation hybrid maps, represent different aspects of the structure, content and behaviour of chromosomes. These properties really need to be integrated with sequence-based maps to fully understand the properties and genomic landmarks that influence genes, mutation and human evolution.

As this book goes to press, the human genome is still unfinished and in the strictest sense it is likely to remain so for several years to come. For example, in April 2002 the human genome draft sequence reached 97.8% coverage, however only 63% of sequence was flagged as finished with 34.8% flagged as draft. The target date for final human sequence completion is 2003. However this may be a moving target, as a combination of contig errors and molecularly intractable regions are likely to continue to keep the genome in at least a partial draft state for many years to come. With this in mind, it is probably pragmatic to assume that the genome will remain unfinished in parts until at least 2005. Mouse genome sequencing is rapidly catching up with human sequencing, with the mouse also projected to finish in 2003. Other mammalian species such as the rat, dog and chimpanzee are further behind, although further genome sequencing will be assisted by existing genomes. The 'pioneer' genome sequences (human and mouse) will be used to span gaps and build contigs by comparison with existing contigs. This approach is already being used to accelerate the mouse and human genome sequencing projects, as both assemblies are being used to span gaps in each respective genome assembly (J. Mullikin, personal communication).

As we are becoming more aware of the difficulties of completing whole genome sequences, the role of physical and genetic maps is changing. Generation of new maps continues to be the first line of study for organisms with poorly characterized genomes. But where the genome sequencing of an organism is advanced, emphasis on maps is shifting to a role in the finishing and QC of existing sequencing maps. With this proviso in mind and with a specific focus on human maps, this chapter will review genetic and physical maps as they are being directly applied and integrated with the human genome and other sequenced mammalian genomes. We will not attempt to cover the full complexity of all forms of maps, or attempt to describe the use of these maps to enable the study of unsequenced organisms. Instead we will review the principles and informatics issues that apply to this area, with a focus on the data which is most likely to be useful to the human geneticist. For example we will examine the use of genetic and physical maps to check the order and orientation of marker maps and genomic contigs. For researchers who wish to construct new genetic and physical maps without sequence data we direct the reader to specialist texts in this research area.

7.1.1 What is a Genome Map?

At the most basic level, a genome map is a collective set of markers with known relative positions. A marker could be any genomic element with a uniquely identifiable sequence or property. Markers can exist in many different forms, such as non-polymorphic sequence tagged sites (STS) which act as a unique anchor or SNPs and short tandem repeats (STR), which act as both unique anchors and markers for differentiation between individuals. Genomic maps are divided into two broad categories. Polymorphic markers are used to construct genetic maps and either polymorphic or non-polymorphic markers are used to construct physical maps.

7.2 GENETIC MAPS

The genetic linkage map is a key concept which gives a fundamental insight into the genetic nature of the genome. Genetic linkage maps inform on more than just order of markers, they also give a measure of the underlying genetic recombination that occurs in a particular chromosomal region. Linkage maps show the relative locations of specific DNA markers along the chromosomes of related individuals. Any inherited physical or molecular characteristic that differs among individuals and is easily detectable is a potential genetic marker, for this reason polymorphic markers, such as SNPs and STRs are particularly suited to genetic map construction as they are plentiful, easy to characterize precisely and amenable to laboratory automation (see Chapter 3 for a review of SNPs and STR markers).

Genetic maps are constructed by evaluating the genotypes of a set of markers in groups of related individuals. This raw mapping data is analysed by software packages, such as MapMaker (Lander *et al.*, 1987; reviewed in Chapter 12) which construct genetic maps by observing how frequently the alleles at any two markers are inherited together. The closer the markers are, the less likely it is that a recombination event will separate the alleles, and the more likely it is that they will be inherited together. Thus, unlike physical maps, the distance between markers on a genetic map is not measured in any kind of physical unit; it is a measure of the recombination frequency between those two markers. This genetic map unit is measured in centimorgans (cM). The distance between two markers would be measured as 1 cM if both markers are separated by recombination on 1% of occasions. Genetic distance has an average correlation with the actual physical distance between markers, on average in humans 1 cM is equivalent to 1 Mb (this ratio varies widely between other species). The 1 cM : 1 Mb ratio is often used as a rule of thumb, but it is important to recognize that this is a genome-wide average and can often diverge significantly from this ratio between different regions of the human genome. The genetic/physical ratio also differs considerably between genders, as recombination frequencies vary between males and females. To overcome these differences, genetic maps typically report distances for each sex and a 'sex-averaged' distance that integrates male and female recombination frequencies.

7.2.1 Human Genetic Maps

A range of genome-wide human genetic maps has now been published at various resolutions. Most genetic maps are based on STR markers, although a genome-wide SNP linkage map has also been published recently (T. C. Matise *et al.*, unpublished data). Most genome-wide linkage maps are constructed with a marker framework spaced at 2.5–10-cM intervals. Denser marker maps have not been widely used for linkage analysis, as the focus of analysis is on a small number of meiotic events observable within a family. These meiotic events do not require a very dense map of markers to find evidence for possible co-segregation of a disease-influencing gene with marker locus alleles. Higher resolution genetic maps have been described, but they are generally restricted to specific chromosomal regions, such as the long arm of chromosome 21 (Lynn *et al.*, 2000), where they have been used to refine initial linkage analysis. Ideally, to be maximally informative, genetic markers need a relatively high level of heterozygosity (>0.6). This provides a high likelihood that a marker (or cluster of SNPs) will be different between any two copies of a chromosome. Markers with lower heterozygosity, for example, SNPs which range in heterozygosity from ~0.1–0.3, need to be used in higher density to give a similar level of information.

The three main genetic maps were developed by Genethon, the Marshfield Institute and the SNP consortium (TSC) (see Table 7.1 for a comparison). The Genethon and Marshfield maps are widely indexed by mapping tools, such as MapViewer and GDB (see below). The newer TSC map is also likely to be available in these tools in the near future.

7.2.2 The Genethon Genetic Linkage Map

The Genethon human linkage map was the first whole genome genetic map to exclusively use STR markers; previous maps were based on less informative RFLPs (which are actually uncharacterized SNPs). The 5264 markers in the Genethon map have a mean heterozygosity of 0.7, which makes it more informative than previous maps. The map was constructed with data from eight CEPH families (comprising 186 meioses) so the fine order of markers is not well resolved, other than by localization within a particular chromosomal region. The map spans a sex-averaged genetic distance of 3699 cM. The average interval size is 1.6 cM, 59% of the map is covered by intervals of 2 cM at most and 1% remains in intervals above 10 cM. The map comprises 2335 positions, of which 2032 could be ordered with an odds ratio of at least 1000 : 1 against alternative orders. This high level of statistical confidence in marker order was subsequently used by DeWan *et al.* (2002), to highlight a number of discrepancies in the order and orientation of clones in the human genome draft assembly. Genethon map data can be accessed at the Genethon website (www.genethon.fr) and the Washington University, St Louis website (www.genlink.wustl.edu/genethon_frame/).

TABLE 7.1 Human Genetic Maps

Map	Genethon	Marshfield	TSC
Marker type	STRs	STRs	SNPs
Marker no.	5264	8325	2679
Av. heterozygosity	0.7	0.68	0.76
Resolution (kb)	1.6 cM	1.3 cM	2.5 cM
Reference	Dib *et al.* (1996)	Broman *et al.* (1998)	Matise *et al.* (unpublished data)

7.2.3 The Marshfield Genetic Linkage Map

The Marshfield genetic linkage map improved on the Genethon map, by offering a larger marker number and a slightly higher resolution. Like the Genethon map, the Marshfield map was constructed with data from eight CEPH families and therefore fine order is still poorly resolved. In particular, markers which are separated by little or no genetic distance generally have no recombination events separating them, and so they are presented in arbitrary order. Accurate ordering information for these markers can be obtained by cross referencing STS marker location with human physical maps, such as RH maps or the human genome sequence itself. The Marshfield database (http://research.marshfieldclinic.org/genetics/), provides a well-documented range of five genome scan marker panels (genome-wide screening sets 6–10), selected from the Marshfield map. These marker panels were initially developed from the first human linkage mapping screening set from the Cooperative Human Linkage Centre (CHLC) (Murray *et al.*, 1994). Each Marshfield marker panel provides a progressively higher density of markers, culminating in set 10 which consists of 405 di, tri and tetra-nucleotide repeat markers with an average spacing of 9 cM. Each marker set is also grouped by allele size so that each panel can be loaded into the same lane or capillary. Primers for marker set 10 are commercially available from Research Genetics, in unlabelled and fluorescent dye-conjugated forms (http://www.resgen.com/).

7.2.4 TSC SNP Linkage Map

Technology developments have brought the cost of SNP genotyping far below the cost of STR genotyping. This has led to calls for the development of a SNP-based linkage map. The only argument against the implementation of such a map is the lower heterozygosity of a single SNP compared to a polymorphic STR (Kruglyak, 1997; see Chapter 8 for a discussion of this issue). Use of single SNPs at similar densities to STRs would essentially be equivalent to the original and less informative RFLP maps. Two related solutions have been proposed to overcome this problem. The first solution is to use a 3–8-fold increase in SNP marker densities to produce an evenly spaced map (Kruglyak, 1997). The second is to use multiple clusters of two to three SNPs in linkage analysis at a similar density to STRs. These SNP clusters provide approximately the same amount of information as an STR in terms of heterozygosity (Goddard and Wijsman, 2002).

Matise *et al.* (unpublished data) used the SNP cluster approach to construct a whole genome SNP linkage map. To do this they selected 666 physically and genetically mapped polymorphic STS anchor loci at 5-cM intervals across the human genome. Ten or more SNPs were then characterized across each STS locus. SNPs were assessed for genotyping success rates, assay quality, allele frequencies (ideally >20%), multi-SNP haplotype heterozygosities (ideally >0.6) and levels of linkage disequilibrium (SNPs in LD with each other were avoided). The three most informative markers per STS locus were then selected to maximize multi-SNP haplotype heterozygosities, to create an informative SNP cluster at each map position. Two thousand SNPs were selected and genotyped in 661 individuals from 48 CEPH reference pedigrees (http://www.cephb.fr/). Linkage maps were constructed without reference to any other mapping or sequence position information. This generated a map with an average resolution of 5 cM; to improve this, a further set of SNPs were identified at half-way points between the SNP clusters loci and were similarly evaluated. The single most informative SNPs at each of these positions were identified ($N = 679$) and genotyped in the CEPH pedigrees. These 'single' SNPs were added to the cluster linkage map to produce a final SNP map with a 2.5-cM resolution.

The construction of this map was supported by the SNP Consortium (TSC), all the data and results are available at the TSC website (http://snp.cshl.org).

7.2.5 SNP-based Haplotype and Linkage Disequilibrium (LD) Maps

As new SNPs arise at different loci and at different points in time, groups of neighbouring SNPs may show distinctive patterns of co-inheritance or LD, which are arranged into distinct haplotypes between individuals. The great abundance of SNPs across the genome creates an opportunity to exploit this haplotypic diversity in association studies by identifying SNPs which capture or 'tag' the majority of common human haplotypes. This enables the construction of very efficient maps, which capture maximal diversity with a minimal number of SNPs. Such haplotype tags have already been used to screen candidate genes. For example, Johnson *et al.* (2001) re-sequenced nine genes to identify common SNP haplotypes among 122 SNPs. Once these haplotypes were defined they were able to define just 34 SNPs or 'haplotype tags' which identified all the haplotypes across the genes. Extension of this principle across the genome would enable the construction of powerful haplotype-based maps which could capture most common haplotype diversity with a minimal number of SNP markers. At the time of going to press, such a map does not exist in the public domain, although at least one company has this data. A public domain genome-wide haplotype/LD map is likely to become available early in 2004 if not sooner.

Some data is already available publicly. Public domain LD or haplotype maps are available for three chromosomes, these have been generated by two distinct methods and consequently the exact nature of the data presented differs between the maps. Orchid Biosciences Inc. in collaboration with the TSC have published a SNP-based map of chromosome 19 which will be available from the TSC website before this book goes to press (Michael Phillips, personal communication); Dawson *et al.* (2002) published a SNP-based LD map of chromosome 22 and Perlegen Inc. published a SNP-based haplotype map of chromosome 21 (Patil *et al.*, 2001). We take a closer look at the Perlegen map data in Chapter 9.

7.3 PHYSICAL MAPS

While genetic maps display the linear order of genes or markers and the recombination between them, they do not give reliable information on the physical distance between markers and genes. By contrast a physical map has an absolute and invariant base-pair scale, which defines the physical distance between markers. Two markers may be very close genetically, i.e. very little recombination occurs between them, but very far apart physically. The difference between genetic and physical maps may seem academic, however if a trait or disease is localized on a physical map between two molecular markers it is important to identify the amount of recombination across the region, to select an appropriately dense panel of markers to detect a genetic association. Conversely if a genetic map places a trait or disease between two molecular markers, it is useful to know if that distance represents 1 kb, 1 Mb or further still, to define the likely number of genes or regulatory regions in the locus.

7.3.1 Cytogenetic Maps

There are many different types of physical maps; the first identified and lowest resolution physical map of the human genome is the cytogenetic map. This type of map is based on the distinctive banding patterns of stained chromosomes. Detailed measurements of these patterns were originally used to define the gross physical size of human chromosomes, and led to the size-based sorting of the autosomal chromosomes from chromosome 1, the largest chromosome, to chromosome 22, the smallest. Unsurprisingly these early efforts at physical mapping were quite inaccurate and prone to distortion by differential contraction, which led to the incorrect ordering of chromosome 19 which is actually slightly smaller than chromosome 20 (Morton, 1991). Use of cytogenetic map locations is still remarkably prevalent, perhaps due to the ease of use of the vocabulary of cytobands, e.g. 1q32, 22q11, etc., to describe and cluster groups of genes and loci. Interestingly the cytobanding recognized by early biologists is not just decorative, but in fact the dark cytobands represent regions of higher average GC content, while light cytobands have a lower average GC content (Nimura and Gojobori, 2002). The region where a transition occurs between a dark and light cytoband is known as an isochore, these regions often show a remarkably increased rate of recombination (Eisenbarth *et al*., 2000). This may make it important to pay special attention to genes and possible regulatory elements in these regions; we specifically address this issue in Chapter 10.

7.3.2 Fluorescence *In Situ* Hybridization (FISH) Mapping

At best a cytogenetic map could be used to locate a DNA fragment to a region of about 10 Mb — the size of a typical chromosome band. Fluorescence *in situ* hybridization (FISH) mapping, is a form of cytogenetic mapping that allows orientation and mapping of DNA sequences to a much higher resolution. Initially FISH resolved markers within 2 Mb, but further development of the FISH method, using chromosomes in interphase when they are less compact, increased map resolution further to around 100 kb. As FISH does not rely on a recombinant map but instead maps a chromosome directly, this has made FISH an important method for the QC of recombinant maps and clone contigs. The level of resolution achieved with interphase FISH, also makes this method directly applicable to the analysis of observable physical traits associated with chromosomal abnormalities, such as prenatal defects or cancer breakpoints. All of these applications are likely to keep the method in regular use well beyond the availability of a complete human genome.

7.3.3 Radiation Hybrid (RH) Mapping

Early physical mapping advanced considerably with the publication of the radiation hybrid (RH) mapping method. Goss and Harris (1975) irradiated human fibroblast chromosomes and fused the resulting fragments with recipient rodent cells. The observed patterns of co-transference of markers in a collection of hybrid cells allowed estimates to be made of linear order and distance between markers by assuming that distant markers are more likely to be separated in different hybrid cell lines than closer markers. The RH mapping technique was refined by Cox *et al*. (1990) who irradiated donor somatic cell hybrids, which contained just a single copy of one human chromosome, and fused the fragments with rodent cells. Several whole genome RH panels were developed in the 1990s which allowed the construction of genome maps containing thousands of STS markers (Gyapay *et al*., 1996; Stewart *et al*., 1997). The human RH map finally reached a high-resolution apex, with the development of the TNG panel (Lunetta *et al*., 1996), which was used to generate an RH map of the human genome consisting of 40,322 STSs (Olivier *et al*., 2001). From the 40,322 STSs mapped to the TNG radiation hybrid panel, only 3604 (9.8%) were absent from the unassembled draft sequence of the human genome.

7.3.4 Human RH-mapping Panels

Three main radiation hybrid panels have been used for mapping STSs and constructing RH maps, each offers a different level of resolution based on the dose of irradiation. The GB4 RH panel (constructed by using 3000 rad of X-rays) and the G3 RH panel (10,000 rad of X-rays) will resolve markers at 1-Mb and 260-kb intervals respectively, both providing a good long-range continuity for mapping (Deloukas *et al.*, 1998). In contrast, the Stanford TNG panel (50,000 rad of X-rays) allows STS resolution down to 60–100 kb with high confidence (Lunetta *et al.*, 1996). The price of this increased resolution is that a large number of STSs need to be scored to produce good long-range continuity. Olivier *et al.* (2001) found a solution to this by using the TNG panel in conjunction with the Stanford G3 panel to produce an RH map with high-resolution and contiguity. Publication of this map saw a shift in the role of human RH-mapping, from a direct role in mapping new genes to a primarily curatorial role to enable the QC and assembly of the human genome.

RH maps provide a marker order confidence supported by LOD (logarithm of the odds ratio of linkage versus no linkage) scores between adjacent markers, coupled with distance measures between markers. Calculation of distance is based on the frequency of breakage between two markers in the radiation hybrid clones which is measured in centiRays (cR). There is a direct linear correlation between cR units and physical distance in kb, which is fairly constant across any given RH panel. The kilobase equivalent of the centiRay unit differs between RH maps. 1 cR on the TNG map corresponds to an average of 2 kb of physical distance, whereas 1 cR on the G3 map corresponds to a physical distance of 24 kb and a distance of 260 kb on the GB4 map. Table 7.2 illustrates the main features of all three panels and Table 7.3 illustrates the main RH maps generated from these panels. RH panels are available from Research Genetics (http://www.resgen.com/).

TABLE 7.2 Human Radiation Hybrid Panels

Panel	GeneBridge4 (GB4)	Stanford G3	Stanford TNG
X-ray dosage	3000 rad	10,000 rad	50,000 rad
Cell lines	93	83	90
Average retention	30%	18%	16%
Av. Frag. Size	10 Mb	4 Mb	800 kb
Resolution	Low	Medium	High
Resolution (kb)	1000	267	60
Reference	Gyapay *et al.* (1996)	Stewart *et al.* (1997)	Lunetta *et al.* (1996)

TABLE 7.3 Human Radiation Hybrid Maps

Map	GeneMap 99-GB4	GeneMap 99-G3	Stanford TNG	NCBI Integrated
Marker panel	GB4	G3	TNG & G3	G3 & GB4
Marker type	STS	STS	STS	STS
Marker no.	45758	7061	40322	23723
Reference	Schuler *et al.* (1996)	Deloukas *et al.* (1998)	Olivier *et al.* (2001)	Agarwala *et al.* (2000)

Comprehensive RH maps generated from these panels can be viewed and integrated with other maps in GDB, MapViewer and other applications (see below). Novel STS markers can be placed on these existing frameworks by PCR, screening STSs against the three RH panels and submitting the results to a web server, several of which are available. For G3 and TNG RH maps Stanford run a server at http://www-shgc.stanford.edu/RH/index.html. The EBI also runs an RH map server which includes all three human panels and also mouse, rat, pig and zebrafish panels (http://corba.ebi.ac.uk/RHdb/RHdb.html). The Whitehead Institute also maintains a GB4 server (http://www-genome.wi.mit.edu/cgi-bin/contig/rhmapper). Data submissions to all three servers are in a binary format to indicate presence or absence of a PCR product in each hybrid bin, e.g. G3.STS1 11000010000 0100000011001100100000110010001000001110100011000000110.

7.4 PHYSICAL CONTIG MAPS

Genetic maps and cytogenetic maps fulfilled many of the short-term goals of the human genome project — to develop low to medium resolution genetic and physical maps of the genome. They have also facilitated longer term goals by assisting in the construction of the more-precise high resolution maps at increasingly finer resolutions needed to organize systematic sequencing efforts (Korenberg *et al.*, 1999). FISH and RH mapping in particular have enabled the development of a complex hierarchy of physical YAC and BAC clone contigs at a range of resolutions (Figure 7.1). These physical maps also became an important framework for positional cloning efforts in the years preceding the availability of a draft human genome. Accurate ordering of YAC and BAC clones (and subsequent

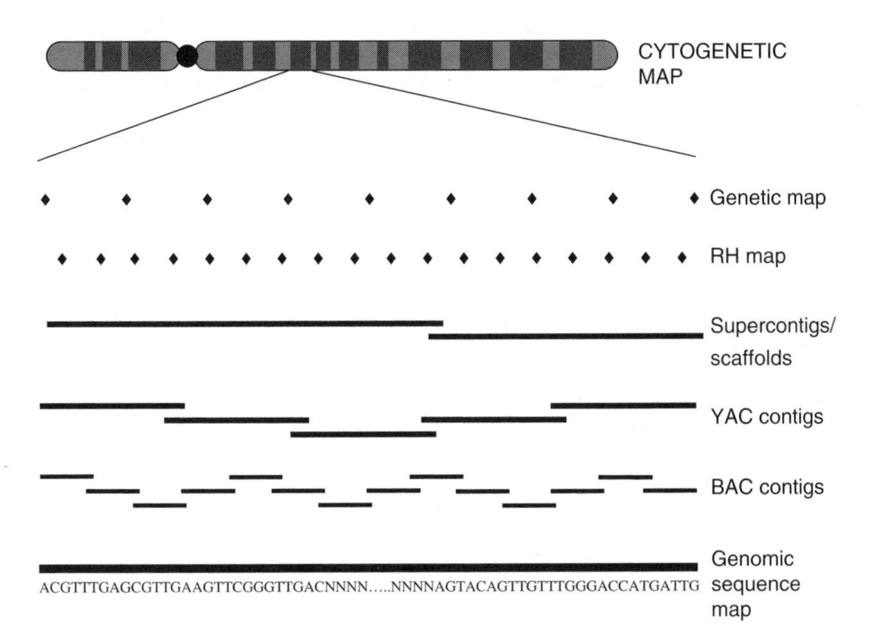

Figure 7.1 Physical and genetic maps used during the sequencing of the human genome. Many different maps were integrated to enable the construction of the framework for human genome sequencing (see Waterston *et al.* (2002) for a review).

shotgun reads) would not have been possible without existing genetic and physical maps which served as a scaffold for orientating, ordering and troubleshooting the human genome sequence assembly.

7.4.1 Yeast Artificial Chromosome (YAC) Maps

Yeast artificial chromosomes (YACs) are the lowest resolution physical clone contig maps, composed of overlapping YAC clones ranging in size from 300 kb–2 Mb. Before YACs were developed, the largest cloning vectors (cosmids) carried inserts of only 20 to 40 kb. YAC methodology drastically reduces the number of clones to be ordered; many YACs span entire human genes, making them a useful resource for further genomic study. The size of YAC inserts can often cause clone instability, which can lead to local rearrangements in the clone, this is the major drawback in the use of YACs for construction of physical contigs and underlines the need to QC YAC contigs with other available genetic and physical maps.

Several whole genome YAC maps are available, including a library of 33,000 YAC clones published by Chumakov et al. (1995). This library and other YAC clones can be obtained from a range of centres which are listed in the CEPH YAC library pages (http://www.cephb.fr/bio/ceph_yac.html).

7.4.2 Bacterial Artificial Chromosome (BAC) Maps

Bacterial artificial chromosomes (BACs), offer a further increase in map resolution, typically ranging in size from 100–300 kb. BAC clones are the primary vehicle of the public human genome sequencing project. Collections of human BACs estimated to represent more than a 10-fold redundancy of the human genome have been used to generate comprehensive BAC maps of the human genome. A minimally redundant set of these BACs have been assembled into physically separate contigs, representing the majority of the human genome. The sequence of these BACs is being determined by shotgun sequencing, where each BAC is digested with restriction enzymes and sub-cloned to generate a library of clones ranging from 0.5–5 kb. These clones are sequenced and assembled to form a complete BAC sequence, which are in turn assembled to form a complete chromosome.

BAC clone data can be accessed in many different ways, either directly from sequencing centres, or alternatively the NCBI have established a Human BAC resource page (http://www.ncbi.nlm.nih.gov/genome/cyto/hbrc.shtml). This page is a useful resource which centralizes information concerning currently available BAC maps and suppliers of BAC clones. Another useful database is GenMapDB (Morley *et al.*, 2001; http://genomics.med.upenn.edu/genmapdb/), which contains over 3000 mapped BAC clones spanning the genome. The database can be searched by map location or accession number. It is also possible to search for BAC clones by using BLAST (http://www.ncbi.nlm.nih.gov/BLAST/) to search the 'HTGS' and 'Genome' divisions of GenBank. BAC sequences can also be accessed indirectly by using tools which show contig information for the draft human genome sequence, e.g. Ensembl, Map Viewer or UCSC human genome browser (see Chapter 5).

7.5 THE ROLE OF PHYSICAL AND GENETIC MAPS IN DRAFT SEQUENCE CURATION

Sequence tagged sites (STSs) are PCR-based anchors used to define a unique genomic sequence in an RH panel, YAC or BAC contig. All that is required to generate a new STS marker is 200–500 bp of unique sequence, this could be a sequence from the 3′ UTR of a transcript or any unique genomic region. Hence STS markers have been extensively identified from characterized genes, expressed sequence tags (ESTs) and random genomic fragments (Schuler *et al.*, 1996). STS markers that include polymorphic sequences, such as microsatellites, are the central integrating force between genetic and physical maps. Common sets of such sequence-based markers can be easily screened and therefore can be used to integrate maps constructed by different mapping methods. RH panels and STS markers will play a critical role in the finishing of the human genome by providing a method to obtain markers from regions of the human genome that may be difficult to clone in conventional vector libraries. Hattori *et al.* (2000) found that up to 10% of certain gene-rich regions of human chromosome 21 were composed of such 'hard-to-clone' DNA. STSs that fail to hit available sequence can be used to screen different DNA libraries to close existing clone gaps in draft genome contigs. High resolution physical maps, such as the TNG map can also be valuable for curating draft genome contigs. Localization of RH markers to working draft sequences provides an independent measure of order and orientation for the clones underlying the draft sequence. Distances between markers can also be used to estimate the physical length of gaps between non-overlapping clones.

7.5.1 Electronic PCR (e-PCR)

Electronic PCR (e-PCR) is an *in silico* equivalent of the laboratory-based STS mapping process (Schuler, 1997; http://www.ncbi.nlm.nih.gov/cgi-bin/STS/nph-sts). The e-PCR tool at the NCBI maps known STSs from the dbSTS, GDB and RHdb databases to a user-submitted sequence. In a directly analogous process to PCR, e-PCR searches for sub-sequences within a query sequence that match known STS PCR primers and are in the correct order, orientation and spacing to be consistent with the PCR product size. These criteria eliminate the possibility of false positives (e.g. hits to psuedogenes or repeat sequences) that occur with other similarity searching methods such as BLAST. Electronic PCR is a valuable tool to assist in the integration of genomic sequence data with existing maps; this can be useful to assist genomic QC and to correlate genetic distances with physical distances. We offer detailed coverage of the use of this and other tools for genomic contig analysis in Chapter 9.

RH maps are playing a critical role in the QC and finishing of the human genome (see below), but once we have a finished genome, these maps may be of limited further use in humans. However, RH maps will continue to be the physical mapping method of choice for other organisms without extensive genome sequence. Human RH maps may also be of some limited use in the construction of comparative maps with other mammalian genomes (Kwitek *et al.*, 2001). But, for the purposes of this chapter, we will focus on the direct integration of genetic and physical maps, with genome sequence as the ultimate integration framework. For consideration of non-human maps we refer the reader to Chapter 6 and other specialist texts.

7.6 THE HUMAN GENOME SEQUENCE – THE ULTIMATE PHYSICAL MAP?

The complete DNA sequence of the human genome will be an accurate physical map resolved down to a single base pair resolution, but we do not have this map as yet. Geneticists will need to work with a draft assembly of the human genome for a somewhat indeterminate number of years, until this task is truly finished. However, the draft genome assembly is still a very valuable asset for genetics, particularly if data are treated with care. With this in mind it is very important to be aware of some of the issues relating to the curation of draft sequences. Genetic and physical maps are one aid in this process.

For example, Olivier *et al.* (2001) used a 40,000-marker RH map to provide an estimate of the size and location of missing sequence in the human genome draft in relation to the existing sequence, and to provide order information for the 15,000 + clones that constitute the human genome working draft. They found that 9.8% of STS markers were absent from the October 2000 draft of the human genome. They suggest that these are likely to represent the 'hard-to-clone' regions of the human genome. Other studies have made similar observations (Hattori *et al.*, 2000) which suggests that a small intractable percentage of the human genome may remain in an unfinished state for longer than we may have anticipated.

Genetic maps are also playing an important role in the QC of human genomic sequence. DeWan *et al.* (2002) compared the genetic order of the Marshfield genome-scan markers (set 9 and 10) with their physical order in the April 2001 public golden path contig and the February 2001 Celera genome assembly. They found inconsistencies in 5 and 2% of the markers in the Celera assembly and the golden path assembly, respectively. The genetic order of these markers was supported with high confidence by a LOD of >3 and most discrepancies were not observed in both contigs, which suggests errors in the physical map order of both genome assemblies. Chromosome-by-chromosome breakdown of this data are available on a website: http://linkage.rockefeller.edu/maps/.

7.7 QC OF GENOMIC DNA – RESOLUTION OF MARKER ORDER AND GAP SIZES

The studies by DeWan *et al.* (2002) and Olivier *et al.* (2001) demonstrate the value of genetic and physical maps in the curation and QC of human genomic sequence contigs. As discussed previously the relationship between genetic distance (cM) and physical distance (Mb) is not uniform, however both genetic and physical mapping methods can resolve marker order to varying degrees of confidence, depending on the map characteristics. Marker or contig order across a locus can be validated by integrating information from different maps. The value and accuracy of different maps is not necessarily hierarchical or directly related to the density of the map. For example, one might assume that a dense RH map, or even a finished sequence map might be more accurate than a less dense genetic map, however as the studies above have shown, this does not always hold true. Using maps in an hierarchical manner may avoid the inevitable discordances between different maps, but this is not necessarily the best order for integration. In some cases for example, YAC STS content data may be more accurate than RH data, or a genetic map may be more reliable than a BAC contig. Both genetic maps and RH maps show a relative confidence in marker order by LOD scores using appropriate maximum likelihood statistical methods (Boehnke *et al.*, 1991). A LOD of 3.0 (odds 1000 : 1) or more is generally accepted as a strong indication of a contiguous relationship between markers. Comparison of LOD scores can help to integrate different data sources in an attempt to reach a consensus. But sometimes, all that can be done *in silico* is to flag up an unresolvable discrepancy between maps to receive special attention in the laboratory. Bioinformatic tools and databases can be a great help in the integration and evaluation of genetic and physical maps. MapViewer, GDB and UDB allow the user to compare and integrate maps; these are described below.

The UCSC human genome browser also provides graphical data on the positions of STS markers on the golden path versus their positions in other maps, including radiation hybrid, sex-averaged genetic (Marshfield), cytogenetic and YAC STS maps at the following URL: http://genome.ucsc.edu/goldenPath/mapPlots/.

It is also possible to view and integrate genetic and physical maps on an *ad hoc* basis using bioinformatics tools, such as Map View at the NCBI (reviewed below).

7.8 TOOLS AND DATABASES FOR MAP ANALYSIS AND INTEGRATION

There are some excellent tools which have recently become available for viewing the human genome sequence. Ensembl and the UCSC human genome browser are shining examples of the kind of biological data integration that geneticists need for their studies. But unfortunately they are lacking in functionality to enable map integration. Other more specialized tools, such as GDB and UDB exist, which allow a user to view and integrate different maps, but unfortunately these have not generally been integrated with the human genome sequence. Fortunately Entrez Map View at the NCBI, is one tool which straddles both the human genome and human genetic maps.

7.8.1 Entrez Map View (http://www.ncbi.nlm.nih.gov/cgi-bin/Entrez/map_search)

In Chapter 5 we reviewed Map View alongside Ensembl and the UCSC human genome browser as a tool for annotation of the human genome sequence. Map View would probably appear lower in most researchers' preferences for this purpose, although it does provide some unique gene annotation information. However, as the name suggests, Map View truly excels in its integration of a wide range of cytogenetic, genetic and physical maps with the NCBI draft and finished sequence contigs. Although this tool is sometimes a little difficult to navigate, once these idiosyncrasies are overcome, Map View becomes a complex and powerful tool.

Map View is an integrated component of the NCBI Entrez system, in the Entrez Genomes division. This division presents a unified graphical view of genetic and physical maps (including sequence maps) for over four vertebrates, including human and mouse. The tools present different genomes at four levels of detail:

- Organism home page — summarizing the resources available for that organism
- Genome View — graphical display of chromosome ideograms and search page
- Map View — presents one or more maps aligned against a master map
- Sequence View — graphically annotates the biological features in a region

7.8.1.1 Searching and Browsing Map View

Map View can be searched with almost any marker, SNP, gene or genomic element either targeted at a chromosome or genome level. Searches at the genome level return a graphic view of the location of the hit with red marks on the chromosome ideograms, this will quickly identify if a query hits multiple regions or chromosomes. A summary of the maps in which the query exists is returned in tabular format at the bottom of the page. This is the essence of the Map View tool — selection of a map from the tabular summary links to a detailed Map View of the corresponding genomic region, with the selected map as the 'master' map. The master map is presented in detail with supporting information, such as LOD scores, cM locations or gene information. To view and integrate the master map with other maps, select the 'maps & options' link at the top of the page. This will summon a pop-up window for Map View configuration. It is possible to select up to eight maps to view alongside the master map, each is presented in a compact view alongside the master map. The alignment between maps is based on common or corresponding objects. Markers or objects shared between maps are indicated by lines connecting the maps. Map View allows the user to zoom and pan into progressively more detailed views.

It is also possible to search and browse Map View by map position or cytoband. This can be achieved from the Map View of a chromosome, by entering a range of interest in the boxes in the side window. A range can be specified in base pairs, cytogenetic bands or between two gene symbols. General chromosomal browsing is possible by clicking on the region of interest in the chromosome thumbnail graphic in the sidebar, or by clicking on a region of interest on the ideograms in the genome view.

Map View is very effective for integration of genetic and physical maps on an *ad hoc* basis. Figure 7.2 shows an integrated view of the Genethon genetic map and the human genome contig for chromosome 3. In this map, the Genethon markers are mapped to sequence and a line is drawn between the marker positions on the two maps. This clearly illustrates some key map integration issues. Firstly several markers in the genetic map are seen to conflict with the order of markers on the sequence (or physical) map. This may be due to an error in either map, so further maps need to be compared to support either order and the LOD scores on the genetic map need to be examined. Figure 7.3 shows such a comparison. The red line traces the Genethon marker, AFMA121WD5, through the Marshfield, GB4, G3 and TNG maps through to the genomic contig level. In this case the marker order is confirmed by each map. Sometimes it may not be possible to conclusively determine which map is 'right', instead further laboratory work may be necessary to resolve marker order. Figure 7.2 also clearly shows the variable relationship between genetic and physical distance. In particular it highlights some of the physical properties of chromosomes, for example the genetic physical distance ratio at the telomere of the P arm of chromosome 3 is very low; the marker AFM234TF4, for example, has a genetic location of 22 cM and physical location of 8 Mb. This illustrates the higher rates of recombination that are often observed in telomere regions (Riethman, 1997). Both figures indicate the presence or absence of each marker in available maps by an array of symbolic green circles at the far right of each marker. This helps to indicate non-specific markers. For example, some markers map to multiple locations in the same chromosome, these are indicated by green circles with a strike through. Other markers map to more than one chromosome, these are indicated by yellow circles and finally some markers, map to multiple chromosomes and multiple locations, indicated by a yellow struck through circles (e.g. AFMA191ZG5 in Figure 7.2).

There are a number of somewhat idiosyncratic features in Map View which might confuse the user. Firstly if a map is viewed in low resolution, it seems to display a somewhat arbitrary selection of markers, the full marker set only becomes visible when the user zooms in. Secondly, if the locus is too large to view in one window it is broken up into pages indicated at the top of the window. This pagination feature can make it slow and difficult to assess a whole locus, but this can be overridden by altering the page size in the configuration window. Setting a page size of 100–200 will allow a very large map to be viewed in a single window, this may take some time to load but it is worth it in the end.

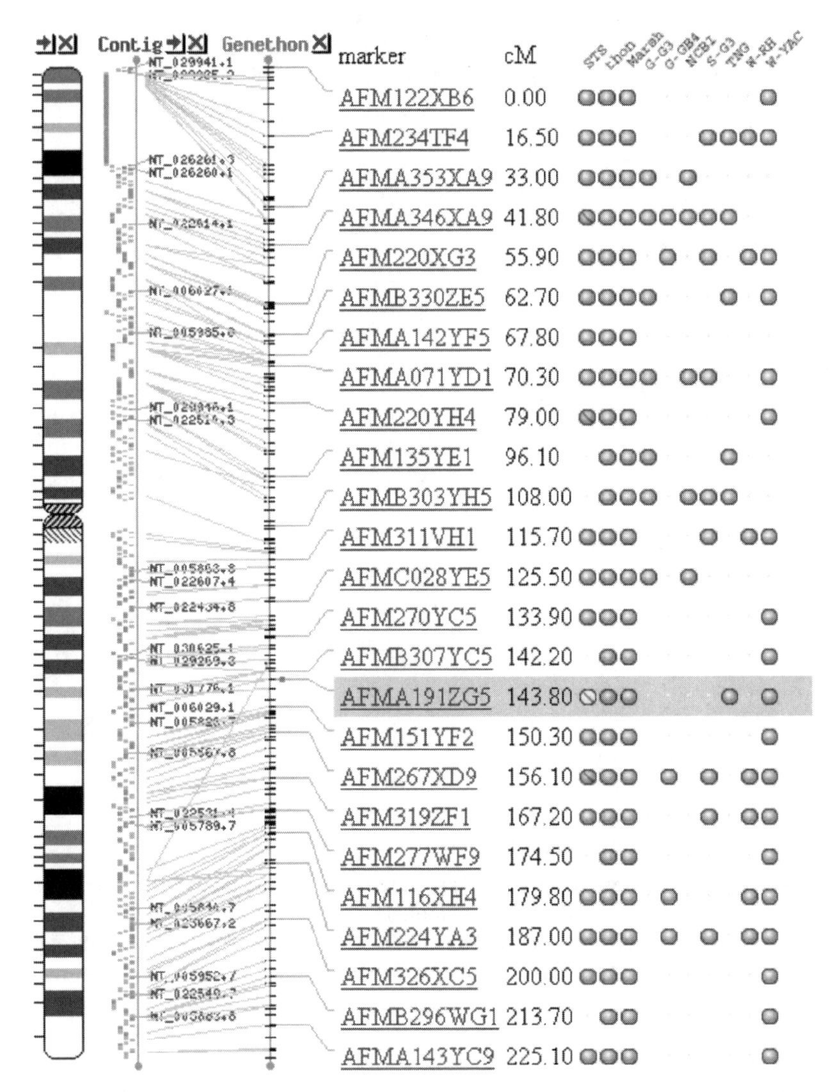

Figure 7.2 Integration of genetic maps and genome contigs. This figure shows an integrated view of the Genethon genetic map and the human genome contig for chromosome 3 generated by the NCBI Map View tool. The Genethon markers are mapped to sequence with a line drawn between the marker positions on the two maps. Lines which cross over show markers which conflict in order between the genetic map and the physical sequence map.

7.8.2 The Genome Database (GDB) (www.gdb.org)

The Genome Database (GDB) was the first web-based graphical interface to the human genome, as such it was a pioneering bioinformatic tool. Now Ensembl, UCSC and Map View present effortless graphical genome views and the GDB graphical interface is starting

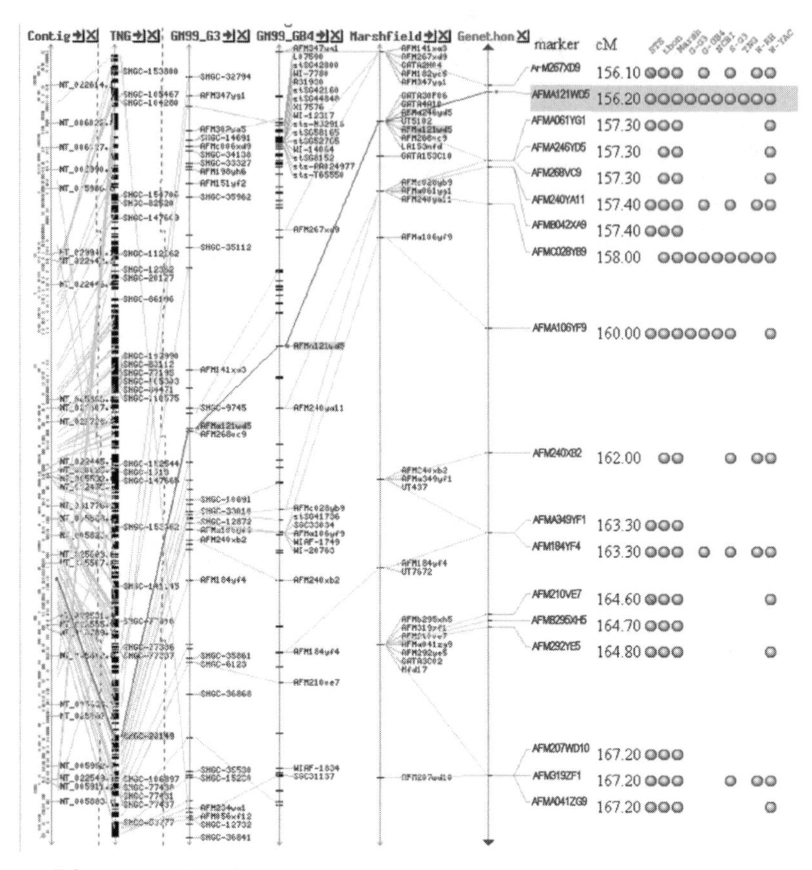

Figure 7.3 Integration of genetic and physical maps with the human genome contig on chromosome 3 using the NCBI Map View tool. The grey line traces the Genethon marker, AFMA121WD5, through though the Marshfield, GB4, G3 and TNG maps through to the genomic contig level. In this case the marker order of this marker is confirmed by each map.

to look a little tired and most of the graphical functionality is covered by Map View. But GDB does have a productive text/table-based search interface which is an improvement on Map View's limited text-based capability. GDB is also a comprehensive source for some forms of genetic data, particularly tandem repeat polymorphisms (it contains over 18,000), and an eclectic range of information on fragile sites, deletions, disease genes and mutations, collected by a mixture of curation and direct submission. This makes GDB a valuable tool for text-based data mining to assist in the construction of marker lists and the identification of marker variables, such as primer and marker sequences.

The text-based search interface is accessible on the front page of the GDB database by following the 'advanced search' link. This interface allows complex queries, for example, it is possible to retrieve all known polymorphic or non-polymorphic markers between two markers or genes. Results are retrieved and ordered based on the genetic distances of the markers, along with a very roughly estimated Mb location (unfortunately actual integration with the human genome draft is currently lacking). As the markers are ordered by genetic distance, the distances are very approximate, with no fine measure of distance or order. It may be necessary to clarify the order with another tool such as Map View.

7.8.3 The Unified Database for Human Genome Mapping (UDB) (http://bioinformatics.weizmann.ac.il/udb/)

The Unified Database for Human Genome Mapping (UDB) is maintained by the Weizmann Institute of Science, Israel. UDB has attempted to create an integrated map based on a diverse range of human genome mapping data retrieved from a number of public databases. The map consists of an integrated hierarchy of genetic, RH, cDNA and YAC maps down to a kilobase resolution, on a scale converted from centiRays (cR) to megabases (Mb). UDB generates its maps using data from the Whitehead/MIT STS map, GeneMap'98, the Stanford TNG map and Genethon maps. The database can be searched in several different ways. An initial search by chromosome number can be narrowed by specification of cytogenetic band, position (in Mb) or marker interval. It is also possible to search by gene or marker name. This gives the estimated location of the gene as well as links to GeneCards and the Genome Database (GDB). The database also displays the estimated boundaries (in Mb) of the cytogenetic bands of any chromosome.

The UDB database is a good starting point for constructing physical or transcript maps across a genomic region. The main benefit of the database is that it eliminates the need to look at a number of different websites and integrates markers from several different maps with genomic contigs from NCBI. Unfortunately UDB is somewhat over zealous in its map integration, sometimes this might cause problems. It assumes an hierarchical value of RH maps over genetic maps and genetic maps over YAC maps which is not necessarily the best order for integration, it may have been better to flag conflicting marker orders for laboratory-based resolution. However as the human genome map solidifies around finished sequence this approach will begin to represent the simplest and most effective use of time and resources.

7.9 CONCLUSIONS

As this chapter has described, there are many tools available to give an integrated view of genetic and physical maps across a defined chromosomal locus. Comparison of the physical and genetic distances between markers can provide a great deal of information about the underlying nature of a locus. Yu *et al.* (2001) compared the genetic and physical distances across the whole genome and found that the genetic/physical distance ratio ranged widely between 0 and 9 cM per Mb. They used this ratio to infer recombination rates and identified several chromosomal regions up to 6 Mb in length with very low or high recombination rates, which they termed recombination 'deserts' and 'jungles', respectively. Linkage disequilibrium (LD) was much more extended in the deserts than in the jungles as higher rates of recombination are likely to reduce the extent of LD.

When sequencing of the human genome is truly complete genetics will become technically much easier. Human map QC may become a distant memory, but presently we are still struggling to study complex phenotypes with draft contigs and incomplete datasets. Every piece of data and data curation may count in this struggle — in Section III of this book we review how physical and genetic map data can come together with literature data, marker data, gene data and comparative organism data to assist genetic studies in the laboratory.

REFERENCES

Agarwala R, Applegate DL, Maglott D, Schuler GD, Schaffer AA. (2000). A fast and scalable radiation hybrid map construction and integration strategy. *Genome Res* **10**: 350–364.

Boehnke M, Lange K, Cox DR. (1991). Statistical methods for multipoint radiation hybrid mapping. *Am J Hum Genet* **49**: 1174–1188.

Broman KW, Murray JC, Sheffield VC, White RL, Weber JL. (1998). Comprehensive human genetic maps: individual and sex-specific variation in recombination. *Am J Hum Genet* **63**: 861–869.

Chumakov IM, Rigault P, Le Gall I, Bellanne-Chantelot C, Billault A, Guillou S, *et al.* (1995). A YAC contig map of the human genome. *Nature* **377**(Suppl.): 175–297.

Cox DR, Burmeister M, Price ER, Kim S, Myers RM. (1990). Radiation hybrid mapping: a somatic cell genetic method for constructing high-resolution maps of mammalian chromosomes. *Science* **250**: 245–250.

Dawson E, Abecasis GR, Bumpstead S, Chen Y, Hunt S, Beare DM, *et al.* (2002). A first-generation linkage disequilibrium map of human chromosome 22. *Nature* (in press).

Deloukas P, Schuler GD, Gyapay G, Beasley EM, Soderlund C, Rodriguez-Tome P, *et al.* (1998). A physical map of 30,000 human genes. *Science* **282**: 744–746.

DeWan AT, Parrado AR, Matise TC, Leal SM. (2002). The map problem: a comparison of genetic and sequence-based physical maps. *Am J Hum Genet* **70**: 101–107.

Dib C, Faure S, Fizames C, Sampson D, Drouot N, Vignal A, *et al.* (1996). A comprehensive genetic map of the human genome based on 5,264 microsatellites. *Nature* **380**: 152–154.

Eisenbarth I, Vogel G, Krone W, Vogel W, Assum G. (2000). An isochore transition in the NF1 gene region coincides with a switch in the extent of linkage disequilibrium. *Am J Hum Genet* **67**: 873–880.

Goddard KA, Wijsman EM. (2002). Characteristics of genetic markers and maps for cost-effective genome screens using diallelic markers. *Genet Epidemiol* **22**: 205–220.

Goss SJ, Harris H. (1975). New method for mapping genes in human chromosomes. *Nature* **255**: 680–684.

Gyapay G, Schmitt K, Fizames C, Jones H, Vega-Czarny N, Spillett D, *et al.* (1996). A radiation hybrid map of the human genome. *Hum Mol Genet* **5**: 339–346.

Hattori M, Fujiyama A, Taylor TD, Watanabe H, Yada T, Park HS, *et al.* (2000). The DNA sequence of human chromosome 21. *Nature* **405**, 311–319.

Johnson GC, Esposito L, Barratt BJ, Smith AN, Heward J, Di Genova G, *et al.* (2001). Haplotype tagging for the identification of common disease genes. *Nature Genet* **29**: 233–237.

Korenberg JR, Chen XN, Sun Z, Shi ZY, Ma S, Vataru E, *et al.* (1999). Human genome anatomy: BACs integrating the genetic and cytogenetic maps for bridging genome and biomedicine. *Genome Res* **9**: 994–1001.

Kruglyak L. (1997). The use of a genetic map of biallelic markers in linkage studies. *Nature Genet* **17**: 21–24.

Kwitek AE, Tonellato PJ, Chen D, Gullings-Handley J, Cheng YS, Twigger S, *et al.* (2001). Automated construction of high-density comparative maps between rat, human, and mouse. *Genome Res* **11**: 1935–1943.

Lander ES, Green P, Abrahamson P, Barlow A, Daly MJ, Lincoln SE, *et al.* (1987). MAPMAKER: an interactive computer package for constructing primary genetic linkage maps of experimental and natural populations. *Genomics* **1**: 174–181.

Lunetta KL, Boehnke M, Lange K, Cox DR. (1996). Selected locus and multiple panel models for radiation hybrid mapping. *Am J Hum Genet* **59**: 717–725.

Lynn A, Kashuk C, Petersen MB, Bailey JA, Cox DR, Antonarakis SE, *et al.* (2000). Patterns of meiotic recombination on the long arm of human chromosome 21. *Genome Res* **10**: 1319–1332.

Morley M, Arcaro M, Burdick J, Yonescu R, Reid T, Kirsch I, *et al.* (2001). GenMapDB: a database of mapped human BAC clones. *Nucleic Acids Res* **29**: 144–147.

Morton NE. (1991). Parameters of the human genome. *Proc Natl Acad Sci USA* **88**: 7474–7476.

Murray JC, Buetow KH, Weber JL, Ludwigsen S, Scherpbier-Heddema T, Manion F, *et al.* (1994). A comprehensive human linkage map with centimorgan density. Cooperative Human Linkage Center (CHLC). *Science* **265**: 2049–2054.

Niimura Y, Gojobori T. (2002). *In silico* chromosome staining: reconstruction of Giemsa bands from the whole human genome sequence. *Proc Natl Acad Sci USA* **99**: 797–802.

Olivier M, Aggarwal A, Allen J, Almendras AA, Bajorek ES, Beasley EM, *et al.* (2001). A high-resolution radiation hybrid map of the human genome draft sequence, *Science* **291**: 1298–1302.

Patil N, Berno AJ, Hinds DA, Barrett WA, Doshi JM, Hacker CR, *et al.* (2001). Blocks of limited haplotype diversity revealed by high-resolution scanning of human chromosome 21. *Science* **294**: 1719–1723.

Riethman H. (1997). Closing in on Telomeric Closure. *Genome Res* **7**: 853–855.

Schuler GD. (1997). Sequence mapping by electronic PCR. *Genome Res* **7**: 541–550.

Schuler GD, Boguski MS, Stewart EA, Stein LD, Gyapay G, Rice K, *et al.* (1996). A gene map of the human genome. *Science* **274**: 540–546.

Stewart EA, McKusick KB, Aggarwal A, Bajorek E, Brady S, Chu A, *et al.* (1997). An STS-based radiation hybrid map of the human genome. *Genome Res* **7**: 422–433.

Waterston RH, Lander ES, Sulston JE. (2002). On the sequencing of the human genome. *Proc Natl Acad Sci USA* **99**: 3712–3716.

Yu A, Zhao C, Fan Y, Jang W, Mungall AJ, Deloukas P, *et al.* (2001). Comparison of human genetic and sequence-based physical maps. *Nature* **409**: 951–953.

SECTION 3

BIOINFORMATICS FOR GENETIC STUDY DESIGN

CHAPTER 8

From Linkage Peak to Culprit Gene: Following Up Linkage Analysis of Complex Phenotypes with Population-based Association Studies

IAN C. GRAY

Discovery Genetics
GlaxoSmithKline Pharmaceuticals Harlow, Essex, UK

8.1 INTRODUCTION

Linkage analysis of complex traits using family-based samples (see Chapter 11) typically results in a number of broad, ill-defined linkage peaks that represent several megabases

Bioinformatics for Geneticists. Edited by M.R. Barnes and I.C. Gray
 ISBNs: 0 470 84393 4; 0 470 84394 2 (PB)

of DNA (see for example Grettarsdottir *et al.*, 2002); beneath the expanse of each peak there may lie a gene (or genes) associated with the disease in question. Under the prior assumption that this preliminary linkage analysis has been completed, the goal of this chapter is to take the investigator through the process of characterizing and narrowing such a region using a population-based approach, with the ultimate aim of identifying candidate genes and testing them directly for association with the disease or trait in question. This is often achieved by testing markers in the linkage interval for differences in allele frequency between case and control cohorts, where the cohorts comprise unrelated individuals (although methods employing family structure, based on the difference in frequency of allele transmission in a large number of small pedigrees are also used — see below; this is covered in more detail in Chapter 11). In general, population-based methods offer large increases in both power and resolution over linkage-based approaches (McGinnis, 2000; Risch 2000; discussed in Chapter 11) and are well suited to the follow-up of preliminary (and often equivocal) linkage results. Examples of the successful application of this two-step linkage-association approach include identification of the involvement of *ApoE* in Alzheimer's disease (Strittmatter *et al.*, 1993) and the recent discovery of the role of *NOD2* in Crohn's disease (Hugot *et al.*, 2001; Ogura *et al.*, 2001). The first part of this chapter focuses on theoretical and practical considerations for good study design, whilst the second part covers a systematic approach to identification of the disease-associated gene, with emphasis on the application of methods, software tools and databases.

8.2 THEORETICAL AND PRACTICAL CONSIDERATIONS

8.2.1 Choice of Study Population

Wherever possible the study population selected for a follow-up analysis of linkage peaks should be derived from the same geographic area as the families used for the original linkage analysis. As the genetic components contributing to complex disease are likely to be varied, there is no guarantee that the predisposing genetic factors in one population will be the same in a second. If we use the term 'study population' in the broadest sense as applied to genetic association studies, a variety of study population structures may be considered. Three of the most common configurations are the case–control cohort, the discordant sib-pair cohort (i.e. one affected and one unaffected sib) and the parent–offspring triad (affected offspring with both parents) cohort. Each of these structures has advantages and disadvantages (for an evaluation of each, see Risch, 2000; Cardon and Bell, 2001).

Case–control cohorts simply consist of one group of individuals (cases) with the disease state and a second group without the disease (controls). Case–control cohorts have the advantage of being more straightforward to collect than the other two structures described above and generally provide more statistical power than similarly sized discordant sib or other nuclear family-based cohorts (McGinnis, 2000; Risch, 2000). However, case–control cohorts are prone to 'population stratification' (or substructure) effects. Population stratification occurs when the cohort under study contains a mix of individuals that can be separated on grounds other than the phenotype under study (most commonly on the basis of geographic origin). This can lead to allele frequency differences in cases and controls that are due to circumstances unrelated to the phenotypic difference under investigation, resulting in erroneous conclusions regarding association between the marker under test and the disease phenotype. Careful selection of individuals for inclusion in disease and control cohorts is necessary to ensure as homogenous a background as possible and therefore avoid stratification. If stratification is suspected it is possible to test for it using randomly selected genetic markers (Devlin and Roeder, 1999; Pritchard and Rosenberg,

1999). It is also important to match the cohorts for phenotypic or environmental variables that may otherwise confound any genetic analysis; for example, hormone replacement therapy (HRT) has a large impact on bone mineral density (BMD) and it would be necessary to account for this in a search for genetic factors influencing BMD using a cohort of post-menopausal women.

Although population homogeneity and well matched cases and controls are preferred, it may be possible to use a cohort even if stratification is present; Pritchard *et al.* (2000) have developed a method for testing for genetic association in the presence of population stratification, by using unlinked markers to make inferences about population substructure and employing this information to test for associations within the identified subpopulations. STRUCTURE and STRAT, software tools for the detection of stratification and testing for genetic association in the presence of stratification can be downloaded from http://pritch.bsd.uchicago.edu/software.html. An alternative approach to correction for population stratification, termed genomic control, measures the degree of variability and magnitude of the test statistics observed at random loci and uses this information to adjust the critical value for significance tests at candidate loci by the appropriate degree (Devlin and Roeder, 1999). However, it should be noted that correction for stratification cannot completely remove the possibility of increased false positive results under all circumstances (Cardon and Bell, 2001; Devlin *et al.*, 2001; Pritchard and Donelly, 2001) and stratification should be avoided where possible.

The main advantage of using study populations that incorporate elements of family structure (e.g. discordant sibs or trios) is that, unlike case–control cohorts, they are immune to population stratification effects. However, as mentioned above, family-based samples are typically more difficult to collect than case–control samples (particularly for late onset diseases) and generally offer less statistical power than the equivalent sized case–control cohort (McGinnis, 2000; Risch, 2000). The remainder of this chapter will focus predominantly on case–control methodology where reference to population structure is necessary; statistical methods for analysing family-based cohorts, such as the transmission disequilibrium (TDT) and sib transmission disequilibrium (S-TDT) tests, together with tools for the analysis of quantitative traits are covered in Chapter 11.

Estimation of required cohort size for a genetic study depends on a number of factors, including the size of the effect of the locus under test, the frequency of the disease-risk conferring allele and genetic nature of this 'risk allele', i.e. recessive, dominant, additive etc. If the causal variant is not being tested directly, the distance between the causal variant and the surrogate marker under test (see Section 8.2.3 below) is also relevant. Most of these factors are unknown prior to the start of the study and the minimum required population size is usually based on assumptions concerning these factors (see McGinnis, 2000; Risch, 2000). In reality, pragmatism typically dictates the available sample size; investigators use the largest obtainable cohort, with the caveat that the available sample may not provide sufficient statistical power to detect effects below a certain magnitude. To detect genetic factors of fairly small or moderate effect, cohorts of a several hundred to a few thousand individuals may be required (McGinnis, 2000; Risch, 2000).

8.2.2 Sequence Characterization at the Locus under Investigation

Following a whole genome linkage scan the investigator is typically faced with several genetic loci of potential involvement in the disease process, the limits of each defined by two genetic markers (usually simple tandem repeats or STRs) spanning several centiMorgans (cM). As 1 cM equates to 1 megabase (Mb) on average, and each Mb contains an estimated average of 15 genes (based on 45,000 genes in the entire 3000-Mb genome; Das *et al.*, 2001), this may represent several thousand kilobases (kb) of DNA and over 100 genes per locus. The first task is to define the locus in the context of the human genome, in order to gain a comprehensive knowledge of genes and further genetic markers in the interval. Until very recently this involved the laborious laboratory process of identifying and ordering genomic clones into contigs and using those contigs as a framework for gene and marker identification. Thankfully locus characterization has become far more straightforward in the wake of the Human Genome Sequencing Project, which provides free access to assembled sequence covering 97% of the human genome at the time of writing, with the goal of complete coverage by 2003. A number of web-based tools are available for exploiting the human genome. These tools are described very briefly in Section 8.3.1 and their practical application is covered in detail in Chapters 5 and 9.

8.2.3 SNPs, Linkage Disequilibrium, Haplotypes and STRs

8.2.3.1 Introduction

In this section we provide a simple introduction to the underlying principles of the detection of genetic association using a population (i.e. non-family)-based approach. The majority of studies of this nature are undertaken using single nucleotide polymorphism (SNP) markers (see Chapter 3). Biallelic SNPs are currently the marker of choice due to their abundance in the human genome and because they are amenable to high throughput genotyping approaches. The other marker system commonly used for genetic studies is the multiallelic STR (see Chapter 3). The paragraphs below on linkage disequilibrium and haplotypes refer mainly to SNPs. The use of STRs for population-based association studies is discussed at the end of this section.

8.2.3.2 Linkage Disequilibrium

A polymorphism associated with a disease state (in the true, rather than statistical, sense) may either directly contribute to the disease process, or may be a surrogate marker which is co-inherited with an adjacent functional variant that contributes to the disease state. This co-inheritance of the surrogate marker with the disease allele can occur to varying degrees and is termed ‘linkage disequilibrium’ (LD). By strict definition, LD is said to be present if co-occurrence of the two polymorphisms happens with a frequency greater than would be expected by chance. A number of measures of LD are used, two of the most commonly employed being Δ and D. Both measures are based on the difference between the observed and expected (assuming independence) number of haplotypes (see below) bearing specified alleles of two markers (see Chapter 11 for a complete explanation of D and Devlin and Risch (1995) for a discussion of D, Δ and other measures of LD). Although by the strict definition given above LD can occur between unlinked variants, for example in the presence of recent population admixture, in the following paragraphs we refer specifically to LD between two linked markers.

Clearly, the greater the extent of LD between two polymorphisms, the larger the chance of detecting the phenotypic influence of one by genotyping the other in a case–control experiment. The degree of LD is dependent on the history of the two adjacent markers and is influenced by the relative times of appearance of the two polymorphisms in the population and the degree of recombination between them. An extreme example would be two polymorphisms that appeared simultaneously on the same chromosome through spontaneous mutation and between which no recombination events have occurred over 2000 generations. During this period, these two linked polymorphisms have attained a population frequency of 20% through chance (random genetic drift) and are in absolute linkage disequilibrium. Imagine an alternative scenario, where a new polymorphism arises adjacent to an ancient polymorphism which has already attained a frequency of 20% over the previous 1000 generations; over the subsequent 1000 generations, there is a high degree of recombination between the markers, eroding the LD (Figure 8.1). Clearly the former case would be more favourable for using one of the markers as a surrogate for detecting the phenotypic influence of the other.

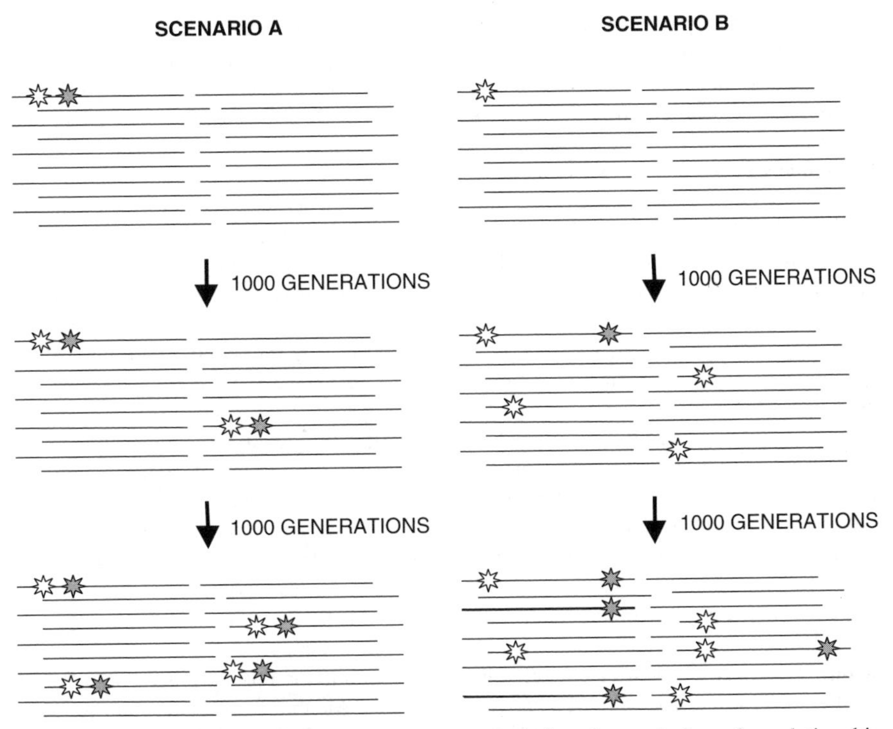

Figure 8.1 Alternative hypothetical scenarios depicting the evolution of a relationship between two SNPs. Identical stretches of DNA within a population are represented by black lines. In scenario A, two adjacent polymorphisms, represented by a white star and a grey star, arise simultaneously and by random drift achieve a population frequency of 0.1 after 1000 generations, increasing to 0.2 after 2000 generations, at which time they are still co-segregating as a tightly linked unit. In scenario B, a lone polymorphism (white star) reaches a frequency of 0.2 after 1000 generations, at which point a new polymorphism (grey star) arises spontaneously, some distance away. Note that although the grey polymorphism only occurs on a background bearing the white polymorphism, the association is less clear-cut than scenario A due to the chromosomes bearing the white polymorphism in the absence of the grey polymorphism. During the subsequent 1000 generations, association between the two polymorphisms is further clouded by recombination between the two SNPs and divergence through random drift. Unfortunately for the genetics investigator, scenario A is idealized and scenario B is more typical.

8.2.3.3 Haplotypes

A haplotype is a string of co-inherited alleles of different markers which are arranged in a successive fashion along a given stretch of DNA, hence each haplotype represents a linear section of DNA rather than the single point corresponding to a single marker. The extent of discernible haplotype length varies widely for different regions of the genome; well-defined haplotypes (characterized by moderate or high LD) are punctuated by regions of extremely low LD, suggesting that the recombination processes, selective pressures and other factors that dictate the degree of LD vary widely in an abrupt fashion across the genome (Goldstein, 2001). Although the length of preserved haplotypes shows dramatic variation from haplotype to haplotype, recent data suggest that the typical length of a discernible haplotypic block is 10–100 kb in the Caucasian population (Daly *et al.*, 2001).

In certain circumstances, statistical analysis of haplotypes is more powerful than single SNP analysis. This is because an SNP usually has only two allelic states, whereas a stretch of DNA can typically be represented by several different haplotypes; the chance that one of the many haplotypes shows strong association with a functional variant (i.e. a variant that influences the phenotype) is higher than the odds of a strong, pure correlation with one of only two possible alleles for a single SNP. In this sense, a series of haplotypes is analogous to a multi-allelic STR marker (although regarded as more stable — see below). Clearly if the functional variant itself is under test, or a polymorphism which shows perfect co-segregation with the functional variant, haplotypic analysis offers no advantage. It should also be noted that haplotype analysis is a double-edged sword and in addition to increasing statistical power has the potential to reduce it by introducing multiple testing and possibly by diluting an association signal due to undetected recombination within the haplotypes.

Haplotypes are usually constructed by comparing the genotypes of closely related individuals at two or more linked markers and identifying groups of alleles which are co-inherited as a set from one generation to the next. However, where no family members are available and the cohort under study consists of a population of unrelated individuals, it is necessary to infer haplotypes and haplotype frequencies using statistical methods. The most common method for the estimation of haplotypes is the expectation-maximization (EM) maximum likelihood estimate (MLE; Excoffier and Slatkin, 1995). The ARLEQUIN software package developed in the Genetics and Biometry Laboratory at the University of Geneva contains an EM algorithm for this purpose. ARLEQUIN can be downloaded from http://lgb.unige.ch/arlequin/. Another popular program for haplotype construction and analysis is EHPLUS (Zhao *et al.*, 2000). EHPLUS can be downloaded from http://www.iop.kcl.ac.uk/IoP/Departments/PsychMed/GEpiBSt/software.stm. Both packages are discussed in detail in Chapter 11. Note that haplotype construction using family inheritance patterns, although more robust than population-based MLE, also typically requires a degree of inference and resulting haplotypes may be probable rather than actual (Hodge *et al.*, 1999). For absolute definition of all haplotypes, it is necessary to physically separate the two copies of each stretch of DNA under analysis, i.e. reduction from a diploid to a haploid state, to allow unmixed analysis of a single haplotype. For very short stretches of DNA (up to approximately 10 kb), this can be achieved by allele-specific PCR (Michalatos-Beloin *et al.*, 1996); for large-scale haplotype construction it is necessary to separate entire chromosomes. This strategy has been successfully employed by the California-based company Perlegen Sciences Inc., who have used a rodent–human somatic cell hybrid technique to physically separate the two copies of human chromosome 21 for haplotype elucidation (Patil *et al.*, 2001; see below). However, most investigators employ the less laborious MLE or family-based inference methods for haplotype construction and accept a certain degree of error or loss of power.

In addition to potentially providing greater power than single markers in subsequent statistical analyses, a knowledge of the haplotypes representing the locus under study is extremely valuable for maximizing efficiency in study design. For example, two markers which always co-segregate (as in Figure 8.1, scenario A) will provide the same information, regardless of which of the two is genotyped, therefore typing both markers is inefficient as the genotype of one can be inferred from the other. Consequently, a detailed knowledge of the haplotypes across the interval theoretically allows a minimum marker set to be identified that will permit the extraction of all haplotypic information (Figure 8.2; see Johnson *et al.*, 2001; Patil *et al.*, 2001). David Clayton of the Medical Research Council Biostatistics Unit, UK has written software (htSNP) to aid the selection of optimum marker sets based on haplotypic information which can be downloaded from http://www-gene.cimr.cam.ac.uk/clayton/software/stata.

Before this optimized marker set can be selected, it is necessary to identify all common SNPs within the area under study and construct haplotypes. The publicly available SNPs catalogued in the SNP database hosted by NCBI (dbSNP: http://www.ncbi.nlm.nih.gov/SNP/) are far too sparse for this purpose at the time of writing (Johnson *et al.*, 2001). Comprehensive identification of all common SNPs in a given interval requires sequencing of a significant number of individuals from the relevant population. For example, sequencing 24 individuals would give a 95% probability of detecting all variants with a minor allele frequency of greater than 5% (Kruglyak and Nickerson, 2001); 5% is a sensible lower cut-off point, as sample size requirements for case–control studies increase dramatically when allele frequencies fall below 5% (Johnson *et al.*, 2001).

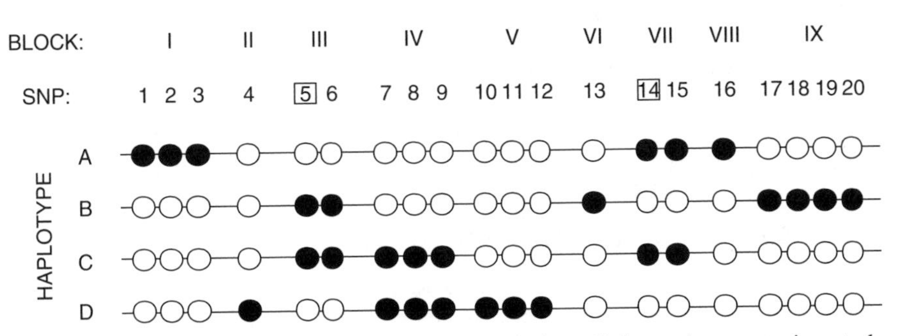

Figure 8.2 Using haplotypic information to maximize efficiency in genotyping study design. Twenty SNPs spanning four haplotypes are shown. Each SNP is represented by a circle; the circle is black or white, depending on the allelic state of the SNP. The SNPs can be grouped into nine blocks—each block contains a group of SNPs with an identical allelic pattern in the four haplotypes. Genotyping all SNPs in any given block is unnecessary, as the genotype of one SNP per block allows the genotypes of the other SNPs in the block to be inferred; for example, genotyping SNP 1 allows the genotypes of SNPs 2 and 3 to be predicted. Moreover, in this simplified example, all four haplotypes can be unambiguously identified by genotyping just two SNPs, 5 and 14 (boxed), yielding a 90% reduction in genotyping compared to a 'blind' strategy (i.e. no knowledge of haplotypic structure).

Obviously it is impractical to sequence a region covering several Mb in 24 individuals. A more realistic approach is the identification of all genes in the interval and sequencing of the coding sequence plus flanking splice sites, together with 1–2 kb of putative promoter (i.e. the region immediately upstream of the transcription start site) and any other known regulatory elements. Although not comprehensive, as unidentified regulatory elements can be intronic or several tens of kilobases away from the genes under their influence (Blackwood and Kadonaga, 1998), this approach offers a good compromise between exhaustive coverage of the locus and practicality. For SNP identification purposes, it may be preferable to use individuals derived from the disease, rather than control, population. This will give a greater chance of detecting rare functional variants (mutations) that are at a higher frequency in the disease population. For example, *NOD2* mutations predisposing to Crohn's disease were recently found to occur at a frequency of 6–12% among cases, but at <5% among controls (Hugot *et al.*, 2001; Ogura *et al.*, 2001).

Having identified the majority of coding and regulatory sequence SNPs with a frequency of greater than 5%, it is necessary to construct haplotypes to allow redundant SNPs to be identified and eliminated from the association study. A subset of 96 individuals from the population under study should be sufficient to detect the majority of haplotypes with a frequency of greater than 5% (B-Rao, 2001; note that studies to date indicate that common haplotypes at any given locus in Caucasian populations are restricted in number and account for the majority of all haplotypes observed; see Daly *et al.*, 2001; Johnson *et al.*, 2001). These haplotypes can then be used as a basis for selecting a minimal SNP set for the full association study. It should be noted, however, that SNPs which suggest a strong possibility of functional consequence (e.g. those that alter residues which are conserved between a number of species, or result in non-conservative amino acid changes; see Chapters 12–14) should not be excluded from analysis and should always be tested individually.

Clearly extensive haplotype information covering the entire human genome for a number of different populations would be an invaluable resource for all research groups undertaking association studies. Perlegen Sciences Inc. has recently released a haplotype map covering the whole of human chromosome 21. Although the number of chromosomes sampled was limited to 48, drawn from a number of different ethnic groups (Patil *et al.*, 2001) this represents a good start in developing a genome-wide haplotype map. Perlegen's haplotype data have been incorporated into the Golden Path Browser (http://genome.ucsc.edu/) and can also be viewed at Perlegen's own website (http://www.perlegen.com/haplotype/).

8.2.3.4 Simple Tandem Repeat Markers (STRs)

STRs (also known as microsatellites — see Chapter 3) were the mainstay of monogenic trait linkage analysis during the 1990s, but are now frequently overlooked following the explosion of interest in SNPs for population-based studies. STRs are out of favour for two main reasons: (i) they are less amenable to cheap, high-throughput genotyping methodology than SNPs and (ii) STRs typically have a much higher mutation rate than SNPs (up to 10^{-3} per meiosis compared with an average of 10^{-9} for SNPs; Ellegren, 2000). It has been suggested that this extreme mutation rate might confound genetic association studies, as a single microsatellite allele may represent an excessive number of haplotypes, having independently arisen on the different haplotypic backgrounds through mutation events (see Moffatt *et al.*, 2000). This may prevent the detection of association between the STR allele and an adjacent polymorphism associated with disease. However, comparison of the entire STR allele frequency distribution profiles for cases and controls may highlight differences which reflect a difference in the frequency of an adjacent disease-associated SNP, due to divergence of the STR profiles associated with SNP allele 1 and SNP allele 2 as a result of frequent STR mutation (Abecasis *et al.*, 2001; Koch *et al.*, 2000). There is also some evidence to suggest that LD can be detected over greater distances with STRs than with SNPs; possibly 10 times as far (Koch *et al.*, 2000), perhaps because in some circumstances STR mutation significantly outstrips recombination at flanking sites. Given the caveat that limited empirical evidence is available at present and the extent of detectable LD is likely to be highly locus and marker specific, inclusion of STRs spaced at intervals of 50–100 kb in a preliminary case–control analysis may assist in identifying regions of potential association within the critical interval that can be prioritized for follow-up with SNPs.

8.2.4 Statistical Analysis

Methods and software for the statistical analysis of both single marker and haplotype data in both a case–control and family-based cohort scenarios are described in detail in Chapter 11. Briefly, a chi-square analysis may be used to test for departure between observed and expected allele frequencies for a biallelic marker in a case–control cohort, while multi-allelic systems (e.g. STRs) may be tested by permutation using software such as CLUMP (Sham and Curtis, 1995a; see Section 8.3.2 below). Family-based samples such as parent–offspring trios and discordant sibs can be analysed using the transmission disequilibrium test (TDT) and associated methods (Spielman *et al.*, 1993; discussed in depth in Chapter 11); although the TDT was originally developed for biallelic markers, an extension of the TDT has been developed for testing multi-allelic markers and haplotypes (Sham and Curtis, 1995b). For case–control studies, haplotypes can be assessed using software such as EHPLUS (see Section 8.2.3 above) which, in addition to haplotype construction, can be used for testing for differences in haplotype frequency between cases and controls (see Chapter 11).

8.3 A PRACTICAL APPROACH TO LOCUS REFINEMENT AND CANDIDATE GENE IDENTIFICATION

Figure 8.3 gives an overview of the practical process of locus refinement, candidate gene selection and testing for phenotype–genotype association using a case–control approach. Each step is described in detail in the following sections.

8.3.1 Sequence Characterization

The most popular web tools for the purpose of human genome sequence characterization are the human genome browser hosted by the National Center for Biotechnology Information (NCBI) at http://www.ncbi.nlm.nih.gov/, the 'Golden Path' genome browser hosted by the University of California, Santa Cruz at http://genome.ucsc.edu/ and the Ensembl human genome browser maintained by the European Bioinformatics Institute (EBI) and the Wellcome Trust Sanger Institute at http://www.ensembl.org/Homo_sapiens/. These browsers are described and reviewed in detail in Chapter 5 and we refer the reader to Chapter 9 for a comprehensive description of methods for defining a locus between two genetic markers at the sequence level using these three tools.

8.3.2 STR Analysis

We suggest that the first step following complete locus characterization should be an attempt to identify regions of potential association within the critical interval using STRs.

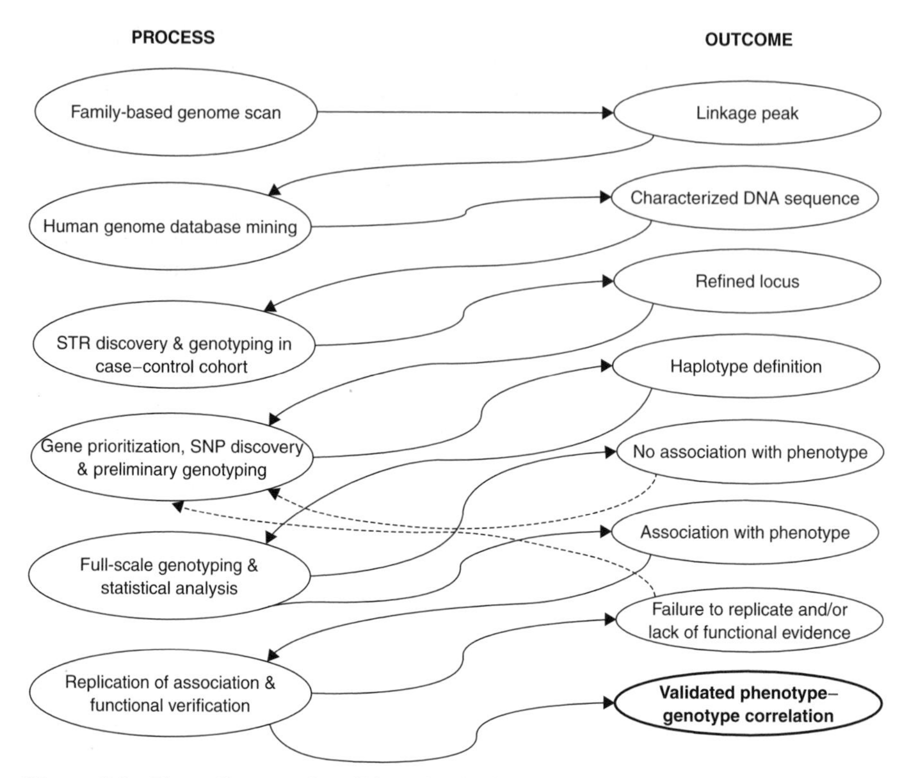

Figure 8.3 Flow diagram describing the logical steps in pinpointing and verifying a gene–phenotype association using a case–control follow-up to a family-based genome scan.

Any regions thus identified can then be prioritized for follow-up with SNPs. It should be noted however, that there is limited empirical data on the use of STRs to detect association in populations and lack of evidence for association with STRs should not deter the investigator from proceeding with a SNP-based association study. STRs can be identified using the tandem repeat finder software at http://c3.biomath.mssm.edu/trf.html (Benson, 1999; see Chapter 9). STR genotyping is typically performed by gel or capillary electrophoresis coupled with a fluorescence detection system (usually an adapted DNA sequencing platform); instrument and software suppliers include Molecular Dynamics, Applied Biosystems and LI-COR Biosciences.

Using a subset of 24 individuals from the control population, test the STRs for polymorphism; aim for a series of polymorphic STRs spaced at 50–100-kb intervals across the critical interval. These markers may then be typed in the entire case and control cohorts and the allele frequency distribution patterns checked for differences between cases and controls in an attempt to pinpoint areas of potential disease association. One of the most popular pieces of software for comparing STR allele frequency distributions is the CLUMP program developed by David Curtis and Pak Sham of the Institute of Psychiatry, London UK (Sham and Curtis, 1995a). CLUMP uses a Monte Carlo simulation to test for departure from expected values for a number of chi-square measures, including the intuitively appealing techniques of considering each allele in turn against the rest and grouping ('clumping') alleles to maximize the chi-squared value. CLUMP is straightforward to use and can be downloaded from http://www.mds.qmw.ac.uk/statgen/dcurtis/software.html.

8.3.3 Gene Selection, SNP Discovery and Haplotype Construction

Genes in the critical interval can be arranged in rank order for analysis, based on biological plausibility with respect to association with the disease under study or other considerations (e.g. pharmaceutical companies may wish to prioritize any tractable drug targets). We suggest sequencing coding and known or putative regulatory regions from each gene in 24 individuals selected randomly from the disease population, as discussed in Section 8.2.3 above, followed by genotyping of all SNPs thus identified in 96 random individuals derived from the control population. ARLEQUIN, EHPLUS (see Section 8.2.3 and Chapter 11) or similar software may then be used to construct haplotypes from this sub-sample, and htSNP (see Section 8.2.3 and Chapter 11) implemented to aid selection of the minimal marker set required for accurate representation of each haplotype. Note that several SNP genotyping platforms are currently available; we will not review them here and the investigator should select the most appropriate system based on cost, robustness and required throughput.

8.3.4 Genotyping and Statistical Analysis

Having selected the optimal SNP set, the whole cohort may now be genotyped. Following genotyping, an EM algorithm can be used again for haplotype construction and haplotype frequency determination (see Chapter 11). It may be beneficial to divide the cohort randomly into two case–control groups for statistical analysis, to allow the possibility of replication of any positive association using the second subset. Haplotype frequency distributions in cases and controls can be compared using CLUMP, as for STRs (see Section 8.3.3) or more specific software tools such as EHPLUS (see Section 8.2.3; discussed in detail in Chapter 11). Individual SNPs can be tested using a chi-square test (see Chapter 11). A test for Hardy–Weinberg equilibrium (HWE) is a useful prior check for ensuring that there is no (or little) population stratification and that each marker is giving the expected genotype distribution for the observed allele frequencies. Expected genotype frequencies are calculated from allele frequencies under the assumption $p^2 + q^2 + 2pq = 1$, where p and q are the allele frequencies and p^2, q^2 and $2pq$ correspond to the frequencies of the three possible genotypic states. The actual genotype frequencies are then tested for departure from the expected frequencies using a chi-square test. The calculation is simple and can be performed by hand or in a Microsoft Excel macro for biallelic markers. Alternatively the ARLEQUIN software suite includes a program for checking HWE for both biallelic and multi-allelic marker systems.

An acceptable p-value threshold for declaring association between a marker and disease is the subject of considerable debate. Clearly a nominal cut-off of $p = 0.05$ is inappropriate where multiple tests have been performed, as this value (or lower) may occur several times by chance. However, standard methods of correction for multiple testing, for example by Bonferroni correction, may be overly stringent (Cardon and Bell, 2001). The authors suggest that the investigator avoids setting thresholds that are excessively rigorous and instead follows up promising leads, adding to the weight of evidence for involvement (or lack of involvement) in the disease process by additional means (see below).

8.3.5 The Burden of Proof – is an Associated Gene Really Involved in the Disease Process?

Unfortunately detection of association between a gene and disease phenotype does not constitute definitive proof that the gene under test is involved in the disease process. Rather, it provides a single piece of evidence to suggest *possible* involvement in the disease process that requires further substantiation. Replication of the association in a second cohort considerably strengthens the argument for involvement; for example the association between the insulin gene and type 1 diabetes has been reproduced a number of times (Bennett and Todd, 1996). However, even in the event of independent replication of results, one should consider the possibility that the replication is due to chance or that the apparent disease association is due to an adjacent gene in LD with the marker under test. If the polymorphism is in protein coding sequence and causes an amino acid change, it may be possible to assess the possible impact on protein function by the nature of the change (conservative or non-conservative), the context in which it occurs (potential disruption of secondary or tertiary protein structure) and the degree of cross-species conservation and conservation within protein families. Conservation may also be used to gauge the potential impact of polymorphisms in putative regulatory elements. These areas are covered in detail in Chapters 12 and 13. However, it should also be remembered that polymorphisms that appear to be innocuous on cursory examination can have functional consequences, for example synonymous coding changes that occur in exonic splicing enhancer (ESE) regions (Liu *et al.*, 2001).

Ultimately it is likely that the investigator will wish to instigate additional laboratory-based experiments to judge the functional effect of the variant in question. These may include gene expression and cell-based reporter assays for putative promoter polymorphisms, functional enzyme or signal transduction assays for amino acid changes and *in vivo* analysis in the mouse using gene knock-out or polymorphism knock-in technology for studies in the context of the whole organism, to name but a small fraction of the available techniques.

8.4 CONCLUSION

In this chapter we have given a basic overview of the process of moving from a large genetic locus to the identification and screening of candidate genes for disease association. More detailed information on all aspects of study design and data analysis can be gleaned from the references cited in the text and further review of the literature and we strongly advise readers to broaden their knowledge beyond the limits of this chapter. Although we have highlighted a number of popular tools and techniques, several other equally valid approaches exist and we encourage investigators to actively seek out and develop further methods for comparison with those presented here. Continual development of new approaches and improvement of existing methodology is a dominant feature of this rapidly moving field; consequently there is a constant need for investigators to keep abreast of new developments to maximize the chances of success.

ACKNOWLEDGEMENTS

Many thanks to Aruna Bansal for a critical reading of the manuscript which resulted in significant improvements to this chapter.

REFERENCES

Abecasis GR, Noguchi E, Heinzmann A, Traherne JA, Bhattacharyya S, Leaves NI, *et al.* (2001). Extent and distribution of linkage disequilibrium in three genomic regions. *Am J Hum Genet* **68**: 191–197.

Bennett ST, Todd JA. (1996). Human type 1 diabetes and the insulin gene: principles of mapping polygenes. *Ann Rev Genet* **30**: 343–370.

Benson G. (1999). Tandem repeats finder: a program to analyze DNA sequences. *Nucleic Acids Res* **27**: 573–580.

Blackwood EM, Kadonaga JT. (1998). Going the distance: a current view of enhancer action. *Science* **281**: 61–63.

B-Rao C. (2001). Sample size considerations in genetic polymorphism studies. *Hum Hered* **52**: 191–200.

Cardon LR, Bell JI. (2001). Association study designs for complex diseases. *Nature Rev Genet* **2**: 91–98.

Daly MJ, Rioux JD, Schaffner SF, Hudson TJ, Lander ES. (2001). High-resolution haplotype structure in the human genome. *Nature Genet* **29**: 229–232.

Das M, Burge CB, Park E, Colinas J, Pelletier J. (2001). Assessment of the total number of human transcription units. *Genomics* **77**: 71–78.

Devlin B, Risch N. (1995). A comparison of linkage disequilibrium measures for fine-scale mapping. *Genomics* **29**: 311–322.

Devlin B, Roeder K. (1999). Genomic control for association studies. *Biometrics* **55**: 997–1004.

Devlin B, Roeder K, Bacanu SA. (2001). Unbiased methods for population-based association studies. *Genet Epidemiol* **21**: 273–284.

Ellegren H. (2000). Microsatellite mutations in the germline: implications for evolutionary inference. *Trends Genet* **16**: 551–558.

Excoffier L, Slatkin M. (1995). Maximum-likelihood estimation of molecular haplotype frequencies in a diploid population. *Mol Biol Evol* **12**: 921–927.

Goldstein DB. (2001). Islands of linkage disequilibrium. *Nature Genet* **29**: 109–211.

Gretarsdottir S, Sveinbjornsdottir S, Jonsson HH, Jakobsson F, Einarsdottir E, Agnarsson U, *et al.* (2002). Localization of a susceptibility gene for common forms of stroke to 5q12. *Am J Hum Genet* **70**: 593–603.

Hodge SE, Boehnke M, Spence MA. (1999). Loss of information due to ambiguous haplotyping of SNPs. *Nature Genet* **21**: 360–361.

Hugot JP, Chamaillard M, Zouali H, Lesage S, Cezard JP, Belaiche J, *et al.* (2001). Association of NOD2 leucine-rich repeat variants with susceptibility to Crohn's disease. *Nature* **411**: 599–603.

Johnson GC, Esposito L, Barratt BJ, Smith AN, Heward J, Di Genova G, *et al.* (2001). Haplotype tagging for the identification of common disease genes. *Nature Genet* **29**: 233–237.

Kruglyak L, Nickerson DA. (2001). Variation is the spice of life. *Nature Genet* **27**: 234–236.

Koch HG, McClay J, Loh EW, Higuchi S, Zhao JH, Sham P, *et al.* (2000). Allele association studies with SSR and SNP markers at known physical distances within a 1-Mb region embracing the ALDH2 locus in the Japanese, demonstrates linkage disequilibrium extending up to 400 kb. *Hum Mol Genet* **9**: 2993–2999.

Liu HX, Cartegni L, Zhang MQ, Krainer AR. (2001). A mechanism for exon skipping caused by nonsense or missense mutations in BRCA1 and other genes. *Nature Genet* **27**: 55–58.

McGinnis R. (2000). General equations for Pt, Ps, and the power of the TDT and the affected-sib-pair test. *Am J Hum Genet* **67**: 1340–1347.

Michalatos-Beloin S, Tishkoff SA, Bentley KL, Kidd KK, Ruano G. (1996). Molecular haplotyping of genetic markers 10 kb apart by allele-specific long-range PCR. *Nucleic Acids Res* **24**: 4841–4843.

Moffatt MF, Traherne JA, Abecasis GR, Cookson WO. (2000). Single nucleotide polymorphism and linkage disequilibrium within the TCR alpha/delta locus. *Hum Mol Genet* **9**: 1011–1019.

Ogura Y, Bonen DK, Inohara N, Nicolae DL, Chen FF, Ramos R, *et al.* (2001). A frameshift mutation in NOD2 associated with susceptibility to Crohn's disease. *Nature* **411**: 603–604.

Patil N, Berno AJ, Hinds DA, Barrett WA, Doshi JM, Hacker CR, *et al.* (2001). Blocks of limited haplotype diversity revealed by high-resolution scanning of human chromosome 21. *Science* **294**: 1719–1723.

Pritchard JK, Donnelly P. (2001). Case–control studies of association in structured or admixed populations. *Theor Popul Biol* **60**: 227–237.

Pritchard JK, Rosenberg NA. (1999). Use of unlinked genetic markers to detect population stratification in association studies. *Am J Hum Genet* **65**: 220–228.

Pritchard JK, Stephens M, Rosenberg NA, Donnelly P. (2000). Association mapping in structured populations. *Am J Hum Genet* **67**: 170–181.

Risch NJ. (2000). Searching for genetic determinants in the new millennium. *Nature* **405**: 847–856.

Sham PC, Curtis D. (1995a). Monte Carlo tests for associations between disease and alleles at highly polymorphic loci. *Ann Hum Genet* **59**: 97–105.

Sham PC, Curtis D. (1995b). An extended transmission/disequilibrium test (TDT) for multi-allele marker loci. *Ann Hum Genet* **59**: 323–336.

Spielman RS, McGinnis RE, Ewens WJ. (1993). Transmission test for linkage disequilibrium: the insulin gene region and insulin-dependent diabetes mellitus (IDDM). *Am J Hum Genet* **52**: 506–516.

Strittmatter WJ, Saunders AM, Schmechel D, Pericak-Vance M, Enghild J, Salvesen GS, *et al.* (1993). Apolipoprotein E: high-avidity binding to beta-amyloid and increased frequency of type 4 allele in late-onset familial Alzheimer disease. *Proc Natl Acad Sci USA* **90**: 1977–1981.

Zhao JH, Curtis D, Sham PC. (2000). Model-free analysis and permutation tests for allelic associations. *Hum Hered* **50**: 133–139.

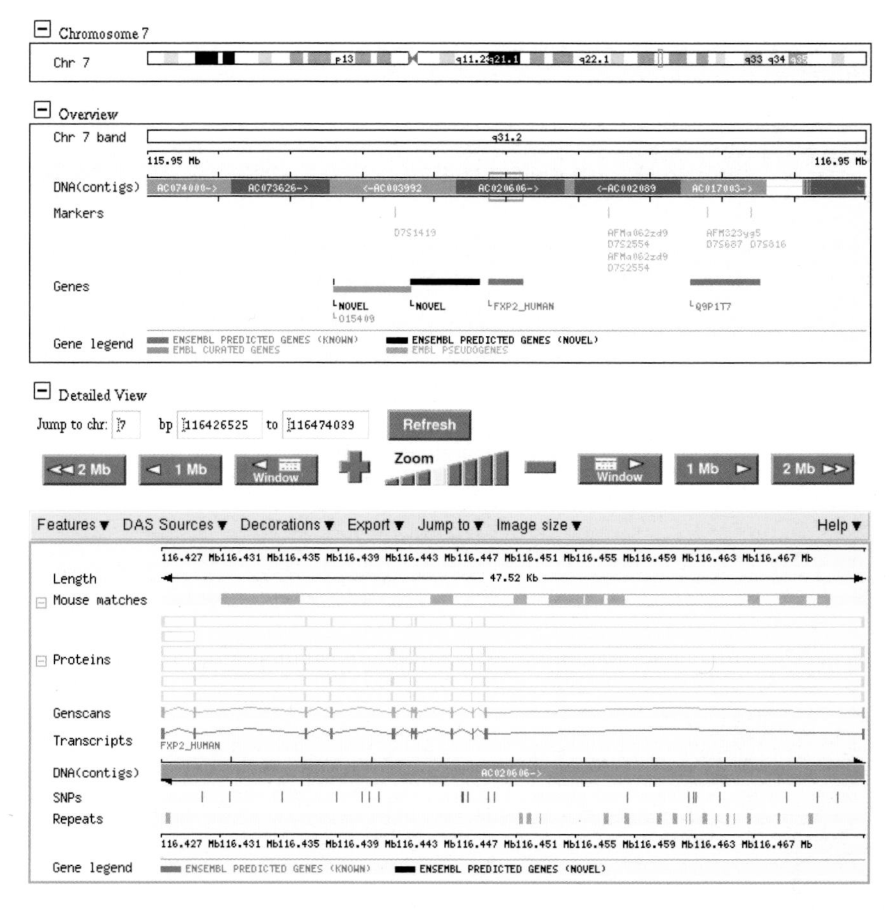

Figure 5.1 The genomic region around the FOXP2 gene according to Ensembl.

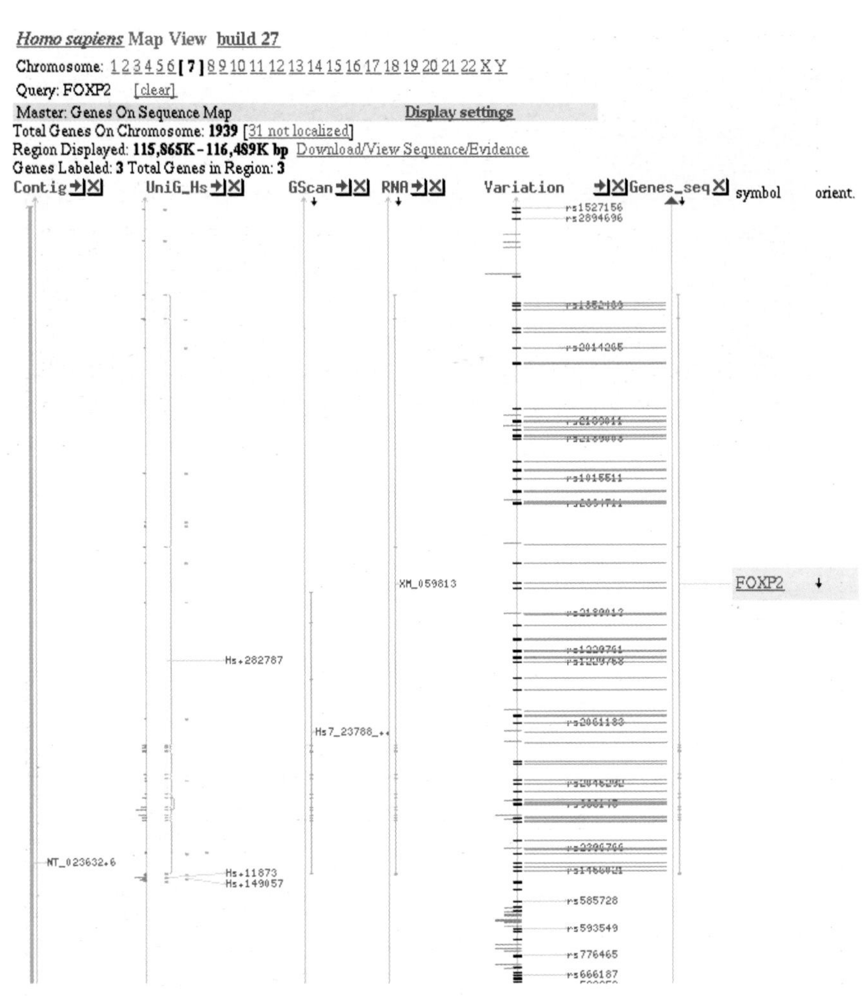

Figure 5.3 The genomic region around the FOXP2 gene according to the NCBI Map Viewer.

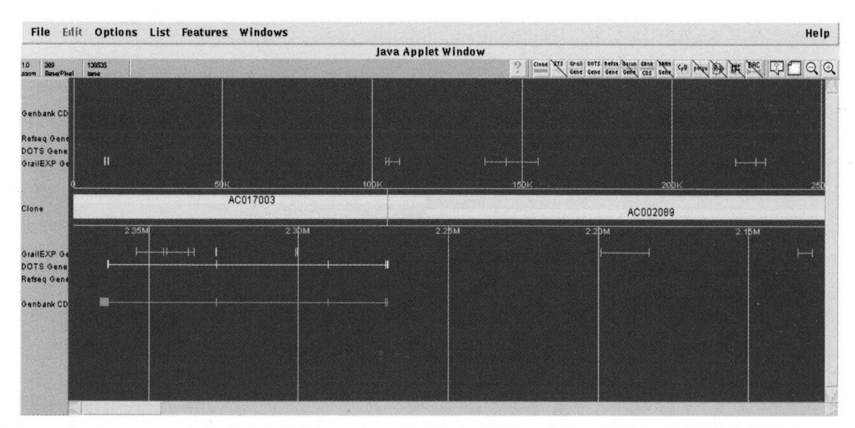

Figure 5.4 The genomic region around the FOXP2 gene according to ORNL Genome Channel.

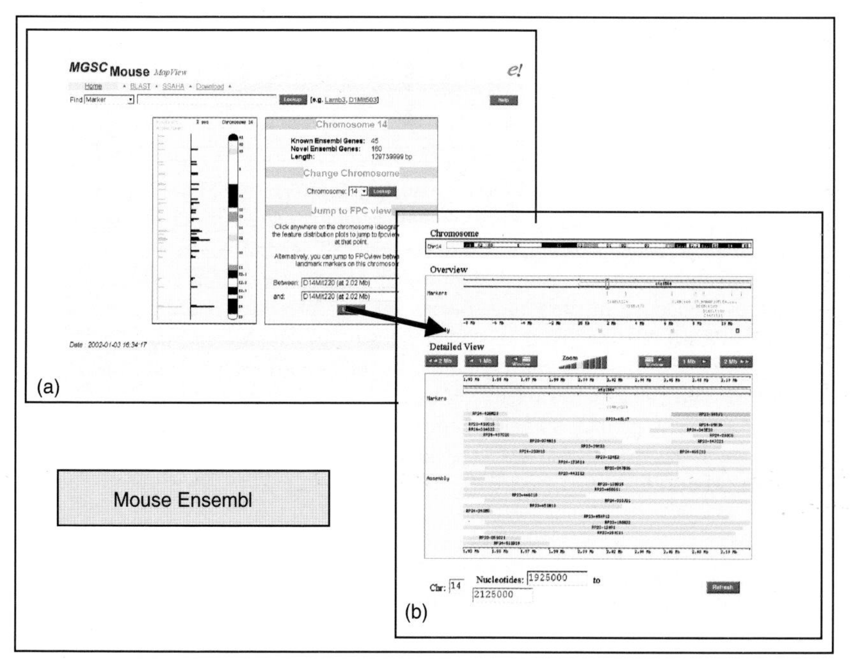

Figure 6.1 Mouse Ensembl. A graphical representation of the clone-based physical map for the proximal end of mouse chromosome 14 from Ensembl. This browser allows users to search for regions of a chromosome between two STS markers and to view the current clone coverage in the selected area. Because the browser is Web-based, users do not have to download and install special software to view the BAC map.

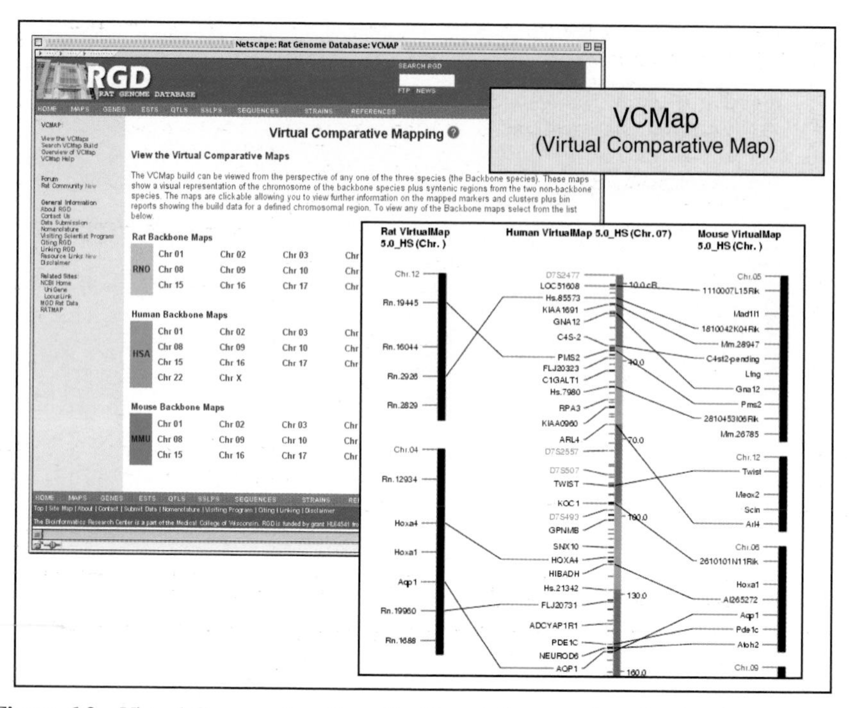

Figure 6.3 Virtual Comparative Map. The Virtual Comparative Map is generated using sequence-based algorithms that predict syntenic regions inferred from homology among mapped sequences. Sequence comparisons between ESTs and cDNAs from human, mouse and rat are combined with Radiation Hybrid map locations to define regions of synteny. Locations for unmapped markers in a species are then predicted based on the map location of the orthologous marker in a syntenic region of another species. The forepanel shows a virtual comparative map using human as the backbone map (centre) and syntenic regions of rat (left) and mouse (right). Mapped genes, UniGenes and STSs are shown, with lines connecting predicted homologues among the species. Data sources for the virtual maps are RGD, NCBI and MGD. The virtual comparative maps are available at http://rgd.mcw.edu/VCMAP/.

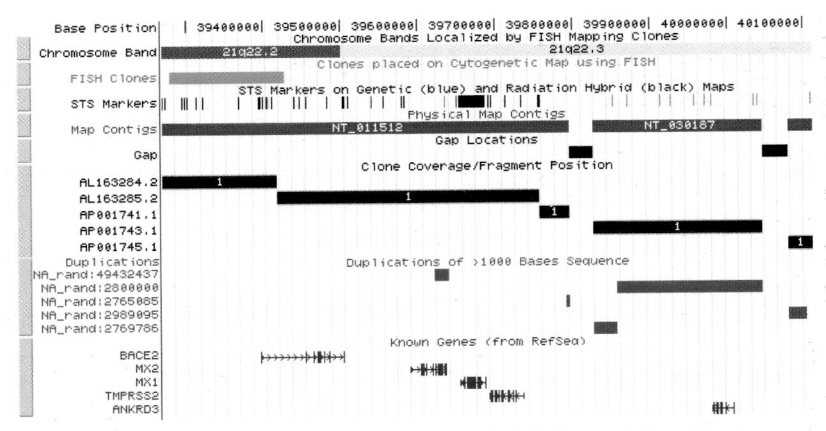

Figure 9.3 Definition of the D21S1245 and D21S1852 interval in genomic sequence using the UCSC human genome browser. The view immediately identifies five known genes, two gaps and several duplications across the region.

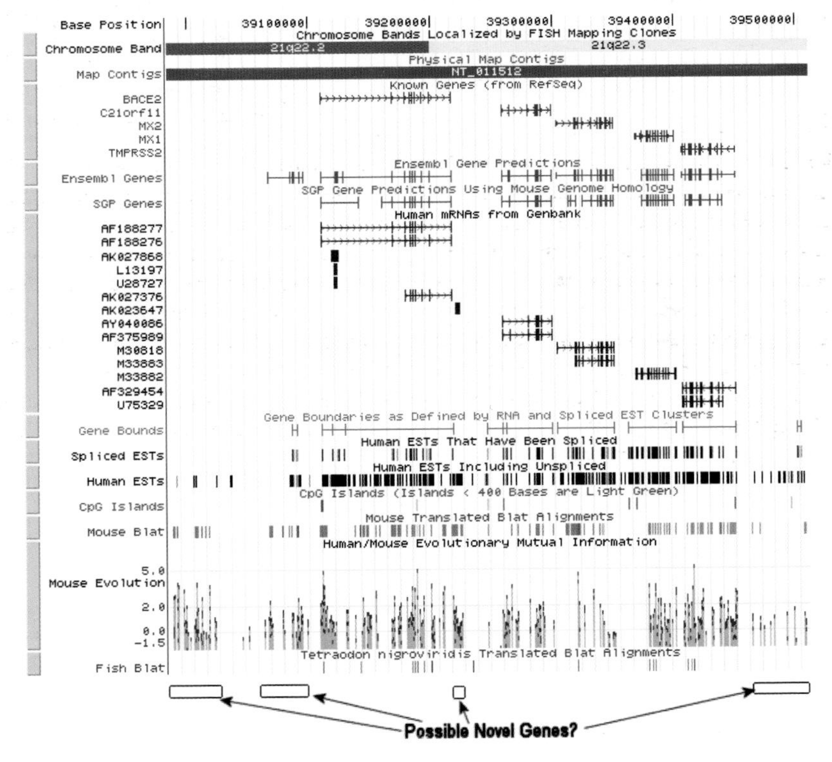

Figure 9.5 Using the UCSC human genome browser to identify known and novel genes. A range of evidence including mRNAs, ESTs and human–mouse homology supports the existence of five known genes and up to four novel genes.

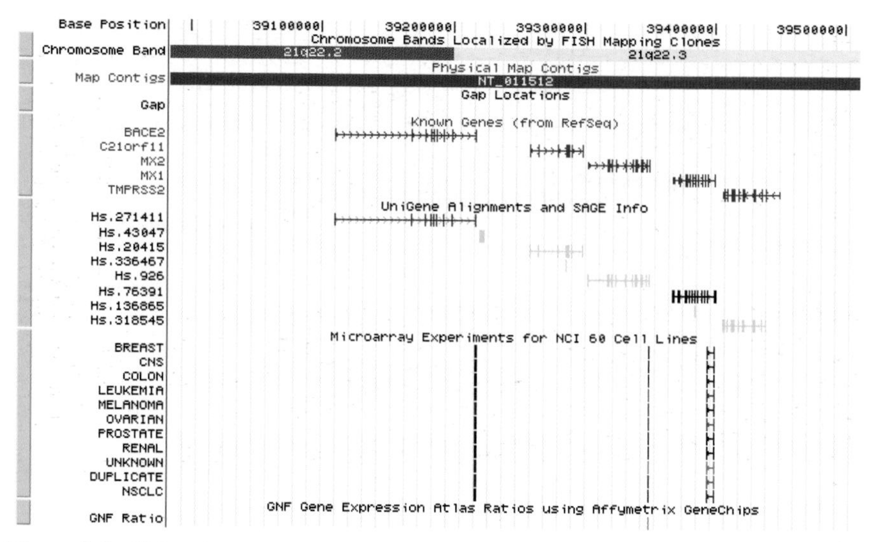

Figure 9.6 Using the UCSC human genome browser to evaluate gene expression across a locus. Four tracks provide information on gene expression. ESTs implicitly measure gene expression as each is derived from a specified tissue source. Unigene clusters link to SAGE expression profiles drawn from the SageMap project. Finally data from two genome-wide microarray projects are presented, the NCI60 cell line project and GNF gene expression atlas ratios.

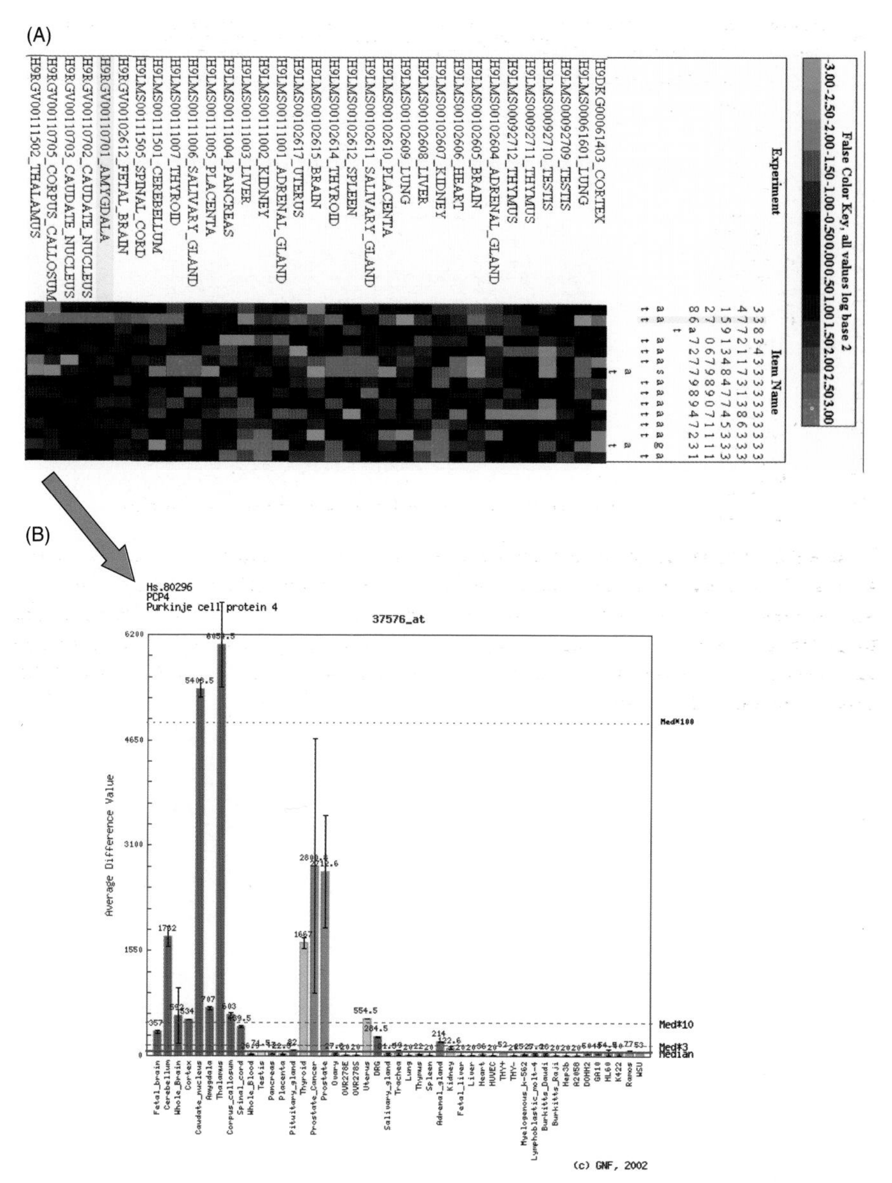

Figure 9.7 GNF gene expression atlas ratios displayed by the UCSC human genome browser. (A) The browser shows a view of the expression profiles of 15 genes across the wider locus. One gene, Purkinje cell protein 4 (PCP4, no. 37576), shows high expression in a wide range of neuronal tissues, including the thalamus. (B) Detailed gene expression profile for the PCP4 gene.

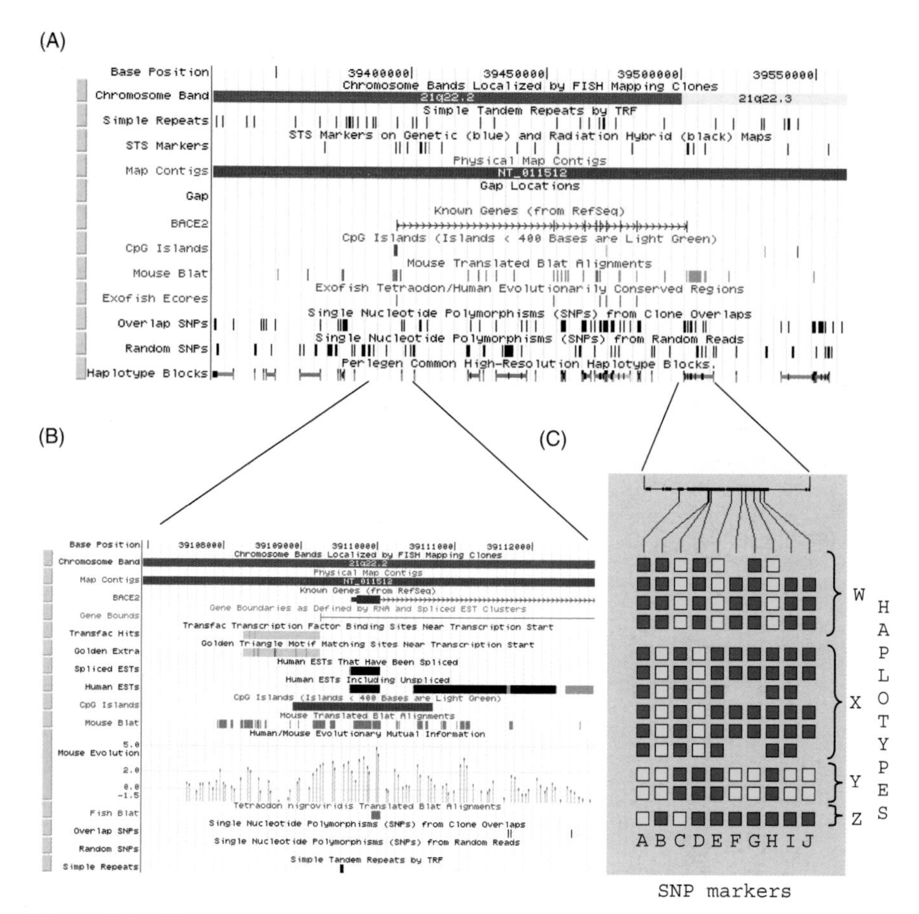

Figure 9.8 SNP visualization in the UCSC human genome browser. (A) A detailed view of the BACE2 gene allows the user view a range of information, including SNP haplotype data. (B) Close viewing of the BACE2 locus allows the user to assess the functional context and genomic conservation of the region surrounding each SNP. (C) A detailed view of a Perlegen haplotype.

CHAPTER 9

Genetic Studies from Genomic Sequence

MICHAEL R. BARNES

GlaxoSmithKline Pharmaceuticals
Harlow, Essex, UK

Bioinformatics for Geneticists. Edited by M.R. Barnes and I.C. Gray
 ISBNs: 0 470 84393 4; 0 470 84394 2 (PB)

9.1 INTRODUCTION

Without risk of hyperbole, the process of definition of a locus or gene in the human genome sequence is probably the single most valuable bioinformatics process that a geneticist can carry out. This immediately places a gene or locus in a wider context. Now we can quickly find out what known genes are in the locus, what evidence exists for novel genes and what markers are available across the locus to study these genes. Digging a little deeper into the data will soon tell us where the genes are expressed and what their biological role is likely to be. We will even gain an insight into some of the common variation that exists across the locus, and also what rare mutations exist in some of the genes. Finally study of the sequence of the locus itself can tell us something about the physical nature of the region. We may even be able to draw some conclusions about the likely genetic nature of the region in terms of recombination and Linkage Disequilibrium (LD). All this is possible before setting foot in the laboratory. But we also need to be aware that there are limitations to this approach and sometimes stepping into the laboratory is the only way to resolve these limitations.

The availability of the golden path is a great advance for genetics—it is too easy to forget the pre-genome era, imprecise genetic localizations are now superseded by absolute genome locations to the nearest base pair. But the golden path must be used carefully; with proper quality checks the data can serve as an invaluable template for genetics, without these checks it can create as many problems as it solves. These caveats should not be ignored. Firstly the golden path is a *draft* dataset composed of hundreds of thousands of fragments of various sizes with many gaps. The order and orientation of the fragments is often not known from the sequencing process itself. In some cases the same part of the genome will be duplicated in more than one fragment. To address the technical challenges of whole genome assembly, the golden path is released as defined 'builds' on a quarterly basis (Lander *et al.*, 2001; reviewed in Chapter 5). This implicitly involves some lag in availability of the most current sequence data. At the time of writing this chapter (March 2002), the December 2001 release of the golden path was in use. It is important to be aware that golden path coordinates can only be compared if both tools are using the same build version of the golden path. Finally, if complete sequence across a locus is critical to a study, additional draft sequence may be available in addition to the material in the golden path; this can be identified by searching the Human Genome BLAST database at the NCBI.

In this chapter we will take a hands-on approach to the application of genomic sequence data to genetics. We will look at the key bioinformatic steps needed to take a genetic study from an initial LOD peak to laboratory genotyping. Figure 9.1 illustrates each step of this process, with genomic sequence as a common thread through every stage.

9.2 DEFINING THE LOCUS

A geneticist may come to be interested in a gene or locus by many routes. The locus could be identified as part of a published genome scan or as part of the scientist's own work; it could be syntenic with a mammalian disease model; it could contain a candidate gene with biological rationale. The possibilities are limitless, but how ever the locus is identified the next steps to define the region in the human genome are similar. Firstly the locus needs to be defined as accurately as possible. In some cases, this may not be easy due to insufficient or unclear data. Taking a complex disease linkage peak as an example, the linkage across the locus may be defined by a broad flat peak, or multiple

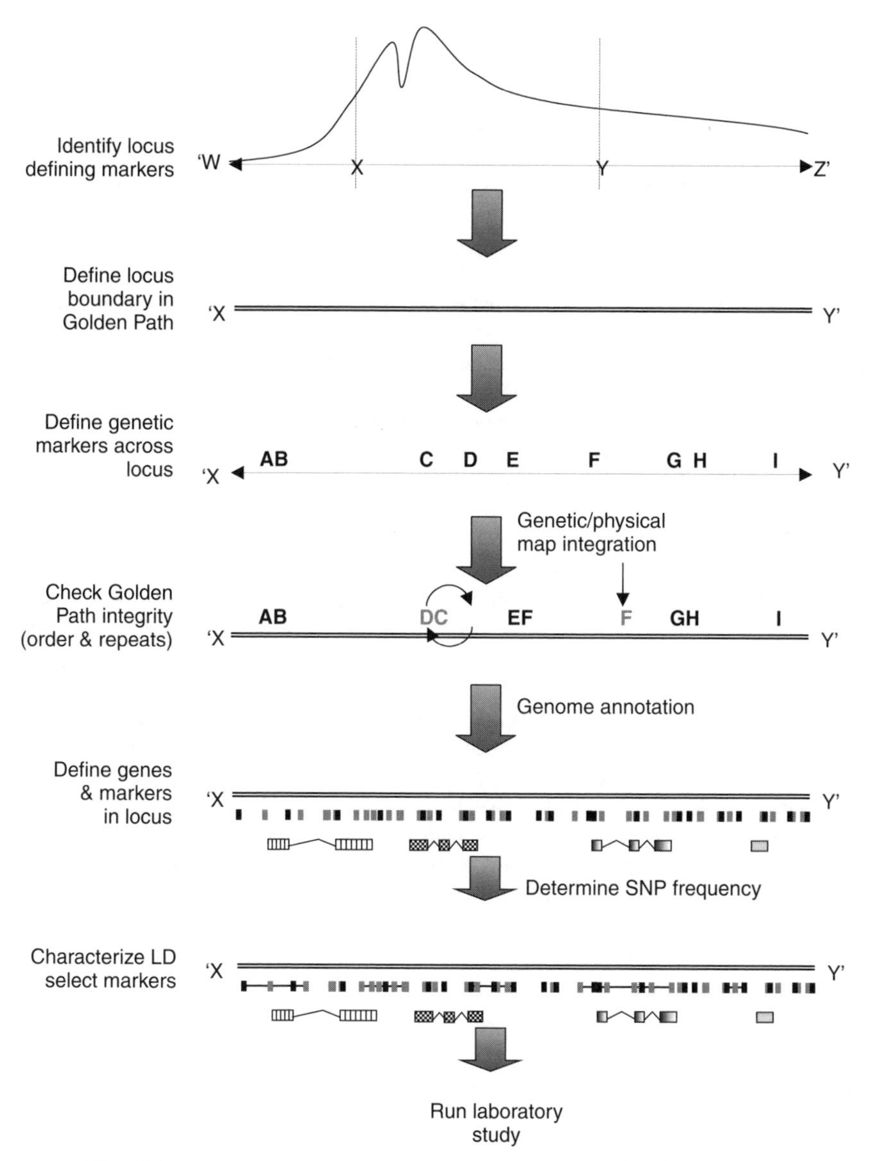

Figure 9.1 Using the Golden Path as a template for genetics. Key bioinformatic steps to take a genetic study from an initial LOD peak to laboratory genotyping are illustrated. The reader should note the role of genomic sequence as a common thread through every stage.

peaks with no well-defined apex. In such cases, it is difficult to define a critical region; unlike monogenic diseases it is not possible to define a region by a clear recombination event between affected and unaffected family members. Instead analysis of complex traits generates an imprecise probabilistic signal based on the increased observance of an allele in affected versus unaffected individuals. Faced with such uncertainties, the best approach is to define a core region within a maximal region, based on LOD score thresholds; this gives some margin for error. Figure 9.2 shows an example of a theoretical complex disease linkage peak. In this case we define an acceptable 'core region' as any region with a LOD score of >3, with a 'maximum region' defined by markers with a LOD >2 or perhaps >1 (respectively a 10- and 100-fold drop in linkage probability). Definition of these regions is necessarily approximate. Where markers do not exactly define a locus it may be necessary to map markers on either side of the locus boundary. So for example in Figure 9.2 the core region would be defined as the region between markers E–K, while the maximum region would be encompassed between markers D and M. If linkage peaks are very flat, approximation to the nearest marker below the threshold might mean including a very large region, in such cases it may be worthwhile extrapolating between markers to identify the most probable region with an estimated LOD above the threshold.

Once the markers delineating the boundaries of the locus have been identified, they need to be mapped onto the human genome to view the full genomic context of the locus. In Chapter 5 Colin Semple reviewed the three primary tools which offer the user an opportunity to localize markers to the draft human genome assembly (the golden path). These tools are Ensembl at the EBI, the UCSC Human Genome Browser (HGB), and NCBI Map View. It is very easy to develop a preference for one or other of these tools, but each tool has its own distinct merits, so for complete characterization of a locus we recommend using all three. In practice this is easy as all three tools use the same coordinates from the same draft version of the golden path allowing reciprocal linking between each tool. Direct comparison between tools allows a second and third opinion across an identical region, which is always a good thing.

In this chapter we will describe some specific case studies which lead the reader through some of the common bioinformatics processes which can help the genetic characterization of genomic loci. In each case we will describe the use of the UCSC HGB, Ensembl or Map View to achieve a specific objective, but the reader should be aware that very similar approaches will produce similar results with each of the other tools unless otherwise mentioned. We will try to highlight the strong points of each tool, where possible with case studies, for an overview of the pros and cons of all three tools see Table 9.1.

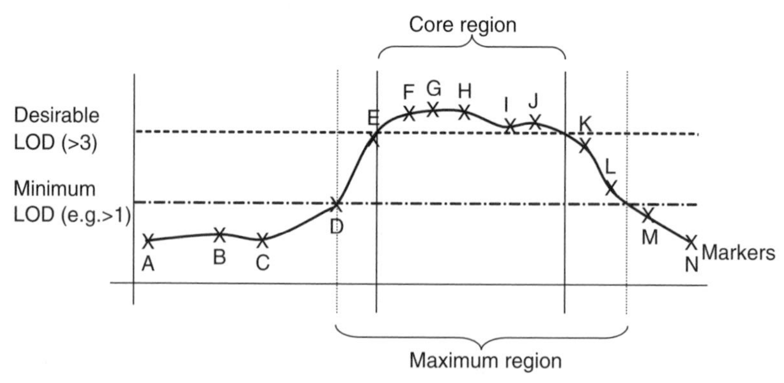

Figure 9.2 Definition of a linkage region by LOD score. In this example of a theoretical complex disease linkage peak, we define an acceptable 'core region' as any region with a LOD score of >3, with a 'maximum region' defined by markers with a LOD >2 or perhaps >1 (respectively a 10- and 100-fold drop in linkage probability).

TABLE 9.1 Pros and Cons of the Three Major Genome Viewers for Genetics

Map View	Ensembl	UCSC HGB
www.ncbi.nlm.nih.gov/cgi-bin/Entrez/map_search	www.Ensembl.org/	genome.ucsc.edu/
Pros	Pros	Pros
Good genetic/RH map integration	Innovative sequence annotation	Innovative sequence annotation
Genetic/physical map focused	Sequence data focused	Sequence data focused
Fully integrated with NCBI tools	Novel gene prediction focused	Good sequence export
Comprehensive genetic markers	Clean innovative interface	Excellent for contig QC
Exclusive data:	Excellent data export	Archives previous golden path annotation/ converts coordinates
• GB4, G3 and TNG RH maps	Distributed annotation (DAS)	Fast BLAT search tool
• Genethon/Marshfield maps	Good integration with mouse and other genomes	Exclusive data:
• Mitelman morbid map	Free source code available	• Chr. 21 haplotype data
• YAC contigs	Exclusive data:	• Fish genome comparison
• FISH clones	• Ensembl gene predictions	• SAGE expression data
	• Detailed gene report	• NCI expression
	• Eponine promoter prediction	• GNF expression data
	• Drosophila genome	• Identifies bridged gaps
	• Zebrafish genome	• Genome duplications
	• Mosquito genome	• Novel tandem repeats
	• Many DAS tracks	
Cons	Cons	Cons
Poor data export	No genetic/RH map integration	No genetic/RH map integration
Limited sequence annotation	No golden path version archive	Data tracks appear and disappear frequently
Linked from Ensembl and HGB but no link back	Accession numbers can be unstable between versions	Newer golden path has less data tracks
Complex and sometimes confounding interface	Not possible to view detailed region >1 Mb	

Finally, the reader should note, that these browsers are subject to frequent change as datasets are updated and software tools improved. It is possible that the steps described here and the data returned may be superseded within months of being written! Nonetheless, the reader should not despair; we hope that the following examples may still be used as a rough overall guide for accessing the required information. Indeed, there are typically many routes to the same data and once familiar with the browser, the user will generally have no problem in retrieving the required datasets.

9.3 CASE STUDY 1: IDENTIFICATION AND EXTRACTION OF A GENOMIC SEQUENCE BETWEEN TWO MARKERS (RECOMMENDED TOOL: UCSC HUMAN GENOME BROWSER (HGB))

In Chapter 2 we reviewed methods for searching the literature for key elements of biological information. A literature search is an important preliminary stage for any study, to define the current state of knowledge in a specific research area. For the purposes of the following case studies imagine we are intending to follow up linkage results to investigate the role of a specific region in bipolar depression. Several linkage studies report a locus for bipolar depression in a region that we define between the genetic marker D21S1245 and the Radiation Hybrid marker D21S1852. To evaluate this region further—to get an idea of its physical and genetic size and number of genes—our first objective should be to locate the markers in the human genome assembly.

Marker localization to the human genome can be achieved either by searching by marker name or by searching directly with the sequence of the marker. The former approach can be problematic as no single tool contains a fully comprehensive index of genetic markers and their aliases. Map View probably contains the most comprehensive list of marker aliases but will not unambiguously localize a marker in genomic sequence. The UCSC HGB is a much more user-friendly tool for this purpose. From the home page select the most current 'browser' from the top left hand menu (for this exercise we used the December 2001 freeze). Type the marker names in the 'position' window, separated by a semicolon (e.g. D21S1245; D21S1852) and submit the request. This returns a 1.04-Mb sequence interval covering the genetic marker D21S1245 and the RH marker D21S1852. Note that in a marker name-based query HGB always returns a larger interval with 100 kb flanking either side of the markers. So in this case the markers D21S1245–D21S1852 actually encompass a 0.84-Mb interval.

The alternative strategy to searching by marker name is to search by marker sequence. This can be a useful technique to use when a marker alias is not found by the genome viewer; in such cases it may be necessary to consult other marker databases, such as GDB or dbSTS to retrieve a marker sequence (see Chapter 3). If all else fails and a marker sequence cannot be found it may be necessary to consult genetic and physical maps to find a neighbouring marker (see Chapter 8). Once a few hundred base pairs of sequence spanning each marker has been found, the sequence between the two marker locations can be identified by using the BLAT sequence search tool at HGB. Select 'BLAT' and enter the DNA sequence spanning the marker. Submit the search and make a note of the genomic position (take the 'start' position for the 5′ marker). Repeat for the second marker (take the 'end' position for the 3′ marker). Return to the genome browser and enter the range spanned by the two markers; this will return the exact genomic interval between the two markers. So for example, a BLAT search with D21S1245 and D21S1852 will return a 0.84-Mb locus without the flanking sequence retrieved by the marker name query. Now that this locus is defined it can be saved for future reference by simply adding a bookmark for the browser page.

9.3.1 Extracting the Genomic Sequence Across the Locus

Either of these approaches will define a genetic locus in the draft human genome sequence; once this has been achieved the sequence can be extracted to provide the ultimate physical map of the region (at 1-bp resolution!). To achieve this select the 'DNA' link in the top tool bar of the HGB, this presents the user with a number of basic options to format the DNA sequence across the selected region. If at this point you are only interested in the DNA sequence, select 'all lower case' and press the submit button. Alternatively, you can select 'lower case repeats' to highlight repeats in the sequence or you can mask them for primer design and other applications. There is also an option to reverse complement the sequence; this is particularly useful if you would like to retrieve a sequence across a gene that is in the reverse orientation in the golden path. If you would like to receive full annotation of the sequence in terms of all the features reported by the HGB then select 'extended case/colour options' and press the submit button. This will take the user to a highly sophisticated annotation interface which allows annotation of almost every available feature on the sequence, with a combination of toggled case, underlining, bold, italics and full colour lettering. This feature can be remarkably useful for preparing figures for publication etc., but bear in mind that the time to retrieve the sequence increases considerably with each added feature. In most cases toggled case annotation of repeats and exons is sufficient, also note that most sequence analysis tools will only maintain upper and lower case annotation, all other annotations (e.g. colour, underlining etc.) will be lost, unless viewed in a rich-text viewer, such as Microsoft Word.

If the sequence across the region is completely finished with no gaps (check the status of the clones in the assembly across the region) then this sequence can be used immediately for further genetic and genomic characterization (see Case study 3), however further QC is needed if the region is still in a draft form or contains gaps, as in the case of our locus between D21S1245 and D21S1852.

9.4 CASE STUDY 2: CHECKING THE INTEGRITY OF A GENOMIC SEQUENCE BETWEEN TWO MARKERS (RECOMMENDED TOOLS: UCSC HGB, NCBI MAP VIEW, NCBI EPCR)

In Figure 9.3 we show the UCSC HGB view of the locus identified between D21S1245 and D21S1852. The HGB interface can be configured to show and hide different datasets, for simplicity we only show tracks with an immediate application to the QC of the genomic sequence.

Now that the genetic locus has been identified in genomic sequence, the next key objective is to check the quality and orientation of the contig across the region. The view of this region immediately identifies two gaps, dividing the region into three contigs. The HGB differentiates between bridged and non-bridged gaps in contigs. Gaps not bridged by any other known physical clone or mRNA are indicated by a black box. If the relative order and orientation of the contigs on either side of the gap is known, then the gap is 'bridged' and indicated by a horizontal white line scored through the black box. This is a valuable feature; contigs between two non-bridged gaps should be evaluated to try to confirm the contig order and orientation. In Figure 9.3 neither gap is bridged, this makes it possible that contig NT_030187 between the two gaps could be incorrectly orientated, or even in the wrong location. This is important information to determine, as this contig constitutes one-third of the entire locus and also contains a complete gene.

9.4.1 Detecting Duplications in Genomic Assemblies

Localized duplications are a common error during genome assembly. Detection of these errors in human genome sequence is complicated by the high number of genuinely duplicated regions, which are estimated at around 3% of the total genome (Bailey *et al.*, 2001).

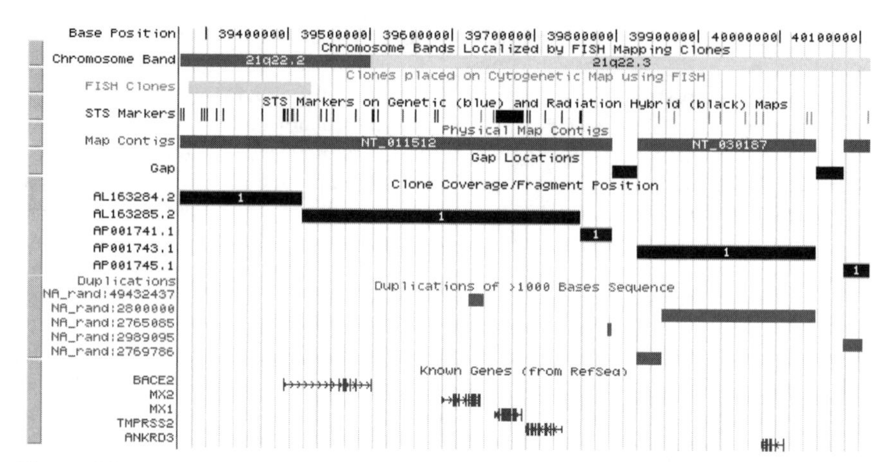

Figure 9.3 Definition of the D21S1245 and D21S1852 interval in genomic sequence using the UCSC human genome browser. The view immediately identifies five known genes, two gaps and several duplications across the region (See Colour Plates).

The HGB interface presents a very useful 'duplication' track. This identifies regions of >1000 bp which are duplicated in other golden path contigs. In this case five duplicated regions are identified across the region; the duplications are coloured red indicating that they have 99% or more similarity. Duplications of 98–99% and 90–98% similarity are shown in yellow and grey respectively. The left hand column reports the location of the duplicate regions. In this case the contigs are all from various 'NA_rand' locations — these are singleton contigs which cannot be placed in a chromosome contig. These are most probably missed overlaps between contigs, so they may not be an overt cause for concern, although this continues to suggest that this region may need careful curation.

The gaps and duplications across this region highlight a need for further QC to validate the sequence assembly before using the sequence as a basis for the construction of a laboratory study. One key *in silico* approach to validate the order and orientation of these contigs is to compare and integrate them with the RH and genetic map frameworks across the human genome. The HGB interface includes links to identical golden path regions in Ensembl and Map View. By selecting the Map View link a view of the region appears in a new window with a default view of the chromosome 21 contig map, unigene clusters and genes. To view the integrated maps, select 'Maps & Options'. Another window will open. In this window select the following pull-down menus, the 'NCBI RH' map, the 'Marshfield' Genetic Map and 'Transcript (RNA)', finally select the 'show connections' tick box and click 'Apply'. The main Map View window will now reload to show an integrated view of the RH maps and genetic maps across the locus. Markers shared between maps are linked by lines, which also show the location of the markers in the human genome contig. The data can also be viewed in a tabular view (with extra information) by selecting the 'Data as Table View' link. Figure 9.4 displays the Map View returned for this region. Examination of the NCBI Integrated RH map supports the correct ordering and orientation of the first contig (NT_011512). The genetic map is also broadly in agreement with the RH map, although there are some conflicts, however comparison with other RH maps supports the RH order for NT_011512 (not shown). Map View also supports the order of the third contig (NT_030188), however no links are drawn to support the

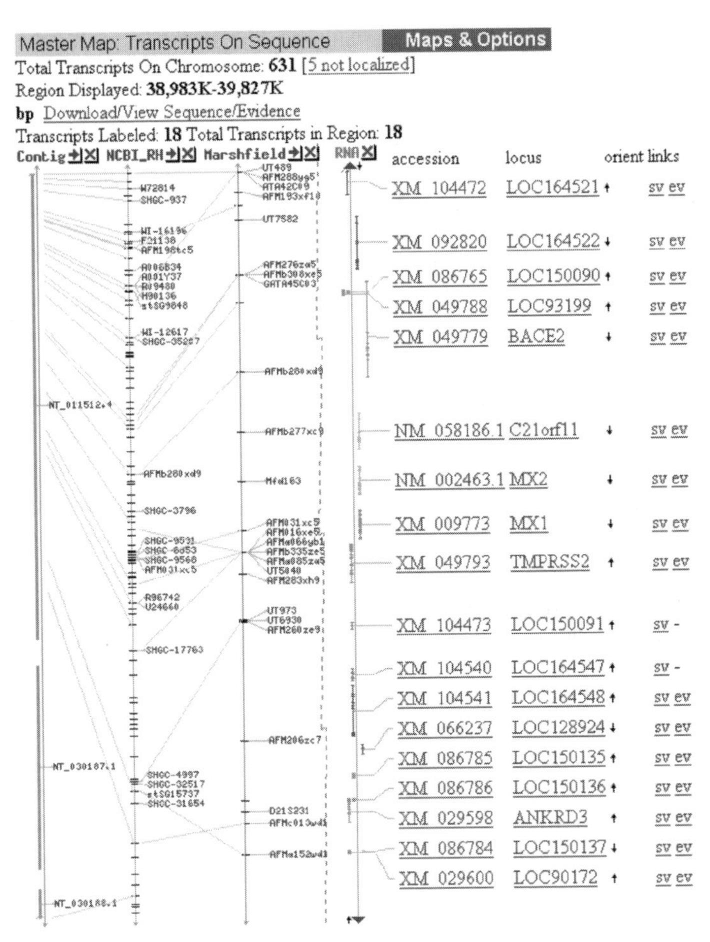

Figure 9.4 Using NCBI Map View to produce an integrated view of genetic and physical map data. Examination of the RH maps and genetic maps support the correct ordering and orientation of the first and third contigs (NT_011512 and NT_030187). No links are drawn to support the order between any of the maps and the second contig (NT_030187) spanned by the gaps, this requires further investigation.

order between the RH map and the second contig (NT_030187) spanned by the gaps. This highlights one problem with Map View: it does not comprehensively localize markers in the human genome draft sequence. In the data reviewed so far there is no firm evidence to place NT_030187 in the region under study. The only solution to this problem is to extract the genomic sequence across the region and screen it directly for matches to STS marker sequences. The tool to achieve this is Electronic PCR (ePCR) at the NCBI (http://www.ncbi.nlm.nih.gov/genome/sts/epcr.cgi). This tool maps known STSs from the dbSTS and RHdb databases to a submitted sequence (Schuler, 1998).

Submission of the 0.84-Mb sequence to the NCBI ePCR server identifies 57 STS markers across the sequence. The contig we need to check, NT_030187, maps from 0.56

TABLE 9.2 RH Marker Order versus Sequence Order across Contig NT_030187

Location in Region (Base Pair)	Marker	TNG cR50000 (LOD)	GeneMap99 GB4 cR3000 (LOD)
489,493–489,804	SHGC-140546	16451 (12.3)	—
558,505	GAP1	—	—
645,681–645,988	SHGC-147821	**16519(8.7)**	—
659,283–659,421	D21S1449	**16508(14.4)**	—
689,527–689,653	stSG46899	—	**222.98(3.00)**
703,493–703,795	D21S356	—	—
711,656–711,819	stSG52786	—	**222.23(2.43)**
767,022–767,332	SHGC-148000	**16583(9.7)**	—
770,359–770,649	WI-20889	**16567(14.6)**	**225.80(3.00)**
771,401–771,520	stSG3262	**16563(17.1)**	**225.21(3.00)**
777,760	GAP2	—	—
843,780–844,059	D21S1852	16617 (11.9)	—

to 0.78 Mb, 10 STS markers span this region. In Table 9.2 we list these markers and note their map order in the other available maps. Integration of maps presents a somewhat confusing picture. Both the TNG and GeneMap99 GB4 maps show local discrepancies in marker order. These results are somewhat inconclusive, RH map resolution is unreliable below 60–100 kb so these localized discrepancies in marker order over 10–30-kb regions may be due to the lack of resolution of the maps. Alternatively it is possible that the finished BAC clone (AP001743) which constitutes most of the NT_030187 contig may contain a sequence rearrangement; further laboratory analysis would be required to confirm this. However, the overall order of the contigs in this region appears to be supported by the integrated maps across the region.

9.5 CASE STUDY 3: DEFINITION OF KNOWN AND NOVEL GENES ACROSS A GENOMIC REGION (RECOMMENDED TOOLS: ENSEMBL AND HGB)

Now that we are (relatively) sure of the order and orientation of the contig across the D21S1245–D21S1852 region, it is important to identify all the known and novel genes in the region, so that they can be evaluated as candidates or to ensure that marker maps across the region are sufficient to detect any genetic effect in genes or regulatory regions. In Chapter 4 we presented a detailed examination of the art of delineating genes in genomic sequence. The UCSC human genome browser and Ensembl are valuable assistants in this process. Both tools run the human genome sequence through sophisticated gene prediction pipelines (Hubbard *et al.*, 2002). These analyses are coupled with a detailed view of supporting evidence for genes, such as ESTs, mouse and fish genome homology, CPG islands and promoter predictions. This wealth of data probably makes further *de novo* gene prediction unnecessary in most cases; improvement on the quality of annotation provided by Ensembl and HGB would require an in-depth understanding of the intricacies of gene prediction, which we cannot hope to impart in this book (see Rogic *et al.* (2001) for an excellent review of this field). Instead we suggest that the user focuses on the available data to build gene models based on existing annotation. Genetics has one advantage over other fields of biology: detection of a genetic effect does not require a completely accurate working model of a gene (although obviously this will help). All that should be needed is an approximate gene model which can be screened by identifying proximal polymorphisms. The only exception to this might be during the analysis of functional polymorphisms where an accurate model of a gene and its regulatory regions may be critical.

For the purposes of our study, we need to identify all known and novel genes across the locus. In Figure 9.5 we show a magnified HGB view of the first contig in the region, NT_011512. Again we have configured the browser to show tracks which are directly applicable to the identification of genes in genomic sequence. Without going into an overt level of detail there are five known genes identified across the region. A number of extra tracks show pieces of evidence which support the known genes and suggest the possible existence of a further four novel genes across the locus which we indicate at the bottom of the figure. Confidence in the identification of novel genes in genomic sequence is in part dependent on the range and nature of supporting evidence. The most convincing single item of evidence is a correctly spliced mRNA transcript, either an EST or whole

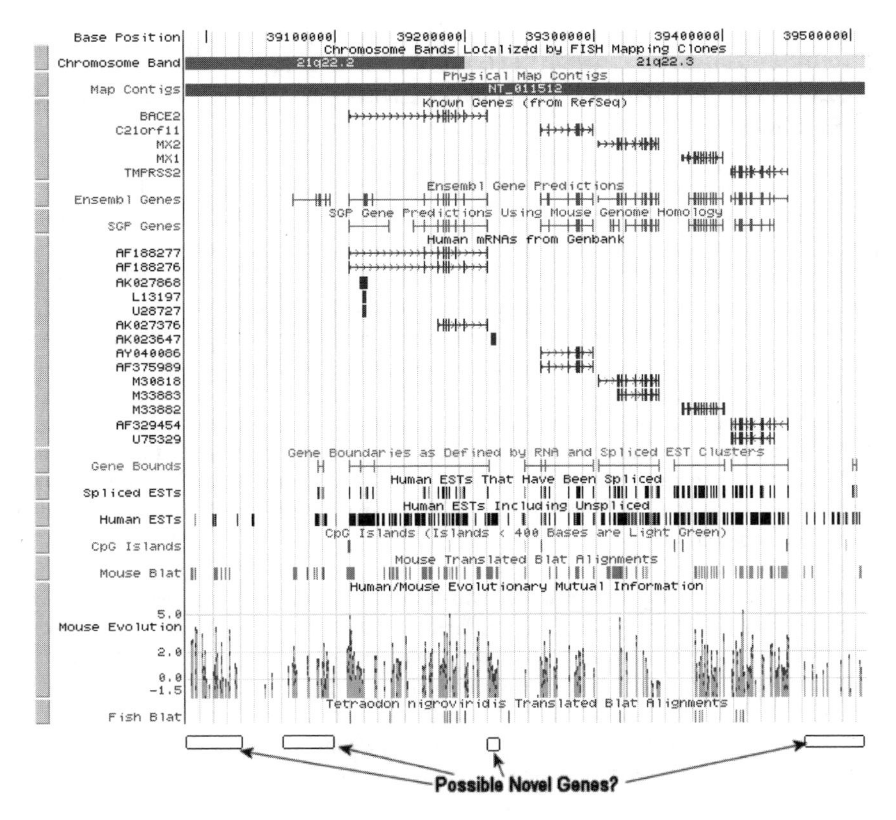

Figure 9.5 Using the UCSC human genome browser to identify known and novel genes. A range of evidence including mRNAs, ESTs and human–mouse homology supports the existence of five known genes and up to four novel genes (See Colour Plates).

transcript. In this figure several human mRNAs from GenBank are identified across the region. These include splice variants and redundant entries of the same transcript. So for example at least two splice variants are apparent for the BACE2 gene (AF188277 and AF188276). There are no novel mRNA transcripts in the GenBank track, however there are a large number of human ESTs which do not appear to map to any known gene. Human ESTs are divided into spliced and unspliced tracks. This is in recognition of the very high number of artefacts that are generated in EST libraries. Spliced ESTs, that is, ESTs which align across exons are much more reliable confirmatory evidence for genes than unspliced ESTs. A final strong source of evidence for genes are the range of tracks which show homology to non-human DNA, including non-human mRNA and comparison with mouse and fish genome sequences. Strong sequence conservation between man and other vertebrates is generally thought to be restricted to coding or regulatory regions. The four putative novel genes across this locus are supported by a range of spliced ESTs, mouse homology and gene prediction. Taken individually each of these pieces of evidence might not be sufficient to reliably support the existence of a novel gene, but taken together they are quite convincing. All that remains is to characterize these novel genes in terms of homology and putative function; we reviewed this process in Chapter 4.

9.6 CASE STUDY 4: CANDIDATE GENE SELECTION – BUILDING BIOLOGICAL RATIONALE AROUND GENES (RECOMMENDED TOOLS: HGB, ENSEMBL)

So far in our study of the D21S1245–D21S1852 region, we have identified our locus in genomic sequence and identified the known and putative novel genes in the region. Further genetic analysis of the region could now take two routes, we could perform further linkage or association studies by defining a suitable set of markers across the region (see Case study 5) or alternatively we could select specific candidate genes for follow-up studies. As we only have nine or 10 genes in our locus it would be quite viable to study each gene, but in most cases a region will contain a much larger number of genes which would make follow-up of each gene an impractical approach. An alternative in such cases would be to prioritize candidate genes based on their biological rationale in the target phenotype or trait. Criteria for biological prioritization of candidate genes are discussed throughout this book. Genes can be prioritized based on a known or putative role in the disease pathway, gene knock-out models, expression in the disease tissue, functional polymorphism and many other criteria.

In our hypothetical study we are looking for a gene with a possible role in bipolar depression, therefore to prioritize our candidate genes, we might first review the literature to search for a link between the candidate genes in the region and this disease pathway. The aetiology of bipolar disorder, like many complex diseases is poorly understood, this makes it difficult to establish a clear biological rationale for any gene in this disorder. Where biological rationale is found it could range from convincing support, such as upregulation of the gene or a related gene or pathway component in a disease model or in a similar phenotype to the most basic support, such as being expressed in a tissue affected by the disease.

Drawing together the complex strands of evidence in the literature is a skill that calls for a good background in biology and ideally a broad understanding of the disease under study. However reliance on literature-based evidence alone can run the risk of over-interpreting tenuous links between genes. This could be a particular problem in the case of poorly understood diseases, where unknown pathways would largely fail to register as a form of rationale. This issue is an argument to support a truly investigative approach to candidate gene identification. The candidate should be in the right place at the right time; beyond this further assumptions may be misleading. Data presented by tools such as Ensembl and the UCSC HGB can provide solid evidence which can help to identify genes which are at least expressed in the tissues affected by the disease. Obviously in the case of bipolar disorder, we are most interested in genes which show evidence of expression in the brain. This will inevitably include a large number of genes. In an analysis of the expression profiles of >33,000 genes, Su *et al.* (2002), found that on average any individual tissue expresses approximately 30–40% of known genes. For candidate gene studies, this implies that 30–40% of genes are likely to be candidates on the basis of expression in the disease tissue (assuming the disease affects only one tissue).

9.6.1 Analysis of Gene Expression

Four tracks in the UCSC HGB provide information about the tissue expression profiles of genes (Figure 9.6). The simplest level of information is provided by ESTs, each is implicitly a measure of gene expression, as each is derived from a specified tissue source. The UniGene track clusters all ESTs which map to a gene. Viewing ESTs from a UniGene record will confirm the expression of a gene in a particular tissue. The number of ESTs represented in each tissue will also give a *very* rough idea of the expression levels of the

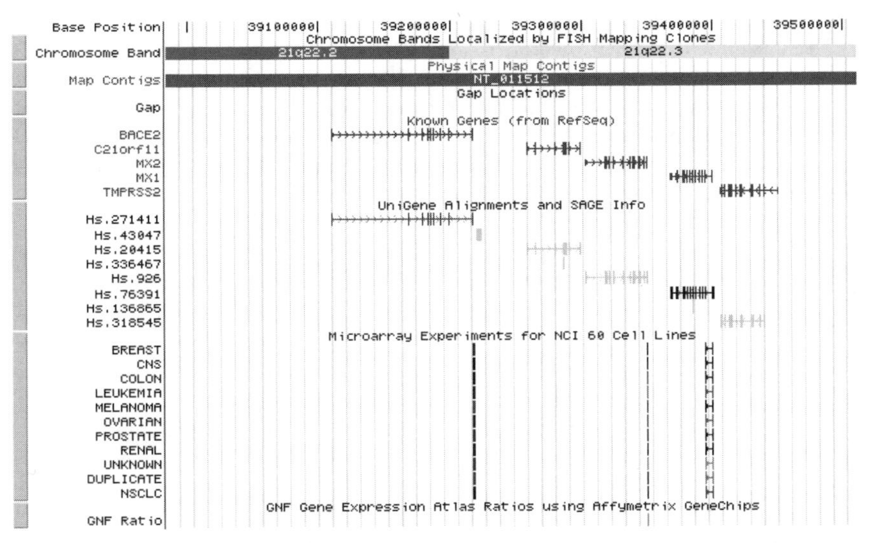

Figure 9.6 Using the UCSC human genome browser to evaluate gene expression across a locus. Four tracks provide information on gene expression. ESTs implicitly measure gene expression as each is derived from a specified tissue source. UniGene clusters link to SAGE expression profiles drawn from the SageMap project. Finally data from two genome-wide microarray projects are presented, the NCI60 cell line project and GNF gene expression atlas ratios (See Colour Plates).

gene, but it will not confirm the absence of a gene in a tissue. The most comprehensive measure of gene expression is also linked from the UniGene track in the HGB interface. Clicking on a UniGene cluster links to SAGE expression profiles drawn from the SageMap project (Lash *et al.*, 2000). Finally the HGB interface also provides two specialist tracks containing data from two genome-wide microarray projects, the NCI60 Cell Line Project (Ross *et al.*, 2000) and GNF Gene Expression Atlas Ratios (Su *et al.*, 2002; Table 9.3).

9.6.2 Serial Analysis of Gene Expression (SAGE)

The HGB UniGene track links to SAGE data for eight of the 10 genes in the D21S1245–D21S1852 interval. SAGE is a quantitative measure of gene expression based on tags in the 3′ UTR of genes (see Chapter 16 for a detailed explanation of this technique). Clicking on a UniGene cluster returns a table of SAGE data for every Unigene cluster contained in the browser window. Selecting the Unigene cluster name will display the SageMap page for the cluster, leading to a so-called 'Electronic Northern', which might more appropriately be called an 'electronic dot-blot', as there is no element of transcript sizing. The summary data from the SageMap of the genes in the region is presented in a tabular form, so if you would rather see a graph across the region, it is quite easy to export this data to a spreadsheet to plot the expression profiles of each gene. SAGE expression data is available in a selected range of brain and neuronal libraries for eight of the genes across the locus. This data identifies expression of BACE2, PAPPA, Novel3, C21orf11, MX2 and MX1 in some of the normal brain tissues in the SAGE libraries. BACE2, Novel3 and MX1 show high expression in a wide range of normal brain tissues, which suggests that these genes may warrant priority as candidates. The TMPRSS1 and Novel2 genes show no evidence of brain expression and so it may be reasonable to reduce the priority of these genes.

The integration of SAGE data across the genome within the HGB interface, makes SAGE data one of the most comprehensive and convenient measures of gene expression across a genetic locus, allowing the user to quickly identify most genes across the locus which are expressed in a range of common tissues. The only real limit to this method is in the number and type of tissues, although these are extensive and growing in numbers (see http://www.ncbi.nlm.nih.gov/SAGE for a list of available tissues).

TABLE 9.3 Comparison of Public Domain Genome-Wide Expression Datasets

Dataset	SAGE	NCI60 Cell Line Project	GNF Gene Expression Atlas
Technology	Serial analysis of gene expression	cDNA microarray (Incyte)	Affymetrix U95a GeneChip microarray
No. of genes	>100 K tags*	~8000	~33 K
Tissues	37	—	33
Cell lines	4	—	—
Induced cell lines	3	—	2
Tumour material (incl. cell lines)	52	60	13
Total tissues	96	60	48
Reference	Lash *et al.* (2000)	Ross *et al.* (2000)	Su *et al.* (2002)

*SAGE tags are redundant across genes.

9.6.3 Microarray Data Tracks

9.6.3.1 The NCI60 Cell Line and GNF Gene Expression Atlas Ratios

The UCSC Human Genome Browser hosts two public domain microarray data tracks. The NCI60 cell line track presents data from a cDNA microarray experiment to assay the expression of more than 8000 genes among 60 tumour-derived cell lines used in the National Cancer Institute's (NCI) anti-cancer drug screens (Ross *et al.*, 2000; http://genome-www.stanford.edu/nci60/). The GNF track shows expression data generated from 46 human tissues and cell lines by the Genomics Institute of the Novartis Research Foundation (GNF) using a u95A Affymetrix GeneChip (Su *et al.*, 2002; http://expression.gnf.org/).

Both the NCI60 and GNF tracks are presented in a similar format, although the experimental details differ. Clicking on a transcript in either track, will bring up a tabular view of all genes in the current browser view, in which each column of coloured boxes represents the variation in transcript levels for a given cDNA across all of the array experiments and each row represents the measured transcript levels for all genes in a sample (Figure 9.7A). The variation in transcript levels for each gene is represented by a colour scale, in which red indicates an increase in transcript levels, and green indicates a decrease in transcript levels. These relative transcript levels are measured in a slightly different way between NCI60 and GNF tracks. In the NCI60 track expression levels are relative to a reference sample of 12 pooled tumour cell lines. In the GNF track the expression levels are relative to the signal of the probe in the particular tissue compared to the median signal of all experiments for the same probe. The saturation of the colour corresponds to the magnitude of transcript variation. A black colour indicates an undetectable change in expression and a grey box indicates missing data (see Su *et al.* (2002) and Ross *et al.* (2000) for a more detailed explanation of this method).

As the NCI60 data focuses on tumour-derived cell lines, it is not well suited for the determination of expression in normal tissues, although obviously this data would be very valuable for studies of cancer genetics. However, the GNF data track presents some very valuable information for complex disease genetics, including a breakdown of gene expression across different regions of the brain. This data is very valuable for candidate prioritization, as certain regions of the brain may have a more significant role in bipolar depression than others. For example, functional neuroimaging studies of bipolar patients have identified the thalamus as a key component of the main neuroanatomic circuitries which are altered in psychiatric illnesses, such as bipolar disorder (Soares and Mann, 1997). This information indicates that expression in the thalamus could help to prioritize candidate genes for analysis.

Only one gene from our core region, MX2, is represented in the GNF dataset, this shows low level expression throughout the different brain regions, with strongest expression in whole blood (data not shown). However if we expand the D21S1245–D21S1852 interval by 1 Mb on either side to include other flanking genes, much more data becomes available. Figure 9.7A shows a view of a selection of the available tissues, including all neuronal tissues for 15 genes across the wider locus. One gene, Purkinje cell protein 4 (PCP4–no. 37576), is immediately apparent with a high level of expression in a wide range of neuronal tissues, including the thalamus. By clicking on the PCP4 gene number in the expression view in HGB, a detailed expression profile of the gene in all available tissues is launched into a new window (Figure 9.7B). This shows that the expression of this gene is primarily limited to the brain, thyroid and prostate glands, with highest levels of expression in the caudate nucleus and thalamus. Obviously this makes PCP4 an interesting candidate gene.

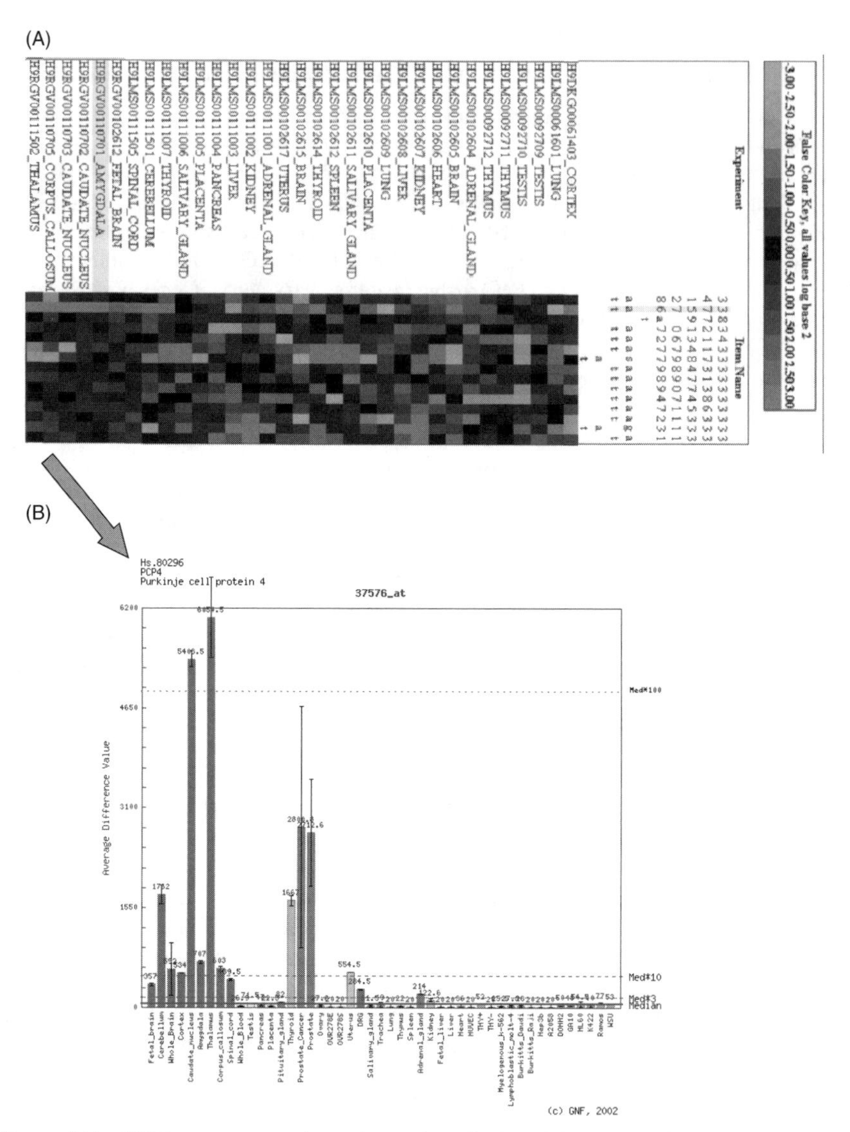

Figure 9.7 GNF gene expression atlas ratios displayed by the UCSC human genome browser. (A) The browser shows a view of the expression profiles of 15 genes across the wider locus. One gene, Purkinje cell protein 4 (PCP4, no. 37576), shows high expression in a wide range of neuronal tissues, including the thalamus. (B) Detailed gene expression profile for the PCP4 gene (See Colour Plates).

The relative cost of microarray technology was a cause of major concern for the academic research community, prompting fears that microarray data would be the preserve of cash-rich industry and biotech. However public domain projects like the NCI60 and GNF gene expression atlas should in part allay these fears. Although these microarray projects currently provide a somewhat limited coverage of human genes, they are complemented by other technologies such as SAGE and both types of data are constantly expanding in the public domain.

9.7 CASE STUDY 5: KNOWN AND NOVEL MARKER IDENTIFICATION (RECOMMENDED TOOLS: ENSEMBL, HGB, MAP VIEW, SNPPER)

Now that we have identified the genes in our locus and established some biological rationale to prioritize them for study, we need to ascertain which markers are available to complete this study. The human genome is a very convenient framework for organization of polymorphism data and so genome viewers are probably the best tools for identifying these polymorphisms.

All the genome viewers maintain SNP annotation across the human genome. Exact numbers of SNPs reported may differ between tools, for example, comparison of the D21S1245–D21S1852 locus between Ensembl, UCSC HGB, Map View and SNPper (see below), identifies 901, 903, 903 and 876 SNPs respectively. These minor discrepancies are a likely result of the different SNP mapping and repeat masking parameters between the tools. Missing a SNP or two across a locus may not be a problem for a large-scale analysis but if candidate gene analysis is the objective, it may be important to identify all variation to enable accurate construction of haplotypes or identification of potentially functional variants.

9.7.1 Identification of Potentially Functional Polymorphisms

Aside from the ordered convenience that genome browsers bring to SNP data, they also place a SNP into a full and diverse genomic context, giving information on nearby genes, transcripts and promoters. Both Ensembl and HGB show genome conservation between human and mouse, while HGB also includes tetradon and fugu (fish) genome conservation. Genome conservation between vertebrates is generally restricted to genes (including undetected genes) and regulatory regions (Aparicio *et al*., 1995). Hence this is a simple but powerful method for identifying SNPs in regions which are potentially functionally conserved. Figures 9.8A shows a detailed HGB view of the BACE2 gene. After close viewing of the locus it is possible to assess the functional context and genomic conservation of the region surrounding each SNP. Where an overlap is unclear it is possible to zoom in to a resolution of just a few hundred base pairs to determine exact locations and overlaps between SNPs and gene features (Figure 9.8B). At the simplest level, identification of potentially functional SNPs is a matter of identifying SNPs which overlap highly conserved regions or putative gene or regulatory features. The UCSC browser presents some detailed information on putative promoter regions, including golden triangle and transfac analyses (see Chapter 12). Once identified, the impact of different alleles can be evaluated by running the alleles through the tool originally used to predict the sequence feature. These could include tools for promoter prediction, splice site prediction or gene prediction. In Chapters 12–14 we describe these analysis approaches in detail. One final point on the functional characterization of SNPs is that we currently know very little about the functional regions of the genome; but we do know that our tools are very limited, and so it is almost impossible to conclude that a SNP is *not* functional. All that we can do is make our best guess from the available evidence.

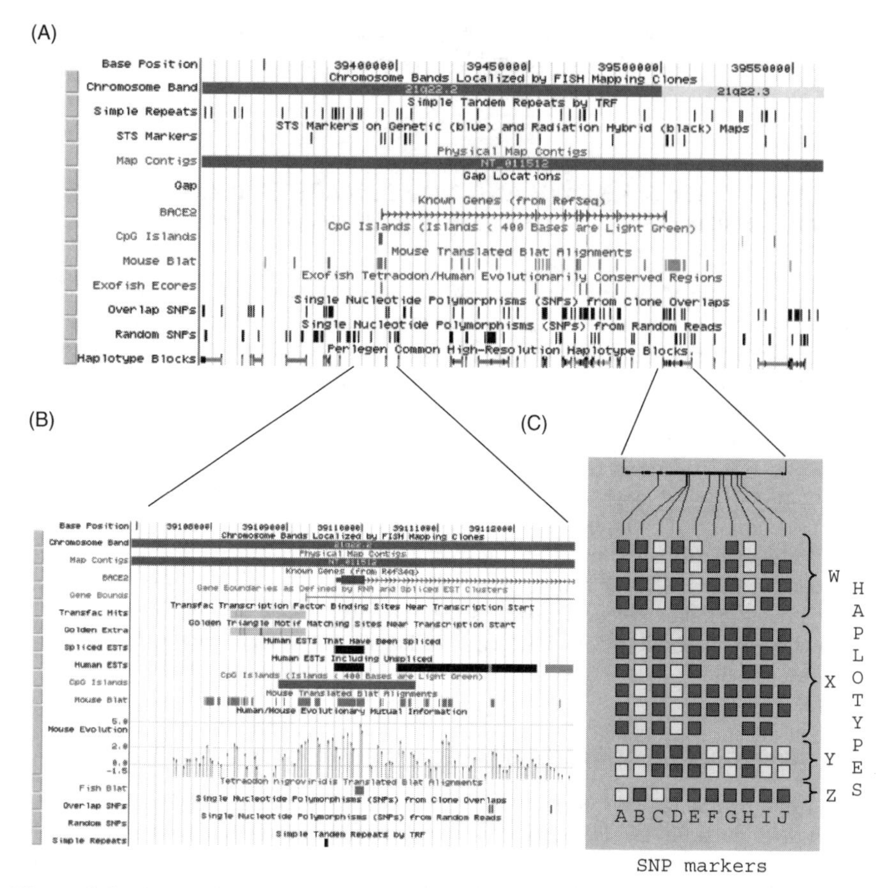

Figure 9.8 SNP visualization in the UCSC human genome browser. (A) A detailed view of the BACE2 gene allows the user to view a range of information, including SNP haplotype data. (B) Close viewing of the BACE2 locus allows the user to assess the functional context and genomic conservation of the region surrounding each SNP. (C) A detailed view of a Perlegen haplotype (See Colour Plates).

9.7.2 Identifying Novel Microsatellites in Sequence Data

The polymorphism data we have identified so far, may be sufficient for a SNP scan across our locus, however in Chapter 8 we reviewed some data to suggest that LD may be detected over greater distances with STRs (Koch *et al.*, 2000). Inclusion of STRs spaced at intervals of 50–100 kb in this locus may assist in narrowing a critical interval to a distance of a few hundred kb in a case–control association study. Known genetic marker maps identify three polymorphic STRs across the 0.84-Mb, D21S1245–D21S1852 interval. To generate a sufficiently dense map, we need 10 to 15 extra markers. Potentially polymorphic STRs can easily be detected across a given region by using Tandem Repeat Finder (Benson, 1999; http://c3.biomath.mssm.edu/trf.html). The UCSC HGB interface already presents output from the Tandem Repeat Finder in the 'simple repeats' track. Figure 9.8B shows an example of this output—a simple repeat in the promoter region of BACE2 is identified by this tool. Use of the Tandem Repeat Finder interface identifies tandem repeats in a submitted sequence over a user-defined repeat unit size range (from 1 to 500 bp). Perfect or close to perfect tandem repeats of greater than 12 repeat units tend to be polymorphic (see Fondon *et al.* (1998) for a review of these methods).

9.7.3 Exporting SNP and Microsatellite Data

If comprehensive SNP coverage is a priority then results from both Ensembl and HGB can be compared and collated. The easiest way to do this is to compare both exported SNP sets in a Microsoft Excel spreadsheet. Both Ensembl and HGB have facilities to export SNP and microsatellite information across a defined locus. In Ensembl the user needs to select the 'export' menu above the detailed analysis window and select 'SNP list'. This allows the user to retrieve information about SNP accessions, golden path location and gene region directly into Microsoft Excel. The UCSC HGB is slightly more complicated, the user needs to select 'TABLES' and then separately select 'Random SNPs' and then repeat with 'Overlap SNPs' from the menu. This produces two similar tab-delimited files. Both files can be loaded into Excel and sorted by golden path location to obtain a non-redundant SNP map across the region.

9.7.4 Construction of Marker Maps

To complete an LD-based association scan across this region, we need to define a sufficiently dense set of markers to detect LD across the region. In the absence of knowledge of the haplotypic diversity of the interval in question, accurate selection of an optimal marker set is not possible. However a framework of markers spaced at 10–30-kb intervals might be appropriate; this is accepted as a reasonable assumption of LD in northern European populations (Ardlie *et al.*, 2002). Informative SNP markers for association studies need to be carefully selected, ideally with an allele frequency of around 25%, and generally no lower than 5%. Lower frequencies would require very large sample sizes to reach a sufficient power to detect association (Johnson *et al.*, 2001; see Chapter 8). Unfortunately this creates a technical problem, most SNPs from dbSNP, are 'candidate SNPs' with no available frequency information (see Chapter 3 for a discussion of this issue). Marth *et al.* (2001) determined the frequency of a large number of candidate SNPs from dbSNP and found that on average, 50% of SNPs assayed showed a frequency of >10%. Considering this success rate it may be necessary to identify SNP markers at 5–15-kb intervals across our region for frequency determination (20 chromosomes should be sufficient to identify the majority of SNPs with a minor allele frequency >10%), before defining a final marker map.

Use of an evenly spaced marker map is a pragmatic approach which assumes evenly spaced LD across the study region. Inevitably this does not always occur; instead LD extends over variable distances. The optimal, but time-consuming approach to map construction across a region is to first determine common haplotypes and use this data to define a minimal set of SNPs ('haplotype tags') that define these common haplotypes. In this study we have one major advantage, the UCSC Human Genome Browser presents information on common SNP haplotypes across the whole of chromosome 21. This data is derived from a study by the Biotech company, Perlegen, Inc. In their study, Patil *et al.* (2001) identified 25,000 SNPs with a frequency >10%, by sequencing 20 haploid copies of chromosome 21 derived from a cell line. They used these SNPs to directly determine common haplotypes. This data is visible in the 'haplotype blocks' track in Figure 9.8A.

Each haplotype block is represented by a blue horizontal line with taller vertical blue bars at the first and last SNPs of each block. The shade of the blue indicates the minimum number of SNPs required to discriminate between haplotype patterns which account for at least 80% of genotyped chromosomes, darker colours indicate fewer SNPs are necessary. Individual SNPs are denoted by smaller black vertical bars. This information is also available at the Perlegen website (http://www.perlegen.com/haplotype). Several haplotype blocks extend over the BACE2 gene region. Clicking on the haplotype bar opens a window displaying the structure of the selected haplotype. In Figure 9.8C we show an example of one of the haplotypes across the BACE2 gene. In this case two SNPs from a total of 10 SNPs (A–J) define four haplotypes (W,X,Y and Z). In this relatively simple haplotype, it is fairly easy to identify the SNP pairs that would 'tag' or distinguish the four haplotypes. However, the HGVbase website also has a tool, 'Tag 'n Tell', to automatically identify haplotype tags (http://hgvbase.cgb.ki.se/). This is particularly useful for identifying tags in larger, complex haplotypes. In the case of the haplotype in Figure 9.8C, we can evaluate the four haplotypes defined by the 10 SNP markers. To format this haplotype for analysis by 'Tag 'n Tell', we need to list the marker IDs (A B C D E F G H I J) on the first line separated by spaces. Then, we introduce each haplotype, defined by a list of alleles (one for each marker) followed by the haplotype name preceded by ':'. All four haplotypes are input on separate lines. Thus the input to the tool is as follows:

INPUT HAPLOTYPES (from Figure 9.8C):

```
A B C D E F G H I J
1 1 2 1 2 1 1 2 1 1 : W
1 2 1 2 1 1 1 1 1 1 : X
2 2 1 1 1 2 2 1 2 2 : Y
2 1 2 1 1 1 1 1 1 1 : Z
```

The 'Tag 'n Tell' program identifies two alternative sets of two tags which will distinguish all four haplotypes. These are SNPs A and B or SNPs A and C:

```
A B                A C
1 1 : W            1 2 : W
1 2 : X      OR    1 1 : X
2 2 : Y            2 1 : Y
2 1 : Z            2 2 : Z
```

This convincingly demonstrates the power and potential economies afforded by use of haplotype data. Unfortunately similar data is not yet available for all chromosomes (LD maps have also been published for chromosomes 19 and 22; see Chapter 7). A publicly funded genome-wide haplotype determination project is in progress so this situation should change fairly quickly.

9.7.5 Identifying 'Candidate SNPs'

Just as the candidate gene approach can complement the locus analysis approach, a complementary approach to regular marker map construction is to screen candidate SNPs across the locus. A candidate SNP is any SNP with a potential for functional effect, this could include non-synonymous SNPs, SNPs in regulatory regions or other functional regions. In fact uncertainty over functional prediction could stretch the definition of a candidate SNP to any SNP within 10 kb of gene. There are many ways to select such SNPs (see Chapters 12 and 13 for details). At the most detailed level, it is possible to identify all available SNPs in genes and putative regulatory regions by eye using human genome browsers to identify overlap between SNPs and features such as promoter regions or comparative genome conservation. This can be a useful and manageable approach for small loci, however, this is not always a practical approach for larger loci. Larger analyses can be facilitated by using the feature 'export facilities' in the human genome browsers. The coordinates of most features can be exported as tab-delimited data, these can be compared in a spreadsheet.

9.7.6 Moving from SNPs to Assays

By using both Ensembl and the UCSC HGB we have been able to identify and export a comprehensive list of SNPs across our locus in a convenient tab-delimited form. But one hurdle stands between our data and a large-scale SNP genotyping experiment. The problem is that the human genome browsers give us a list of SNP accession numbers, but not a primer design-ready list of SNP sequences. Until very recently the only web based approach to obtaining these sequences was to access each SNP individually! Fortunately both dbSNP and SNPper, now incorporate features which allow the user to export a list of SNPs with flanking sequence. SNPper (Riva and Kohane, 2001) maps RefSNPs and genes to the golden path in a very similar way to the genome browsers (but without a graphical interface), allowing SNP searching by gene or SNP name or by golden path position. This makes SNPper completely compatible with the coordinates generated by Ensembl and HGB (and so this tool can also be used at an earlier stage for SNP data mining). SNPper also produces a very effective gene report. The 'Annotated' link in the gene report displays a very informative SNP report which positions SNPs in the context of introns, exons and other gene features, including a mark-up of non-synonymous SNPs across a gene.

The great strength of SNPper lies in its data export and manipulation features. At the SNP report level, SNPs can be sent directly to automatic primer design through Primer3 (which allows the entry of multiple SNP sequences). At a whole gene level or even at a locus level, SNP sets can be defined and refined and e-mailed to the user in an Excel spreadsheet with SNP names in the first column and flanking sequences in the second, ready for primer design. This is a very useful function which is not currently offered by any other tool.

9.8 CASE STUDY 6: GENETIC/PHYSICAL LOCUS CHARACTERIZATION AND MARKER PANEL DESIGN (RECOMMENDED TOOLS: ENSEMBL, HGB AND MAP VIEW)

The wealth of data presented so far in this chapter has enabled us to define our locus to a level of detail that would allow us to complete a quite effective genetic study. However, before finalizing the requirements for this study it may be useful to spend some time characterizing the actual genetic and physical characteristics of the locus.

Ideally we would like to establish a detailed LD map across the D21S1245–D21S1852 interval. The Perlegen haplotype data presented in the HGB interface goes a long way towards this objective. However the haplotype coverage across this region is not particularly complete. Extended haplotypes only cover 40–50% of the interval, leaving large gaps across the region (Figure 9.9). These gaps may simply reflect insufficient coverage

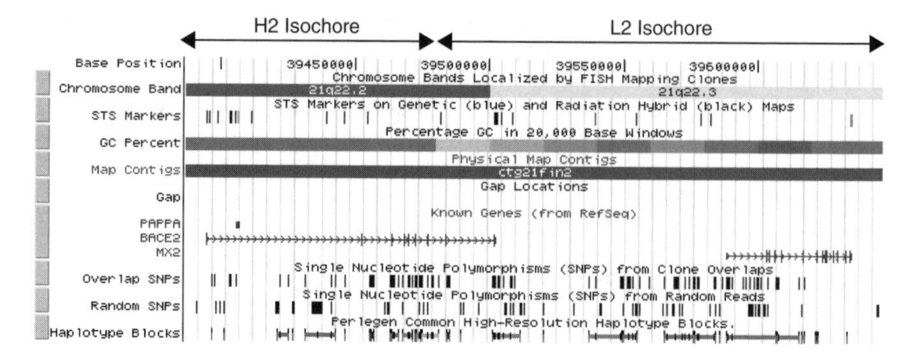

Figure 9.9 A putative isochore boundary in the BACE2 gene.

of the region by Perlegen SNPs. Alternatively LD across the gaps may actually be limited. A number of factors are known to influence LD (see Ardlie *et al.* (2002) for an excellent review). One of the most influential is the recombination frequency across the region. As we showed in Chapter 8, comparison of the physical and genetic distances between markers can give a direct measure of recombination frequency. In a comparison of Marshfield maps with the golden path, Yu *et al.* (2001) found that the genome-wide genetic/physical distance ratio ranged between 0 to 9 cM per Mb. They used this ratio to infer recombination rates and identified several chromosomal regions up to 6 Mb in length with very low or high recombination rates. They termed these recombination 'deserts' and 'jungles', respectively. LD was much more extended in recombination 'deserts' than in 'jungles' as higher rates of recombination reduced the extent of LD. This is an interesting approach, although its major drawback is the low resolution of genetic maps, this makes it very difficult to draw accurate conclusions about recombination over ranges of less than 1 Mb. There are only two genetic markers in the D21S1245–D21S1852 interval, so it is not possible to draw conclusions on recombination rate across this locus, analysis would need to encompass a much wider region.

9.8.1 Analysis of GC Ratio and Identification of Isochore Boundaries

Beyond the analysis of genetic and physical ratios, even simpler measures can give clues to the nature of recombination in a locus. GC content across a locus also has a weak influence on recombination rates. Lower GC ratios generally correspond to lower recombination rates (Yu *et al.*, 2001). There is cytogenetic evidence for this phenomenon, analyses have shown that meiotic crossovers are seen more frequently in GC-rich R and T bands than in GC-poor G bands (Holmquist, 1992). This observation directly relates recombination frequency with the gross Giemsa banding of chromosomes. These bands were believed to be made up of tracts of DNA with homogeneous GC content, known as isochores. Isochores are divided into two classes the GC-rich H2 and H3 isochores and the GC-poor L1, L2 and H1 isochores (Bernardi, 2000). Interestingly our region between D21S1245–D21S1852 spans the 21q22.2–q22.3 cytoband. The region also shows a clear shift from high GC (average 50%) to low GC (average 40%) in the region of the cytoband boundary (high GC is indicated by dark grey in Figure 9.9, lower GC is indicated by lighter shades of grey). This region is a putative isochore boundary, between an H2 and L1 isochore. It is difficult to determine if there is a significantly different extent of

LD between the putative L1 and H2 isochore regions. The markers within the first eight exons of the BACE2 gene do not generally show LD over a greater distance than 5 kb, while markers in the L1 region near the MX2 gene show LD over longer distances up to 10–15 kb (Figure 9.9). In a study of the NF1 gene, Eisenbarth *et al.* (2000), found a marked reduction in LD, which coincided with an L1 to H2 isochore boundary in the NF1 gene.

This analysis of the D21S1245–D21S1852 interval, may seem somewhat esoteric. We are just starting to understand how the physical properties of chromosomes affect their genetic properties. Undoubtedly, our understanding of these issues is still limited, but a pragmatic approach to marker map design across this region might be to establish a baseline marker density across the entire region (say at 1SNP/10 kb), once this has been reached within the budget of the project, then supplemental markers could be placed in regions with a higher predicted recombination rate.

9.9 CONCLUSIONS

In this chapter we have reviewed the key steps in the design and construction of genetic studies using genomic sequence as a template. When sequencing of the human genome is finally complete and as studies of the genome become more and more precise, much of genetics as we know it today may become an increasingly *in silico* process. Ten years ago who might have believed that the details of the genetic study process would have changed so dramatically (although the principles remain the same). As the genome wave continues to roll towards us, we may be looking at much more intelligently designed genetic studies, with maps which account for local recombination, LD and a detailed knowledge of genes and regulatory regions. And 10 years further on, perhaps we will look back again and marvel again at how much things have changed?

REFERENCES

Aparicio S, Morrison A, Gould A, Gilthorpe J, Chaudhuri C, Rigby P, *et al.* (1995). Detecting conserved regulatory elements with the model genome of the Japanese puffer fish, *Fugu rubripes*. *Proc Natl Acad Sci USA* **92**: 1684–1688.

Ardlie KG, Kruglyak L, Seielstad M. (2002). Patterns of linkage disequilibrium in the human genome. *Nature Rev Genet* **3**: 299–309.

Bailey JA, Yavor AM, Massa HF, Trask BJ, Eichler EE. (2001). Segmental duplications: organization and impact within the current human genome project assembly. *Genome Res* **11**: 1005–1017.

Benson G. (1999). Tandem repeats finder: a program to analyze DNA sequences. *Nucleic Acids Res* **27**: 573–580.

Bernardi G. (2000). Isochores and the evolutionary genomics of vertebrates. *Gene* **241**: 3–17.

Eisenbarth I, Vogel G, Krone W, Vogel W, Assum G. (2000). An isochore transition in the NF1 gene region coincides with a switch in the extent of linkage disequilibrium. *Am J Hum Genet* **67**: 873–880.

Fondon JW III, Mele GM, Brezinschek RI, Cummings D, Pande A, Wren J, *et al.* (1998). Computerized polymorphic marker identification: experimental validation and a predicted human polymorphism catalog. *Proc Natl Acad Sci USA* **95**: 7514–7519.

Johnson GC, Esposito L, Barratt BJ, Smith AN, Heward J, Di Genova G, *et al.* (2001). Haplotype tagging for the identification of common disease genes. *Nature Genet* **29**: 233–237.

Holmquist GP. (1992). Chromosome bands, their chromatin flavors, and their functional features. *Am J Hum Genet* **51**: 17–37.

Hubbard T, Barker D, Birney E, Cameron G, Chen Y, Clark L, *et al.* (2002). The Ensembl genome database project. *Nucleic Acids Res* **30**: 38–41.

Koch HG, McClay J, Loh EW, Higuchi S, Zhao JH, Sham P, *et al.* (2000). Allele association studies with SSR and SNP markers at known physical distances within a 1-Mb region embracing the ALDH2 locus in the Japanese, demonstrates linkage disequilibrium extending up to 400 kb. *Hum Mol Genet* **9**: 2993–2999.

Lander ES, Linton LM, Birren B, Nusbaum C, Zody MC, Baldwin J, *et al.* (2001). Initial sequencing and analysis of the human genome. *Nature* **409**: 860–921.

Lash AE, Tolstoshev CM, Wagner L, Schuler GD, Strausberg RL, Riggins GJ, *et al.* (2000). SAGEmap: a public gene expression resource. *Genome Res* **10**: 1051–1060.

Marth G, Yeh R, Minton M, Donaldson R, Li Q, Duan S, *et al.* (2001). Single-nucleotide polymorphisms in the public domain: how useful are they? *Nature Genet* **27**: 371–372.

Patil N, Berno AJ, Hinds DA, Barrett WA, Doshi JM, Hacker CR, *et al.* (2001). Blocks of limited haplotype diversity revealed by high-resolution scanning of human chromosome 21. *Science* **294**: 1719–1723.

Riva AA, Kohane IS. (2001). A web-based tool to retrieve human genome polymorphisms from public databases. *Proc AMIA Symp* 558–562.

Rogic S, Mackworth AK, Ouellette FBF. (2001). Evaluation of gene-finding programs on mammalian sequences. *Genome Res* **11**: 817–832.

Ross DT, Scherf U, Eisen MB, Perou CM, Rees C, Spellman P, *et al.* (2000). Systematic variation in gene expression patterns in human cancer cell lines. *Nature Genet* **24**: 227–235.

Schuler GD. (1998). Electronic PCR: bridging the gap between genome mapping and genome sequencing. *Trends Biotechnol* **16**: 456–459.

Soares JC, Mann JJ. (1997). The anatomy of mood disorders — review of structural neuroimaging studies. *Biol Psychiatry* **41**: 86–106.

Su AI, Cooke MP, Ching KA, Hakak Y, Walker JR, Wiltshire T, *et al.* (2002). Large-scale analysis of the human and mouse transcriptomes. *Proc Natl Acad Sci USA* **99**: 4465–4470.

Yu A, Zhao C, Fan Y, Jang W, Mungall AJ, Deloukas P, *et al.* (2001). Comparison of human genetic and sequence-based physical maps. *Nature* **409**: 951–953.

CHAPTER 10

SNP Discovery and PCR-based Assay Design: From *In Silico* Data to the Laboratory Experiment

ELLEN VIEUX[1], GABOR MARTH[2] AND PUI KWOK[3]

[1]*Washington University School of Medicine, St. Louis, MO, USA*

[2]*National Center for Biotechnology Information, Bethesda, MD, USA*

[3]*Cardiovascular Research Institute, University of California, San Francisco, USA*

Bioinformatics for Geneticists. Edited by M.R. Barnes and I.C. Gray
 ISBNs: 0 470 84393 4; 0 470 84394 2 (PB)

10.1 INTRODUCTION

Single nucleotide polymorphisms (SNPs) are the most abundant form of DNA sequence variation in the human genome. It is widely believed that a significant fraction of SNPs contribute to our susceptibility to various diseases. In order to identify the SNPs associated with diseases, however, many groups are pursuing a case–control mapping strategy that requires a large number of SNP markers distributed throughout the human genome. Once a set of genes is implicated in a disease (either by genetic mapping or by obtaining biological evidence), the candidate genes are scanned for sequence variations that are likely to alter the genes' function. Therefore, identifying single base-pair changes, in a global or targeted fashion, is extremely important in genome research.

The central public polymorphism database, dbSNP (Sherry *et al.*, 2002), serves as an archival repository of nucleotide sequence variations. An important subset of these data, nearly 100,000 SNPs in transcribed regions, were found by analysing clusters of expressed sequence tags (ESTs) (Buetow *et al.*, 1999, 2001; Irizarry *et al.*, 2000) or by aligning ESTs to the human reference sequence (Marth *et al.*, 1999). The vast majority of genomic SNPs (single base pair variations found by analysing genomic sequence clones without regard to whether they represent exonic DNA) were discovered in sequences from restricted genome representation libraries (Altshuler *et al.*, 2000), random shotgun reads aligned to genome sequence (Sachidanandam *et al.*, 2001), and in the overlapping sections of the large-insert clones (mainly bacterial artificial chromosome, or BAC) that make up the public human reference genome (Tallion-Miller *et al.*, 1998). Because most sequences of these comparisons involved a small number of chromosomes (typically two), this collection of SNPs is enriched for common variants. Experimental characterization of these polymorphisms demonstrates that many of them occur at a high frequency in independently chosen samples, and often segregate in all or most human populations (Marth *et al.*, 2001). By the same argument, many rare polymorphisms, including those that cause noticeable but rare phenotypic effects, are likely to be absent from this set. The identification of rare phenotypic mutations will require significantly higher sample sizes and may only be possible by the cross-comparison between large samples of affected patients and those of controls (see Halushka *et al.* (1999) for an example of such a study).

Because the numbers are extremely large and the need for identifying SNPs in a timely fashion is great, computer tools are indispensable in the SNP discovery process. Fundamentally, one identifies a SNP by comparing two or more sequences from the same region on the chromosome. This can be done quite easily if the DNA sequence quality is high and the sequence data are derived from cloned DNA because each clone comes from a single copy of one of the two chromosomes in the diploid human cell. There are no unambiguous bases in regions where the data quality is high. In the case of identifying SNPs in targeted regions in the genome, one amplifies genomic DNA by PCR and sequences the PCR products derived from different individuals. In this situation, SNP discovery is complicated by the fact that the same regions on both chromosomes in the diploid cell are amplified by PCR and some bases will be heterozygous in one or more individuals. A good computer tool will be able to identify a SNP even when only heterozygotes and homozygotes of just one of the two alleles are present in the samples sequenced. This is not a trivial problem because the commonly used dye terminator-based DNA sequencing methods yield peaks of uneven heights at the polymorphic sites and the base-calling algorithm will frequently miscall the base at these sites in the sequences of heterozygous individuals.

In this chapter, we will survey the computer tools used in global and targeted SNP discovery and PCR-based assay design. Instead of describing the mechanics of how to use these bioinformatic tools, we refer the reader to the primary literature and the excellent documentations of these tools and concentrate on explaining the approaches these tools take and the limitations (if any) they may have.

10.2 SNP IDENTIFICATION

Computational discovery of polymorphisms in sequence data usually follows a four-step procedure. First, sequences of high similarity in multiple individuals are identified, usually with a BLAST (Altschul *et al.*, 1990) similarity search. To avoid spurious similarity due to known human repeats, sequences are masked for high copy number repetitive elements with REPEATMASKER (Arian Smit, unpublished data). Still, the possibility exists that the sequences originate from regions of as yet uncharacterized chromosomal duplications (Lander *et al.*, 2001). Inclusion of a second, paralogue-filtering step into the procedure can reduce false positive SNP predictions arising from comparing paralogous sequence copies. Following this step, false predictions due to paralogy were as low as 0.2% of the data collected through pooled SNP characterization in the Kwok laboratory (unpublished data).

The third step is the construction of a base-wise multiple alignment of the sequences. In the general case, this is a computationally expensive task. Aligning expressed sequences is even more complicated because of exon–intron punctuation and possible alternative splice variants. In the case of human data one can organize fragmentary sequences on top of the nearly complete reference sequence (Lander *et al.*, 2001). This approach was shown to work well for discovering SNPs in clusters of cDNA sequences (Marth *et al.*, 1999).

In the fourth (and final) step, sequences in the precise, base-to-base multiple alignment are scanned for nucleotide differences. Because of the possibility of sequencing errors, not every mismatch is a polymorphic site. Discrimination between true polymorphism and sequencing error uses statistical tools based on measures of sequence accuracy, or base quality values (Ewing and Green, 1998; Ewing *et al.*, 1998). Each SNP prediction is accompanied by a measure of confidence. Accurate confidence values permit one to use the highest number of candidates with an acceptable false positive rate.

Both commercial and academically developed programs are available for use in SNP detection. Some methods use sequence quality data to eliminate false positives due to poor sequencing quality. Others incorporate expected mutation rates to distinguish true SNPs. The most prominent methods of detecting SNPs are PolyBayes, PolyPhred, and Sequencher. Other methods incorporate neighbourhood quality standard (NQS) generated by Phred (Ewing *et al.*, 1998) to determine the quality of the data surrounding the SNP (Altshuler *et al.*, 2000; Mullikin *et al.*, 2000). PolyPhred and PolyBayes are freely available to academic groups, while Sequencher is produced by Gene Codes Corporation (URL: www.genecodes.com). Other companies have developed software based on the same principles. Typically, these products either offer a built-in graphical interface, or use an external, licensable interface program (such as CONSED). They can be used for both comparison of short known regions, or long shotgun regions, and are extremely useful when searching known regions of interest for novel SNPs.

10.2.1 PolyBayes

The POLYBAYES program was developed for *de novo* SNP discovery in non-ambiguous (clonal) sequence data (Marth *et al.*, 1999). The SNP detection algorithm employs a Bayesian approach to combine prior knowledge (such as average polymorphism rate or expected transition to transversion ratio) with the base calls and base quality values of the sequences in the multiple sequence alignment. Each SNP prediction comes with a

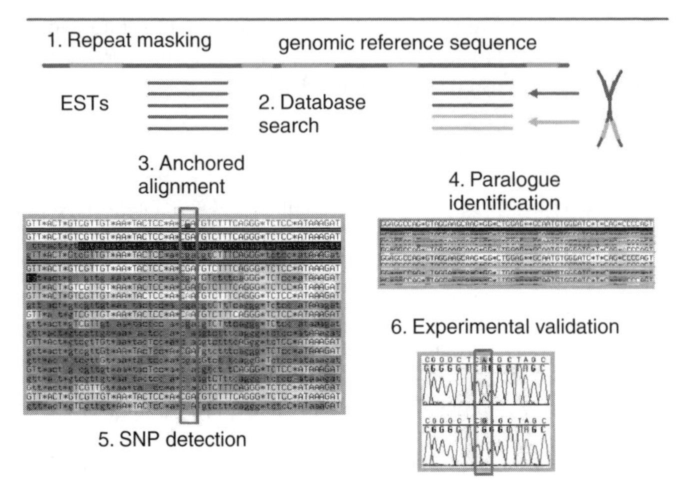

Figure 10.1 The POLYBAYES SNP discovery tool applied to EST-Mining. (1) The genome sequence (BAC clone or assembled sequence contig) is masked for known, large copy number human repeats. (2) Using the BLAST similarity search tool, expressed sequences in the public database (dbEST) that match the genomic sequence are identified. (3) Matching ESTs are aligned to the genomic sequence using an anchored alignment approach. (4) Possible paralogous ESTs are identified and discarded. (5) The multiple alignment is scanned for polymorphic sites. (6) Candidates are validated by sequencing in independent, population-specific DNA pools.

predicted true positive rate, or 'SNP score', which have been shown to be accurate (Marth *et al.*, 2001). Figure 10.1 illustrates an example of using POLYBAYES for SNP discovery in ESTs aligned to genome sequence. POLYBAYES has been used to discover SNPs in overlapping regions of human BACs (Marth *et al.*, 1999), in *C. elegans* (Wicks *et al.*, 2001) and Drosophila (Berger *et al.*, 2001).

10.2.2 PolyPhred

PolyPhred was developed to be used with Phred, Phrap, and CONSED to identify candidate SNPs in sequence trace data (Ewing and Green, 1998; Gordon *et al.*, 1998; Table 10.1). In Consed, coloured marks in the sequence alignment are used to indicate candidate SNPs as well as confidence in the variations base call. The accuracy of the calls by PolyPhred has been tested using previously screened mitochondrial DNA. The results show that this software exhibits over 95% accuracy depending on the quality of the sequence traces (Nickerson *et al.*, 1998). The primary use of PolyPhred is to identify SNPs in PCR-amplified data as it can detect heterozygous sequence peaks, and is thus widely employed in sequence-based genotyping applications.

10.2.3 Sequencher

Sequencher is a tool developed by GeneCodes for sequence alignment, annotation, editing and mutation identification. Although it is a commercial product, a free demo version is available (www.genecodes.com/features/html). Sequencher can be used with automated sequencers such as ABI, Pharmacia/ALF, LI-COR and VISTRA. GeneCodes continues to

TABLE 10.1 Tools and Related Resources for Primer Design

SNP detection tools	
Sequencher	http://www.genecodes.com/features.htm
PolyPhred	http://droog.mbt.washington.edu/PolyPhred.html
POLYBAYES	http://www.genome.wustl.edu/gsc/polybayes/
Repeat masking tools	
RepeatMasker	www.genome.washington.edu/uwgc/analysistools/repeatmask.html
MaskerAid	http://sapiens.wustl.edu/maskeraid/
Primer design tools	
Primer3	http://www-genome.wi.mit.edu/cgi-bin/primer/primer3_www.cgi
TSC primer db	ftp://snp.cshl.org/pub/SNP/.
Primer design tips	http://www.alkami.com/primers/refdsgn.htm.
Tools for sequence extraction and manipulation	
SNPper	http://bio.chip.org:8080/bio/.
UCSC HGB	http://genome.ucsc.edu/index.html

add new tools and functionality to the software. The most appealing part of the software is the graphical interface, which is intuitive and easy to use on multiple platforms.

10.2.4 Non-sequencing Methods

Several groups have explored non-sequencing methods for SNP discovery. Among the most promising of these techniques is the use of high density DNA chips (Dong *et al.*, 2001). Variations of this method have been used to scan for genome wide SNPs (Wang *et al.*, 1998), in mitochondrial DNA (Chee *et al.*, 1996), and to scan all of chromosome 21 (Patil *et al.*, 2001). Methods for SNP scanning using DNA chips vary considerably in design (for a review see Draghici *et al.*, 2001).

10.3 PCR PRIMER DESIGN

A large number of candidate SNPs exists in public databases. Key to taking advantage of this resource is the ability to design PCR assays to amplify these loci uniquely and SNP genotyping assays for genetic studies under standardized conditions. Genetic researchers wanting to validate and assay SNPs are faced with the need for high throughput primer design. Manual picking of primers is time consuming, and some automated tools only allow for submission of one sequence at a time. There are many tools available over the web as well as software. In addition, some commercial companies offer genotyping assay design by order for their customers and a few assays are available through public databases.

10.3.1 Tools

Currently there is no standard method for calculating the annealing temperature (TM) of primers. Although many tools have been developed to determine the annealing temperature, their results vary. Furthermore, many of these programs use different entropy and enthalpy tables in their TM calculations, leading to further discrepancies (Owczarzy *et al.*, 1997). Despite these variances most of these tools will work and one program that has become a standard is Primer3 (http://www-genome.wi.mit.edu/genome_software/other/primer3.html). A comprehensive review of primer picking and TM predicting tools can be found at http://www.alkami.com/primers/refdsgn.htm.

Primer3 is a standard because it is freely available and easy to use. It is particularly useful for high throughput design because it can determine primers for multiple sequences at once. Some of its particular strengths are its many useful and well-documented options, its easily parsed output and its simple command line interface. Primer3 can be used for the design of both PCR primers and internal sequencing primers. Although Primer3 allows for individual SNP position targets and target lengths to be set for each sequence, if the data is highly varied in position and length it is possible to avoid setting parameters for each SNP by pre-formatting the data. The SNP sequence retrieval option on SNPper (see below) is a tool that can provide this uniformly formatted flanking sequence.

10.3.2 Custom Primer Design Services

Although primer design may be carried out in-house, many companies as well as public databases are offering high throughput design as part of their product support. Sequenom is one such example. Sequenom is in the process of making primers available through a site called RealSNP (www.RealSNP.com). Applied Biosystems is another company providing primer design through their 'Assay by Design™' Genomic Assay Service. The researcher provides the sequence, while Applied Biosystems designs and test all assays. These designs are optimized for Taqman™ assays (http://www.appliedbiosystems.com/). There is no charge if an assay cannot be designed. Perkin-Elmer will also be providing SNP-specific assays through their website.

10.3.3 Public Databases

Primers generated by The SNP Consortium (TSC) Allele Frequency Project are available via ftp (Table 10.1). These primers have been released, by some of the groups, to the public with the assistance of TSC. It should be noted that these primers have been generated by separate groups via different methods and for specific experimental conditions. The NCBI's dbSNP database also contains primer designs for some of the SNP entries, but these have not been specifically designed for SNP validation (Sherry *et al.*, 2002). In addition to these public databases the Kwok laboratory has over 980,000 assays designed for sequencing of PCR products, for the specific purpose of pooled allele frequency determination. The Kwok laboratory also maintains over 1,400,000 SNP genotyping assays designed for single base extension using fluorescence polarization detection (available at http://snp.wustl.edu).

10.4 BROADER PCR ASSAY DESIGN ISSUES

SNP assay methods have three major components: (1) allelic discrimination methods, (2) reaction formats and (3) detection methods. Each area presents different challenges during SNP assay design. The most important consideration for assay design is the method of allelic discrimination. These methods vary greatly. For example four main methods of allelic discrimination are allele-specific hybridization, primer extension (includes single base extension), ligation and invasive cleavage. The reaction formats are either homogeneous reactions or solid phase reactions and the detection methods currently use product light emissions, product–mass measurements and electrical property changes in the product (Kwok, 2001).

In some cases the critical parameters that apply to one technique will not apply to others. However, almost all SNP genotyping assay techniques use PCR to amplify DNA. In order to design the correct primers one must first determine the method of assay. However, there are some basic guidelines used when designing primers for genomic sequence. All designs require obtaining sequence, repeat masking, setting experimental and design parameters, picking primers and formatting the information.

10.4.1 Obtaining Sequence

The flanking sequences for each SNP can be obtained from a variety of sources. For known SNPs two public databases, dbSNP and SNPper (Riva and Kohane, 2001; Table 10.1) provide a method for obtaining sequence. SNPper is run by Harvard's Children's Hospital Informatics Program (CHIP). Both dbSNP and SNPper offer batch query modes and return sequence in FASTA format. In the case of single SNP analysis, SNPper provides a link to the Primer3 website which will import the retrieved sequence into Primer3 and analyse it using default values. For SNPs that can be uniquely mapped SNPper can provide up to 1000 bases on either side of the SNP. It should be noted that at this time SNPper does not contain the most recent uniquely mapped SNPs in dbSNP.

10.4.2 Repeat Masking

A large amount of the genome consists of repeated regions or low complexity DNA. It is important to avoid selecting primers from these regions in order to avoid amplification of multiple products. Masking the repeats or making repeated sequence unavailable to the automated primer-picking programs prevents most unwanted amplification. A commonly used program for masking is RepeatMasker (see Table 10.1). A new resource that improves upon RepeatMasker is MaskerAid (Table 10.1), which increases the speed of masking more than 30-fold (Bedell *et al.*, 2000). Default parameters in RepeatMasker will mask known repeat regions with Ns. RepeatMasker accepts FASTA files, and returns the sequence in the same format. Ready masked sequence can also be obtained from some of the public databases. dbSNP provides sequence in FASTA format with low-complexity sequence in lower case, while the University of California Santa Cruz Human Genome Browser (UCSC HGB) has options to save repeats as either Ns or lower case. However, this format is problematic when trying to represent the start and stop exons and introns on UCSC HGB, because lower case can also be used to represent introns. In some cases two files may be required to represent the masked and unmasked forms of sequence.

Masking repeats can only be accomplished in known repeat regions with current resources. However, there remain repeat regions of the genome that have not yet been identified. By using pooled sequencing, it is possible to identify regions that have duplicated and subsequently diverged. These can be identified by the presence of a large number of apparent SNPs that are all 50% in frequency, as shown in Figure 10.2. When designing SNP specific primers within PCR products, for example for a single base extension assay, the RepeatMasking stage is not necessary.

10.4.3 Setting Experimental and Design Parameters

If a large number of SNP candidates are to be assayed it is more efficient to eliminate the experiments that are less likely to be successful *in vitro* during the *in silico* design stage. Stringent design parameters allow for a first level of screening when designing primers.

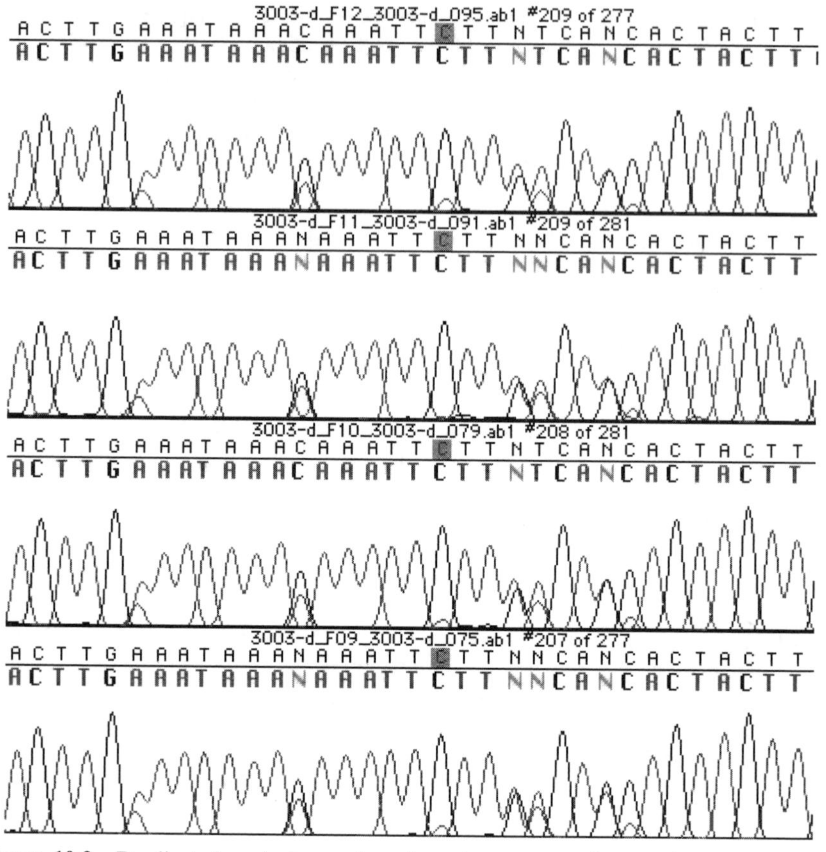

Figure 10.2 Duplicated and diverged regions in genome characterized by multiple 'SNPs' at 50% frequency in each pool.

Primer design programs such as Primer3 allow for input of both experimental parameters and primer structure parameters. The second level of screening can be done after candidate primers are chosen by primer selection programs to determine if the primers are likely to work.

Some suggestions for optimizing design parameters for the best experimental results can be found in *PCR Applications* (Beasley *et al.*, 1999). In general most primer design methods work (see http://gsu.med.ohio-state.edu/primer_design/sld001.htm for a detailed guide). However, under stringent experimental conditions optimized design parameters can decrease the level of experimental failures.

10.5 PRIMER SELECTION

In most design programs primer picking is preformed by the software. The primers are picked according to the specified parameters. If more than one primer set is returned some post processing will be required to select the most appropriate pair. Post processing can also be necessary for techniques such as pooled sequencing, where selection of a sequencing primer from the PCR primers is required.

10.5.1 Design Specific to Pooled Sequencing

Pooled sequencing uses sequencing to observe the frequency of a SNP in a group of individuals in one reaction. The candidate SNP and its flanking sequence is amplified from pools of DNA each containing individuals and a single reference individual. After sequencing of the PCR products is preformed using fluorescent dye-terminators, the sequence traces are aligned, allowing the allele frequencies to be estimated (Kwok *et al.*, 1994). Pooling DNA in this way prior to PCR amplification and estimating allele frequencies by subsequent quantitation of trace peak heights yields considerable time and cost savings.

There are several steps to designing pooled sequencing reactions. This method of design is carried out on a UNIX-based system, using RepeatMasker and Primer3. Repeats are masked before choosing PCR primers. Sequence that is not masked is retained for post processing. The input for Primer3 is set according to the optimized parameters (Beasley *et al.*, 1999) with a few optimizations. The optimizations are most important in the placement of the primers relative to the SNP. The primers are not allowed closer than 25 bases to the SNP, but are close enough to use one of the PCR primers for sequencing. After running Primer3 the results are processed to select for the best sequencing primer based on criteria to optimize experimental performance. These criteria are (1) the sequencing primer should be 100 bases from the target and (2) there should be no poly As or Ts greater than eight bases and no poly GTs or CAs greater than 10 pairs between the primer and the SNP. This design has been shown to work with less than 3% experimental failure and allows for the primers to be far enough from the SNP that the sequence is of high quality around the target as shown in Figure 10.3. During the design process as many as half of the SNPs fail to meet the design criteria, but this failure is at far lower cost than laboratory-based trial and error (Vieux *et al.*, 2002).

10.5.2 Design Specific to Single Base Extension (SBE) Reactions

SBE requires a primer that abuts the SNP under test. The primer is then extended by a single base, usually a labelled ddNTP (Hsu *et al.*, 2001). By using two different labels for the ddNTPs representing the two possible alleles, the allelic state of the SNP can be determined. SNP-specific SBE primer design can be undertaken using many of the same tools as pooled sequencing primer design. Both require repeats to be masked before designing PCR primers. The SNP-specific primers are chosen using non-masked sequence. The PCR product sizes can be smaller than for sequencing for all of the single base extension reactions. The SBE primer should not hang over the end of the PCR product and the PCR primers should not overlap with the SBE primer. In Primer3 the primers can overlap the target so it is important to give a SNP a large enough target area to prevent the overlap of primers. When choosing parameters and methods for SBE primers it is important to remember that different methods can have different primer requirements.

We have found that picking the shortest primer from 16–40 bases which has a TM between 60–65 degrees works well. In order to calculate TM for a small number of SBE primers it is possible to use free tools on the web. For high throughput design the best option is to solve TM equations after determining which set of entropy and enthalpy tables work best for the relevant method (Owczarzy *et al.*, 1997), and picking the shortest primer in the defined range. Further optimization can be achieved by picking the SBE primer with the least amount of secondary structure, and fewest runs of poly As and Ts.

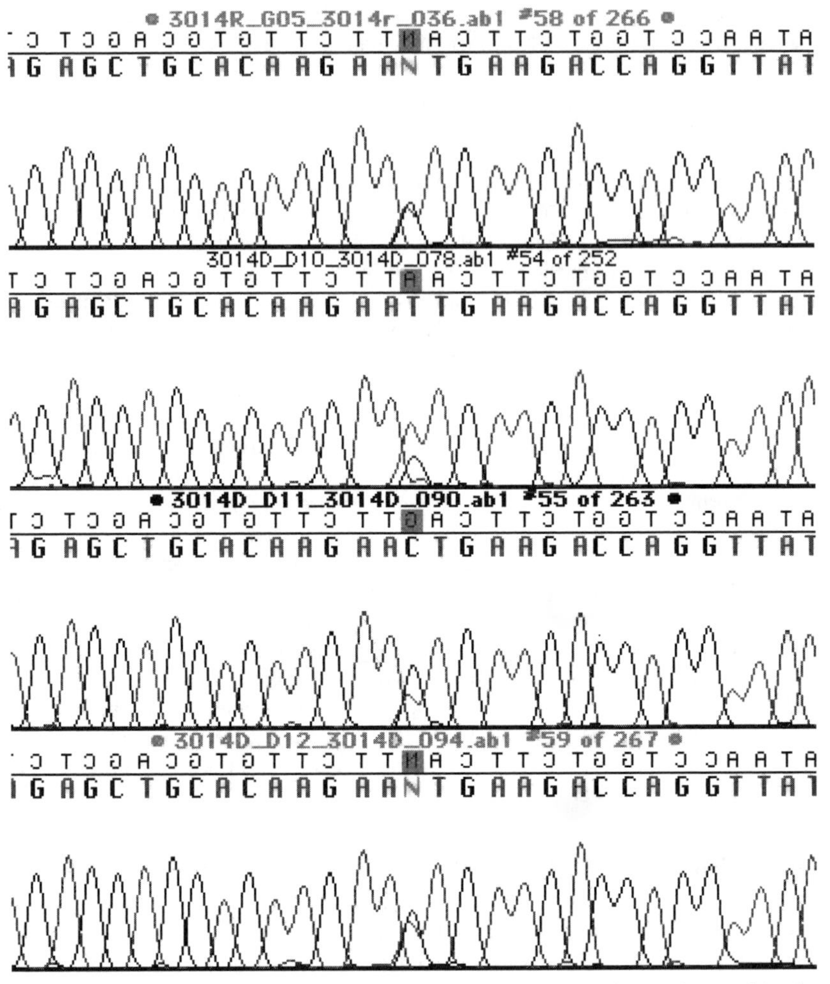

Figure 10.3 A clean pooled sequencing assay result shows a clear polymorphic site at the position of the candidate SNP.

10.6 PROBLEMS RELATED TO SNP ASSAY VALIDATION

As with any experimental design, assays for validation of candidate SNPs require attention to detail. Problems arise that are not always obvious or clearly stated in the documentation associated with the tools being used. Some problems are easy to overcome, while others cannot yet be solved.

With the completion of a final version of the human genome assembly a number of problems will be resolved, while inherent challenges will remain. There are many errors due to incorrect physical map order, gaps in physical map data and incorrect assembly (DeWan *et al.*, 2002). These errors lead to SNPs mapping to multiple locations, incorrect haplotypes and difficulty in identifying paralogues. However, SNP locations are continually amended as assemblies are progressively corrected. Map locations will continue to change until the Human Genome Project is complete. This can cause difficulties in the analysis of data and obtaining guide sequence. Another difficulty with the unfinished map is unidentified paralogues. A SNP can appear to map to a unique position, when it is actually an artefact generated from an unknown paralogue of the original reference sequence. An example of such an artefact generated by pooled sequencing data is shown in Figure 10.2.

Guide sequence is provided for known SNPs through dbSNP, TSC and SNPper. The first two sites only provide a small amount of flanking sequence in their database for any given SNP. This can lead to failure in the design of PCR primers due to limited sequence information. SNPper provides far more flanking sequence by mapping the SNP location and retrieving guide sequence from the human genome assembly at the UCSC (Table 10.1).

Other problems are inherent when working with DNA and the current technologies. Long runs of a single nucleotide can cause sequencing reactions to fail, while insertion/deletion events can cause problems with sequencing and with SNP allelic discrimination methods such as allele-specific hybridization, primer extension (including SBE), ligation and invasive cleavage. These problems may only be solved with new technologies for SNP characterization.

10.7 CONCLUSION

Given the large number of SNPs in the human genome and the potential for large-scale experimentation, bioinformatics tools are essential for SNP discovery and genotyping assay development. The tools for comparing cloned (and hence homozygous) sequences are well developed and have proven useful. However, tools for comparing genomic sequences amplified by PCR, which are often heterozygous, still have room for computational and technical improvements. Following SNP discovery, there are many assay methods for genotyping, but none can satisfy all requirements. The basic methods for assay design are well defined, but specific optimizations are different for each method. With technology improvements, some of the current problems in SNP assay design will be solved, resulting in a reduction in the number of SNPs that are refractory to successful assay design. But for now design optimization using currently available tools and careful interpretation of subsequent results will provide assays and allele frequencies for a large portion of the SNPs currently available.

REFERENCES

Altschul SF, Gish W, Miller W, Myers EW, Lipman DJ. (1990). Basic local alignment search tool. *J Mol Biol* **215**: 403–410.

Altshuler D, Pollara VJ, Cowles CR, Van Etten WJ, Baldwin J, Linton L, *et al.* (2000). An SNP map of the human genome generated by reduced representation shotgun sequencing. *Nature* **407**: 513–516.

Beasley EM, Myers RM, Cox DR, Lazzeroni LC. (1999). Statistical refinement of primer design parameters. In Michael DHG, Innis A, Sninsky JJ. (Eds), *PCR Applications*. Academic Press: New York, pp. 55–71.

Bedell J, Korff I, Gish W. (2000). MaskerAid: a performance enhancement to RepeatMasker. *Bioinformatics* **16**: 1040–1042.

Berger J, Suzuki T, Senti K, Stubbs J, Schaffner G, Cickson BJ. (2001). Genetic mapping with SNP markers in Drosophila. *Nature Genet* **29**: 475–481.

Buetow KH, Edmonson M, MacDonald R, Clifford R, Yip P, Kelley J, *et al.* (2001). High-throughput development and characterization of a genomewide collection of gene-based single nucleotide polymorphism markers by chip-based matrix-assisted laser desorption/ionization time-of-flight mass spectrometry. *Proc Natl Acad Sci USA* **98**: 581–584.

Buetow KH, Edmonson MN, Cassidy AB. (1999). Reliable identification of large numbers of candidate SNPs from public EST data. *Nature Genet* **21**: 323–325.

Chee M, Yang R, Hubbell E, Berno A, Huang XC, Stern D, *et al.* (1996). Accessing genetic information with high density DNA arrays. *Science* **274**: 610–626.

DeWan AT, Parrado AR, Matise TC, Leal SM. (2002). The map problem: a comparison of genetic and sequence-based physical maps. *Am J Hum Genet* **70**: 101–107.

Dong S, Wang E, Hsie L, Cao Y, Chen X, Gingeras TR. (2001). Flexible use of high-density oligonucleotide arrays for single-nucleotide polymorphism discovery and validation. *Genome Res* **11**: 1418–1424.

Draghici S, Kulin A, Hoff B, Shams S. (2001). Experimental design, analysis of variance and slide quality assessment in gene expression arrays. *Curr Opin Drug Discov Devel* **4**: 332–337.

Ewing B, Hillier L, Wendl MC, Green P. (1998). Base-calling of automated sequencer traces using phred. I. Accuracy assessment. *Genome Res* **8**: 175–185.

Ewing B, Green P. (1998). Base-calling of automated sequencer traces using phred. II. Error probabilities. *Genome Res* **8**: 186–194.

Gordon D, Abajian C, Green P. (1998). Consed: a graphical tool for sequence finishing. *Genome Res* **8**: 195–202.

Halushka MK, Fan JB, Bentley K, Hsie L, Shen N, Weder A, *et al.* (1999). Patterns of screening of single-nucleotide polymorphisms in candidate genes for blood-pressure homeostasis. *Nature Genet* **22**: 239–247.

Hsu TM, Chen X, Duan S, Miller RD, Kwok PY. (2001). Universal SNP genotyping assay with fluorescence polarization detection. *Biotechniques* **31**: 560, 562, 564–568, *passim*.

Irizarry K, Kustanovich V, Li C, Brown N, Nelson S, Wong W, *et al.* (2000). Genome-wide analysis of single-nucleotide polymorphisms in human expressed sequences. *Nature Genet* **26**: 233–236.

Kwok P-Y. (2001). Methods for genotyping single nucleotide polymorphisms. *Annu Rev Genomics Hum Genet* **2**: 235–258.

Kwok P-Y, Carlson C, Yager TD, Ankener W, Nickerson DA. (1994). Comparative analysis of human DNA variations by fluorescence-based sequencing of PCR products. *Genomics* **23**: 138–144.

Lander ES, Linton LM, Birren B, Nusbaum C, Zody MC, Baldwin J, *et al.* (2001). Initial sequencing and analysis of the human genome. *Nature* **409**, 860–921.

Marth GT, Korf I, Yandell MD, Yeh RT, Gu Z, Zakeri H, *et al.* (1999). A general approach to single-nucleotide polymorphism discovery. *Nature Genet* **23**, 453–456.

Marth G, Yeh R, Minton M, Donaldson R, Li Q, Duan S, *et al.* (2001). Single-nucleotide polymorphisms in the public domain: how useful are they? *Nature Genet* **27**, 371–372.

Mullikin JC, Hunt SE, Cole CG, Mortimore BJ, Rice CM, Burton J, *et al.* (2000). An SNP map of human chromosome 22. *Nature* **407**: 516–520.

Nickerson DA, Rieder MJ, Taylor SL, Tobe VO. (1998). Automating the identification of DNA variations using quality-based fluorescence re-sequencing: analysis of the human mitochondrial genome. *Nucleic Acids Res* **26**: 967–973.

Owczarzy R, Vallone PM, Gallo FJ, Paner TM, Lane MJ, Benight AS. (1997). Predicting sequence-dependent melting stability of short duplex DNA oligomeres. *Biopolymers* **44**: 217–239.

Patil N, Berno AJ, Hinds DA, Barrett WA, Doshi JM, Hacker CR, *et al.* (2001). Blocks of limited haplotype diversity revealed by high-resolution scanning of human chromosome 21. *Science* **294**: 1719–1723.

Riva AA, Kohane IS. (2001). A web-based tool to retrieve human genome polymorphisms from public databases. *Proc AMIA Symp* 558–562.

Sachidanandam R, Weissman D, Schmidt SC, Kakol JM, Stein LD, Marth G, *et al.* (2001). A map of human genome sequence variation containing 1.42 million single nucleotide polymorphisms. *Nature* **409**: 928–933.

Sherry ST, Ward MH, Kholdov M, Baker J, Phan L, Smigielski EM, *et al.* (2002). dbSNP: the NCBI database of genetic variation. *Nucleic Acids Res* **29**: 308–311.

Tallion-Miller P, Gu Z, Li Q, Hiller L, Kwok PY. (1998). Overlapping genomic sequences: a treasure trove of single-nucleotide polymorphisms. *Genome Res* **8**: 748–754.

Vieux EF, Kwok P-Y, Miller RD. (2002). Primer design for PCR and sequencing in high-throughput analysis of SNPs. *Biotechniques* (in press).

Wang DG, Fan J-B, Siao C-J, Berno A, Young P, Sapolsky R, *et al.* (1998). Large-scale identification, mapping, and genotyping of single-nucleotide polymorphisms in the human genome. *Science* **280**: 1077–1081.

Wicks SR, Yeh RT, Gish WR, Waterston RH, Plasterk RHA. (2001). Rapid gene mapping in *Caenorhabditis elegans* using a high density polymorphism map. *Nature Genet* **28**: 160–164.

CHAPTER 11

Tools for Statistical Analysis of Genetic Data

ARUNA BANSAL[1], PETER R. BOYD[2] and RALPH MCGINNIS[1]

[1]*GlaxoSmithKline, Population Genetics*
New Frontiers Science Park (North)
Third Avenue, Harlow, Essex CM19 5AW, UK

[2]*GlaxoSmithKline, Population Genetics*
Medicines Research Centre, Gunnels Wood Road
Stevenage, Herts SG1 2NY, UK

Bioinformatics for Geneticists. Edited by M.R. Barnes and I.C. Gray
 ISBNs: 0 470 84393 4; 0 470 84394 2 (PB)

11.1 INTRODUCTION

The focus of this chapter is on methods that aid in the identification of genetic variants that influence a trait of interest. The trait may be a biological measurement, possibly indicating risk of disease or it may be the response to an environmental stimulus such as a drug. Techniques such as linkage analysis and association analysis are central to the process. These methods are described and corresponding software is reviewed, with worked examples to show how they can be applied. The majority of tools covered may be downloaded, together with full documentation, by following links at http://linkage.rockefeller.edu. Web addresses for the few exceptions are provided in the text. Almost all are available free of charge.

11.2 LINKAGE ANALYSIS

Linkage analysis is applied in the early stages of gene localization and is one means by which an initial, often broad, chromosomal interval of interest is defined. It is a process of tracking the inheritance pattern of genetic markers with the inheritance pattern of a disease or trait. Disease linkage manifests as a marker allele being inherited in diseased individuals more often than would be expected under independent assortment.

Linkage analysis may be parametric to test whether the inheritance pattern of the trait fits a specific model of inheritance or it may be non-parametric (model-free). The former is more powerful under a correctly specified model and is most informative for large, multiply affected pedigrees. The latter is more powerful when the mode of inheritance is unknown, as in complex trait analysis for which small pedigrees are often ascertained.

11.2.1 Parametric Linkage Analysis

By the parametric approach (and in certain non-parametric cases), evidence of linkage is measured by the LOD score (Morton, 1955). It proceeds by an assessment of the recombination fraction, often denoted by theta (θ). Theta is the probability of a recombination event between the two loci of interest and as such it is a function of distance. Two unlinked loci are given by $\theta = 0.5$ and the closer a pair of loci, the lower their recombination fraction. The LOD may be expressed as follows, using L to denote likelihood.

$$LOD = \log_{10} \frac{L(\theta = \hat{\theta})}{L(\theta = 0.5)}$$

The likelihood in the numerator is based upon the maximum likelihood estimate of the recombination fraction, derived from the data. It is compared to that calculated under the null hypothesis of no linkage ($\theta = 0.5$). A high LOD score is thus consistent with the presence of linkage. Due to the computational complexity of the likelihood calculation, software for exact parametric linkage analysis is constrained either by pedigree size or by the number of markers included in the calculation.

The software VITESSE (O'Connell and Weeks, 1995) allows rapid, exact parametric linkage analysis of very extended pedigrees. At the expense of some speed, an alternative, FASTLINK (Cottingham *et al.*, 1993), allows the analysis of large pedigrees that also contain loops (marriages between related individuals). Both VITESSE and FASTLINK are based on an earlier program, LINKAGE (Lathrop *et al.*, 1984) and are available for UNIX, VMS and PC(DOS) systems. Using these pieces of software, analysis is typically conducted by means of a sliding window of one, two or four markers along the chromosome, although larger windows are also possible.

Parametric linkage analysis in more moderately-sized pedigrees is commonly carried out using the software GENEHUNTER (Kruglyak *et al.*, 1996). It is written in C, to be run on UNIX and uses a command-line interface. A major feature of this program is that it allows the rapid, simultaneous analysis of dozens of markers (often an entire chromosome) in a multipoint fashion, thereby providing increased power over single-marker analyses when map positions are known (Fulker and Cardon, 1994; Holmans and Clayton, 1995; Olson, 1995). In order to accommodate uncertainty in marker ordering, an option to perform single marker tests is also available. On most platforms, pedigrees up to size $2n - f = 16$ may be analysed by GENEHUNTER, where n is the number of non-founders (those with parents included in the pedigree), and f is the number of founders. This limit is important to consider, because larger pedigrees are automatically trimmed until they fall within it, leading to possible information loss. Results are stored graphically in postscript files for easy interpretation and presentation.

11.2.2 Non-parametric (Model-free) Linkage Analysis

Non-parametric linkage (NPL) analysis does not allow direct estimation of the recombination fraction, but one source of multiple testing — that derived from examining multiple models — is removed. The general principle is that relatives who share similar trait values will exhibit increased sharing of alleles at markers that are linked to a trait locus (see Holmans (2001) for a review of the method).

Allele sharing may be defined as identical by state (IBS) or identical by descent (IBD). Two alleles are IBS if they have the same DNA sequence. They are IBD if, in addition to being IBS, they are descended from (and are copies of) the same ancestral allele (Sham, 1998). A statistical test is performed to compare the observed degree of sharing to that expected under the assumption that the marker and the trait are not linked. While the test statistic may take the form of a chi-squared, normal or F statistic, often it is transformed to allow it to be expressed in LOD units.

NPL analysis often examines IBD or IBS allele sharing in sets of affected sib-pairs (ASPs), in which both siblings exhibit the trait of interest. In the absence of linkage, ASPs are expected to share zero, one or two alleles IBD, with probabilities 0.25, 0.5 and 0.25 respectively. The presence of linkage to a tested marker leads to a departure from these proportions which may be detected by means of a χ^2 test (Cudworth and Woodrow, 1975). Another model-free test, the mean test, tests the null hypothesis that the proportion of IBD allele-sharing equals 0.5. The latter is implemented in the programs SAGE (1999) and SIBPAIR (Terwilliger, 1996), allowing for larger sibships and cases where IBD status cannot be determined unequivocally.

MAPMAKER/SIBS (Kruglyak and Lander, 1995) is a piece of software widely used to test for linkage in sibling data. It was originally written as a stand-alone program, but its functionality and commands have now also been fully incorporated into GENEHUNTER (Kruglyak *et al.*, 1996) whose algorithms are similar. It accommodates both qualitative and quantitative data for either autosomal or sex-linked chromosomes and again, it allows large numbers of markers to be examined jointly.

For dichotomous trait data, a likelihood ratio (LR) test, analogous to the LOD score above is constructed in MAPMAKER/SIBS. The LR is a test for comparing two models in which the parameters of one model (the reduced model), form a subset of the parameters of the other (the full model). It has many genetic applications and may be expressed as follows, where L denotes likelihood.

$$LR = 2\log_e \frac{L_{full}}{L_{reduced}}$$

It is asymptotically distributed as a χ^2, with degrees of freedom equal to the difference in the number of parameters between the two models. In the current context, the numerator is calculated under maximum likelihood estimates of allele sharing proportions and the denominator is calculated assuming random segregation (Risch, 1990a, b). This LR test is also implemented in other software including SPLINK (Holmans and Clayton, 1995), and ASPEX (Hinds and Risch, 1996).

In the case of quantitative trait (QT) data, a test based on the Wilcoxon rank-sum test is available in MAPMAKER/SIBS. It is broadly applicable, as it makes no assumptions concerning the distribution of phenotypic effects. Alternatively, if the sib-pair QT differences are normally distributed, then the original Haseman–Elston method (1972), also implemented, may be applied with greater power. In this test, the squared QT differences between pairs of siblings are regressed on the proportion of alleles that each pair is estimated to share IBD. It is also implemented in SIBPAL2, part of SAGE (1999).

For pedigrees larger than sibships, there is an 'NPL' option in GENEHUNTER, but it was shown to be conservative (Kong and Cox, 1997). Alternatives include the modified version, GENEHUNTER-PLUS (Kong and Cox, 1997) and MERLIN (Abecasis *et al.*, 2002), which also incorporates this modification. The latter is a C++ program for UNIX, again with a command-line interface. It offers further improvements in computational speed and reduction in memory constraints, making it more suited to very dense genetic maps. It has the attractive properties of incorporating error detection routines to improve power, and simulation routines to estimate p-values. Graphical output is not however, currently provided.

For normally distributed quantitative traits (or those capable of being transformed to normality), variance component analysis represents a powerful approach to the study of pedigrees of any size (Amos, 1994; Blangero and Almasy, 1996; Goldgar, 1990). The variance component approach to linkage analysis assumes that the joint distribution of the data for a family depends only on means, variances and covariances. The variance of the phenotype is decomposed into (a) components due to linkage to individual marker locations and (b) residual polygenic and environmental components. Familial covariances are modelled in terms of a maximum of two parameters: an additive genetic-variance component and a dominant genetic-variance component, each estimated from the data. The method is implemented in SOLAR (Blangero and Almasy, 1996), in which the size of each effect may be estimated and tested by an LR test. This is a powerful approach and a major advantage is its scope for incorporating into models the effects of covariates, epistasis and gene–environment interaction. For highly complex problems, Markov Chain Monte Carlo Methods are also available, as implemented for example in LOKI (Heath, 1997) and BLOCK (Jensen *et al.*, 1995). When the parameter set is large however, the computational burden of these methods can be prohibitive.

11.2.3 Example: MAPMAKER/SIBS (Kruglyak and Lander, 1995)

11.2.3.1 Data Import

The current example follows a format originally designed for MAPMAKER/SIBS, but now also accommodated by GENEHUNTER. The input files match rather closely what has become known as 'LINKAGE format' due to the software in which it was first introduced (Lathrop *et al.*, 1984; Terwilliger and Ott, 1994). Two files are required, namely a pedigree file and a map file. In the current example, a genetic trait has been simulated for 200 sibships, and the files have been named *regionA.ped* and *regionA.loc* respectively. For the analysis of a quantitative trait, a third file is also required, called for our purposes, *test.pheno*.

The file *regionA.ped* takes the following form where, for simplicity, only a single marker, genotyped in two families has been presented.

70	8699	0	0	2	0	0	0
70	8698	0	0	1	0	0	0
70	2230	8698	8699	2	2	1	2
70	2231	8698	8699	2	2	2	2
75	8787	0	0	2	0	0	0
75	8786	0	0	1	0	0	0
75	2238	8786	8787	2	2	2	2
75	2239	8786	8787	2	2	2	2

The columns are as follows: kindred ID, individual ID, father's ID, mother's ID, sex (1 = male, 2 = female), affection status (1 = unaffected, 2 = affected), genotype. In practice, multiple (paired) columns of genotypes would be included, in map order, for each individual. Missing values are denoted by a zero.

This file therefore provides pedigree structure information, genotypes and, in the case of dichotomous traits, phenotype. For liability class data, an additional liability class column may be included after the affection status column and this is described in more detail in the manual.

For the current example, quantitative trait data is loaded separately using *test.pheno* (not shown). This file contains, on the first line, a count of the number of traits in the file. All subsequent lines take the space-separated form: kindred ID, individual ID and phenotype(s). Only sibling phenotypes should be included.

Lastly, the file *regionA.loc* lists the marker details in map order. Here, the 'internal' format is described, but LINKAGE format is also supported. The first line provides a count of the number of markers in the file, and is followed by a blank line. Subsequent lines are in six line blocks as follows. The first line has the marker name and number of alleles; the second has the allele labels; the third has the allele frequencies for each label; the fourth is blank; the fifth is the distance to the next marker; the sixth is blank. If the distances are all below 0.5, they are assumed to be recombination fractions, otherwise they are assumed to be distances in cM. The following is an example of a map file, say *regionA.loc* with just the first two markers shown for the sake of brevity.

```
29

MARKER1 6
1 2 3 4 5 6
.01 .95 .01 .01 .01 .01

5.1

MARKER2 10
1 2 3 4 5 6 7 8 9 10
.114988 .110626 .070579 .218874 .250991 .141158 .028549
.062649 .000793 .000793
```

Note that the program tends to crash if inter-marker distances less than 0.1 cM are provided. This should therefore be used as the lower bound even in the case of apparently recombinationally inseparable markers.

11.2.3.2 NPL Analysis for a Quantitative Trait

The following sequence of UNIX commands may be used.

```
load markers regionA.loc
prepare pedigrees regionA.ped
y
test.pheno
increment step 10
scan
p
nonparametric
1
np.out
np.ps
q
```

The process is as follows. The first step is to import the marker and pedigree data that are stored, respectively in *regionA.loc* and *regionA.ped*. You are then asked whether you wish to import additional phenotypic data. Upon typing *y* (yes), you are prompted for a filename, in this case, *test.pheno*. Increment step 10 specifies that linkage is to be assessed at 10 equally spaced points in each marker interval.

The *scan* command computes the full multi-point probability that two sibs share zero, one or two alleles identical by descent (IBD) with the given map and allele frequencies. You are asked whether to include affected (*a*) or phenotyped pairs (*p*). The latter (*p*) allows NPL analysis to follow. Non-parametric linkage analysis is to be applied to trait 1, with numerical output to be piped to *np.out* and graphical output to be stored in *np.ps* (Figure 11.1, below). Note that if only one trait exists in the phenotype file then the 1 above is not required.

As shown in Figure 11.1, a Z-score provides the measure of linkage and in this case evidence peaks close to marker 22. Localization cannot however be assumed to be precise and separation of at least 10 cM may be seen between studies (Hauser and Boehnke, 1997). It is therefore usual to construct a support interval around a strong linkage signal (Conneally *et al.*, 1985). For example, having converted to LOD units, a 1-unit support interval is the interval that includes all (possibly disjoint) map positions with LOD score less than 1 LOD unit below the peak score. A conservative approach is to adopt a 1.5 to 2 LOD support interval. All points within the support interval are considered to be of interest.

A determination of information content in MAPMAKER/SIBS allows a representation of the amount of IBD information extracted by the genotype data, as plotted along the chromosome. Dips in the graph allow regions to be highlighted in which the typing of additional markers could be beneficial. The following commands are applied.

```
scan
a
infomap
info.out
info.ps
```

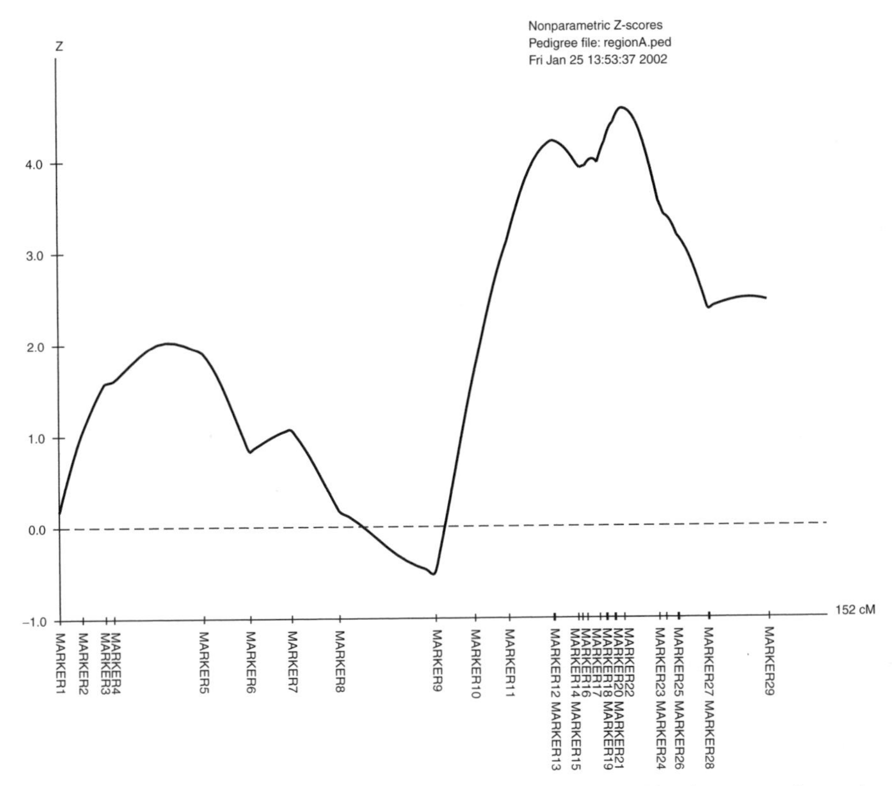

Figure 11.1 Postscript output from MAPMAKER/SIBS. This is *np.ps* from the example run.

A *scan* of affected pairs (*a*) is conducted and *infomap* is requested. The filenames ensure that numerical output is stored in *info.out* and a graphical representation is saved as *info.ps* (Figure 11.2). In this example the large gaps between markers 4 and 5 and between markers 8 and 9 manifest as troughs in the Information Content graph.

Another useful option (not shown) is that the IBD distribution can be output as a text file using the command *dump ibd.* This is a very rapid means of generating IBD probabilities for sibships and, after re-formatting, the output may be used to generate input files for other software such as QTDT (Abecasis *et al*., 2000), to be discussed later. Another piece of software, SimWalk2 (Sobel and Lange, 1996) will generate IBD probabilities for a wider range of family structures, but in the case of sibships it is slower than MAPMAKER/SIBS.

11.3 ASSOCIATION ANALYSIS

Association analysis may be regarded as a test for the presence of a difference in allele frequency between cases and controls. A difference does not necessarily imply causality in disease, as many factors, including population history and ethnic make-up may yield this effect. In a well-designed study, however, evidence of association provides a flag

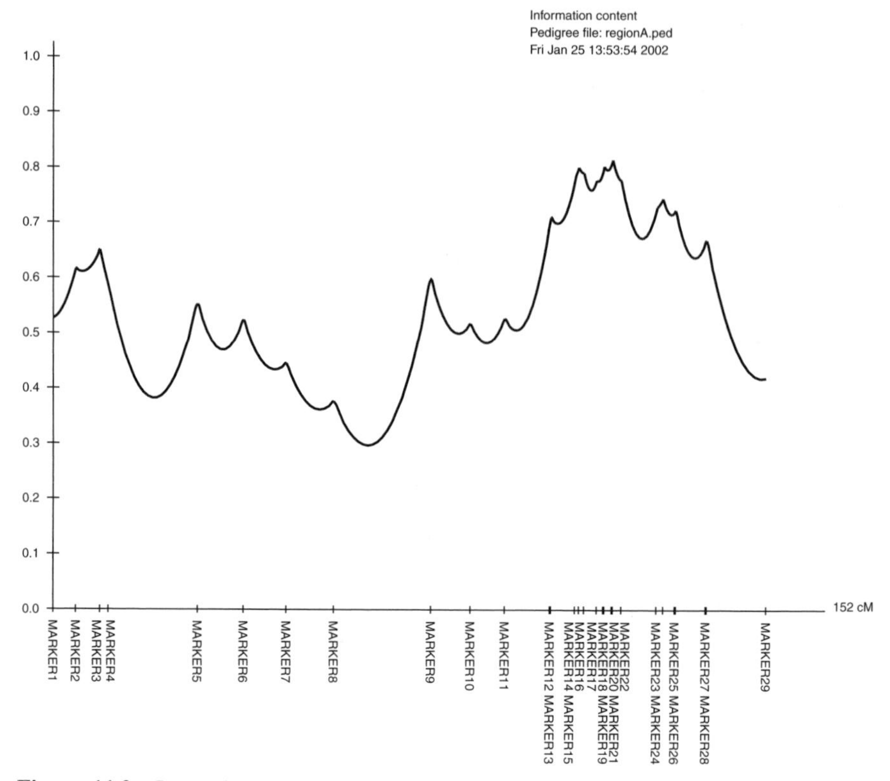

Figure 11.2 Postscript output from MAPMAKER/SIBS. This is *info.ps* from the example run.

for further study. In some instances it is due to the marker being physically close to the causal variant.

Association testing for case–control or population data is often carried out using general (non-genetic) statistical software packages, such as SAS or S-PLUS. A χ^2 test is applied to a contingency table, in which case/control status is tabulated by frequencies of either genotypes or alleles. The test takes the usual form,

$$\chi = \sum \frac{(Obs - Exp)^2}{Exp}$$

where *Obs* and *Exp* are the observed and expected frequencies respectively, and the sum is taken over all cells in the table. The number of degrees of freedom is $(r - 1)$ $(c - 1)$, where r is the number of rows, and c is the number of columns in the table. Equivalently, logistic regression can be applied, using disease status as the dependent variable and alleles or genotypes as the independent variables (see Clayton (2001) for a detailed review of the method). The remaining sections of this chapter all involve applications and extensions of the traditional association test.

11.3.1 Transmission Disequilibrium Tests

In recent years, there has been an upsurge in interest in family-based testing owing to the concern that ethnic mismatching of non-family cases and controls (population stratification) can sometimes yield false positive evidence of association. In particular, the transmission/disequilibrium test or TDT (Spielman *et al.*, 1993) has gained prominence as a test of linkage in the presence of association that does not give false evidence of linkage due to population stratification. The TDT is applied by counting alleles transmitted from heterozygous parents to one or more affected children in nuclear families. The alleles *not* transmitted to affected children may be regarded as control alleles, perfectly ethnically matched to the 'case' alleles seen in the affected children. The test takes the form of a McNemar's test, which, under the null hypothesis of no linkage, follows a χ^2 distribution with one degree of freedom. The TDT is also a valid test for association, but only when applied to alleles transmitted from heterozygous parents to just one affected child per family.

Assuming a diallelic locus, let b denote the counts of heterozygous parent-to-offspring transmissions in which allele 1 goes to an affected child, while allele 2 is not transmitted. Let c denote the counts of transmissions the other way around, in which allele 2 is inherited in an affected child, while allele 1 is not transmitted. The test takes the following form:

$$\chi_1^2 = \frac{(b-c)^2}{(b+c)}$$

A number of groups have focused on generalizing the TDT to quantitative traits or to designs in which parental genotypes are not available. The sib-TDT or S-TDT (Spielman and Ewens, 1998) does not use parental genotypes and, like the original TDT, it is not prone to false positives due to population stratification. For association testing, the S-TDT requires that the data in each family consist of at least one affected and one unaffected sibling, each with different marker genotypes. This test and the original TDT are widely implemented, for example in the Java-based program TDT/S-TDT (Spielman and Ewens, 1996, 1998).

Multi-allelic markers may be tested using ANALYZE (Terwilliger, 1995). This has the advantage of taking LINKAGE format files as input and so provides a natural follow-up to a genome scan. It does however require that LINKAGE (Lathrop *et al.*, 1984) be installed on your system. Other software able to handle multi-allelic markers includes ETDT (Sham and Curtis, 1995) and GASSOC (Schaid, 1996).

For quantitative traits, a major development was the release of QTDT (Abacasis *et al.*, 2000), software which allows TDT testing under a variance components framework. It is applicable to sibships with or without parental genotypes and incorporates a broad range of quantitative trait tests—those proposed by Rabinowitz (1997), Allison (1997), Monks *et al.* (1998), Fulker *et al.* (1999) and Abecasis *et al.* (2000). It is written in C++, to be run on UNIX and has a command-line interface. Its input files are based on LINKAGE format, but in addition, one input file of IBD probabilities must be prepared in advance. QTDT assumes the IBD format generated by the programs SimWalk2 (Sobel and Lange, 1996) and MERLIN (Abecasis *et al.*, 2002). Covariates may also be modelled, but should be kept to a minimum in order to maintain performance.

11.4 HAPLOTYPE RECONSTRUCTION

A haplotype is a string of consecutive alleles lying on the same chromosome. Each individual therefore has a pair of haplotypes for any chromosomal interval — one inherited from the paternal side and one inherited maternally. In statistical genetics, their importance lies in the fact that tests of association may be applied to haplotypes instead of single loci. This may yield increased power if the variant of interest is not being tested directly or if adjacent loci are contributing to a single effect (see Clark *et al.*, 1998; Nickerson *et al.*, 1998). Haplotypes can be inferred from the genotypes of parents or other family members (Weeks *et al.*, 1995) or by laboratory methods (Clark 1990; Nickerson *et al.*, 1998). Often, however, they are estimated by means of the Expectation–Maximization (EM) algorithm (Dempster *et al.*, 1977; Excoffier and Slatkin, 1995; Hawley and Kidd, 1995; Little and Rubin, 1987; Long *et al.*, 1995).

The EM algorithm is a method that aims to provide maximum likelihood parameter estimates in the presence of incomplete data. In the case of haplotype frequency estimation, it proceeds as follows (Schneider *et al.*, 2000).

1. An initial set of plausible haplotype frequencies is assigned — for example the product of the relevant allele frequencies may be used.
2. The E-step: assuming Hardy–Weinberg equilibrium, the haplotype frequencies are used to estimate the expected frequencies of ordered genotypes.
3. The M-step: the expected genotype frequencies are used as weights to produce improved estimates of haplotype frequencies.
4. Steps 2 and 3 are repeated until the haplotype frequencies reach equilibrium.

Note that, as with other iterative techniques, it is wise to compare the results of multiple starting points as the EM algorithm may converge to a local, rather than global, optimum. It is not always reasonable to assume that the maximum likelihood haplotype configuration has been reached.

Software written specifically for haplotype analysis includes EHPLUS (Zhao *et al.*, 2000), a reworked and extended version of the earlier program EH (Xie and Ott, 1993). It is written in C and is available in both UNIX and PC versions. EHPLUS can be applied to either case–control data or data assumed to come from a random-mating population. It accommodates large numbers of haplotypes and incorporates a companion program, PMPLUS, which will reformat genotype data ready for use. Estimated haplotypes and their frequencies are output and may be subjected to association tests. Permutation features allow the calculation of empirical p-values for these.

Further software for sophisticated haplotype analysis is available from ftp://ftp-gene.cimr.cam.ac.uk/software/clayton/. Resources include SNPHAP, a program that uses the EM algorithm to estimate haplotype frequencies for large numbers of diallelic markers using genotype data. Another program, TDTHAP (Clayton and Jones, 1999) allows the TDT to be applied to extended haplotypes. STATA routines to aid SNP selection by haplotype tagging (Johnson *et al.*, 2001) are available in ftp://ftp-gene.cimr.cam.ac.uk/software/clayton/stata/htSNP/.

Haplotype reconstruction from family data can be achieved by using SimWalk2 (Sobel and Lange, 1996). The derived haplotypes may then be imported to a pedigree-drawing package such as Cyrillic (Chapman, 1990) for viewing recombinants in positional cloning for example. MERLIN (Abecasis *et al.*, 2002) and GENEHUNTER (Kruglyak *et al.*, 1996) also output haplotypes estimated from family data. Another piece of software,

TRANSMIT (Clayton, 1999) allows association testing of family-based haplotypes. All of these programs allow for missing parental genotypes.

11.4.1 Example: EHPLUS and PMPLUS (Zhao *et al.*, 2000)

11.4.1.1 Data Import

PMPLUS requires two input files, namely a parameter file and a data file. The data file contains for each individual, subject ID, subject status (0 = control, 1 = case) and genotypes listed as either pairs of numbered alleles or as numerical genotype codes. A data file with three markers takes the following form:

```
[Subject ID] [Status] [1a] [1b]   [2a] [2b]   [3a] [3b]
   or
[Subject ID] [Status] [1] [2] [3]
```

where *[1a]* and *[1b]* are the alleles of the first genotype or, alternatively, *[1]* alone represents the first genotype. Currently, the compiled limits are 15 alleles, 30 markers and 800 subjects. Note also that whereas subject IDs with a decimal point (e.g. '20.1') work well, more complex IDs containing several dashes and decimal points may lead to erroneous output.

The parameter file consists of five lines of space-delimited integer values, and it defines the tests to be carried out. The following parameter file *(hapfrest.par)* may be used to estimate haplotype frequencies:

```
3 0 0 0
2 2 2
0 0
1 1 1
1 1 1
```

The four values on line 1 specify the number of markers in the data file, whether to perform a marker–marker or case–control analysis (0 or 1, respectively), whether case–control status is to be permuted (0 = no, 1 = yes) and the number of permutations to perform. Line 2 gives the number of alleles for each marker in the data file. The first value on line 3 specifies whether genotypes in the data file are pairs of alleles or numbered genotypes (0 or 1, respectively), while the second value specifies whether screen output is suppressed or shown (0 or 1). Line 4 has a 1 for each marker to be included in the analysis; zero otherwise. Line 5 assigns each marker to one of two blocks (0 or 1), if required in a marker–marker analysis.

11.4.2 Estimating Haplotype Frequencies

Firstly, PMPLUS is run by typing the following:

```
>pmplus hapfrest.par hapfrest.dat hapfrest.out
```

Here *hapfrest.out* is an output file named by the user and created by PMPLUS to record chi-squared statistics and associated p-values for the specified analysis. A second output file named *ehplus.dat* is also generated, in which the contents of *hapfrest.dat* have been converted into EHPLUS format ready for estimation of haplotypes.

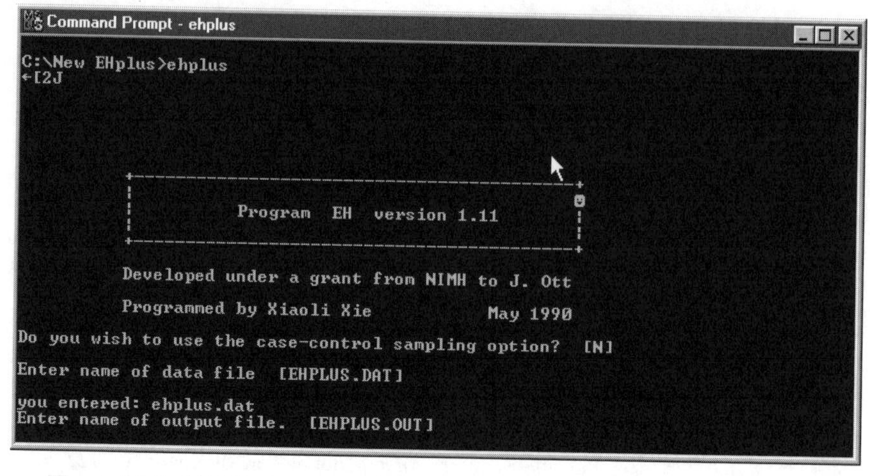

Figure 11.3 EHPLUS interface, as used for estimating haplotype frequencies.

```
---------------------------------
# of Typed Individuals: 200

There are 8 Possible Haplotypes of These 3 Loci.
They are Listed Below, with their Estimated Frequencies:
-----------------------------------------------------------
| Allele   Allele   Allele  |      Haplotype Frequency      |
|   at       at       at    |                               |
| Locus 1  Locus 2  Locus 3 |  Independent   w/Association  |
-----------------------------------------------------------   Observed   Expected   Freeman-Tukey Z:
    1        1        1         0.253194      0.220146          88.06     101.28       -1.33
    1        1        2         0.011931      0.044928          17.97       4.77        4.11
    1        2        1         0.248181      0.259925         103.97      99.27        0.49
    1        2        2         0.011694      0.000001           0.00       4.68       -3.42
    2        1        1         0.229081      0.239855          95.94      91.63        0.47
    2        1        2         0.010794      0.000071           0.03       4.32       -3.09
    2        2        1         0.224544      0.235074          94.03      89.82        0.46
    2        2        2         0.010581      0.000000           0.00       4.23       -3.23
-----------------------------------------------------------
# of Iterations = 20

                                      df    Ln(L)     Chi-square
-------------------------------------------------------------
H0: No Association                     3   -474.91       0.00
H1: Allelic Associations Allowed       7   -466.63      16.57
```

Figure 11.4 Haplotype frequency output from EHPLUS — named *ehplus.out*.

Haplotype estimation is carried out by typing >*ehplus* to invoke the program and then pressing the <*CarriageReturn*> three times to accept the default options provided. The process, as seen using the PC(DOS) version, is shown in Figure 11.3. The output file, *ehplus.out*, shown in Figure 11.4, contains the estimated haplotype frequencies (see column labelled *w/Association*) as well as log likelihoods for the null and alternative hypotheses of *No Association* between the markers and *Allelic Associations Allowed* between the markers. Such inter-marker association is termed linkage disequilibrium and shall be the topic of the next section of this chapter.

11.4.3 Haplotype-based Association Testing

In order to test 2-point haplotypes for association to a disease, the parameter file, *hapfrest.par* was modified to produce the following parameter file (*cscntcom.par*):

```
3 1 1 100
2 2 2
0 0
1 1 0
0 0 0
.01 0 1 1
```

The second, third and fourth entries on line 1 of this parameter file now specify that a case–control analysis is to be performed and significance determined by permuting case–control status 100 times. Furthermore, other entries specify that only the first two markers are to be included (line 4), and that genotypes are *not* to be permuted (line 5). This time a sixth line is included, applicable only to case–control analyses. This line contains four possibly non-integer values that define a model of the mode of inheritance of the trait. The first value specifies the assumed disease allele frequency; the following three are penetrant estimates, resulting from zero, one or two copies of the disease allele respectively. In the current example, the model assigns a 0.01 allele frequency and a fully penetrant, pure dominant mode of inheritance. The new data file (*cscntcom.dat*) contains the simulated genotypes of both cases and controls, formatted as described previously. PMPLUS is executed as follows:

```
>pmplus cscntcom.par cscntcom.dat cscntcom.out.
```

When PMPLUS is instructed to perform permutations, EHPLUS is automatically invoked at the end of each PMPLUS run (i.e. following each permutation of the dataset) and thus program control passes back and forth between the two programs until the permutations are complete. Since PMPLUS permutes the data via the *ehplus.dat* input file, the final EHPLUS output file (*ehplus.out*) does not have meaningful haplotype frequency estimates. These are based on permuted, rather than real data.

The output produced by PMPLUS (in this case *cscntcom.out*) contains the key analysis results. These are the χ^2 values and permutation-derived p-values obtained under five sets of assumptions as follows: (1) under the user-specified disease model, (2) under a Mendelian recessive model, (3) under a Mendelian dominant model, (4) by maximizing the log likelihood ratio over multiple disease models and (5) by a non-parametric 'homogeneity' test, to compare log likelihoods calculated from pooling cases and controls and considering them separately. The fifth test is completely non-parametric, while the others are constrained by the population prevalence of disease implied by the user-specified disease model. Figure 11.5 shows the results of evaluating the 2-point haplotype for association with the simulated disease. Note that p-values below 0.0001 are rounded down to zero.

11.5 LINKAGE DISEQUILIBRIUM

Linkage disequilibrium (LD) is a lack of independence, in the statistical sense, between the alleles at two loci. LD exists between two linked loci when particular alleles at these loci occur on the same haplotype more often than would be expected by chance alone. This phenomenon can provide valuable information in locating disease variants from marker data, as a marker in LD with the causal variant provides a flag for its location. LD information also provides a means by which the efficiency of high-density marker maps can be increased. If markers are in strong LD with each other, there is an argument for genotyping only a subset of them.

```
Chi-squared statistic for user-specified model = 23.76, df=3, p=0.0000
Chi-squared statistic for recessive model      = 20.58, df=3, p=0.0001
Chi-squared statistic for dominant model       = 23.76, df=3, p=0.0000
Chi-squared statistic for model-free analysis  = 23.76, df=4, p=0.0001
Chi-squared statistic for heterogeneity model  = 20.38, df=3, p=0.0001

Random number seed = 3000
Number of replicates = 100

User-specified model chi-squared statistic (23.76) was reached 0 times
Recessive model chi-squared statistic (20.58) was reached 0 times
Dominant model chi-squared statistic (23.76) was reached 0 times
Model-free chi-squared statistic (23.76) was reached 0 times
Heterogeneity model chi-squared statistic (20.38) was reached 0 times

Empirical p-values for these statistics are as follows:
T1 - User specified model:         P-value = 0.0000
T2 - Mendelian recessive model: P-value = 0.0000
T3 - Mendelian dominant model:  P-value = 0.0000
T4 - Model-free analysis:          P-value = 0.0000
T5 - Heterogeneity model:          P-value = 0.0000
```

Figure 11.5 Output of haplotype-based association testing in EHPLUS.

The extent of pair-wise LD may be measured by the value D, as follows (Lewontin, 1964). Assume two diallelic loci are linked and let p_{ij} be the proportion of chromosomes that have allele i at the first locus and allele j at the second locus. For example, p_{12} is the frequency of the haplotype with allele 1 at the first locus and allele 2 at the second locus. The disequilibrium coefficient D is the difference between the observed haplotype frequency p_{12} and the haplotype frequency expected under linkage equilibrium, the latter being the product of the two allele frequencies, say p_{1+} and p_{+2}. It may be written as follows:

$$D = p_{12} - p_{1+}p_{+2}$$

Another commonly quoted measure of LD is D' (Lewontin, 1964). This is a normalized form, with numerator equal to D and denominator equal to the absolute maximum D that could be achieved given the allele frequencies at the two loci. Many other valid measures of pair-wise LD exist and have been reviewed elsewhere (Devlin and Risch, 1995; Hedrick, 1987).

As noted above, EHPLUS can perform tests of LD among a group of markers. The complete set of pair-wise tests for the group, together with D and D' values, can be achieved in a single step using software such as Arlequin (Schneider *et al.*, 2000). This is a C++ program available for PC(Win), Linux and MacOS systems. The statistical significance of observed LD is estimated for phase known (haplotype) data by means of a Fisher's Exact Test. For phase unknown data, a likelihood ratio test is applied. An alternative tool is GDA (Lewis and Zaykin, 2001), the PC(Win) companion program to the book, *Genetic Data Analysis II* (Weir, 1996). Both are well documented and perform a broad range of population genetic tests.

The software, GOLD (Abecasis and Cookson, 2000), available for PC(Win), is another program that will calculate D and D', and it is noteworthy in that it can output them in graphical form. For each marker pair, the pair-wise disequilibrium statistics are colour coded (bright red to dark blue) and plotted. The output is valuable for presentation purposes and provides a useful summary of the properties of dense maps. The software takes haplotype estimates as input and, in the case of family data, these must be reconstructed using software such as SimWalk2 (Sobel and Lange, 1996) prior to use. Case–control data is not well supported by GOLD, which relies for this purpose upon a limited interface to the software, EH (Xie and Ott, 1993).

Other methods of estimating LD include the Moment Method, applicable to newly-formed populations under certain assumptions concerning the evolutionary process (Hastabacka *et al.*, 1992; Kaplan *et al.*, 1995; Lehesjoki *et al.*, 1993). Maximum likelihood methods have also been explored (Hill and Weir, 1994; Kaplan *et al.*, 1995). Composite likelihood methods were proposed to evaluate the information from multiple pairs of loci simultaneously. Examples of software for the composite likelihood approach include DMAP (Devlin *et al.*, 1996) and ALLASS (Collins and Morton, 1998). The latter uses the Malecot isolation by distance equation and has the advantage of accommodating multiple founder mutations. Each method however relies upon population assumptions and may suffer reduced power when these are not met.

11.5.1 Example: Arlequin (Schneider *et al.*, 2000)

11.5.1.1 Data Import

Arlequin categorizes data into five groups, namely DNA sequences, RFLP data, microsatellite data, allele frequency data and standard data. The latter assumes that different alleles are mutationally equidistant from each other, as is the case with SNP data. Data can be loaded in two ways, by importing a project file, or by using the Project Wizard, to guide you through the creation of a project. Figure 11.6 shows the Arlequin interface in Windows NT, having selected the import screen. As shown, a number of data formats may be read in, and converted by selecting Arlequin as the Target format. LINKAGE format is not however, supported.

With the objective of testing for LD between five markers, the current example may be regarded as a Standard data project. The data and the parameters of the project are shown below in an Arlequin format, for which the filename extension *.arp* is required. The first *[Profile]* section describes the data before it is listed in the second, *[Data]* section. Comments are included, preceded by '#' and these are ignored by the program.

```
[Profile]               # first describe the data for this project

Title = 'Simulated data for five genetic markers'
      NbSamples = 1     # Number of study populations in the project.
  DataType = STANDARD
      GenotypicData = 1    # 1= yes; 0 = no (i.e. haplotypic)
  LocusSeparator = WHITESPACE
      GameticPhase = 0     # 1 = yes; 0 = no (i.e. phase unknown)
      RecessiveData = 0    # 1 = yes; 0 = no (i.e. codominant alleles)
      RecessiveAllele = null # because RecessiveData = 0
      MissingData  = '.'   # the missing data code
[Data]                     # next list the data points

[[Samples]]

         SampleName = 'Simulation 1'
         SampleSize = 200 #200 individuals are in the study set
         SampleData = {
```

```
CONFIG1 34 1 1 1 1 2 # The first genotype combination is labelled CONFIG1
           2 1 2 1 2 # 34 individuals have this set of five genotypes
CONFIG2 14 2 1 1 1 2
           2 1 1 1 2
CONFIG3 9  1 1 1 1 2 # 9 individuals have this set of five genotypes
           1 2 2 1 2
```

Subsequent lines of data follow the same paired format and the final line consists of a '}' symbol. This project file is specific to the problem in hand, namely phase-unknown genotype data. Variations exist for other data types and are described in detail in the user manual. It can be seen that genotypes are written with one allele directly below the other allele. This allows a mechanism for inputting phase-known data, for which each line represents a haplotype. In our case, the phase is unknown, so the relative orderings of the alleles are ignored.

Upon successful import, a 'Project' is created by Arlequin. It is remembered by the system and can be recalled at a later date. Its details can be viewed by selecting the menu

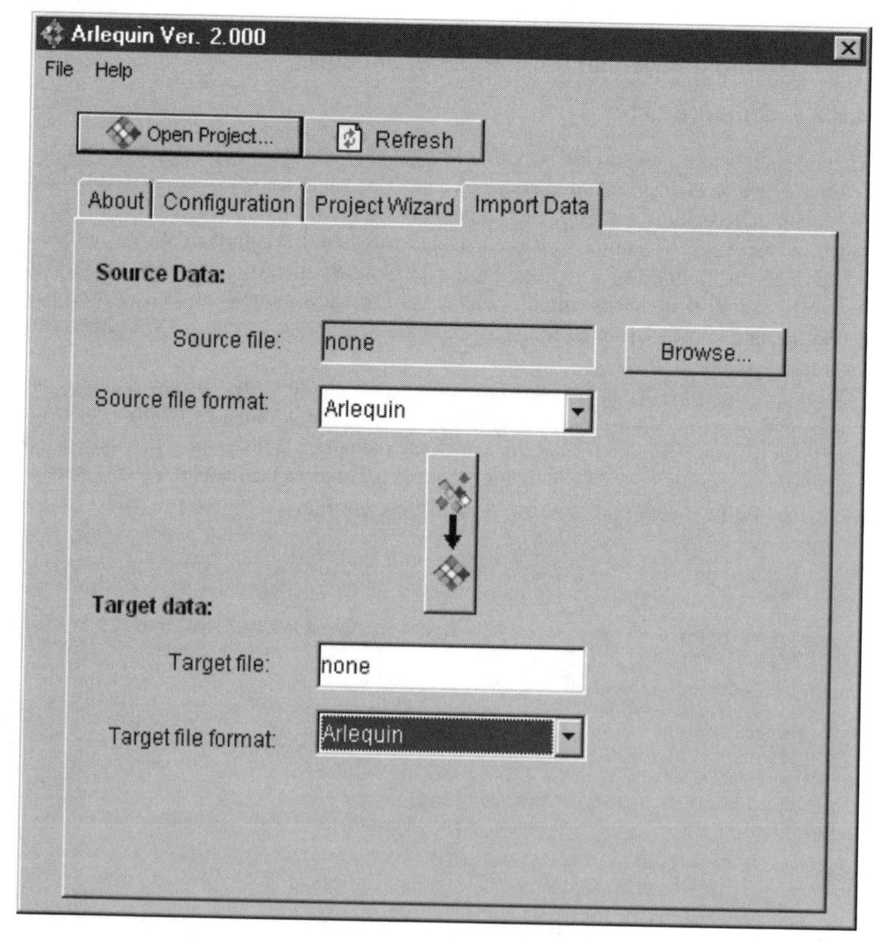

Figure 11.6 Arlequin Screen. Initiating an analysis run.

items Project>View Project Info. The following analyses are then performed by making selections in the launch pad dialogue box.

11.5.2 Linkage Disequilibrium Analysis of Genotypes with Unknown Phase

An LR test statistic, denoted by S, is used to test for LD between a pair of loci when phase is unknown (Slatkin and Excoffier, 1996). It compares the likelihood of a model assuming linkage equilibrium to that of a model allowing linkage disequilibrium. Asymptotically, this statistic follows a χ^2 distribution, but to allow for small sample size or the study of markers with large numbers of alleles, Arlequin also uses a permutation procedure to test for significance.

The analysis screen is given in Figure 11.7. The procedure is as follows:

1. Click on the *Calculation Settings* tab
2. Click to the left of the folder *Linkage disequilibrium*

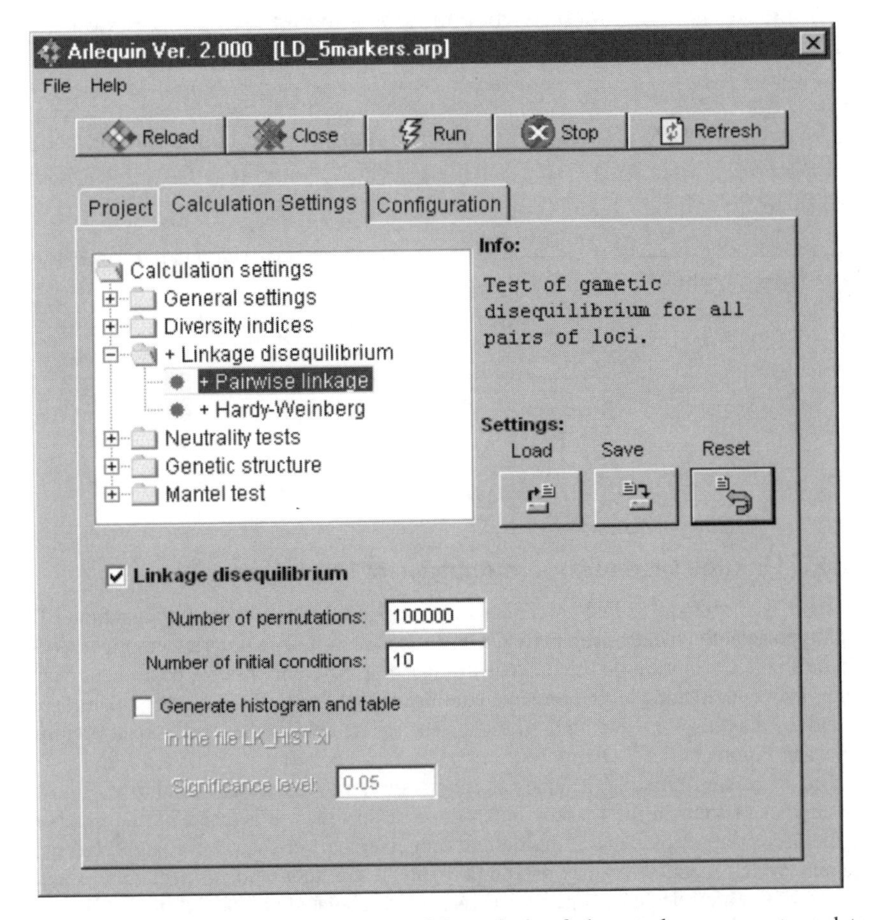

Figure 11.7 Arlequin Screen. Setting up LD analysis of phase-unknown genotype data.

3. Select *Pairwise linkage*
4. Select the *Linkage Disequilibrium* box below the settings window. A plus sign will appear
5. Input parameter values. Default values are given in Figure 11.7. However in the manual, it is recommended that 16,000 permutations be conducted to establish significance and the EM be applied to at least 100 initial conditions
6. Click on the *Run* button

Detailed output is written to an HTML result file, in a sub-directory of that containing the input. First, parameter settings are stated, then for each locus pair, there is a listing of the log-likelihoods under the null and alternative hypotheses, a p-value determined by permutation, the χ^2 test statistic and its corresponding (asymptotic) p-value. Lastly, a table is provided, in which a '+' sign denotes nominal evidence of a departure from linkage equilibrium. This allows the results to be scanned rapidly by eye. Samples of the output are shown below.

```
Pair(0, 1)
     LnLHood LD: -302.71677       LnLHood LE: -319.36838
     Exact P = 0.00000 +- 0.00000 (16002 permutations done)
Chi-square test value = 33.30322 (P  = 0.00000, 1 d.f.)
Pair(0, 2)
     LnLHood LD: -420.27411       LnLHood LE: -420.70319
     Exact P = 0.36164 +- 0.00381 (16002 permutations done)
Chi-square test value  = 0.85815 (P  = 0.35426, 1 d.f.)
```

(and so on)

```
Table of significant linkage disequilibrium (significance
level  = 0.0500):

Locus # |  0|  1|  2|  3|  4|
------------------------------------------
       0|   *   +   -   -   -
       1|   +   *   +   -   -
       2|   -   +   *   -   -
       3|   -   -   -   *   -
       4|   -   -   -   -   *
```

11.5.3 Linkage Disequilibrium Analysis of Haplotypes

Arlequin uses a modified Fisher's Exact test, as opposed to the LR test, to examine LD in haplotype data. Such data is given by $GameticPhase = 1$. The program employs Markov Chain Monte Carlo sampling to explore the space of different possible contingency tables rather than enumerating all the possible contingency tables. In this case, the LD measures, D and D' may also be generated. The analysis screen reflects these additional options as shown in Figure 11.8.

The process of initiating the analysis is very similar to that described above. This time, the number of steps in the Markov chain must be specified, together with the number of de-memorization steps. Again, the default values are lower than those suggested in the manual, which mentions values of 100,000 and 'a few thousand' respectively. If the D and D' boxes are selected, all pair-wise values are tabulated and output in HTML format as well as in a file called *LD_DIS.XL*, ready for inputting to MS Excel.

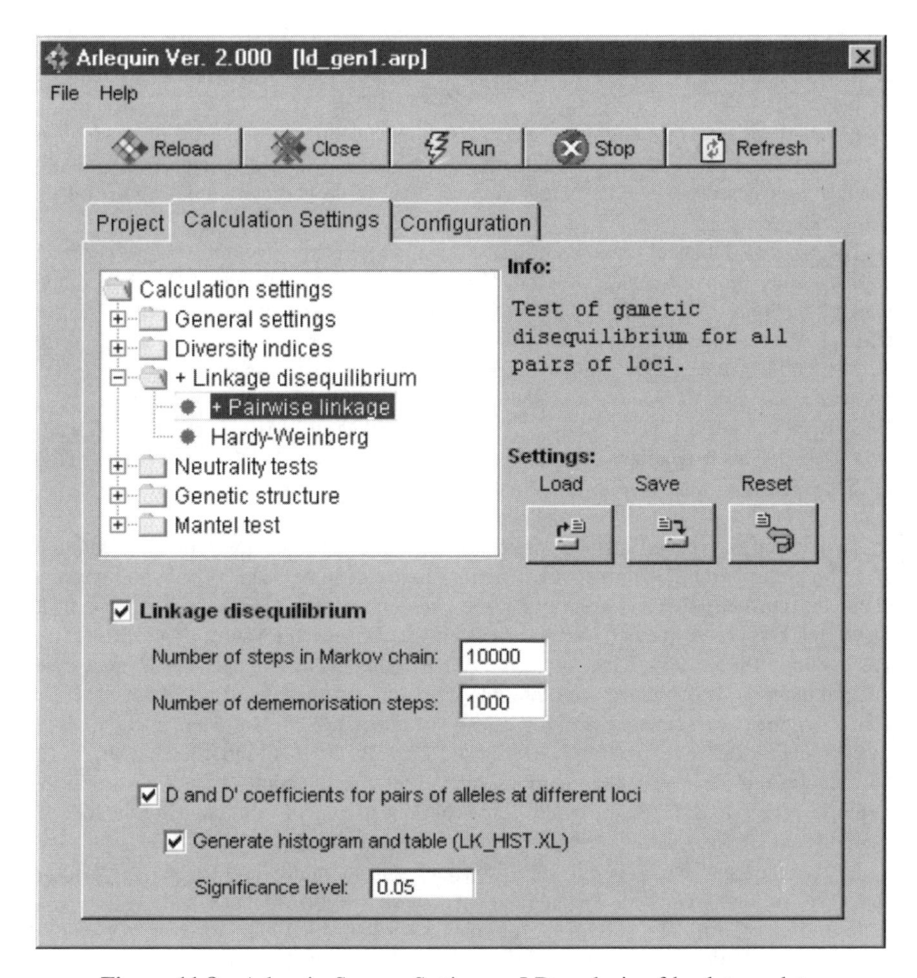

Figure 11.8 Arlequin Screen. Setting up LD analysis of haplotype data.

11.6 QUANTITATIVE TRAIT LOCUS (QTL) MAPPING IN EXPERIMENTAL CROSSES

In contrast to human studies, in which variances of phenotypic differences are used to establish the presence of linkage, QTL mapping in experimental crosses involves comparing means of progeny inheriting specific parental alleles. This is simpler and more powerful (Kruglyak and Lander, 1995). It can be achieved by any of a number of standard statistical methods, such as t-tests, analysis of variance (ANOVA), Wilcoxon rank-sum and regression techniques. Again, missing data can be accommodated by an application of the EM algorithm.

Of the very broad array of possible diploid crosses, the following are particularly common. They are derived from a pair of divergent inbred lines in which the genotypes at the majority of loci are homozygous and distinct, say *aa* and *bb* for a particular locus in the two lines respectively. The filial F_1 generation results from crossing these two lines to produce individuals with heterozygous genotype *ab*. In the backcross (BC) design, F_1 is crossed with one of the parent strains. For example, in the case of a cross with the *aa* parent, half the offspring produced are *ab* and half are *aa*. In the filial F_2 design, the F_1 is selfed, or two F_1 individuals are crossed so that offspring are *aa, ab,* and *bb* in the ratio 1:2:1. Lastly, in the recombinant inbred line (RIL), each F_2 enters individually a single seed descent-inbreeding programme so that all progeny are homozygous for the chosen allele.

The original statistical framework for QTL mapping in experimental crosses was based upon a marker-by-marker analysis. Of particular relevance to sparse maps however, simple interval mapping (IM or SIM) allows the evaluation of any position within a marker interval. The maximum likelihood approach to IM proceeds by calculation of a LOD score (Lander and Botstein, 1989). Similarly, and with lower computational burden, least squares regression achieves the same goal (Haley and Knott, 1992; Martinez and Curnow, 1992). IM may be carried out using a range of software, including MAPMAKER/QTL (Lander *et al.*, 1987). This may appeal to regular users of MAPMAKER/SIBS or GENEHUNTER, as the syntax is similar. It relies upon data pre-processing in MAPMAKER/EXP (Lander *et al.*, 1987) and allows simple graphical output.

Two newer and related methods are Composite Interval Mapping (CIM) and Multiple QTL Mapping (MQM). Both involve performing a genome scan by moving stepwise along the chromosome and testing for the presence of the QTL using a pre-defined set of markers as co-factors (Jansen 1992, 1993; Jansen and Stam, 1994; Kao *et al.*, 1999; Zeng, 1993, 1994; Zeng *et al.*, 1999). In other words, in the sparse map case, interval mapping is combined with multiple regression on markers. This approach allows you to control, to some extent, for effects of other QTLs. Software such as QTL Cartographer (Basten *et al.*, 1994, 1997) and PLABQTL (Utz and Melchinger, 1996; http://probe.nalusda.gove:8000/otherdocs/jqtl/) allow the selection of such co-factors by stepwise regression. These programs offer options that will automatically include or exclude background markers according to user-defined criteria.

Lastly, Bayesian methods allow the consideration of multiple QTLs, QTL positions and QTL strengths (Jansen, 1996; Satagopan *et al.*, 1996; Sillanpaa and Arjas, 1998; Uimari *et al.*, 1996). The software Multimapper (Sillanpaa, 1998), for example, allows the automatic building of models of multiple QTLs within the same linkage group. It is designed to work as a companion program to QTL Cartographer (Basten *et al.*, 1994, 1997) and allows a more detailed follow-up of regions of interest. As with other Markov Chain Monte Carlo methods, however, this approach is computer intensive and may suffer from problems of convergence to a local, rather than global, optimum or of lack of convergence if run for a short time.

Ten of the most prominent pieces of software for QTL mapping are reviewed in greater detail by Manley and Olson (1999). The majority will perform IM and CIM for backcross, filial F_2 and recombinant inbred lines. Cordell (2002) provides worked examples of the usage of three of them, MAPMAKER/QTL, QTL Cartographer and another piece of software, MapQTL (van Ooijen and Maliepaard, 1996a, b).

A major limitation of QTL mapping using inbred lines is the broad, ill-defined nature of the resulting linkage peaks, which typically span tens of centiMorgans even if large numbers of progeny are analysed (for example see Farmer *et al.*, 2001). This is a consequence of the multifactorial nature of quantitative traits, which results in an inability to identify unequivocal recombinants that precisely delineate a critical genetic interval, in contrast with monogenic phenotypes. Subsequent attempts to narrow a locus by, for example, successive rounds of backcrossing are often frustrated by the dilution or loss of unlinked genetic co-factors that are required for trait manifestation. In the future, QTL mapping using genetically heterogeneous stocks may gain in prominence (Mott *et al*., 2000). Talbot *et al*. (1999) were able to achieve a mapping resolution of less than 1 cM by the study of heterogeneous stocks from eight known inbred mouse progenitor strains that had been intercrossed over 30–60 generations. The group has released software called HAPPY (Mott *et al*., 2000) which requires knowledge of the ancestral alleles in the inbred founders, together with the genotypes and phenotypes in the final generation. It will then apply variance component methods to test for linkage to the QTL.

11.6.1 Example: Map Manager QTX (Manley *et al*., 2001)

Map Manager QTX is available for both MacOS and PC(Win). It has no licence fee and was selected here due to the usefulness of its graphic user interface. It has both IM and CIM capability and can reformat data for use in other important software such as QTL Cartographer. Interval mapping is based on the Haley and Knott (1992) procedure, and CIM is achieved by adding background loci. Significance can be assessed by permutation (Churchill and Doerge, 1994).

The genotype data may derive from inbred or non-inbred stock and options are provided for a variety of experimental designs. Extensive documentation can be downloaded in either pdf or Hypertext formats. The *Tutorial* is especially helpful; but readers should be aware that its files are somewhat inconspicuously tucked in with *Sample Data* files, rather than being included in the Map Manager QTX Manual.

For the current example, genotype data was downloaded from the Mouse Genome Database (2001). Specifically, it consists of mouse chromosome 1 genotypes from the Copeland–Jenkins backcross, and a selected subset of 10 markers spanning the entire ~ 100-cM length of the chromosome. Marker *En1* is located near the middle of the chromosome, between markers *Col6a3* and *D1Fcr15*, and it was used to simulate the quantitative trait (QT) for the 193 backcross mice. Homozygotes (denoted as *b*) at *En1* received a QT value of 50 ± 20 (mean $\pm$ SD) while heterozygotes (*s*) at *En1* received a QT value of 100 ± 20. *En1* was then removed from the dataset and Map Manager QTX was used to analyse QT association with the remaining nine markers as shown below.

11.6.1.1 Data Import

Map Manager QTX is launched by a mouse click on the Map Manager icon (*QTXb13.exe*), thus opening the main menu. The genotype data (alternatively termed 'Phenotype data' by Map Manager QTX) is imported by selecting *File>Import>Text*. The name of each marker and the genotypes (phenotypes) of the cross progeny are imported as a single line of text. The marker name is separated from the genotypes by a tab character but the genotypes, each represented as above by a single letter, can be given as either an unbroken string of characters or space-separated. In our case, the first two lines of input therefore took the following form (with missing genotypes given by a hyphen):

```
Actn3<tab>sssbbbbbsbsbsbsssbbsbbsbbbbssbsbbsbsb-bbb-ssss
  <CarriageReturn>
Laf4<tab>-sbbbb--sb-------bb--bsbb-bbsb--s-bbbbbsbssb-
  bs<CarriageReturn>
```

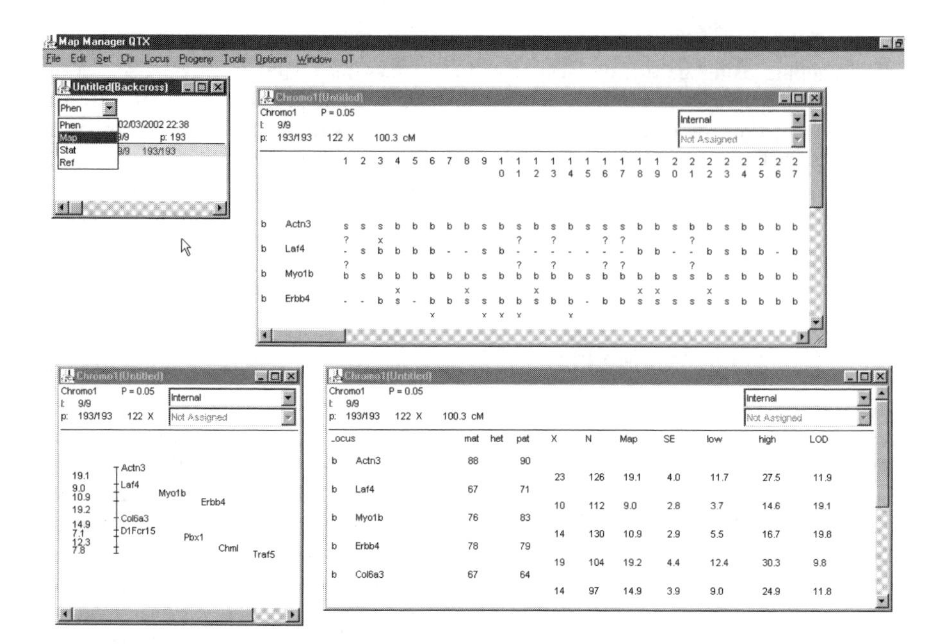

Figure 11.9 Screens in Map Manager QTX. The dataset window (upper left), the Phenotype window (upper right), the Map window (lower left) and the Statistics window (lower right). Genotypes with permission from Mouse Genome Database (2001).

QT Links report for set Backcross(Backcross)
Using quantitative trait QTrait
Chi Square Stats, P = 0.05
Additive regression model
With no control for other QTLs.
182 to 193 informative progeny from Progeny Order Internal

Add highlighted Locus to background for Trait QTrait

Chr	Locus	Stat	%	P	Add
Chromo 1	:: Laf4	13.6	6	0.00022	-18.52
Chromo 1	:: Myo1b	23.9	11	0.00000	-23.54
Chromo 1	:: Erbb4	50.6	23	0.00000	-33.49
Chromo 1	:: Col6a3	143.5	52	0.00000	-52.25
Chromo 1	:: D1Fcr15	99.8	40	0.00000	-44.62
Chromo 1	:: Pbx1	58.7	26	0.00000	-34.83
Chromo 1	:: Chml	23.7	11	0.00000	-23.28
Chromo 1	:: Traf5	6.4	3	0.01151	-12.49

Figure 11.10 Output from single marker association testing in Map Manager QTX: The 'Links Report'. 'Add' denotes the additive regression coefficient for the association. Genotypes with permission from Mouse Genome Database (2001).

Quantitative trait data are then read in from a second text file via *File>Import>Trait Text*. The format is almost identical, except that the name of the trait replaces marker name and the trait value for each mouse must be separated from adjacent values by at least one space. Again, the name of the quantitative trait and all of the values for cross progeny must be in a single line of text.

Successful import of a text genotype file produces a small pop-up window (the *dataset* window), as shown in Figure 11.9, top left. Within it is a menu allowing selection of *Phen, Map, Stat* or *Ref*. Selecting one of these options and double-clicking on a chromosome name in the *dataset* window, produces the chosen window as shown in Figure 11.9. The *Phenotype* window (top right) displays the marker names on the left side of the window, with one column for each member of the progeny. The body of the *Phenotype* window shows the genotype at each locus and also indicates locations of recombination

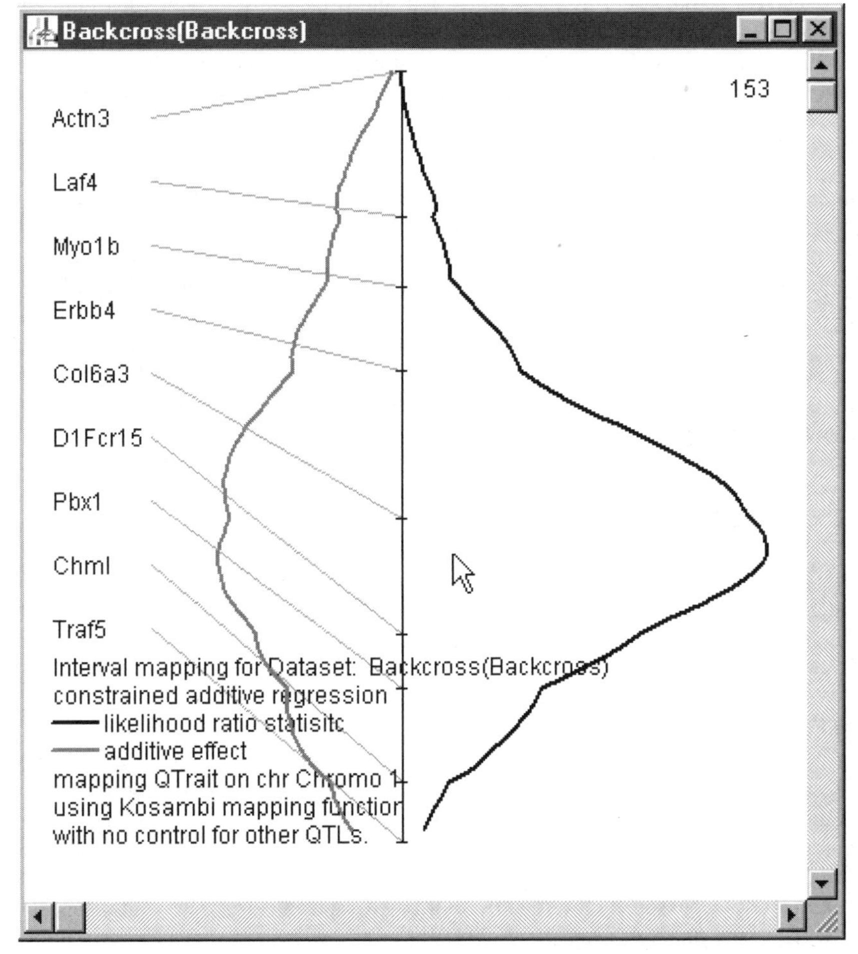

Figure 11.11 Output from Map Manager QTX. Results of interval mapping across nine markers. Genotypes with permission from Mouse Genome Database (2001).

events with an X. Pairs of question marks denote the possible locations of crossovers whose more precise location cannot be specified due to missing genotype data. The *Map* window (bottom left), shows a genetic map with estimated cM distance between markers, and the *Statistics* window (bottom right) summarizes useful numerical information, such as the number of recombination events between adjacent markers and LOD evidence for linkage.

11.6.1.2 Single marker association

Testing for association between an individual marker and a quantitative trait is accomplished by first selecting a p-value cut-off in the *Main* menu under *Options>Search&Linkage criteria*, and then choosing *QT>Links Report* in the *Main* menu. This produces a window allowing the user to select both the name of the quantitative trait to test and the background QTLs to be included in the analysis.

Figure 11.10 shows the table or *Links Report* that was produced by testing each of the nine markers in our panel for association with the simulated trait. Note that only eight markers appear in the table, as one marker did not meet the $p < 0.05$ criterion. Note also that marker *Col6a3* is highlighted as giving the strongest association and therefore as being the best marker to include as a background QTL in analyses of other chromosomal loci.

11.6.1.3 Simple Interval Mapping

Simple interval mapping of a QT across a series of markers is accomplished by choosing *QT>Interval Mapping* from the *Main* menu. This produces a window which again allows the user to specify the trait to be analysed and whether any background QTLs are to be included in the analysis. Once options in this window are specified, Map Manager QTX produces a table and a figure displaying the Interval Mapping results. Figure 11.11 shows the result of interval mapping our simulated trait across the nine markers on mouse chromosome 1. As indicated by the position of the cursor, the peak of the likelihood ratio statistic falls very close to the true location of the simulated QT locus, between markers *Col6a3* and *D1Fcr15*.

ACKNOWLEDGEMENTS

The authors wish to thank Heather Cordell, Dmitri Zaykin, Clive Bowman, Meg Ehm and Leonid Kruglyak for helpful discussions, advice and support during the writing of this chapter.

REFERENCES

Abecasis GR, Cardon LR, Cookson WO. (2000). A general test of association for quantitative traits in nuclear families. *Am J Hum Genet* **66**: 279–292.

Abecasis GR, Cherny SS, Cookson WO, Cardon LR. (2002). Merlin—rapid analysis of dense genetic maps using sparse gene flow trees. *Nature Genet.* **30**: 97–101.

Abecasis GR, Cookson WOC. (2000). GOLD—Graphical Overview of Linkage Disequilibrium. *Bioinformatics* **16**: 182–183.

Allison DB. (1997). Transmission-disequilibrium tests for quantitative traits. *Am J Hum Genet* **60**: 676–690.

Amos CI. (1994). Robust variance components approach for assessing genetic linkage in pedigrees. *Am J Hum Genet* **54**: 535–543.

Basten C, Weir BS, Zeng Z-B. (1994). Zmap—a QTL cartographer. In Smith C, Gavora JS, Benkel B, Chesnais J, Fairfull W, Gibson JP, Kennedy BW, Burnside EB. *Proceedings of the 5th World Congress on Genetics Applied to Livestock Production: Computing Strategies and Software* vol. 22. pp. 65–66. (On-line publication)

Basten C, Weir BS, Zeng Z-B. (1997). *QTL Cartographer: A Reference Manual and Tutorial for QTL Mapping*. Department of Statistics, North Carolina State University: Raleight, NC. (http://statgen.ncsu.edu/qtlcart/).

Blangero J, Almasy L. (1996). *SOLAR: Sequential Oligogenic Linkage Analysis Routines*. Technical notes no. 6, Population Genetics Laboratory, Southwest Foundation for Biomedical Research: San Antonio, TX.

Chapman CJ. (1990). A visual interface to computer programs for linkage analysis. *Am J Med Genet* **36**: 155–160.

Churchill GA, Doerge RW. (1994). Empirical threshold values for quantitative trait mapping. *Genetics* **138**: 963–971.

Clark AG. (1990). Inference of haplotypes from PCR amplified samples of diploid populations. *Mol Biol Evol* **7**: 111–122.

Clark AG, Weiss KM, Nickerson DA, Taylor SL, Buchanan A, Stengard J, *et al.* (1998). Haplotype structure and population genetic inferences from nucleotide-sequence variation in human lipoprotein lipase. *Am J Hum Genet* **63**: 595–612.

Clayton D. (1999). A generalization of the transmission/disequilibrium test for uncertain-haplotype transmission. *Am J Hum Genet* **65**: 1170–1177.

Clayton D. (2001). Population association. In Balding DJ, Bishop M, Cannings C. (Eds), *Handbook of Statistical Genetics*. John Wiley: Chichester, pp. 519–540.

Clayton D, Jones HB. (1999). Transmission/disequilibrium tests for extended marker haplotypes. *Am J Hum Genet* **65**: 1161–1169.

Collins A, Morton NE. (1998). Mapping a disease locus by allelic association. *Proc Natl Acad Sci USA* **95**: 1741–1745.

Conneally PM, Edwards JH, Kidd KK, Lalouel J-M, Morton NE, Ott J, *et al.* (1985). Report of the committee on methods of linkage analysis and reporting. *Cytogenet Cell Genet* **40**: 356–359.

Cordell HJ. (2002). Diabetes in the NOD mouse. In Camp N, Cox A. (Eds), *Quantitative Trait Loci: Methods and Protocols*. Humana Press: pp. 165–198.

Cottingham RW Jr, Idury RM, Schaffer AA. (1993). Fast sequential genetic linkage computation. *Am J Hum Genet* **53**: 252–263.

Cudworth AG, Woodrow JC. (1975). Evidence for HLA-linked genes in 'juvenile' diabetes mellitus. *Br Med J* **3**: 133–135.

Dempster AP, Laird NM, Rubin DB. (1977). Maximum likelihood from incomplete data via the EM algorithm. *J Roy Stat Soc* **B39**: 1–38.

Devlin B, Risch N. (1995). A comparison of linkage disequilibrium measures for fine-scale mapping. *Genomics* **29**: 311–322.

Devlin B, Risch N, Roeder K. (1996). Disequilibrium mapping: composite likelihood for pairwise disequilibrium. *Genomics* **36**: 1–16.

Excoffier L, Slatkin M. (1995). Maximum-likelihood estimation of molecular haplotype frequencies in a diploid population. *Mol Biol Evol* **12**: 921–927.

Farmer MA, Sundberg JP, Bristol IJ, Churchill GA, Li R, Elson CO, *et al.* (2001). A major quantitative trait locus on chromosome 3 controls colitis severity in IL-10-deficient mice. *Proc Natl Acad Sci USA* **98**: 13820–13825.

Fulker DW, Cardon LR. (1994). A sib-pair approach to interval mapping of quantitative trait loci. *Am J Hum Genet* **54**: 1092–1103.

Fulker DW, Cherny SS, Sham PC, Hewitt JK. (1999). Combined linkage and association sib-pair analysis for quantitative traits. *Am J Hum Genet* **64**: 259–267.

Goldgar DE. (1990). Multipoint analysis of human quantitative genetic variation. *Am J Hum Genet* **47**: 957–967.

Haley CS, Knott SA. (1992). A simple regression method for mapping quantitative trait loci in line crosses using flanking markers. *J Hered* **69**: 315–324.

Haseman JK, Elston RC. (1972). The investigation of linkage between a quantitative trait and a marker locus. *Behav Genet* **2**: 3–19.

Hastabacka J, de la Chapelle A, Kaitila I, Sistonen P, Weaver A, Lander E. (1992). Linkage disequilibrium mapping in isolated founder populations: Diastrophic dysplasia in Finland. *Nature Genet* **2**: 204–211.

Hauser ER, Boehnke M. (1997). Confirmation of linkage results in affected-sib-pair linkage analysis for complex genetic traits. *Am J Hum Genet* **61**: A278.

Hawley ME, Kidd KK. (1995). HAPLO: a program using the EM algorithm to estimate the frequencies of multi-site haplotypes. *J Hered* **86**: 409–411.

Heath S. (1997). Markov chain segregation and linkage analysis for oligogenic models. *Am J Hum Genet* **61**: 748–760.

Hedrick PW. (1987). Gametic disequilibrium measures: Proceed with caution. *Genetics* **117**: 331–341.

Hill WG, Weir BS. (1994). Maximum-likelihood estimation of gene location by linkage disequilibrium. *Am J Hum Genet* **54**: 705–714.

Hinds D, Risch N. (1996). The ASPEX package: affected sib-pair mapping ftp://lahmed.stanford.edu/pub/aspex.

Holmans P. (2001). Nonparametric linkage. In Balding DJ, Bishop M, Cannings C. (Eds), *Handbook of Statistical Genetics*. John Wiley: Chichester, pp. 487–505.

Holmans P, Clayton D. (1995). Efficiency of typing unaffected relatives in an affected sib-pair linkage study with single locus and multiple tightly-linked markers. *Am J Hum Genet* **57**: 1221–1232.

Jansen RC. (1992). A general mixture model for mapping quantitative trait loci by using molecular markers. *Theoretical and Applied Genetics* **85**: 252–260.

Jansen RC. (1993). Interval mapping of multiple quantitative trait loci. *Genetics* **135**: 205–211.

Jansen RC. (1996). A general Monte Carlo method for mapping multiple quantitative trait loci. *Genetics* **142**: 305–311.

Jansen RC, Stam P. (1994). High resolution of quantitative traits into multiple loci via interval mapping. *Genetics* **136**: 1447–1455.

Jensen CS, Kong A, Kjaerulff KM. (1995). Blocking–Gibbs sampling in very large probabilistic expert systems. *Int J Hum Computer Studies* 647–666.

Johnson GC, Esposito L, Barratt BJ, Smith AN, Heward J, Di Genova G, *et al.* (2001). Haplotype tagging for the identification of common disease genes. *Nature Genet* **29**: 233–237.

Kao CH, Zeng Z-B, Teasdale RD. (1999). Multiple interval mapping for quantitative trait loci. *Genetics* **152**: 1203–1216.

Kaplan N, Hill WG, Weir BS. (1995). Likelihood methods for locating disease genes in nonequilibrium populations. *Am J Hum Genet* **56**: 18–32.

Kong A, Cox NJ. (1997). Allele sharing models: LOD scores and accurate linkage tests. *Am J Hum Genet* **61**: 1179–1188.

Kruglyak L, Lander ES. (1995). Complete multipoint sib-pair analysis of qualitative and quantitative traits. *Am J Hum Genet* **57**: 439–454.

Kruglyak L, Daly MJ, Reeve-Daly MP, Lander ES. (1996). Parametric and nonparametric linkage analysis: a unified multipoint approach. *Am J Hum Genet* **58**: 1347–1363.

Lander ES, Botstein D. (1989). Mapping Mendelian factors underlying quantitative traits using RFLP linkage maps. *Genetics* **121**: 185–199.

Lander ES, Green P, Abrahamson J, Barlow A, Daly M, Lincoln SE, *et al.* (1987). MAPMAKER: an interactive computer package for constructing primary genetic linkage maps of experimental and natural populations. *Genomics* **1**: 174–181.

Lathrop GM, Lalouel JM, Julier C, Ott J. (1984). Strategies for multilocus linkage analysis in humans. *Proc Natl Acad Sci USA* **81**: 3443–3446.

Lehesjoki A-E, Koskiniemi M, Norio R, Tirrito S, Sistonen P, Lander E, *et al.* (1993). Localization of the EPM1 gene for progressive myoclonus epilepsy on chromosome 21: Linkage disequilibrium allows high resolution mapping. *Hum Mol Genet* **2**: 1229–1234.

Lewis PO, Zaykin D. (2001). Genetic Data Analysis: Computer program for the analysis of allelic data. Version 1.0 (d16c). Free program distributed by the authors over the internet from http://lewis.eeb.uconn.edu/lewishome/software.html.

Lewontin RC. (1964). The interaction of selection and linkage I. General considerations; heterotic models. *Genetics* **49**: 49–67.

Little RJA, Rubin DB. (1987). *Statistical Analysis with Missing Data*. Wiley: New York.

Long JC, Williams RC, Urbanek M. (1995). An E-M algorithm and testing strategy for multiple locus haplotypes. *Am J Hum Genet* **56**: 799–810.

Manly KF, Olson JM. (1999). Overview of QTL mapping software and introduction to map manager QT. *Mamm Genome* **10**: 327–334.

Manly KF, Cudmore RH, Meer JM. (2001). Map Manager QTX, cross-platform software for genetic mapping. *Mamm Genome* **12**: 930–932.

Martinez O, Curnow RN. (1992). Estimating the locations and the sizes of the effects of quantitative trait loci using flanking markers. *Theor Appl Genet* **85**: 480–488.

Monks SA, Kaplan NL, Weir BS. (1998). A comparative study of sibship tests of linkage and/or association. *Am J Hum Genet* **63**: 1507–1516.

Morton NE. (1955). Sequential tests for the detection of linkage. *Am J Hum Genet* **7**: 277–318.

Mott R, Talbot C, Turri M, Collins AC, Flint J. (2000). A new method for fine mapping quantitative trait loci in outbred animal stocks. *Proc Natl Acad Sci USA* **97**: 12649–12654.

Mouse Genome Database (MGD). (2001). Mouse Genome Informatics Web Site, The Jackson Laboratory, Bar Harbor, Maine. World Wide Web (URL: http://www.informatics.jax.org/).

Nickerson DA, Taylor SL, Weiss KM, Clark AG, Hutchinson RG, Stengard J, *et al.* (1998). DNA sequence diversity in a 9.7-kb region of the human lipoprotein lipase gene. *Nature Genet* **19**: 233–240.

O'Connell JR, Weeks DE. (1995). The VITESSE algorithm for rapid exact multilocus linkage analysis via genotype set-recoding and fuzzy inheritance. *Nature Genet* **11**: 402–408.

Olson JM. (1995). Multipoint linkage analysis using sib pairs: an interval mapping approach for dichotomous outcomes. *Am J Hum Genet* **56**: 788–798.

Rabinowitz D. (1997). A transmission disequilibrium test for quantitative trait loci. *Hum Hered* **47**: 342–350.

Risch N. (1990a). Linkage strategies for genetically complex traits II. The power of affected relative pairs. *Am J Hum Genet* **46**: 229–241.

Risch N. (1990b). Linkage strategies for genetically complex traits: III. The effect of marker polymorphism on analysis of affected relative pairs. *Am J Hum Genet* **46**: 242–253.

SAGE (1999). *Statistical Analysis for Genetic Epidemiology*, Release 4.0. Department of Epidemiology and Biostatistics, Rammelkamp Center for Education and Research, MetroHealth campus, Case Western Reserve University: Cleveland, OH.

Satagopan JM, Yandell BS, Newton MA, Osborn TC. (1996). A Bayesian approach to detect quantitative trait loci using Markov chain Monte Carlo. *Genetics* **144**: 805–816.

Schaid DJ. (1996). General score tests for associations of genetic markers with disease using cases and their parents. *Genet Epidemiol* **13**: 423–449.

Schneider S, Roessli D, Excoffier L. (2000). *Arlequin ver. 2.000: A software for population genetics data analysis*. Genetics and Biometry Laboratory, University of Geneva: Geneva, Switzerland.

Sham PC. (1998). *Statistics in Human Genetics*. Arnold Publishers: London; John Wiley and Sons Inc.: New York.

Sham PC, Curtis D. (1995). An extended transmission/disequilibrium test (TDT) for multi-allele marker loci. *Ann Hum Genet* **59**: 323–336.

Sillanpaa MJ (1998). Multimapper Reference Manual. http://www.RNL.Helsinki.F1/~mjs/.

Sillanpaa MJ, Arjas E. (1998). Bayesian mapping of multiple quantitative trait loci from incomplete inbred line cross data. *Genetics* **148**: 1373–1388.

Slatkin M, Excoffier L. (1996). Testing for linkage disequilibrium in genotypic data using the EM algorithm. *Heredity* **76**: 377–383.

Sobel E, Lange K. (1996). Descent graphs in pedigree analysis: applications to haplotyping, location scores, and marker sharing statistics. *Am J Hum Genet* **58**: 1323–1337.

Spielman RS, Ewens WJ. (1996). The TDT and other family-based tests for linkage disequilibrium and association. *Am J Hum Genet* **59**: 983–989.

Spielman RS, Ewens WJ. (1998). A sibship test for linkage in the presence of association: the sib-transmission/disequilibrium test. *Am J Hum Genet* **62**: 450–458.

Spielman RS, McGinnis RE, Ewens WJ. (1993). Transmission test for linkage disequilibrium: the insulin gene region and insulin-dependent diabetes mellitus. *Am J Hum Genet* **52**: 506–516.

Talbot CJ, Nicod A, Cherny SS, Fulker DW, Collins AC, Flint J. (1999). High-resolution mapping of quantitative trait loci in outbred mice. *Nature Genet* **21**: 305–308.

Terwilliger JD. (1995). A powerful likelihood method for the analysis of linkage disequilibrium between trait loci and one or more polymorphic marker loci. *Am J Hum Genet* **56**: 777–787.

Terwilliger JD. (1996). Program SIBPAIR — sib pair analysis on nuclear families. ftp://linkage.cpmc.columbia.edu.

Terwilliger JD, Ott J. (1994). *Handbook of Human Genetic Linkage*. Johns Hopkins: Baltimore.

Uimari P, Thaller G, Hoeschele I. (1996). The use of multiple markers in a Bayesian method for mapping quantitative trait loci. *Genetics* **143**: 1831–1842.

Utz HF, Melchinger AE. (1996). PLABQTL: a program for composite interval mapping of QTL. *J Quant Trait Loci* **2**, http://probe.nalusda.gove:8000/otherdocs/jqtl/.

van Ooijen JW, Maliepaard C. (1996a). MapQTL version 3.0: software for the calculation of QTL positions on genetic maps. Plant Genome IV abstracts. http://probe.nalusda.gov:3000/otherdocs/pg/pg4/abstracts/p316.html.

van Ooijen JW, Maliepaard C. (1996b). MapQTL version 3.0: software for the calculation of QTL positions on genetic maps. CPRO-DLLO: Wageningen, ISBN90-73771-23-4.

Weeks DE, Sobel E, O'Connell JR, Lange K. (1995). Computer programs for multilocus haplotyping of general pedigrees. *Am J Hum Genet* **56**: 1506–1507.

Weir BS. (1996). *Genetic Data Analysis II*. Sinauer Associates Inc Publishers: Sunderland, MA, USA.

Xie X, Ott J. (1993). Testing linkage disequilibrium between a disease gene and marker loci. *Am J Hum Genet* **53**: 1107.

Zeng Z-B. (1993). Theoretical basis for separation of multiple linked gene effects in mapping quantitative trait loci. *Proc Natl Acad Sci USA* **90**: 10972–10976.

Zeng Z-B. (1994). Precision mapping of quantitative trait loci. *Genetics* **136**: 1457–1468.

Zeng Z-B, Kao CH, Basten CJ. (1999). Estimating the genetic architecture of quantitative traits. *Genet Res* **74**: 279–289.

Zhao JH, Curtis D, Sham PC. (2000). Model-free analysis and permutation tests for allelic associations. *Hum Hered* **50**: 133–139.

SECTION 4

BIOLOGICAL SEQUENCE ANALYSIS AND CHARACTERIZATION

CHAPTER 12

Predictive Functional Analysis of Polymorphisms: An Overview

MICHAEL R. BARNES

Genetic Bioinformatics
GlaxoSmithKline Pharmaceuticals
Harlow, Essex, UK

Bioinformatics for Geneticists. Edited by M.R. Barnes and I.C. Gray
 ISBNs: 0 470 84393 4; 0 470 84394 2 (PB)

12.1 INTRODUCTION

Human genetic disease is generally characterized by a profound range of phenotypic variability manifested in variable age of onset, severity, organ specific pathology and response to drug therapy. The causes underlying this variability are likely to be equally diverse, influenced by differing levels of genetic and environmental modifiers. The vast majority of human genetic variants are likely to be neutral in effect, but some may cause or modify disease phenotypes. The challenge for bioinformatics is to identify the genetic variants which are most likely to show a non-neutral allelic effect. Geneticists studying complex disease are already seeking to identify these genetic determinants by genetic association of phenotypes with markers. The literature is now replete with reported associations, but moving from associated marker to disease allele is proving to be very difficult. So why are we so unsuccessful in making this transition? Disregarding false positive associations (which may make up the bulk of reported associations to date!) it may be that the diverse effects of genetic variation are helping disease alleles to elude us. Genetic variation can cause disease at any number of stages between promotion of gene transcription to post-translational modification of protein products. Many geneticists have chosen to focus their efforts on the most obvious form of variation — non-synonymous coding variation in genes. While this category of variation is undoubtedly likely to contribute considerably to human disease, this may overlook many equally important categories of variation in the genome, namely the effects of variation on gene transcription, temporal and spatial expression, transcript stability and splicing.

Clearly all polymorphisms are not equal. Analysis of polymorphism distribution across the human genome shows significant variations in polymorphism density and allele frequency distribution. Chakravarti (1999) showed an immediate difference between the density of SNPs in exonic regions and intragenic and intronic regions. SNPs occurred at 1.2-kb average intervals in coding regions and 0.9-kb intervals in intragenic and intronic regions. These differences point to different selection intensities in the genome, particularly in protein coding regions, where SNPs may result in alteration of amino acid sequences (non-synonymous SNPs (nsSNPs)) or the alteration of gene regulatory sequences. These observations are intuitive — natural selection is obviously likely to be strongest across gene regions, essentially encapsulating the objective of genetics — to identify non-neutral alleles with a role in disease.

So how should we go about identifying disease alleles? One approach used to identify disease mutations is to directly screen strong candidate genes for mutations present in affected but not unaffected family members. This approach is very useful in the study of monogenic diseases and cancers, where transmission of the disease allele can generally be demonstrated to be restricted to affected individuals/tissues. But in the case of complex disease the odds of identifying disease alleles by population screening of candidate genes would seem to be very high and proving their role is problematic as disease alleles are likely to be present in cases and controls. Instead we detect common marker alleles in LD with rarer disease alleles. This methodical approach to disease gene hunting localizes disease alleles rather than actually identifying them directly, the next step is to identify the disease allele from a range of alleles in LD with the associated marker. To conclusively identify this allele a functional mechanism for the allele in the disease needs to be identified.

12.1.1 Moving from Associated Genes to Disease Genes

Many potential associations have been reported between markers and disease phenotypes. Aside from the potential for false positive association, magnitude of effect in complex disease is also a problem. There may be a few gene variants with major effects, but generally complex disease is very heterogeneous and polygenic, it therefore follows that studies of single gene variants will be inconclusive and inconsistent — this is just something we have to work with. We may also find a bewildering array of complex disease genes with somewhat indirect roles in disease, such as modifier genes and redundant genes, that have many effects on phenotype. Understanding the mode of action of these associated alleles will help in determining how susceptibility genes may give rise to a multifactorial phenotype. Bioinformatics may be critical in this process. Follow-up studies need to be designed to ask the right questions, to ensure that the right candidates are tested and to confirm the biological role of positive associations. It may also be necessary to attempt to characterize polymorphisms with a potential functional impact, to help to identify the molecular mechanisms by a combination of bioinformatics and laboratory follow-up. Many of these informatics approaches are similar to the approaches originally used to identify candidates, but by necessity these analyses benefit from a far more detailed approach as in-depth analyses transfer to in-depth laboratory investigation.

Moving from an 'associated gene' to a 'disease gene' is not a purely academic objective. Genetics may sometimes be our only insight into the nature of a disease, such insights may help us to restore the normal function of disease genes in patients, develop drugs and better still it may help prevent disease in the first place. Better diagnosis and treatments are also prospects afforded by better understanding of the pathology of disease. A validated 'disease gene' is one of the most tangible progressions towards this end.

12.1.2 Candidate Polymorphisms

To turn the arguments for association analysis on their head, there is also theory that suggests that the direct identification of disease alleles may not be entirely futile. The common disease/common variant (cd/cv) hypothesis predicts that the genetic risk for common diseases will often be due to disease-predisposing alleles with relatively high frequencies (Reich and Lander, 2001). There is not enough evidence to prove or disprove this hypothesis, however several examples of common disease variants have been identified, some of which are listed in Table 12.1, the allele frequency of these variants in the public databases is also listed.

The possibility that many disease alleles may be common, presents an intriguing challenge for genetics (and bioinformatics), if the cd/cv hypothesis holds true, then a substantial number of disease alleles may already be present in polymorphism databases or the human genome sequence. These might be termed 'candidate polymorphisms'. To extend this idea, just as genes with a putative biological role in disease are often prioritized for genetic association analysis, 'candidate polymorphisms' can be prioritized based on a predicted effect on the structure and function of regulatory regions, genes, transcripts or proteins. Thus selection of candidate polymorphisms is an extension of the candidate gene selection process — but in this case a link needs to be established between a predicted functional allelic effect and a target phenotype. As discussed earlier, DNA polymorphism can impact almost any biological process. Much of the literature in this area

TABLE 12.1 Disease Alleles Supporting the Common Disease/Common Variant Hypothesis

Gene (Allele)	Minor Allele Freq. (In dbSNP)	Disease/Trait Association	OMIM Review
APOE ε4	16% (14%)	Alzheimer's and cardiovascular disease	107741
Factor V^{leiden} R506Q	2–7% (ND)	Deep vein thrombosis	227400
KCNJ11 E23K	14% (25%)	Type II diabetes	600937
COMT V158M	0.1–62% (45%)	Catechol drug pharmacogenetics	116790

has focused on the most obvious form of variation — non-synonymous changes in coding regions of genes. Alterations in amino acid sequences have accounted for a great number of diseases. Coding variants may impact protein folding, active sites, protein–protein interactions, protein solubility or stability. But the effects of DNA polymorphism are by no means restricted to coding regions, variants in regulatory regions may alter the consensus of transcription factor binding sites or promoter elements; variants in the untranslated regions (UTR) of mRNA may alter mRNA stability; variants in the introns and silent variants in exons may alter splicing efficiency.

Approaches for evaluating the potential functional effects of DNA polymorphisms are almost limitless, but there are very few tools designed specifically for this task. Instead almost any bioinformatics tool which makes a prediction based on a DNA or protein sequence can be commandeered to analyse polymorphisms — simply by analysing wild-type and mutant sequences and looking for an alteration in predicted outcome by the tool. Polymorphisms can also be evaluated at a simple level by looking at physical considerations of the properties of genes and proteins or they can be evaluated in the context of a variant within a family of homologous or orthologous genes or proteins.

12.2 PRINCIPLES OF PREDICTIVE FUNCTIONAL ANALYSIS OF POLYMORPHISMS

Faced with the extreme diversity of disease, analysis of polymorphism data calls for equally diverse methods to assess functional effects that might lead to these phenotypes. The complex arrangements that regulate gene transcription, translation and function are all potential mechanisms through which disease could act and so analysis of potential disease alleles needs to evaluate almost every eventuality. Figure 12.1 illustrates the logical decision-making process that needs to be applied to the analysis of polymorphisms and mutations. The tools and approaches for the analysis of variation are completely dependent on the location of the variant within a gene or regulatory region. Many of these questions can be answered very quickly using genomic viewers such as Ensembl or the UCSC human genome browser (see Chapter 5 for a tutorial on these tools). Placing a polymorphism in full genomic context is useful to evaluate variants in terms of location

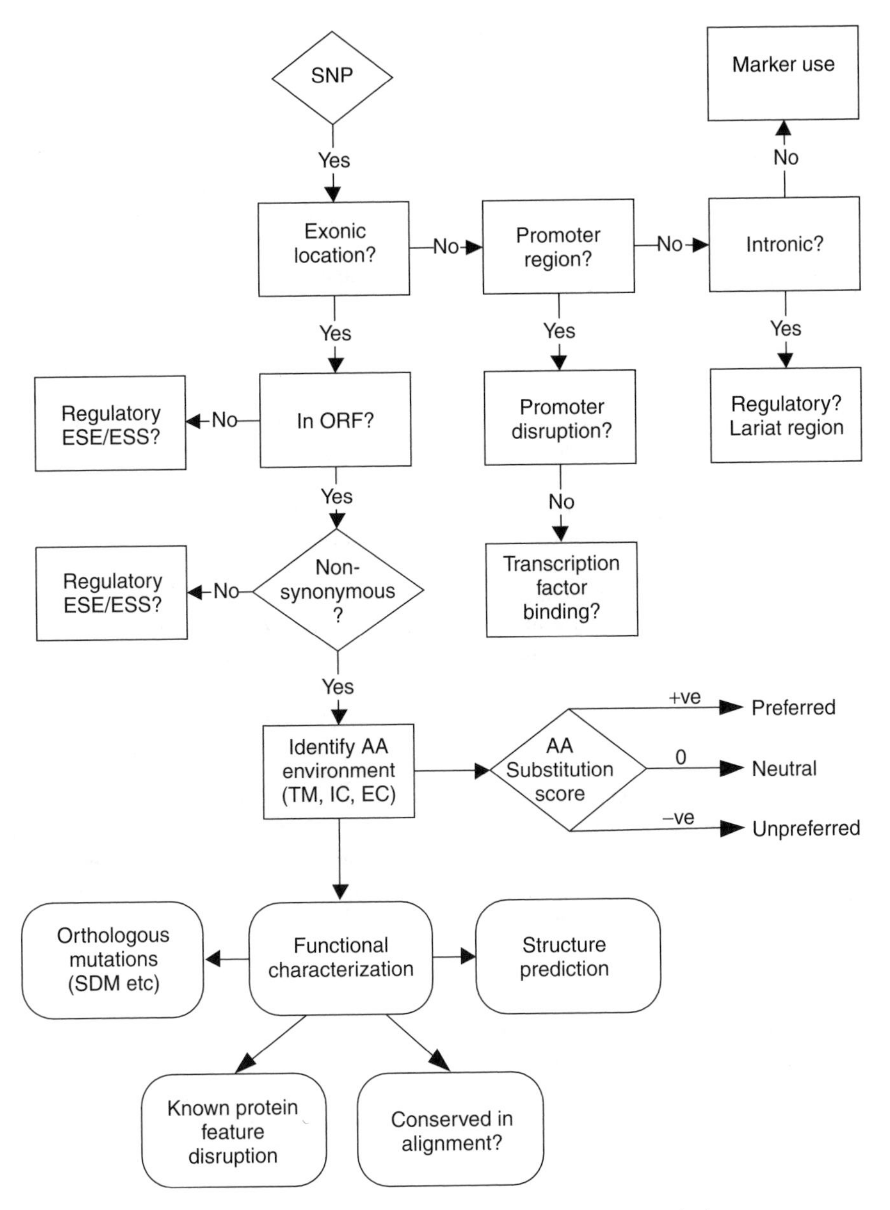

Figure 12.1 A decision tree for polymorphism analysis.

within or near genes (exonic, coding, UTR, intronic, promoter region) and other functionally significant features, such as CPG islands, repeat regions or recombination hotspots. Once approximate localization is achieved, specific questions need to be asked to place the polymorphism in a specific genic or intergenic region. This will help to narrow down the potential range of functional effects attributable to a variant, which will in turn help to identify the appropriate laboratory follow-up approach to evaluate function. Tables 12.2

TABLE 12.2 Functional Polymorphisms in Genes and Gene Regulatory Sequences

Location	Gene/Disease	Mechanism
Transcription factor binding	TNF in cerebral malaria	−376A SNP introduces OCT1 binding site-altering TNF expression, associated with four-fold increased susceptibility to cerebral malaria. (Knight *et al.*, 1999)
Promoter	CYP2D6	Common — 48T > G substitution disrupts the TATA box of the CYP2D6 promoter, causing 50% reduction in expression. (Pitarque *et al.*, 2001)
Promoter	RANTES in HIV progression	−28G mutation increases transcription of the RANTES gene slowing HIV-1 disease progression (Liu *et al.*, 1999)
cis-regulatory element	Bruton's tyrosine kinase in X-linked agammaglo-bulinemia	+5G/A (intron 1) shows reduced BTK transcriptional activity, suggesting a novel *cis*-acting element, involved in BTK downregulation but not splicing (Jo *et al.*, 2001)
Lariat region	HNF-4alpha	NIDDM-associated C/T substitution in polypyrimidine tract in intron 1b in an important *cis*-acting element directing intron removal (lariat region) (Sakurai *et al.*, 2000)
Splice donor/acceptor sites	ATP7A in Menke disease	Mutation in donor splice site of exon 6 of ATP7A causes a lethal disorder of copper metabolism (Moller *et al.*, 2000)
Cryptic donor/acceptor sites	β-glucuronidase gene (GUSB) in MPS VII	A 2-bp intronic deletion creates a new donor splice site activating a cryptic exon in intron 8 (Vervoort *et al.*, 1998)
Exonic splicing enhancers (ESE)	BRCA1 in breast cancer	Both silent and nonsense exonic point mutations were demonstrated to disrupt splicing in BRCA1 with differing phenotypic penetrance (Liu *et al.*, 2001)
Intronic splicing enhancers (ISE)	Alpha galactosidase in Fabry disease	G > A transversion within 4 bp of splice acceptor results in greatly increased alternative splicing (Ishii *et al.*, 2002)
Exonic splicing silencers (ESS)	CD45 in multiple sclerosis	Silent C77G disrupts ESS that inhibits the use of the 5′ exon four splice sites (Lynch and Weiss, 2001)
Intronic splicing silencers (ISS)	TAU in dementia with parkinsonism	Mutations in TAU intron 11 ISS cause disease by altering exon 10 splicing (D'Souza and Schellenberg, 2000)

TABLE 12.2 *(continued)*

Location	Gene/Disease	Mechanism
Polyadenylation signal	FOXP3 in IPEX syndrome	A→G transition within the polyadenylation signal leads to unstable mRNA with 5.1 kb extra UTR (Bennett *et al.*, 2001)

TABLE 12.3 Tools for Functional Analysis of Gene Regulation and Splicing

Tool	URL
Promoter prediction	
NNPP	http://www.fruitfly.org/seq_tools/promoter.html
CorePromoter	http://sciclio.cshl.org/genefinder/CPROMOTER/
Promoter Scan II	http://www.molbiol.ox.ac.uk/promoterscan.htm
Orange	http://wwwiti.cs.uni-magdeburg.de/~grabe/orange/
Transcription factor binding site prediction	
TRANSFAC	http://transfac.gbf.de/TRANSFAC/
FastM/ModelInspector	http://genomatix.gsf.de/cgi-bin/fastm2/fastm.pl
TESS	http://www.cbil.upenn.edu/tess/
TFSEARCH	http://www.cbrc.jp/research/db/TFSEARCH.html
Splice site prediction	
NETGENE	http://genome.cbs.dtu.dk/services/NetGene2/
Splice Site Prediction	http://www.fruitfly.org/seq_tools/splice.html
SpliceProximalCheck	http://industry.ebi.ac.uk/~thanaraj/SpliceProximalCheck.html
Gene prediction and ORF finding	
Genscan	http://genes.mit.edu/GENSCAN.html
Genie	http://www.fruitfly.org/seq_tools/genie.html
ORF Finder	http://www.ncbi.nlm.nih.gov/gorf/gorf.html
Detection of novel regulatory elements and comparative genome analysis	
PipMaker	http://bio.cse.psu.edu/pipmaker/
TRES	http://bioportal.bic.nus.edu.sg/tres/
Improbizer	http://www.soe.ucsc.edu/~kent/improbizer/
Regulatory Vista	http://www-gsd.lbl.gov/vista/rVistaInput.html
Integrated platforms for gene, promoter and splice site prediction	
Webgene	http://www.itba.mi.cnr.it/webgene/
BCM Gene Finder	http://dot.imgen.bcm.tmc.edu:9331/gene-finder/gf.html

and 12.3 illustrate some carefully selected examples of non-coding polymorphisms in genes and transcripts, these publications were specifically selected as each also includes a detailed laboratory based follow-up to evaluate each form of polymorphism. We refer the reader to these publications as a potential guide to assist in laboratory investigation.

12.2.1 Defining the Boundaries of Normal Function in Genes and Gene Products

Beyond the general localization of variants that general bioinformatics tools, such as Ensembl, can afford, there is a further more detailed context to many known regulatory elements in genes and gene regulatory regions. Our knowledge of these elements is still very sparse, but certain elements are relatively well defined. Many of these elements have been defined by mutations in severe Mendelian phenotypes. By definition this suggests that many elements which may have moderate effects on gene function are less likely to have been identified as they are less likely to have come to the attention of physicians. In the case of complex disease it may be very difficult to distinguish genuine disease susceptibility alleles from the normal spectrum of variability in human individuals.

12.2.2 A Decision Tree for Polymorphism Analysis

The first step in our decision tree for polymorphism analysis (Figure 12.1) is a simple question — is the polymorphism located in an exon? Answering this accurately may not always be simple or even possible with only *in silico* resources. As we have already seen in the previous section, delineation of genes is really the key step in all subsequent analyses, once we know the location of a gene all other functional elements fall into place based on their location in and around genes. In Chapter 4 we presented a detailed examination of the art of delineating genes, including methods for extending sequences to identify the true boundaries of a gene, not just its coding region. This activity may seem superfluous in the 'post genome' era, but the fact is that we still know very little about the full diversity of genes and the vast majority of genes are still incompletely characterized. Gene prediction and gene cloning has generally focused on the open reading frame — the protein coding sequence (ORF/CDS) of genes. For the most part UTR sequences have

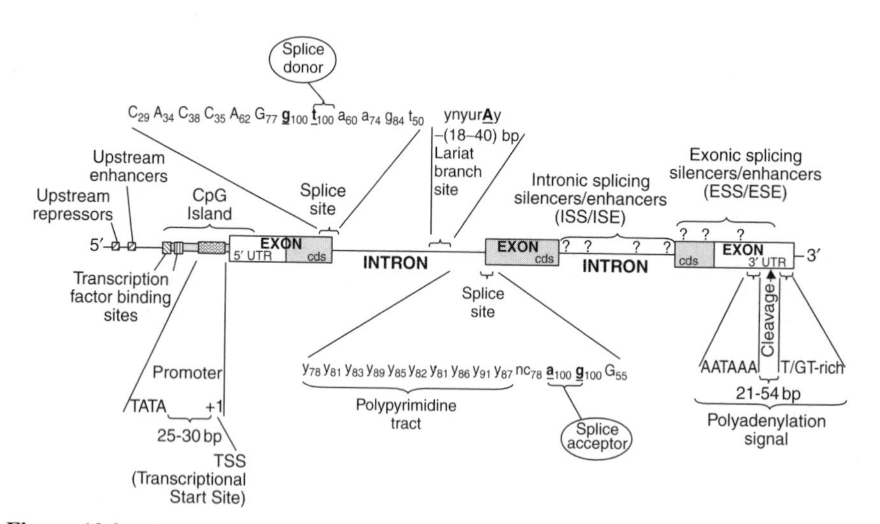

Figure 12.2 The anatomy of a gene. This figure illustrates some of the key regulatory regions which control the transcription, splicing and post-transcriptional processing of genes and transcripts. Polymorphisms in these regions should be investigated for functional effects.

been neglected in the rush to find an ORF and a protein. In the case of polymorphism analysis, these sequences should not be overlooked as the extreme 5′ and 3′ limits of UTR sequence delineate the true boundaries of genes. This delineation of gene boundaries is illustrated in a canonical gene model in Figure 12.2. As the model shows, most of the known regulatory elements in genes are localized to specific regions based on the location of the exons. So for example, the promoter region is generally located in a 1–2-kb region immediately upstream of the 5′ UTR and splice regulatory elements flank intron/exon boundaries. Many of these regulatory regions were first identified in Mendelian disorders and now some are also being identified in complex phenotypes. Table 12.2 lists some of the disease mutations and polymorphisms that have helped to shape our knowledge of this complex area.

12.3 THE ANATOMY OF PROMOTER REGIONS AND REGULATORY ELEMENTS

Prediction of eukaryotic promoters from genomic sequence remains one of the most challenging tasks for bioinformatics. The biggest problem is over-prediction; current methods will on average predict promoter elements at 1-kb intervals across a given genomic sequence. This is in stark contrast to the estimated average 40–50-kb distance of functional promoters in the human genome (Reese *et al.*, 2000). Although it is possible that some of these predicted promoters may be expressed cryptically, the vast majority of predictions are likely to be false positives. To avoid these false predictions it is essential to provide promoter prediction tools with the appropriate sequence region, that is, the region immediately upstream of the gene transcriptional start site (TSS). It is important to define the TSS accurately; it is certainly insufficient to simply take the sequence upstream from the start codon as 5′ UTR can often span additional 5′ exons in higher eukaryotes (Reese *et al.*, 2000). As Uwe Ohler of the Drosophila genome project so eloquently stated, 'without a clear idea of the TSS location we may well be looking for a needle in the wrong haystack' (Ohler, 2000). If we can identify the TSS, the majority of RNA polymerase promoter elements are likely to be located within 150 bp, although some may be more distant so it may be important to analyse 2 kb or more upstream, particularly when the full extent of the 5′ UTR or TSS is not well defined.

Once a potential TSS has been identified there are many tools which can be applied to identify promoter elements and transcription factor binding sites. The human genome browsers (UCSC and Ensembl) are the single most valuable resources for the analysis of promoters and regulatory elements. Specifically, Ensembl annotates putative promoter regions using the Eponine tool. The UCSC browser annotates known transcription factor binding sites from the Transfac database and novel predicted regulatory elements in the 'golden triangle' track (see Section 12.6.2 below). These are very useful for rapid evaluation of the location of variants in relation to these features, although this data needs to be used with caution as whole genome analyses may over-predict or overlook evidence for alternative gene models. The analysis approaches for promoter and transcription binding site analysis are reviewed thoroughly in Chapter 13.

Characterization of gene promoters and regulatory regions is not only valuable for functional analysis of polymorphisms, but it can also provide important information about the regulatory cues that govern the expression of a gene, which may be valuable for pathway expansion to assist in the elucidation of the function of candidate genes and disease-associated genes.

12.4 THE ANATOMY OF GENES

12.4.1 Gene Splicing

Alternative splicing is an important mechanism for regulation of gene expression which can also expand the coding capacity of a single gene to allow production of different protein isoforms, which can have very different functions. The recent completion of the human genome draft has given an interesting new insight into this form of gene regulation. Despite initial estimates of a human gene complement of > 100 K genes, direct analysis of the sequence suggests that humans may only have 30–40 K genes, which is only a two- to three-fold gene increase over invertebrates (Aparicio, 2000). Indeed, extrapolation of results from an analysis of alternatively spliced transcripts from chromosomes 22 and 19 have led to estimates that at least 59% of human genes are alternatively spliced (Lander *et al.*, 2001). This highlights the probable significance of post-transcriptional modifications such as alternative splicing as an alternative means by which to express the full phenotypic complexity of vertebrates without a very large number of genes.

A much simpler organism has given us a glimpse of the possibilities of splicing as a mechanism to generate phenotypic complexity. The drosophila homologue of the human Down syndrome cell adhesion molecule (DSCAM) has 115 exons, 20 of which are constitutively spliced and 95 of which are alternatively spliced (Schmucker *et al.*, 2000). The alternatively spliced exons are organized into four clusters, with 12 alternative versions of exon 4, 48 versions of exon 6, 33 versions of exon 9 and two versions of exon 17. These clusters of alternative exons code for 38,016 related but distinct protein isoforms!

12.4.2 Splicing Mechanisms, Human Disease and Functional Analysis

The remarkable diversity of potential proteins produced from the DSCAM gene, gives us some idea of the tight regulation of alternative splicing that must be in place to not only regulate the choice of each version of a particular exon, but also to exclude all other versions of the exon once one version has been selected. Regulation of splicing is mediated by the spliceosome, a complex network of small nuclear ribonucleoprotein (snRNP) complexes and members of the serine/arginine-rich (SR) protein family. At its most basic level, pre-mRNA splicing involves precise removal of introns to form mature mRNA with an intact open reading frame (ORF). Correct splicing requires exon recognition with accurate cleavage and rejoining at the exon boundaries designated by the invariant intronic GT and AG dinucleotides, respectively known as the splice donor and splice acceptor sites (Figure 12.2). Other more variable consensus motifs have been identified in adjacent locations to the donor and acceptor sites, including a weak exonic 'CACCAG' consensus flanking the splice donor site, an intronic polypyrimidine- (Y : C or T) rich tract flanking the splice acceptor site and a weakly conserved intronic 'YNYURAY' consensus 18–40 bp from the acceptor site, which acts as a branch site for lariat formation (Figure 12.2). Other regulatory motifs are known to be involved in splicing, including exonic splicing enhancers (ESE) and intronic splicing enhancers (ISE), both of which promote exon recognition, and exonic and intronic splicing silencers (ESS and ISS, respectively), which have an opposite action, inhibiting the recognition of exons. DNA recognition motifs for splicing enhancers and silencers are generally quite degenerate. The degeneracy of these consensus recognition motifs points to fairly promiscuous binding by SR proteins. These interactions can also explain the use of alternative and inefficient splice sites, which may be influenced by competitive binding of SR proteins and hnRNP determined by the relative ratio of hnRNP to SR proteins in the nucleus. A natural stimulus that influences the ratio of these proteins is genotoxic stress, which can lead to the often observed phenomenon of differential splicing in tumours and other disease states (Hastings and Krainer, 2001).

Mutations affecting mRNA splicing are a common cause of Mendelian disorders, 10–15% of Mendelian disease mutations affect pre-mRNA splicing (Human Gene Mutation Database, Cardiff). These mutations can be divided into two subclasses according to their position and effect on the splicing pattern. Subclass I (60% of the splicing mutations) includes mutations in the invariant splice-site sequences, which completely abolish exon recognition. Subclass II includes mutations in the variant motifs, which can lead to both aberrantly and correctly spliced transcripts, by either weakening or strengthening exon-recognition motifs. Subclass II also includes intronic mutations, which generate cryptic donor or acceptor sites and can lead to partial inclusion of intronic sequences. These Mendelian disease mutations have helped to define our understanding of splicing mechanisms. Considering the proven complexity of splicing in the human genome (Lander *et al.*, 2001), it seems reasonable to expect splicing abnormality to play a significant role in complex diseases, but examples are rare. This is explained in part by the power of family-based mutations, the inheritance of which can be traced between affected and unaffected relatives. It is difficult to determine similar causality for a population-based polymorphism.

12.4.3 Functional Analysis of Polymorphisms in Putative Splicing Elements

If taken individually, there are many sequences within the human genome that match the consensus motifs for splice sites, but most of them are not used. In order to function, splice sites need appropriately arranged positive (ESEs and ISEs) and negative (ESSs, and ISSs) *cis*-acting sequence elements. These *cis*-acting arrangements of regulatory elements can be both activated and deactivated by DNA sequence polymorphisms. DNA polymorphism at the invariant splice acceptor (AG) and donor (GT) sites, are generally associated with severe diseases and so, are likely to be correspondingly rare. But, as we have seen, recognition motifs for some of the elements that make up the larger splice site consensus are very variable, so splice site prediction from undefined genomic sequence is still imprecise at the best of times. Bioinformatics tools can fare rather better when applied to known genes with known intron/exon boundaries — this information can be used to carry out reasonably accurate evaluations of the impact of polymorphisms in putative splice regions. There are several tools which will predict the location of splice sites in genomic sequence, all match and score the query sequence against a probability matrix built from known splice sites (see Table 12.3). These tools can be used to evaluate the effect of splice region polymorphisms on the strength of splice site prediction by alternatively running wild-type and mutant alleles. As with any other bioinformatics prediction tool it is always worth running predictions on other available tools to look for a consensus between different prediction methods. These tools can also be used to evaluate the propensity of an exon to undergo alternative splicing. For example an unusually low splice site score may indicate that aberrant splicing may be more likely at a particular exon compared to exons with higher splice site scores. The phase of the donor and acceptor sites also needs to be taken into account in these calculations. Coding exons exist in three phases 0, 1 and 2, based on the codon location of the splice sites, if alternative donor or acceptor sites are in unmatched phases then a frameshift mutation will occur.

Splice site prediction tools will generally predict the functional impact of a polymorphism within close vicinity of a splice donor or acceptor site, although they will not predict the functional effect of polymorphisms in other elements such as lariat branch sites. Definition of consensus motifs for these elements (Figure 12.2) makes it reasonably easy to assess the potential functional impact of polymorphisms in these gene regions by simply inspecting the location of a polymorphism in relation to the consensus motif. As with all functional predictions laboratory investigation is required to confirm the hypothesis.

Other *cis*-regulatory elements, such as ESE, ESS, ISE and ISS sites are very poorly defined and may be located in almost any location within exons and introns. There are currently no available bioinformatics tools to generally predict the locations of these regulatory elements. Some specific elements, *cis*-regulatory elements, have been defined in specific genes, but these do not form a consensus sequence to search other genes. One of the only possible approaches for *in silico* analysis of such elements is to use comparative genome data to look for evolutionarily conserved regions, particularly between distant species, e.g. comparison of Human/Fugu (fish) genomes. Although there may be some value in these approaches, confirmation of *cis*-regulatory elements really needs to be achieved by laboratory methods (see D'Souza and Schellenberg (2000) for a description of such methods).

12.4.4 Polyadenylation Signals

Polyadenylation of eukaryotic mRNA occurs in the nucleus after cleavage of the precursor-RNA. Several signals are known which determine the site of cleavage and subsequent polyadenylation, the most well known is a canonical hexanucleotide (AAUAAA) signal 20–50 bp from the 3′ end of the pre-RNA, this works with a downstream U/GU-rich element which is believed to regulate the complex of proteins necessary to complete 3′ processing (Pauws *et al.*, 2001). The specific site of cleavage of pre-RNA is located between these regulatory elements and is determined by the nucleotide composition of the cleavage region with the following nucleotide preference $A > U > C >> G$. In a study of 9625 known human genes Pauws *et al.* (2001) found that 44% of human genes regularly used more than one cleavage site, resulting in the generation of slightly different mRNA species.

Mutations in the canonical AAUAAA polyadenylation signal have been shown to disrupt normal generation of polyadenylated transcripts (Bennett *et al.*, 2001). This signal is needed for both cleavage and polyadenylation in eukaryotes, and failure to polyadenylate will prevent maturation of mRNA from nuclear RNA (Wahle and Keller, 1992). The complete aggregate of elements that make up the polyadenylation signal including the U/GU-rich region may not be universally required for processing (Graber *et al.*, 1999). Single nucleotide variations in this region cannot be conclusively identified as functional although any polymorphism in this region might be considered a candidate for further consideration.

12.4.5 Analysis of mRNA Transcript Polymorphism

The potential functional effects of genetic polymorphism can extend beyond a direct effect on the genomic organization and regulation of genes. Messenger RNA is far more than a simple coded message acting as an intermediary between genes and proteins. mRNA molecules have different fates related to structural features embedded in discrete regions of the molecule. The processing, localization, translation or degradation of a given mRNA may vary considerably, depending upon the environment in which it is expressed. Figure 12.3 illustrates a simplified model of an mRNA molecule, indicating the

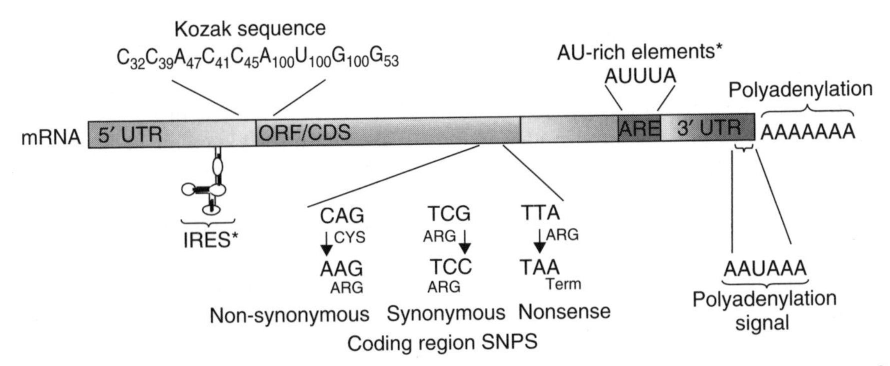

Figure 12.3 The anatomy of an mRNA transcript. This figure illustrates some of the key regulatory and structural elements that control the translation, stability and post-transcriptional processing of mRNA transcripts. Polymorphisms in these regions should be investigated for functional effects.

key features and regulatory motifs that could potentially be disrupted by polymorphism. At the most basic level an mRNA molecule consists of a protein coding, open reading frame (ORF), flanked by 5′ and 3′ UTR. Most polymorphism analysis in the literature has tended to focus on the coding sequence of genes, but there is evidence to suggest that UTR sequences also serve important roles in the function of mRNA. At the risk of generalizing, 5′ UTR sequences are important as they are known to accommodate the translational machinery, while there is accumulating evidence that strongly implicates the 3′ UTR in the regulation of gene expression. In Table 12.4 we highlight some examples of polymorphisms which impact mRNA transcripts.

12.4.6 Initiation of Translation

If a gene is known, the ORF will probably be well defined, but if a novel transcript is being studied the ORF needs to be identified. Again we refer the reader to Chapter 4 which contains details on the extension of mRNA transcripts and ORF finding procedures. The accepted convention is that the initiator codon will be the first inframe AUG encoding the largest open reading frame in the transcript. There is evidence of a scanning mechanism for initiation of translation; the initiator codon generally conforms to a 'CCACCaugG' consensus motif known as the Kozak sequence (Kozak, 1996). However, Peri and Pandey (2001) and others have recently reappraised this convention and actually found that more than 40% of known transcripts contain inframe AUG codons upstream of the actual initiator codon, some of which conform more closely to the Kozak motif than the authentic initiator codon. Their revised Kozak consensus '$C_{32}C_{39}A_{47}C_{41}C_{45}A_{100}U_{100}G_{100}G_{53}$' was much weaker. These observations have cast some doubt on the validity of the scanning mechanism for initiation of translation, some have argued that the frequent occurrence of AUG codons upstream of the putative initiator codon, may indicate misassignment of the initiator codon or cDNA library anomalies (Kozak, 2000), others point to the empirical increase in gene expression measured in the laboratory when initiator codons conforming to the Kozak consensus are compared to other sequences. This debate may never resolve conclusively and it seems certain that the mechanism for translation initiation is still not fully understood.

TABLE 12.4 Functional Non-Coding Polymorphisms in mRNA Transcripts

Location	Gene/Disease	Mechanism
Internal ribosome entry segment (IRES)	Proto-oncogene c-myc in multiple myeloma	C–T mutation in the c-myc-IRES causes aberrant translational regulation of c-myc, enhanced binding of protein factors and enhanced initiation of translation leading to oncogenesis (Chappell *et al.*, 2000)
Kozak initiation sequence	Platelet glycoprotein Ib-alpha (GP1BA) in ischaemic stroke	C/T polymorphism at the −5 position from the initiator ATG codon of the GP1BA gene is located within the 'Kozak' consensus nucleotide sequence. The presence of a C at this position significantly increases the efficiency of expression of the GPIb/V/IX complex (Afshar-Kharghan *et al.*, 1999)
Anti-termination mutation and 3′ UTR stability determinants	Alpha-globin in alpha-thalassemia	UAA to CAA to anti-termination mutation allows translation to proceed into the 3′ UTR which masks stability determinants to substantially decrease mRNA half-life (Conne *et al.*, 2000)
UTR stability	Protein tyrosine phosphatase-1B (PTP1B)	1484insG in 3′ UTR causes PTP1B over-expression leading to insulin resistance (Di Paola *et al.*, 2002)

There are some examples of polymorphisms in Kozak sequences that appear to have a direct bearing in human disease. Kaski *et al.* (1996) reported a T > C SNP with an 8–17% minor allele frequency at the −5 position from the initiator ATG codon of the GP1BA gene. This SNP is located within the most 5′ (and weakest) part of the Kozak consensus sequence. The cytosine (C) allele at this position conforms more closely to the consensus and subsequent studies of the SNP found that it was associated with increased expression of the receptor on the cell membrane, both in transfected cells and in the platelets of individuals carrying the allele. The polymorphism was also associated with cardiovascular disease susceptibility (Afshar-Kharghan *et al.*, 1999).

An alternative mechanism for translation initiation has been identified which does not obey the 'first AUG rule', this involves cap-independent internal ribosome binding mediated by a Y-shaped secondary structure, denoted the Internal Ribosome Entry Site (IRES), located in the 5′ UTR of 5–10% of human mRNA molecules (see Le and Maizel, (1997) for a review of these elements). IRES elements are complex stem loop structures, there is no reliable sequence consensus to allow prediction of the possible functional effects of polymorphisms in these elements instead this needs to be attempted by the use of RNA secondary structure prediction tools such as MFOLD (see below).

12.4.7 mRNA Secondary Structure Stability

While we have already established that nucleotide variants in mRNA can alter or create sequence elements directing splicing, processing or translation of mRNA, variants may also influence mRNA synthesis, folding, maturation, transport and degradation. Many of these diverse biological processes are strongly dependent on mRNA secondary structure. Secondary structure is essentially determined by ribonucleotide sequence and so folding of mRNA is also likely to be influenced by SNPs and other forms of variation at any location in a transcript. Shen *et al.* (1999) studied two common silent SNPs in the coding regions of two essential genes — a U1013C transition in human alanyl tRNA synthetase (AARS) and a U1674C transition in the human replication protein A 70-kDa subunit (RPA70). The minor allele frequency was 0.49 for the AARS U allele and 0.15 for the RPA70 C allele. Using structural mapping and structure-based targeting strategies they demonstrated that both SNPs had marked effects on the structural folds of the mRNAs, suggesting phenotypic consequences of SNPs in mRNA structural motifs.

RNA stability is an intriguing disease mechanism, unfortunately beyond this and a handful of other published studies (see Conne *et al.* (2000) for a review), the true extent of detectable differences in mRNA folding caused by polymorphism is quite unknown, this may reflect the difficulties involved in studying such mutational effects *in vitro*.

There are several tools which can help to construct *in silico* secondary-structure models of polymorphic mRNA alleles. One of the best tools is MFOLD (M. Zuker, Washington University, St. Louis, MO), this is maintained on the Zuker laboratory homepage which also contains an excellent range of RNA secondary structure-related resources (http://bioinfo.math.rpi.edu/~zukerm/rna/). MFOLD will construct a number of possible models based on all structural permutations of a user-submitted mRNA sequence. Submission of mutant and wild-type mRNA alleles to this tool will give the user a fairly good indication of whether an allele could alter mRNA secondary structure. This can help to prioritize alleles for laboratory-based investigation of mRNA stability studies.

12.4.8 Regulatory Control of mRNA Processing and Translation

Beyond splicing and promoter based regulation, mRNAs are also tightly controlled by regulatory elements in their 5′ and 3′ untranslated regions (Figure 12.3). Proteins that bind to these sites are key players in controlling mRNA stability, localization and translational efficiency. Consensus motifs have been identified for many of these factors, usually corresponding to short oligonucleotide tracts, which generally fold in specific secondary structures, which are protein binding sites for various regulatory proteins. Some of these regulatory signals tend to be protein family specific, while others have a more general effect on diverse mRNAs. AU-rich elements (AREs) are the largest class of *cis*-acting 3′ UTR-located regulatory molecules that control the cytoplasmic half-life of a variety of mRNA molecules. One main class of these regulatory elements consists of pentanucleotide sequences (AUUUA) in the 3′ UTR of transcripts encoding oncoproteins, cytokines and growth and transcription factors. Many RNA-binding proteins, mostly members of the highly conserved ELAV family, recognize and bind AREs (Chen and Shyu, 1995). Defective functioning of AREs can lead to the abnormal stabilization of mRNA, this forms the basis of several human diseases, including mantle cell lymphoma, neuroblastoma, immune and several inflammatory diseases. Polymorphisms which disrupt AU-rich motifs in a 3′ UTR sequence may be worth evaluation as potentially functional polymorphisms. Some databases to assist in the identification of these motifs are described below.

12.4.9 Tools and Databases to Assist mRNA Analysis

To assist in the analysis of diverse and often family specific regulatory elements, such as ARE elements, Pesole *et al.* (2000) have developed UTRdb, a specialized non-redundant database of 5′ and 3′ untranslated sequences of eukaryotic mRNAs (http://bighost.area.ba.cnr.it/BIG/UTRHome/). In March 2002, UTRdb contained 39,527 non-redundant human entries; these are enriched with specialized information absent from primary databases including the presence of RNA regulatory motifs with experimental proof of a functional role. It is possible to BLAST search the database for the presence of annotated functional motifs in a query sequence.

Jacobs *et al.* (2002) have also developed Transterm, a curated database of mRNA elements that control translation (http://uther.otago.ac.nz/Transterm.html). This database examines the context of initiation codons for conformation with the Kozak consensus and also contains a range of mRNA regulatory elements from a broad range of species. Access is provided via a web browser in several different ways: a user-defined sequence can be searched against motifs in the database or elements can be entered by the user to search specific sections of the database (e.g. coding regions or 3′ flanking regions or the 3′ UTRs) or the user's sequence. All elements defined in Transterm have associated biological descriptions with references.

12.5 PSEUDOGENES AND REGULATORY MRNA

As a final word on the analysis of mRNA transcripts, it is important to be aware that not all mRNAs are intended to be translated. Some genes may produce transcripts that are truncated or retain an intron or are otherwise configured in a way that precludes translation. It is difficult to clarify the role of some of these transcripts; where a transcript has multiple premature termination codons, it is likely to be a pseudogene, others may have no obvious open reading frames, these may also be pseudogenes or they may be regulatory mRNA molecules. Several non-coding RNA (ncRNA) molecules have been described which act as riboregulators with a direct influence on post-transcriptional regulation of gene expression (see Erdmann *et al.* (2001) for a comprehensive review of the properties of regulatory mRNA). Analysis of polymorphisms in these molecules is difficult as they are very poorly defined in terms of functionality.

12.6 ANALYSIS OF NOVEL REGULATORY ELEMENTS AND MOTIFS IN NUCLEOTIDE SEQUENCES

It is very likely that our current knowledge of regulatory elements in the human genome is quite superficial. In terms of transcription factors alone, the TRANSFAC database contains a redundant set of 2263 profiles for vertebrate binding sites (Heinemeyer *et al.*, 1999), yet the first pass analysis of the human genome has identified over 4000 proteins with a putative DNA binding role (Venter *et al.*, 2001). This is likely to be an underestimate. Geneticists are working at the vanguard of efforts to close the gap between our current understanding and the full complexity of human gene regulation. Genetics has already contributed greatly to the identification of new regulatory elements by the identification of regulatory mutations and polymorphisms.

In this chapter we have reviewed a number of regulatory mechanisms and motifs in DNA sequences, including motifs in promoter regions, splice sites, introns and transcripts. Functional analysis of polymorphisms located in the consensus sequences identified for some of these elements may be an important indicator of a potential functional effect. However, despite advances in bioinformatic tools, predictive functional analysis of sequence polymorphism is still difficult to validate without laboratory follow-up. Even with the benefit of laboratory verification, identification of deleterious alleles can be laborious and the results of analyses do not always hold true between *in vitro* and *in vivo* environments. In a sense evolution is an *in vivo* experiment on a grand scale and so Sydney Brenner (2000) and others have proposed the concept of 'inverse genetics' to cover the use of information recovered from different genomes to inform on function. Brenner suggested comparing genomes to highlight conserved areas 'in a vast sea of randomness'. This is an elegant approach for the characterization of polymorphisms. Characterization by conventional genetics demands analysis of large sample numbers, complex *in vitro* analysis or laborious transgenic approaches. In the case of inverse genetics, evolution and time have already done the work in a long-term 'experiment' which would be impossible to match in the laboratory.

Inverse genetics also has a wider application — analysis of a single promoter sequence will often identify many putative regulatory elements by chance alone. However, simultaneous analysis of many evolutionarily-related but diverse promoter sequences will clearly identify known and novel conserved motifs which are more likely to be functionally important to a particular family of genes. This approach known as phylogenetic footprinting, has been used to successfully elucidate many common regulatory modules (Gumucio *et al.*, 1996). Kleiman *et al.* (1998) used a similar approach to identify a novel potential element in the polyadenylation regulatory apparatus, a TG deletion (deltaTG) in the 3′ UTR of the *HEXB* gene, 7 bp upstream from the polyadenylation signal. The deltaTG HEXB allele, which occurred at a 10% frequency, showed 30% lower enzymatic activities compared to WT individuals. Polyacrylamide gel electrophoresis analysis of the allele revealed that the 3′ UTR of the HEXB gene had an irregular structure. After studying a large range of eukaryotic mRNAs, including human, mouse and cat *HEXB* genes they found that the TG dinucleotide was part of a conserved sequence (TGTTTT) immersed in an A/T-rich region observed in more than 40% of mRNAs analysed. This study clearly illustrates how effective bioinformatic analysis of mRNA processing signals may require more than sequence analysis of known regulatory motifs; clearly tools are needed to identify novel regulatory elements. The web-based TRES tool is an example of a tool to assist in the identification of such novel elements.

12.6.1 TRES (http://bioportal.bic.nus.edu.sg/tres/)

TRES can be used to compare as many as 20 nucleotide sequences. The tool is multifunctional, it can either be used to identify conserved sequence motifs between submitted sequences or alternatively it can be used to identify known transcription factor binding sites shared between sequences using nucleotide frequency distribution matrices described in the TRANSFAC database (Heinemeyer *et al.*, 1999). This approach is not just applicable to evolutionarily-related sequences it can also be used to study unrelated sequences which may share similar regulatory cues, such as genes which show similar patterns of gene expression.

TRES also has another versatile search mode which allows detection of palindromic motifs or inverted repeats shared between sequences. These have unique features of dyad symmetry which can form hairpins or loops to facilitate protein binding in homo- or heterodimer form. Many transcription factors have palindromic recognition sequences and bind as dimmers; these motifs may be important to allow greater regulatory diversity from a limited number of transcription factors (Lamb and McKnight, 1991).

Although TRES is generally focused on the identification of transcription factor binding sites and promoter elements, the sequence motif identification facilities of the tool also make it suitable for the identification of other motifs in non-coding sequences including UTR sequences and intronic sequences.

12.6.2 Improbizer

Improbizer was developed at the UCSC; the tool searches for motifs in DNA or RNA sequences that occur with an improbable frequency; that is greater than might be expected to occur by chance alone. Probabilities are estimated using the expectation maximization (EM) algorithm (Jim Kent, personal communication; for more details see http://www.soe.ucsc.edu/~kent/improbizer/improbizer.html).

Improbizer is available as a web interface, this allows the analysis of multiple sequences (up to 100 can be entered) for common motifs between sequences. Improbizer has also been used to annotate a large number of predicted promoter regions in the UCSC human genome browser (see Chapter 5). This data is presented as the so-called 'golden triangle' track. Kent and colleagues adopted this name to describe the process they called 'Regulatory region Triangulation' (J. Kent and D. Haussler, personal communication). This approach combines cDNA, genomic DNA and microarray data to locate and characterize regulatory regions in the human genome. The method identified a large set of putative transcription start sites by aligning G-cap selected ESTs (which represent 5′ ends of transcripts) and other cDNA data to the human genome using BLAT. This data was compared with regions conserved between the human and mouse genomes with BLASTZ. Finally to complete the 'triangulation' process, they clustered Affymetrix microarray data to find co-regulated clusters of genes; once identified the promoter sequences were analysed using Improbizer. The highly novel data generated by this analysis is a valuable resource for the evaluation of polymorphisms in regulatory regions.

12.7 FUNCTIONAL ANALYSIS ON NON-SYNONYMOUS CODING POLYMORPHISMS

The huge diversity of protein molecules makes it very difficult to provide a generic model of a protein. Returning to our decision tree for polymorphism analysis (Figure 12.1), the consequences of an amino acid substitution are first and foremost defined by the environment in which the amino acid exists. Different cellular locations can have very different chemical environments which can have diverse effects on the properties of amino acids. The cellular location of proteins can be divided at the simplest level between intracellular, extracellular or transmembrane environments. The latter location is the most complex as amino acids in transmembrane proteins can be exposed to all three cellular environments, depending upon the topology of the protein and the location of the particular amino acid. Environments will also differ in extracellular and intracellular proteins, depending on the location of the residue within the protein. Amino acid residues may be buried in a protein core or exposed on the protein surface. Once the environment of an amino acid has been defined, different matrices are available to evaluate and score amino acid changes. For reference we have provided four amino acid substitution matrices in Appendix II. These matrices can be used to evaluate amino acid changes in extracellular, intracellular and transmembrane proteins; where the location of the protein is unknown, a matrix for 'all proteins' is also available. Preferred (conservative) substitutions have positive scores, neutral substitutions have a zero score and unpreferred (non-conservative) substitutions are scored negatively. These matrices are another application of 'inverse genetics' and are constructed by observing the propensity for exchange of one amino acid for another based on

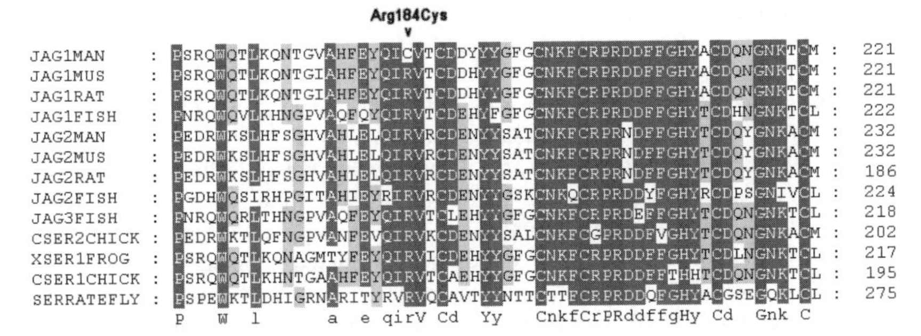

Figure 12.4 Functional evaluation of an Arg184Cys mutation in the Jagged protein family. Arg184Cys causes Alagille syndrome (OMIM 118450). Alignment of the mutated human amino acid sequence with vertebrate and invertebrate orthologues and homologues in the Jagged family identifies the Arg184 residue in a highly conserved position throughout this gene family. A mutation to a cysteine at this position would be expected to lead to the aberrant formation of disulphide bonds with other cysteine residues in the Jagged protein, this is likely to have a disruptive effect on the structure of the Jagged1 protein.

TABLE 12.5 Tools for Functional Analysis of Amino Acid Polymorphisms

Sequence manipulation and translation	
Sequence Manipulation Suite	http://www.bioinformatics.org/sms/
Amino acid properties	
Properties of amino acids	http://www.russell.embl-heidelberg.de/aas/
Secondary structure prediction	
TMPRED	http://www.ch.embnet.org/software/TMPRED_form.html
SOSUI	http://sosui.proteome.bio.tuat.ac.jp/sosuiframe0.html
TMHMM	http://www.cbs.dtu.dk/services/TMHMM/
PREDICTPROTEIN	http://www.embl-heidelberg.de/predictprotein/
GPCRdb 7TM plots (Snake plots for most 7TMs)	http://www.gpcr.org/7tm/seq/snakes.html
Tertiary structure prediction and visualization	
Swiss-Model	http://expasy.hcuge.ch/swissmod/SWISS-MODEL.html
SCOP	http://scop.mrc-lmb.cam.ac.uk/scop/
Identification of functional motifs	
INTERPRO	http://www.ebi.ac.uk/interpro/scan.html
PROSITE	http://www.ebi.ac.uk/searches/prosite.html
PFAM	http://www.sanger.ac.uk/Software/Pfam/
NetPhos (serine, threonine and tyrosine phosphorylation)	http://www.cbs.dtu.dk/services/NetPhos/
NetOGlyc (O-glycosylation)	http://www.cbs.dtu.dk/services/NetOGlyc/
NetNGlyc (N-glycosylation)	http://www.cbs.dtu.dk/services/NetNGlyc/
SIGNALP (signal peptide prediction)	http://www.cbs.dtu.dk/services/SignalP/
Swissprot (functional annotation)	http://www.expasy.ch/cgi-bin/sprot-search-ful

comparison of very large sets of related proteins (see Chapter 14 and www.russell.embl-heidelberg.de/aas for more details). Defining the environment of an amino acid may be relatively straightforward if the protein is known, by looking at existing protein annotation or better still a known tertiary structure. Beyond the cellular environment of a variant there are many other important characteristics of an amino acid that need to be evaluated. These include the context of an amino acid within known protein features and the conservation of the amino acid position in an alignment of related proteins. Figure 12.4 shows an example of an evaluation of a mutation in Jagged1, a ligand for the Notch receptor family. Krantz *et al.* (1998) identified an Arg184Cys missense mutation in patients with Alagille syndrome (OMIM 118450). In terms of amino acid substitutions, Arg > Cys is very non-conservative (the extracellular substitution matrix score for this change is — 5). Alignment of the mutated human amino acid sequence with vertebrate and invertebrate orthologues and homologues in the Jagged family identifies the Arg184 residue as a highly conserved position throughout this gene family. A mutation to a cysteine at this position would be expected to lead to the aberrant formation of disulphide bonds with other cysteine residues in the Jagged protein, this is likely to have a disruptive effect on the structure of the Jagged1 protein, presumably leading to the Alagille syndrome phenotype (see Chapter 14 for a description of the effects of inappropriate disulphide bond formation).

There are many different sources of protein annotation and tools to evaluate the impact of substitutions in known and predicted protein features, some of the best are listed in Table 12.5. The protein analysis approaches underlying these tools are comprehensively reviewed in Chapter 14.

12.8 A NOTE OF CAUTION ON THE PRIORITIZATION OF *IN SILICO* PREDICTIONS FOR FURTHER LABORATORY INVESTIGATION

Just as the complexity of genes, transcripts and proteins are virtually limitless, so too are the possibilities for developing functional hypotheses. If every aspect of the analyses explored in this chapter were examined in any single polymorphism, it would probably be possible to assign a *potential* deleterious function to almost every one. But clearly the human genome does not contain millions of potentially deleterious mutations (thousands maybe, but not millions!), so it is important to treat *in silico* predictions with caution. If a polymorphism shows genetic association with a phenotype it is important to first consider if the polymorphism is causal or in LD with a causal mutation. Hypotheses need to be constructed and tested in the laboratory. For example if a polymorphism is predicted to impact splicing, then *in vitro* analysis methods need to be employed to investigate evidence for alternative transcripts.

12.9 CONCLUSIONS

In this chapter we have taken an overview of some of the approaches for predictive functional analysis of polymorphisms in genes, proteins and regulatory regions. These methods can be applied equally at the candidate identification stage or at later stages to assist in the progression of associated genes to disease genes. The chapter has also examined the role of bioinformatics in the formulation of laboratory-based investigation for confirmation of functional predictions. As we have shown there are very few tools specifically designed to evaluate the impact of polymorphisms on gene and protein function. Instead functional prediction of the potential impact of variation requires a very good grasp of the full gamut of bioinformatics tools used for predicting the properties and structure of genes, proteins and regulatory regions. This huge range of applications makes polymorphism analysis one of the most difficult bioinformatics activities to get right. The complexity of some analysis areas are worthy of special attention, particularly the analysis of polymorphisms in gene regulatory regions and protein sequences. To address some of these highly specialized analysis issues, Tom Werner presents a detailed examination of gene regulatory sequence analysis (Chapter 13) and Rob Russell and Matthew Betts present on tools and principles of protein analysis (Chapter 14).

REFERENCES

Afshar-Kharghan V, Li CQ, Khoshnevis-Asl M, Lopez JA. (1999). Kozak sequence polymorphism of the glycoprotein (GP) Ib-alpha gene is a major determinant of the plasma membrane levels of the platelet GP Ib-IX-V complex. *Blood* **94**: 186–191.

Aparicio SA. (2000). How to count human genes. *Nature Genet* **B25**: 129–130.

Bennett CL, Brunkow ME, Ramsdell F, O'Briant KC, Zhu Q, Fuleihan RL, *et al.* (2001). A rare polyadenylation signal mutation of the FOXP3 gene (AAUAAA→AAUGAA) leads to the IPEX syndrome. *Immunogenetics* **53**: 435–439.

Brenner S. (2000). Inverse genetics. *Curr Biol* **10**: R649.

Chakravarti A. (1999). Population genetics — making sense out of sequence. *Nature Genet* **21** (Suppl.): 56–60.

Chappell SA, LeQuesne JP, Paulin FE, deSchoolmeester ML, Stoneley M, Soutar RL, *et al.* (2000). A mutation in the c-myc-IRES leads to enhanced internal ribosome entry in multiple myeloma: a novel mechanism of oncogene de-regulation. *Oncogene* **19**: 4437–4440.

Chen CY, Shyu AB. (1995). AU-rich elements: characterization and importance in mRNA degradation. *Trends Biochem Sci* **20**: 465–470.

Conne B, Stutz A, Vassalli JD. (2000). The 3′ untranslated region of messenger RNA: a molecular 'hotspot' for pathology? *Nature Med* **6**: 637–641.

Di Paola R, Frittitta L, Miscio G, Bozzali M, Baratta R, Centra M, *et al.* (2002). A variation in 3prime prime or minute UTR of hPTP1B increases specific gene expression and associates with insulin resistance. *Am J Hum Genet* **70**: 806–812.

D'Souza I, Schellenberg GD. (2000). Determinants of 4-repeat tau expression. Coordination between enhancing and inhibitory splicing sequences for exon 10 inclusion. *J Biol Chem* **275**: 17700–17709.

Erdmann VA, Barciszewska MZ, Hochberg A, de Groot N, Barciszewski J. (2001). Regulatory RNAs. *Cell Mol Life Sci* **58**: 960–977.

Graber JH, Cantor CR, Mohr SC, Smith TF. (1999). *In silico* detection of control signals: mRNA 3′-end-processing sequences in diverse species. *Proc Natl Acad Sci USA* **96**: 14055–14060.

Gumucio DL, Shelton DA, Zhu W, Millinoff D, Gray T, Bock JH, *et al.* (1996). Evolutionary strategies for the elucidation of cis and trans factors that regulate the developmental switching programs of the beta-like globin genes. *Mol Phylogenet Evol* **5**: 18–32.

Hastings ML, Krainer AR. (2001). Pre-mRNA splicing in the new millennium. *Curr Opin Cell Biol* **13**: 302–309.

Heinemeyer T, Chen X, Karas H, Kel AE, Kel OV, Liebich I, *et al.* (1999). Expanding the TRANSFAC database towards an expert system of regulatory molecular mechanisms. *Nucleic Acids Res* **27**: 318–322.

Ishii S, Nakao S, Minamikawa-Tachino R, Desnick RJ, Fan JQ. (2002). Alternative splicing in the alpha-galactosidase A gene: increased exon inclusion results in the Fabry cardiac phenotype. *Am J Hum Genet* **70**: 994–1002.

Jacobs GH, Rackham O, Stockwell PA, Tate W, Brown CM. (2002). Transterm: a database of mRNAs and translational control elements. *Nucleic Acids Res* **30**: 310–311.

Jo EK, Kanegane H, Nonoyama S, Tsukada S, Lee JH, Lim K, *et al.* (2001). Characterization of mutations, including a novel regulatory defect in the first intron in Bruton's tyrosine kinase gene from seven Korean X-linked agammaglobulinemia families. *J. Immunol* **167**: 4038–4045.

Kaski S, Kekomaki R, Partanen J. (1996). Systemic screening for genetic polymorphism in human platelet glycoprotein Ib-alpha. *Immunogenetics* **44**: 170–176.

Kleiman FE, Ramirez AO, Dodelson de Kremer R, Gravel RA, Argarana CE. (1998). A frequent TG deletion near the polyadenylation signal of the human HEXB gene: occurrence of an irregular DNA structure and conserved nucleotide sequence motif in the 3′ untranslated region. *Hum Mut* **12**: 320–329.

Knight JC, Udalova I, Hill AV, Greenwood BM, Peshu N, Marsh K, *et al.* (1999). A polymorphism that affects OCT-1 binding to the TNF promoter region is associated with severe malaria. *Nature Genet* **22**: 145–150.

Kozak M. (1996). Interpreting cDNA sequences: some insights from studies on translation. *Mamm Genome* **7**: 563–574.

Kozak M. (2000). Do the 5′ untranslated domains of human cDNAs challenge the rules for initiation of translation (or is it vice versa)? **70**: 396–406.

Krantz ID, Colliton RP, Genin A, Rand EB, Li L, Piccoli DA, *et al.* (1998). Spectrum and frequency of Jagged1 (JAG1) mutations in Alagille syndrome patients and their families. *Am J Hum Genet* **62**: 1361–1369.

Lamb P, McKnight SL. (1991). Diversity and specificity in transcription regulation: the benefits of heterotypic dimerization. *Trends Biochem Sci* **16**: 417–422.

Lander ES, Linton LM, Birren B, Nusbaum C, Zody MC, Baldwin J, *et al.* (2001). Initial sequencing and analysis of the human genome. *Nature* **409**: 860–921.

Le SY, Maizel JV Jr, (1997). A common RNA structural motif involved in the internal initiation of translation of cellular mRNAs. *Nucleic Acids Res* **25**: 362–369.

Liu H, Chao D, Nakayama EE, Taguchi H, Goto M, Xin X, *et al.* (1999). Polymorphism in RANTES chemokine promoter affects HIV-1 disease progression. *Proc Natl Acad Sci USA* **96**: 4581–4585.

Liu HX, Cartegni L, Zhang MQ, Krainer AR. (2001). A mechanism for exon skipping caused by nonsense or missense mutations in BRCA1 and other genes. *Nature Genet* **27**: 55–58.

Lynch KW, Weiss A. (2001). A CD45 polymorphism associated with multiple sclerosis disrupts an exonic splicing silencer. *J Biol Chem* **276**: 24341–24347.

Moller LB, Tumer Z, Lund C, Petersen C, Cole T, Hanusch R, *et al.* (2000). Similar splice site mutations of the ATP7A gene lead to different phenotypes: classical Menkes disease or occipital horn syndrome. *Am J Hum Genet* **66**: 1211–1220.

Ohler U. (2000). Promoter prediction on a genomic scale: The Adh experience. *Genome Res* **10**: 539–542.

Pauws E, van Kampen AH, van de Graaf SA, de Vijlder JJ, Ris-Stalpers C. (2001). Heterogeneity in polyadenylation cleavage sites in mammalian mRNA sequences: implications for SAGE analysis. *Nucleic Acids Res* **29**: 1690–1694.

Peri S, Pandey A. (2001). A reassessment of the translation initiation codon in vertebrates. *Trends Genet* **17**: 685–687.

Pesole G, Grillo G, Larizza A, Liuni S. (2000). The untranslated regions of eukaryotic mRNAs: structure, function, evolution and bioinformatic tools for their analysis. *Brief Bioinform* **3**: 236–249.

Pitarque M, von Richter O, Oke B, Berkkan H, Oscarson M, Ingelman-Sundberg M. (2001). Identification of a single nucleotide polymorphism in the TATA box of the CYP2A6 gene: impairment of its promoter activity. *Biochem Biophys Res Commun* **284**: 455–460.

Reese MG, Hartzell G, Harris NL, Ohler U, Abril JF, Lewis SE. (2000). Genome annotation assessment in *Drosophila melanogaster*. *Genome Res* **10**: 483–501.

Reich DE, Lander ES. (2001). On the allelic spectrum of human disease. *Trends Genet* **17**: 502–510.

Sakurai K, Seki N, Fujii R, Yagui K, Tokuyama Y, Shimada F, *et al.* (2000). Mutations in the hepatocyte nuclear factor-4alpha gene in Japanese with non-insulin-dependent diabetes: a nucleotide substitution in the polypyrimidine tract of intron 1b. *Horm Metab Res* **32**: 316–320.

Schmucker D, Clemens JC, Shu H, Worby CA, Xiao J, Muda M, *et al.* (2000). Drosophila Dscam is an axon guidance receptor exhibiting extraordinary molecular diversity. *Cell* **101**: 671–684.

Shen LX, Basilion JP, Stantoon VP Jr. (1999). Single-nucleotide polymorphisms can cause different structural folds of mRNA. *Proc Natl Acad Sci USA* **96**: 7871–7876.

Venter JC, Adams MD, Myers EW, Li PW, Mural RJ, Sutton GG, *et al.* (2001). The sequence of the human genome. *Science* **291**: 1304–1351.

Vervoort R, Gitzelmann R, Lissens W, Liebaers I. (1998). A mutation (IVS8 + 0.6kbdelTC) creating a new donor splice site activates a cryptic exon in an Alu-element in intron 8 of the human beta-glucuronidase gene. *Hum Genet* **103**: 686–693.

Wahle E, Keller W. (1992). The biochemistry of 3-end cleavage and polyadenylation of messenger RNA precursors. *Annu Rev Biochem* **61**: 419–440.

CHAPTER 13

Functional *In Silico* Analysis of Non-coding SNPs

THOMAS WERNER

Genomatix Software GmbH
Munich, Germany

13.1 INTRODUCTION

The total amount of nucleotides within the human genome was found to be well within the expected range of about 3 billion base pairs. That was no big surprise since the physical size of the genome could already be measured fairly accurately by biophysical means well before sequencing became possible. However, the number of genes turned out to be surprisingly low, especially after the gene counts of *Drosophila melanogaster* and *Caenorhabtis elegans* were released (Adams *et al.*, 2000; The *C. elegans* Sequencing

Bioinformatics for Geneticists. Edited by M.R. Barnes and I.C. Gray
 ISBNs: 0 470 84393 4; 0 470 84394 2 (PB)

Consortium, 1998). It was common sense that humans should have at least double or triple the amount of genes as compared to those much simpler organisms. However, despite high expectations and correspondingly high initial estimates, the number of genes to be expected within the human genome decreased constantly. There is still no final answer but current estimates converge somewhere between 30,000 and 40,000 (Venter *et al*., 2001). This leads to a lot of questions as to where the huge differences between species will be found in the genomes, if not in gene numbers. It is also quite obvious from those numbers that only about 2–3% of the human genome is expected to encode proteins. Even disregarding the 40% repetitive sequences present in the human genomic sequence, this leaves more than half of the genomic sequence in search of a function.

Of course, encoding proteins is just one of the many known functions of the genome. There are three very prominent additional tasks that must be fulfilled by the genome. The first one is to maintain some physical ordered structure of the genomic sequence, which is a prerequisite for everything else. A hopeless tangle of 3 billion base pairs would most likely interfere severely with gene expression as well as with DNA replication.

The second task that has to be faithfully fulfilled over a lifetime in any organism is the correct replication of the genomic information to allow cell divisions. And last but not least, gene expression itself involves much more than synthesis of an RNA copy of the coding parts of the genome. The correct regulation both of transcription as well as DNA replication in space and time is probably the most crucial part of life for any organism. No cell let alone a multicellular organism, can develop or survive without perfect control over gene expression (control of replication is just one of the consequences of controlled gene expression).

Here the genome has to fulfil a formidable task. The information encoded within the genome can be regarded as invariant regardless of the few mutations that occur continuously within a living cell (most are either repaired or eliminated by selection). This view also includes Single Nucleotide Polymorphisms (SNPs) because most SNPs are frozen in evolution and very few arise during the lifespan of an individual organism. Survival of such mutations becomes most prominently visible in allelic differences where there appears to be more than one solution for a functional sequence. Development and differentiations are examples of extremely complex and linked programmes that have to be fulfilled in an exact time-frame. Nevertheless, this is the easy part for the genome as both of these processes are deterministic and every step is clear from the very beginning with very little variation included.

In contrast, every organism encounters a variety of unexpected environmental stimuli (availability of food resources, climate conditions, interactions with other organisms such as predators or competing species). The static genome must provide *a priori* all information suitable to react appropriately to such external challenges. This requires an enormous amount of 'conditional programming' within the genetic code, most of which is not directly manifest in protein sequences. This is probably the major reason why regulatory sequences appear to occupy almost five to 10 times more genomic sequence than coding regions (this estimate includes all regulatory sequences not only transcriptional regulation). For the very same reason most of this chapter will focus on regulatory aspects. The allelic differences (something in the region of one nucleotide in 1000) also called Single Nucleotide Polymorphisms SNPs (pronounced 'Snips') may have no effect under normal conditions but can make a huge difference in the case of changing conditions. Such changes do not need to be external. A developing tumour can bring a dramatic change to the physiology of an organism and genetic predisposition is linked to a large extent to allelic differences. However, the effect of any individual SNP or sets of SNPs (e.g. haplotypes) depends critically on the local context within which the SNP is located. Therefore, functional analysis or estimation of SNPs requires detailed understanding of the functional features and sequence regions within the genome.

13.2 GENERAL STRUCTURE OF CHROMATIN-ASSOCIATED DNA

DNA is complexed with histone proteins forming nucleosomes which are distributed along most of the genomic DNA like beads on a string. This structure is then organized in chromosomal loops and such loops are known to form solenoids (Daban and Bermudez, 1998). Solenoids in turn form the chromosomal fibres already visible by microscopy. For the purpose of this chapter the most relevant structures are all within a chromosomal loop. Therefore, the schematic organization of a chromosomal loop will be used as a framework for the explanation of all further components.

Repetitive DNA of retroposon origin is ubiquitously found throughout the genome. As we learned from the first published chromosomal sequence (chromosome 22, Dunham *et al.*, 1999) about 40% of the human genomic DNA consists of repetitive DNA, most prominent among these are ALU repeat sequences. They have been named after the restriction enzyme (ALU I) which generates a characteristic satellite band in digests of genomic DNA. ALUs belong to the class of short interspersed elements (SINEs) which are short retroposon sequences of only about 300 nucleotides in length. Another class of repetitive DNA is the long interspersed elements (LINEs) which reach up to 7 kb in length and include retroviral sequences (Smit, 1999).

A chromatin loop is the region of chromosomal DNA located between two contact points of the DNA with a protein framework within the nucleus, the so-called nuclear matrix. These contact points are marked in the genomic DNA as Matrix/Scaffold Attachment Regions (S/MARs). Association of DNA with this nuclear matrix is a prerequisite for transcription of nucleosomal DNA (Bode *et al.*, 2000). S/MARs are themselves complex structures not yet fully understood at the molecular level. There is an excellent review on chromatin domains and prediction of MAR sequences by Boulikas (1995) explaining S/MARs and their elements in detail. There are currently two methods to detect S/MAR elements in genomic sequences, the first is MARFinder by Kramer (1996) (http://www.ncgr.org/MAR-search/). The second method is SMARTest developed by Genomatix Software GmbH, Munich and available for academic researchers free of charge from http://www.genomatix.de (Frisch *et al.*, 2002).

Enhancers are regulatory regions found within chromosomal loops that can significantly boost the level of transcription from a responsive promoter regardless of their orientation and distance with respect to the promoter within the same chromatin loop. Currently, there is no way to detect enhancers in general by *in silico* methods. However, at least a subclass of enhancers is organized in a very similar manner to that of promoters, i.e. they also contain frameworks of transcription factors (Gailus-Durner *et al.*, 2000). In cases where enhancers share modules with promoters it is possible to find them via the module. However, since an isolated module match is no proof of either a promoter or an enhancer, experimental verification is still mandatory. Silencers are basically identical to enhancers and follow the same requirements but exert a negative effect on promoter activities. Enhancers and silencers often show a similar internal organization as promoters (Werner, 1999).

13.3 GENERAL FUNCTIONS OF REGULATORY REGIONS

The biological functionality of regulatory regions is generally not a property evenly spread over the regulatory region in total. Functional units are usually defined by a combination of defined stretches that can be delimited and possess an intrinsic functional property (e.g. binding of a protein or a curved DNA structure). Several functionally similar types of these stretches of DNA are already known and will be referred to as *elements*. Those elements are neither restricted to regulatory regions nor individually sufficient for the regulatory function of a promoter or enhancer. The function of the complete regulatory region is composed of the functions of the individual elements either in an additive manner (independent elements) or by synergistic effects (modules) (Werner, 1999). With respect to SNPs it is important to view regulatory DNA in a similar way as we regard coding genes: there are short stretches of immediate functional importance (e.g. exons or regulatory elements) and there are much larger regions with either unknown or more implicit functions (introns, 'spacer' DNA in regulatory regions). As a direct consequence of this sort of discontinuous organization of functional sequences the potential effects of SNPs can vary dramatically (from none to lethal) depending which part of the sequence the SNP is located in.

13.4 TRANSCRIPTION FACTOR BINDING SITES (TF-SITES)

Binding sites for specific proteins are most important among regulatory elements. They consist of about 10 to 30 nucleotides, not all of which are equally important for protein binding, a reminder of the somewhat fractal properties of genomic sequences, i.e. a binding site looks like a tiny copy of a promoter, which in turn looks like a small copy of a gene, etc. Individual protein binding sites may vary in part of their sequence, even if they bind to the same protein. There are nucleotides which are in contact with the protein in a sequence-specific manner ('recognition exons'), which usually represent the best-conserved areas of a binding site. Different nucleotides are involved in more non-specific contacts to the DNA backbone (i.e. not sequence specific as they do not involve the bases A, G, C or T), and there are internal 'spacers' ('introns') which are not in contact with the protein at all. All in all, protein binding sites exhibit enough sequence conservation to allow for the detection of candidates by a variety of sequence similarity-based approaches. There have been many attempts to collect TF binding sites (Wingender *et al.*, 2001), as well as several developments in the location of TF binding sites in genomic sequences (Chen *et al.*, 1995; Prestridge, 1996; Quandt *et al.*, 1995). However, potential binding sites can be found almost anywhere in the genome and are not restricted to regulatory regions. Quite a number of binding sites outside regulatory regions are also known to bind their respective binding proteins (e.g. Kodadek, 1998). Therefore the abundance of predicted binding sites is not just a shortcoming of the detection algorithms but reflects biological reality although currently hard to interpret in functional terms.

13.5 STRUCTURAL ELEMENTS

Secondary structures are mostly known for RNAs and proteins but they also play important roles in promoters (e.g. Bates *et al.*, 2001). Potential secondary structures can be easily determined. For an excellent start point see Michael Zuckers homepage (Table 13.1).

TABLE 13.1 Useful URLs for Regulatory SNP Analysis

RNA secondary structure prediction	
Homepage of M. Zucker	http://bioinfo.math.rpi.edu/~zukerm/
Pattern definition:	
CoreSearch (ftp)	ariane.gsf.de/pub/unix/coresearch_1.2.tar.Z
CONSENSUS	http://bioweb.pasteur.fr/seqanal/interfaces/consensus-simple.html
S/MAR detection	
MARFinder	http://www.ncgr.org/MAR-search/
SMARTest	http://www.genomatix.de
UTR analysis	
UTR database	http://bighost-area.ba.cnr.it/BIG/BioWWW/#UTRdb
Genomatix tools	
(free to academic users)	
SMARTest	http://www.genomatix.de/free_services/
PromoterInspector	
MatInspector professional	
GEMS Launcher	
ELDorado	
Sequence tools	

There is a plethora of tools now available on the web. However, Michael Zucker has been one of the most important pioneers in the field, so starting from his page would be a good choice. Secondary structures are also often not conserved in primary nucleotide sequence but are subject to strong positional correlation within the structure. There is also always a trade-off between best and fastest structure prediction. Some algorithms dive deep into energy calculations to provide the best possible structure for one RNA while others do a much more rudimentary analysis, which can be applied to many sequences within the same time-frame as a single in-depth analysis would take. It is impossible to decide in advance which approach will be most suitable for any problem. A few experiments using different methods will be called for.

13.6 ORGANIZATIONAL PRINCIPLES OF REGULATORY REGIONS

Regulatory regions are not just statistical collections of the regulatory elements introduced above. Therefore, it is necessary to understand at least some basic organizational features of regulatory regions in order to understand the different consequences SNPs can have. Eukaryotic polymerase II promoters will serve as examples as they appear to be the currently best-studied regulatory regions. The TF-sites within promoters (and likewise most other regulatory sequences) do not show any obvious general patterns with respect to location and orientation within the promoter sequences. TF binding sites can be found virtually everywhere in promoters but in individual promoters possible locations are much more restricted. A closer look reveals that the function of a TF binding site often depends on the relative location and especially on the sequence context of the binding site.

The context of a TF-site is one of the major determinants of its role in transcription control. However, the context is not merely a few nucleotides around the binding site, which would be more like an extension of the binding site rather than a context. More important is the context of other TF binding sites located at some distance that are often grouped together and such functional groups have been described in many cases. A systematic attempt to collect synergistic or antagonistic pairs of TF binding sites has been made with the COMPEL database (Kel *et al.*, 1995, Heinemeyer *et al.*, 1998). In many cases, a specific promoter function (e.g. a tissue-specific silencer) will require more than two sites simultaneously. Such groups of promoter subunits consisting of several TF binding sites that carry a specific function independent of the promoter, will be referred to as *promoter modules*. This is a definition at the molecular level, which is more specific than the definition recently given by Arnone and Davidson (1997) requiring only the presence of the sites within a loosely defined DNA region. Within a molecular promoter module both sequential order and distance can be crucial for function indicating that these modules may be the critical determinants of a promoter rather than individual binding sites. However, promoters can contain several modules that may use overlapping sets of binding sites. Therefore, the conserved context of a particular binding site cannot be determined from the primary sequence without additional information about the modular structure (Figure 13.1).

The peculiar property of promoter modules to function only as intact units has an important consequence for the effects SNPs can have in promoters. A SNP inactivating a single TF binding site can in fact destroy the function of a complete module, which will not be obvious from inspection of the individual binding site in which the SNP was detected. A SNP affecting another binding site of the same factor within the same promoter may have a quite different effect if this binding site is either part of another module or has no direct function at all. Therefore, identification of binding sites affected by SNPs is not enough to estimate the functional consequences of regulatory SNPs.

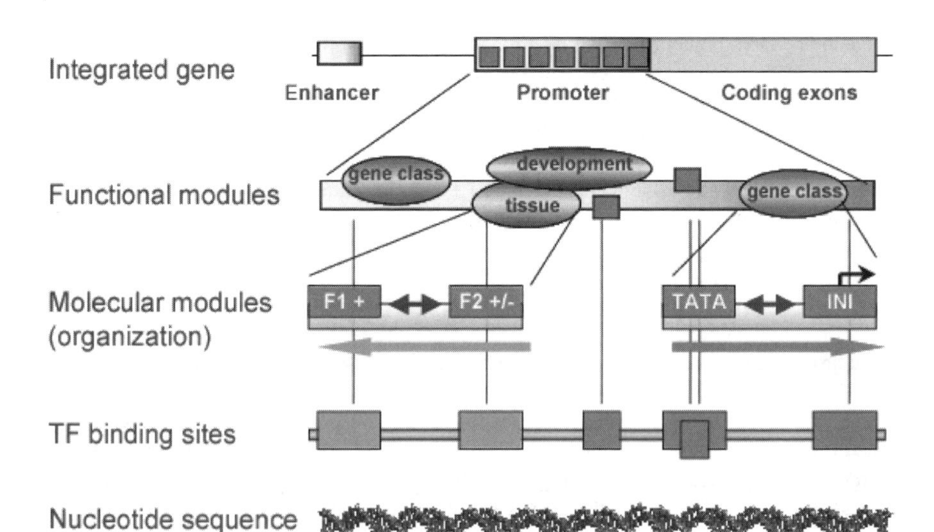

Figure 13.1 Hierarchical structure of a polymerase II promoter (schematic). Oval shapes indicate modules without defined internal structure; rectangular boxes in the lower part of the figure indicate transcription factor binding sites. Note that a direct assignment of individual binding sites to functional modules is only possible for molecular modules. The arrows below molecular modules indicate strand orientation of the modules.

Similar logic holds for insertions/deletions in promoter sequences (e.g. polymorphic microsatellite sequences). The quite variable distances found between elements in functionally related promoters could be interpreted as the result of insertion/deletion events, although it is hard to find clear evidence for this. Insertions or deletions affecting the organization of modules may well interfere with function. However, as most of the promoter functions are crucial for the function of the whole gene, such deleterious mutations are most likely selected against in evolution. Therefore, the variability in spacing seen in present time sequences should be considered neutral with respect to function unless there is direct evidence to the contrary.

13.7 RNA PROCESSING

Genomic regulation does not stop at the level of the genomic DNA sequence. RNAs contain important regulatory signals of their own, among them are transport signals that direct the RNAs to specific subcellular locations and RNA instability signals that mark RNAs for rapid destruction unless specifically protected. Most of these signals reside within the 5′ and 3′ untranslated terminal regions of the mRNAs (UTRs), which have been collected in a specialized database (Pesole *et al.*, 1996, see Table 13.1 for URL). This database is an excellent representation of the knowledge about UTRs as reported in the literature. However, there were no efforts made to complete 5′ or 3′ UTRs in case they were reported incomplete. Because RNAs are faithful copies of the genomic sequences (except for RNA editing) all of these signals can also be directly studied in the genomic DNA.

Removal of intronic sequences is one of the most important processing steps of primary transcripts (splicing). As became quite clear in recent years splicing is governed by complex and discontinuous signals located within exons as well as introns (Kramer, 1996). The set-up of complete splice signals resembles the general set-up of promoters and enhancers to a large extent. There are splice enhancers, splice donor and acceptor sites, branch point sequences as well as some less well defined accessory sequences within introns. Again the organizational context of splicing elements is most likely the most important factor determining biological function.

13.8 SNPs IN REGULATORY REGIONS

SNPs are to be found all over the genome and as a mere consequence of the amount of sequence a considerable number of SNPs are expected to be located within regulatory sequences. As discussed above the potential effects of SNPs on gene regulation depend on the location of SNPs with respect to the regulatory elements. SNPs located in non-functional spacer DNA (if anything like that exists) will not affect regulation in all likelihood, whereas a SNP destroying the binding site of a crucial transcription factor can alter transcription of a gene quite dramatically.

13.8.1 Examples for Regulatory SNPs

SNPs can influence TF binding sites in three different ways. A binding site can be destroyed by loss of binding affinity due to the SNP. The opposite effect is also possible, that is, generation of a new binding site within a regulatory sequence. A combination of these two events would result in an altered binding site that might have switched specificity to another protein. There are examples with well-established effects on gene regulation for TF binding sites being deleted as well as created by SNPs.

The RANTES gene encodes a chemokine involved in immune signalling. Unfortunately, RANTES expression is also involved in supporting HIV-1 infection, which of course is detrimental for the individual. There is one mutation (SNP) known in the RANTES promoter that destroys a potential c-myb binding site immediately upstream of the TATA box. This mutation has the astonishing effect of delaying the CD4 depletion by HIV-1 infection although it has no effect on the infection itself. However, it was found that this mutation increases the transcription of the RANTES chemokine gene, demonstrating nicely that SNPs *per se* are not determined to be positive or negative (Lui *et al.*, 1999). The positive effect of this SNP only becomes apparent upon HIV-1 infection.

Cystic fibrosis is a devastating disease caused by a defective protein which results in a dramatically shortened lifetime for the sufferers. In one case a SNP generated a new binding site for the transcription factor YY1 in the promoter of the gene already affected by a mutation in exon 11, which caused the disease. The effect of this new YY1 binding site was over-expression of the (not completely) defective protein via attracting additional protein(s) to the promoter complex, which reduced the symptoms of the disease via a gene dosage effect (Romey *et al.*, 2000).

The 'bottom line' of all these examples is that SNPs affected transcriptional control elements that were actually involved in the gene transcription of the respective genes.

13.9 EVALUATION OF NON-CODING SNPs

In a case where a SNP is located within the coding sequence of a gene it is very simple to find out whether this is a silent exchange or not, just from the triplet code. In the case where there is an amino acid exchange it is much less obvious whether the exchange will affect protein function or not. If there is no known example for the particular exchange there is no way to predict the functional consequences solely from the sequence.

SNPs in regulatory regions are always difficult to assess for two reasons: the first is simply to find out whether a SNP is located inside regulatory regions at all. Given that locating regulatory regions is much more difficult than locating coding sequences, this is no trivial task. However, even if this prerequisite can be satisfied there remains the question of whether the SNP affects any regulatory elements. This requires knowledge or at least a well-supported hypothesis about the regulatory elements relevant for the regulation of the gene in question. Since this information cannot be directly derived from the nucleotide sequence of a promoter for example, this requires additional efforts to locate relevant regulatory elements.

Once it has been established that a SNP is located within a putative regulatory element, it is possible to evaluate the primary effect of the SNP on this regulatory element. The primary effect is the change in binding affinity or specificity of a TF binding site or the change in the stability of a secondary structure. This kind of information can be derived from a comparison of the wild-type sequence with the SNP-containing sequence and usually a fair estimate of the resulting changes can be made.

Unfortunately, even that information is only part of the answer. The real question is whether the SNP-induced change has functional consequences on regulation, which is not necessarily a consequence of a single altered regulatory element. Therefore, it is also necessary to ascertain the relevant context of the affected elements, e.g. whether a TF binding site is part of a transcriptional module or not.

13.10 SNPs AND REGULATORY NETWORKS

Although SNPs are always necessarily located within one gene or one regulatory region and they can have pleiotropic effects. SNPs affecting the protein sequence or the regulation of transcription factors can influence the expression of many target genes of this particular factor and lead to pronounced systemic effects. This is achieved via the regulatory networks in which the affected transcription factor participates. A functional correlation of a SNP with an observed phenotype can be established generally by epidemiological studies. However, this kind of correlation does not reveal any data about the molecular mechanisms behind the correlation. The molecular link between the regulatory SNP and the observed phenotype is finally established on the level of regulatory networks by tracking relevant transcriptional modules. However, reconstruction of regulatory networks on the molecular level is far from easy with current tools and may turn out to be (still) a futile effort in many cases. Where it works, it provides the final answer not only to the effect of the SNP but also reveals the molecular mechanisms behind the phenotype. Due to that enormous gain in insight as well as the therapeutic possibilities, it is well justified to invest a significant effort into the elucidation of the pertinent regulatory networks even if the chances of success are sometimes slim.

13.11 SNPs MAY AFFECT THE EXPRESSION OF A GENE ONLY IN SPECIFIC TISSUES

In addition, SNPs are always necessarily present in all tissues and their effects may vary dramatically. As outlined above many promoter modules are active under specific conditions only, such as when they are stimulated by a signalling pathway or only in particular cell or tissue types. Consequently, a SNP affecting a specific module will only show an effect under conditions where the corresponding module is active. Therefore, lack of association of a promoter SNP with an observable phenotype in cell culture experiments only excludes a functional effect in the particular cell type under the specific conditions used in the experiment. The very same 'silent' SNP may have a clear effect either in the same cells under different conditions or in other cells/tissues which have not been tested. Therefore, *in vitro* results are only conclusive in the case of positive results, while negative results are of limited value.

13.12 *IN SILICO* DETECTION AND EVALUATION OF REGULATORY SNPs

So far we have established the basic factors and requirements of how to attack the problem of regulatory SNPs. Now it is time to detail the strategies that will help to elucidate the different levels of SNP-caused effects in a practical approach.

Figure 13.2 shows an overview of the general strategy to analyse SNPs for potential regulatory effects. In brief SNPs are first mapped onto predetermined regulatory regions (in this example, promoters). The gene annotation is used to identify promoters of interest as well as to include pre-existing knowledge about functional aspects of these promoters, e.g. promoter elements already known to be involved in promoter function. Then relevant transcriptional elements are identified either from knowledge databases or by comparative analyses of sets of functionally related promoters (detailed below). At this point it is

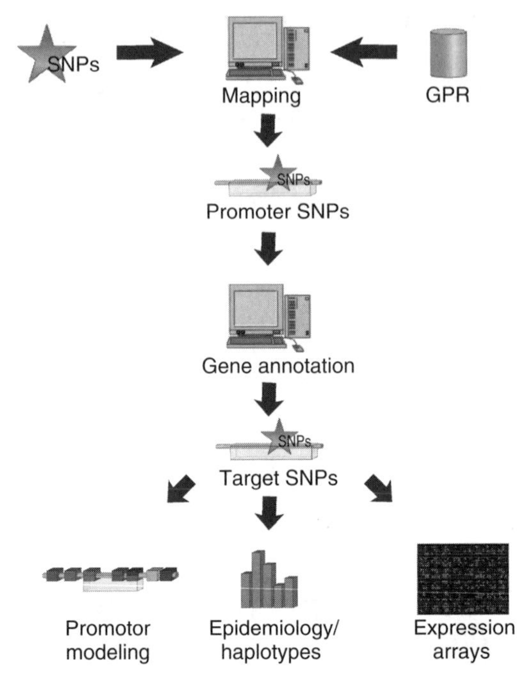

Figure 13.2 Strategy to qualify genomic SNPs as relevant regulatory SNPs. GPR (Genomatix Promoter Resource now part of ElDorado®). The transparent light grey box symbolizes a promoter region; darker boxes indicate transcription factor binding sites. (See note added in proof).

possible to determine whether the SNP is located within a relevant transcription factor binding site and the effect on binding affinity can be calculated. The comparative sequence analysis may also have revealed promoter modules facilitating estimation of regulatory effects. However, since a promoter can contain several independent functional modules it may be necessary to re-analyse the promoter in another functional context to focus on the effect of a particular SNP.

How to implement such a general strategy for real life application? This will be shown in the example below of Genomatix sequence analysis tools that were especially developed to facilitate this kind of approach.

13.13 GETTING PROMOTER SEQUENCES

Promoter sequences can be derived from the literature (about 10 to 20% of genomic promoters), by promoter prediction (about 50% of the genomic promoters) or by mapping of 5′-complete mRNAs (up to the Transcription Start Site, TSS) to the genomic sequence. In the last case the promoter is the sequence containing about 100 bp of the mRNA and a region of about 500–600 nucleotides immediately upstream of the TSS. So far there are also only about 10% of the mRNAs available as 5′-complete sequence. In summary it is possible to obtain about 50 to 70% of the human promoters by a combination of these approaches.

Of course, there are resources as well as methods claiming to be able to provide up to 90% of human promoters. However, the problem is that higher sensitivity is always achieved at the cost of lower specificity and a promoter collection containing all true promoters burdened by a high amount of false positives is absolutely useless for the regulatory SNP analysis, while half of the promoters with few false positives are sufficient to obtain useful results, although only for a subset of regulatory SNPs.

There is no best way to deal with the results of promoter prediction. For example, while it is quite popular to let the user play with some sort of cryptic scoring/quality Genomatix PromoterInspector does not have such a parameter. A promoter cannot be predicted with more or less scoring. The point is that every method will also produce false positives at any threshold. Any specificity given is only valid using exactly the corresponding parameters. A much better way to strengthen the likelihood of a true prediction is to check for additional evidence, e.g. a gene annotation or prediction that indicates the same region as the potential promoter. However, for a significant part of the human genome there is no alternative to prediction at this time.

Genomatix provides the Genomatix Promoter Resource (GPR) for this purpose (containing predictions for about 50% of all human promoters with a false positive rate of less than 15%, Scherf *et al.*, 2000, 2001) complemented by mRNA mapping as well as promoters extracted from the experimental literature. This is by no means the most complete collection of promoters available but most likely the one with the least false positives. GPR is a product of Genomatix that requires licensing. However, the software tool used to generate GPR (PromoterInspector) is available (with some restrictions) to academic scientists free of charge (URL see Table 13.1).

This resource can be used to map SNPs to the promoter regions (Genomatix software can do this automatically high throughput). Based on the gene annotation corresponding to these promoters, genes of interest with SNPs inside the promoter regions can be selected for further analysis.

At this point there are two possible strategies to evaluate the functional importance of the SNPs in question. A straightforward approach involves epidemiological data connecting the SNP to a phenotype by statistical coupling as can be seen in haplotype studies (e.g. Judson *et al.*, 2000, Stephens *et al.*, 2001). If such data are available the analysis can directly proceed to the identification of the transcription factor binding sites affected and the consequences for binding affinities (see below).

13.14 IDENTIFICATION OF RELEVANT REGULATORY ELEMENTS

If no such epidemiological data are available it is necessary to first select binding sites which are likely to be involved in promoter function, because direct analysis for potential binding sites usually yields about a 10 times excess of potential binding sites. Selection can be carried out with Genomatix software (GEMS Launcher) and is based on the principle of evolutionary conservation of functional binding sites in promoters. There are two ways to assess such functional conservation. The first is to compare promoters from orthologous genes in several species (e.g. man, mouse and dog or another non-rodent mammal). Of course mouse, rat and hamster might be related too closely to reveal a useful pattern. Such an analysis usually results in a conserved framework of about three to eight binding sites, which can be directly used for further evaluation. An example of such an analysis was the determination of the general mammalian actin promoter model (Frech *et al.*, 1998).

Another approach is horizontal conservation derived from sets of genes within the same organisms that are coupled functionally, e.g. by co-expression. The strategy has already been outlined in detail in Werner (2001).

Although not immediately evident, the difference between co-regulation and mere co-expression is of great importance for this strategy. Co-regulated genes usually share partial promoter features, so-called modules responsible for the observed co-regulation. Co-expressed genes which just show up at the same time but are not co-regulated do not necessarily share such modules. Therefore, they may interfere with comparative sequence analysis and should be removed first. There are several possible ways to focus co-regulated rather than co-expressed genes.

The basic idea of transcription event-oriented clustering is to include additional information beyond the mere expression level into the clustering process. One way to do this is to use multiple time points. Different pathways might resemble each other in expression level of genes for a limited period of time but separate at other time points. Clustering of genes based on time profiles of gene expression is leading more directly towards identification of the underlying mechanisms than clustering based on expression levels at a single time point. Groups of genes suitable for clustering can be derived from pathway information or directly from expression arrays. Again the GPR can be used to locate the corresponding promoters for human genes (or PromoterInspector to analyse other mammalian genomic sequences for promoters, Scherf *et al.*, 2000).

Comparative sequence analysis of such a set of co-expressed genes usually reveals the promoter module responsible for the observed co-expression and not a complete model. If the same promoter is analysed in combination with different sets of expression-related genes, different modules may be found. For example, analysis of a set of promoters expressed during glucose starvation may reveal a different module within the promoters than analysis of one of these promoters in a context of growth factor-induced genes.

13.15 ESTIMATION OF FUNCTIONAL CONSEQUENCES OF REGULATORY SNPs

The selection of conserved binding sites either from the analysis of orthologous promoters or from sets of co-expressed genes can be directly evaluated for SNP-induced differences in binding affinities, e.g. by applying the MatInspector program (also integrated in the GEMS Launcher, Frech *et al.*, 1997; Quandt *et al.*, 1995). If a site from this selected set is affected, GEMS Launcher can directly determine the resulting change in binding affinity for the protein at least in a qualitative manner.

If promoters from co-expressed genes were used, there are also indications about module structures, indicating which signal response might be affected by the SNP. In this manner it is possible to formulate a detailed hypothesis about the functional consequence of a particular SNP. However, final proof will only come from an experiment. The bioinformatics analysis will provide exact guiding of how to set up a decisive experiment.

The analysis for conserved binding sites as well as the evaluation of binding affinity changes will be fully automatic in a new software package Genomatix is releasing in 2002. So far, these steps remain interactive and may require a substantial amount of interactive work in some cases.

13.15.1 Limitations of this Approach

The up-side of the approach is that the tedious work of defining the promoter framework has only to be done once. After that an unlimited number of SNPs hitting this promoter can be evaluated automatically.

Unfortunately, this strategy has more limitations than the amount of work required. If a SNP hits the promoter outside of any binding site belonging to an identified framework, there is no guarantee that this will be a silent mutation as promoter frameworks are usually incomplete and there is always the possibility that a binding site, unknown so far, may be affected. In such cases additional experimental validation of the SNP is required or orthologous or co-expressed promoters have to be analysed for unknown but conserved patterns that might correspond to new binding sites. There is also software available to carry out such analyses (e.g. Stormo and Hartzell, 1989, Wolfertstetter *et al.*, 1996) (CONSENSUS and CoreSearch, see Table 13.1 for URLs).

13.16 CONCLUSION

In summary, SNPs located within known elements of frameworks can be evaluated for potential functional consequences, while SNPs located in a region not assigned to any functional framework remain unresolved in the absence of additional data. Fortunately, the information about regulatory sequences which accumulates during the analysis of individual SNPs remains valid for additional SNP analyses. Therefore, this new kind of genomic analysis may have a steep learning curve in the beginning but will gradually develop into a very powerful high throughput system in the near future.

The Evaluation of regulatory SNPs described in 13.12 to 13.15 and in Figure 13.2 down to the level of target SNPs has been carried out genome-wide in the meantime and is all available as part of the ElDorado system.

REFERENCES

Adams MD, Celniker SE, Holt RA, Evans CA, Gocayne JD, Amanatides PG, *et al.* (2000). The genome sequence of *Drosophila melanogaster*. *Science* **287**: 2185–2195.

Arnone MI, Davidson EH. (1997). The hardwiring of development: organization and function of genomic regulatory systems. *Development* **124**: 1851–1864.

Bates MD, Schatzman LC, Harvey RP, Potter SS. (2001). Two CCAAT boxes in a novel inverted repeat motif are required for Hlx homeobox gene expression. *Biochim Biophys Acta* **1519**: 96–105.

Bode J, Benham C, Knopp A, Mielke C. (2000). Transcriptional augmentation: modulation of gene expression by scaffold/matrix-attached regions (S/MAR elements). *Crit Rev Eukaryot Gene Expr* **10**: 73–90.

Boulikas T. (1995). Chromatin domains and prediction of MAR sequences. *Int Rev Cytol* **162A**: 279–388.

Chen QK, Hertz GZ, Stormo GD. (1995). MATRIX SEARCH 1.0: a computer program that scans DNA sequences for transcriptional elements using a database of weight matrices. *Comp Appl Biosci* **11**: 563–566.

Daban JR, Bermudez A. (1998). Interdigitated solenoid model for compact chromatin fibres. *Biochemistry* **37**: 4299–4304.

Dunham I, Shimizu N, Roe BA, Chissoe S, Hunt AR, Collins JE, *et al.* (1999). The DNA sequence of human chromosome 22. *Nature* **402**: 489–495.

Frech K, Quandt K, Werner T. (1997). Software for the analysis of DNA sequence elements of transcription. *Comp Appl Biosci* **13**: 89–97.

Frech K, Quandt K, Werner, T. (1998). Muscle actin genes: a first step towards computational classification of tissue specific promoters. *In Silico Biol* **1**: 5.

Frisch M, Frech K, Klingenhoff A, Quandt K, Liebich I, Werner T. (2002). *In Silico* prediction of matrix attachment regions in large genomic sequences. *Genome Res* **12**: 349–354.

Gailus-Durner V, Scherf M, Werner T. (2000). Experimental data of a single promoter can be used for *in silico* detection of genes with related regulation in absence of sequence similarity. *Mammal Genome* **12**: 67–72.

Heinemeyer T, Wingender E, Reuter I, Hermjakob H, Kel AE, Kel OV, *et al.* (1998). Databases on transcriptional regulation: TRANSFAC, TRRD and COMPEL. *Nucleic Acids Res* **26**: 362–367.

Judson R, Stephens JC, Windemuth A. (2000). The predictive power of haplotypes in clinical response. *Pharmacogenomics* **1**: 5–16.

Kel OV, Romaschenko AG, Kel AE, Wingender E, Kolchanov NA. (1995). A compilation of composite regulatory elements affecting gene transcription in vertebrates. *Nucleic Acids Res* **23**: 4097–4103.

Kodadek T. (1998). Mechanistic parallels between DNA replication, recombination and transcription. *Trends Biochem Sci* **23**: 79–83.

Kramer A. (1996). The structure and function of proteins involved in mammalian pre-mRNA splicing. *Annu Rev Biochem* **65**: 367–409.

Kramer JA, Singh GB, Krawetz SA. (2000). Computer-assisted search for sites of nuclear matrix attachment *Genomics* **35**: 273.

Liu H, Chao D, Nakayama EE, Taguchi H, Goto M, Xin X, *et al.* (1999). Polymorphism in RANTES chemokine promoter affects HIV-1 disease progression. *Proc Natl Acad Sci USA* **96**: 4581–4585.

Pesole G, Grillo G, Liuni S. (1996). Databases of mRNA untranslated regions for metazoa. *Comput Chem* **20**: 141–144.

Prestridge DS. (1996). SIGNAL SCAN 4.0: additional databases and sequence formats. *Comp Appl Biosci* **12**: 157–160.

Quandt K, Frech K, Karas H, Wingender E, Werner T. (1995). Matlnd and Matlnspector: new fast and versatile tools for detection of consensus matches in nucleotide sequence data. *Nucleic Acids Res* **23**: 4878–4884.

Romey MC, Pallares-Ruiz N, Mange A, Mettling C, Peytavi R, Demaille J, *et al.* (2000). A naturally occurring sequence variation that creates a YY1 element is associated with increased cystic fibrosis transmembrane conductance regulator gene expression. *J Biol Chem* **275**: 3561–3567.

Scherf M, Klingenhoff A, Werner T. (2000). Highly specific localization of promoter regions in large genomic sequences by PromoterInspector—a novel context analysis approach. *J Mol Biol* **297**: 599–606.

Scherf M, Klingenhoff A, Frech K, Quandt K, Schneider R, Grote K, *et al.* (2001). First pass annotation of promoters on human chromosome 22. *Genome Res* **11**, 333–340.

Smit AF. (1999). Interspersed repeats and other mementos of transposable elements in mammalian genomes. *Curr Opin Genet Dev* **6**: 657–663.

Stephens JC, Schneider JA, Tanguay DA, Choi J, Acharya T, Stanley SE, *et al.* (2001). Haplotype variation and linkage disequilibrium in 313 human genes. *Science* **293**: 489–493.

Stormo GD, Hartzell III GW. (1989). Identifying protein-binding sites from unaligned DNA fragments. *Proc Natl Acad Sci USA* **86**: 1183–1187.

The *C. elegans* Sequencing Consortium. (1998). Genome sequence of the nematode *C. elegans*: a platform for investigating biology. *Science* **282**: 2012–2018.

Venter JC, Adams MD, Myers EW, Li PW, Mural RJ, Sutton GG, *et al.* (2001). The sequence of the human genome. *Science* **291**: 1304–1351.

Werner T. (1999). Identification and characterization of promoters in eukaryotic DNA sequences. *Mammal Genome* **10**: 168–175.

Werner T. (2001). Cluster analysis and promoter modelling as bioinformatics tools for the identification of target genes from expression array data. *Pharmacogenomics* **2**: 25–36.

Wingender E, Chen X, Fricke E, Geffers R, Hehl R, Liebich I, *et al.* (2001). The TRANSFAC system on gene expression regulation. *Nucleic Acids Res* **29**: 281–283.

Wolfertstetter F, Frech K, Herrmann G, Werner T. (1996). Identification of functional elements in unaligned nucleic acid sequences by a novel tuple search algorithm. *Comp Appl Biosci* **12**: 71–80.

CHAPTER 14

Amino Acid Properties and Consequences of Substitutions

MATTHEW J. BETTS[1] and ROBERT B. RUSSELL[2]

[1] *Bioinformatics*
deCODE genetics, Sturlugötu 8
101 Reykjavık, Iceland

[2] *Structural & Computational Biology Programme*
EMBL, Meyerhofstrasse 1
69117 Heidelberg, Germany

Bioinformatics for Geneticists. Edited by M.R. Barnes and I.C. Gray
 ISBNs: 0 470 84393 4; 0 470 84394 2 (PB)

14.1 INTRODUCTION

Since the earliest protein sequences and structures were determined, it has been clear that the positioning and properties of amino acids are key to understanding many biological processes. For example, the first protein structure, haemoglobin provided a molecular explanation for the genetic disease sickle cell anaemia. A single nucleotide mutation leads to a substitution of glutamate in normal individuals with valine in those who suffer the disease. The substitution leads to a lower solubility of the deoxygenated form of haemoglobin and it is thought that this causes the molecules to form long fibres within blood cells which leads to the unusual sickle-shaped cells that give the disease its name.

Haemoglobin is just one of many examples now known where single mutations can have drastic consequences for protein structure, function and associated phenotype. The current availability of thousands or even millions of DNA and protein sequences means that we now have knowledge of many mutations, either naturally occurring or synthetic. Mutations can occur within one species, or between species at a wide variety of evolutionary distances. Whether mutations cause diseases or have subtle or drastic effects on protein function is often unknown.

The aim of this chapter is to give some guidance as to how to interpret mutations that occur within genes that encode for proteins. Both authors of this chapter have been approached previously by geneticists who want help interpreting mutations through the use of protein sequence and structure information. This chapter is an attempt to summarize our thought processes when giving such help. Specifically, we discuss the nature of mutations and the properties of amino acids in a variety of different protein contexts. The hope is that this discussion will help in anticipating or interpreting the effect that a particular amino acid change will have on protein structure and function. We will first highlight features of proteins that are relevant to considering mutations: cellular environments, three-dimensional structure and evolution. Then we will discuss classifications of the amino acids based on evolutionary, chemical or structural principles, and the role for amino acids of different classes in protein structure and function in different contexts. Last, we will review several studies of mutations, including naturally-occurring variations, SNPs, site-directed mutations, mutations that allow adaptive evolution and post-translational modification.

14.2 PROTEIN FEATURES RELEVANT TO AMINO ACID BEHAVIOUR

It is beyond the scope of this chapter to discuss the basic principles of proteins, since this can be gleaned from any introductory biochemistry text-book. However, a number of general principles of proteins are important to place any mutation in the correct context.

14.2.1 Protein Environments

A feature of key importance is cellular location. Different parts of cells can have very different chemical environments with the consequence that many amino acids behave differently. The biggest difference is between *soluble* proteins and *membrane* proteins. Whereas soluble proteins tend to be surrounded by water molecules, membrane proteins are surrounded by lipids. Roughly speaking this means that these two classes behave in an 'inside-out' fashion relative to each other. Soluble proteins tend to have polar or hydrophilic residues on their surfaces, whereas membrane proteins tend to have hydrophobic residues on the surface that interact with the membrane.

Soluble proteins also come in several flavours. The biggest difference is between those that are *extracellular* and those that are *cytosolic* (or *intracellular*). The cytosol is quite different from the more aqueous environment outside the cell; the density of proteins and other molecules effects the behaviour of some amino acids quite drastically, the foremost among these being cysteine. Outside the cell, cysteines in proximity to one another can be *oxidized* to form disulphide bonds, sulphur–sulphur covalent linkages that are important for protein folding and stability. However, the reducing environment inside the cell makes the formation of these bonds very difficult; in fact they are so rare as to warrant special attention.

Cells also contain numerous compartments, the organelles, which can also have slightly different environments from each other. Proteins in the nucleus often interact with DNA, meaning they contain different preferences for amino acids on their surfaces (e.g. positive amino acids or those containing amides most suitable for interacting with the negatively charged phosphate backbone). Some organelles such as mitochondria or chloroplasts are quite similar to the cytosol, while others, such as lysosomes or Golgi apparati are more akin to the extracellular environment. It is important to consider the likely cellular location of any protein before considering the consequences of amino acid substitutions.

A detailed hierarchical description of cellular location is one of the three main branches of the classification provided by the Gene Ontology Consortium (Ashburner *et al.*, 2000), the others being 'molecular function' and 'biological process'. The widespread adoption of this vocabulary by sequence databases and others should enable more sophisticated investigation of the factors governing the various roles of proteins.

14.2.2 Protein Structure

Proteins themselves also contain different microenvironments. For soluble proteins, the surface lies at the interface with water and thus tends to contain more polar or charged amino acids than one finds in the core of the protein, which is more likely to comprise hydrophobic amino acids. Proteins also contain regions that are directly involved in protein function, such as active sites or binding sites, in addition to regions that are less critical to the protein function and where mutations are likely to have fewer consequences. We will discuss many specific roles for particular amino acids in protein structures in the sections below, but it is important to remember that the context of any amino acid can vary greatly depending on its location in the protein structure.

14.2.3 Protein Evolution

Proteins are nearly always members of homologous families. Knowledge about the family a protein belongs in will generally give insights into the possible function, but several things should be considered. Two processes can give rise to homologous protein families: *speciation* or *duplication*. Proteins related by speciation only are referred to as *orthologues*, and as the name suggests, these proteins have the same function in different species. Proteins related by duplications are referred to as *paralogues*. Successive rounds of speciation and intra-genomic duplication can lead to confusing situations where it becomes difficult to say whether paralogy or orthology applies.

To be maintained in a genome over time, paralogous proteins are likely to evolve different functions (or have a dominant negative phenotype and so resist decay by point mutation (Gibson and Spring, 1998)). Differences in function can range from subtle differences in substrate (e.g. malate versus lactate dehydrogenases), to only weak similarities in molecular function (e.g. hydrolases) to complete differences in cellular location and function (e.g. an intracellular signalling domain homologous to a secreted growth factor (Schoorlemmer and Goldfarb, 2001)). At the other extreme, the molecular function may be identical, but the cellular function may be altered, as in the case of enzymes with differing tissue specificities.

Similarity in molecular function generally correlates with sequence identity. Mouse and human proteins with sequence identities in excess of 85% are likely to be orthologues, provided there are no other proteins with higher sequence identity in either organism. Orthology between more distantly related species (e.g. human and yeast) is harder to assess, since the evolutionary distance between organisms can make it virtually impossible to distinguish orthologues from paralogues using simple measures of sequence similarity. An operational definition of orthology can sometimes be used, for example if the two proteins are each other's best match in their respective genomes. However there is no substitute for constructing a phylogenetic tree of the protein family, to identify which sequences are related by speciation events. Assignment of orthology and paralogy is perhaps the best way of determining likely equivalences of function. Unfortunately, complete genomes are unavailable for most organisms. Some rough rules of thumb can be used: function is often conserved down to 40% protein sequence identity, with the broad functional class being conserved to 25% identity (Wilson *et al.*, 2000).

When considering a mutation, it is important to consider how conserved the position is within other homologous proteins. Conservation across all homologues (paralogues and orthologues) should be considered carefully. These amino acids are likely to play key structural roles or a role in a common functional theme (i.e. catalytic mechanism). Other amino acids may play key roles only in the particular orthologous group (i.e. they may confer specificity to a substrate), thus meaning they vary when considering all homologues.

14.2.4 Protein Function

Protein function is key to any understanding of the consequences of amino acid substitution. Enzymes, such as trypsin (Figure 14.1), tend to have highly conserved active sites involving a handful of polar residues. In contrast, proteins that function primarily only to interact with other proteins, such as fibroblast growth factors (Figure 14.2), interact over a large surface, with virtually any amino acid being important in mediating the interaction (Plotnikov *et al.*, 1999). In other cases, multiple functions make the situation even more confusing, for example a protein kinase (Hanks *et al.*, 1988) can both catalyse a phosphorylation event and bind specifically to another protein, such as cyclin (Jeffrey *et al.*, 1995).

It is not possible to discuss all of the possible functional themes here, but we emphasize that functional information, if known, should be considered whenever studying the effects of substitution.

14.2.5 Post-translational Modification

Although there are only 20 possible types of amino acid that can be incorporated into a protein sequence upon translation of DNA, there are many more variations that can occur

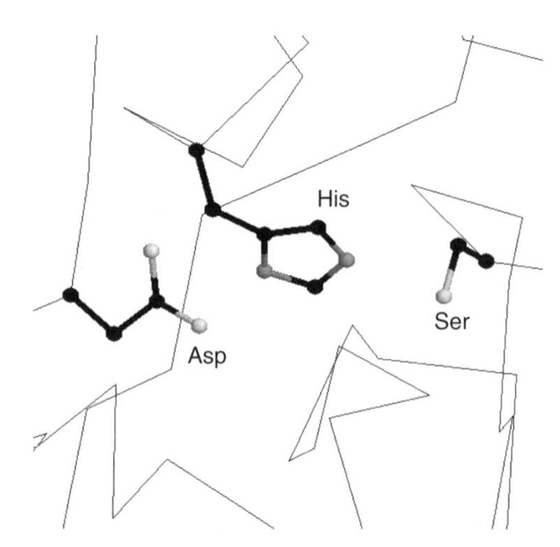

Figure 14.1 RasMol (Sayle and Milner-White, 1995) figure showing the catalytic Asp-His-Ser triad in trypsin (PDB code 1mct; Berman *et al.*, 2000).

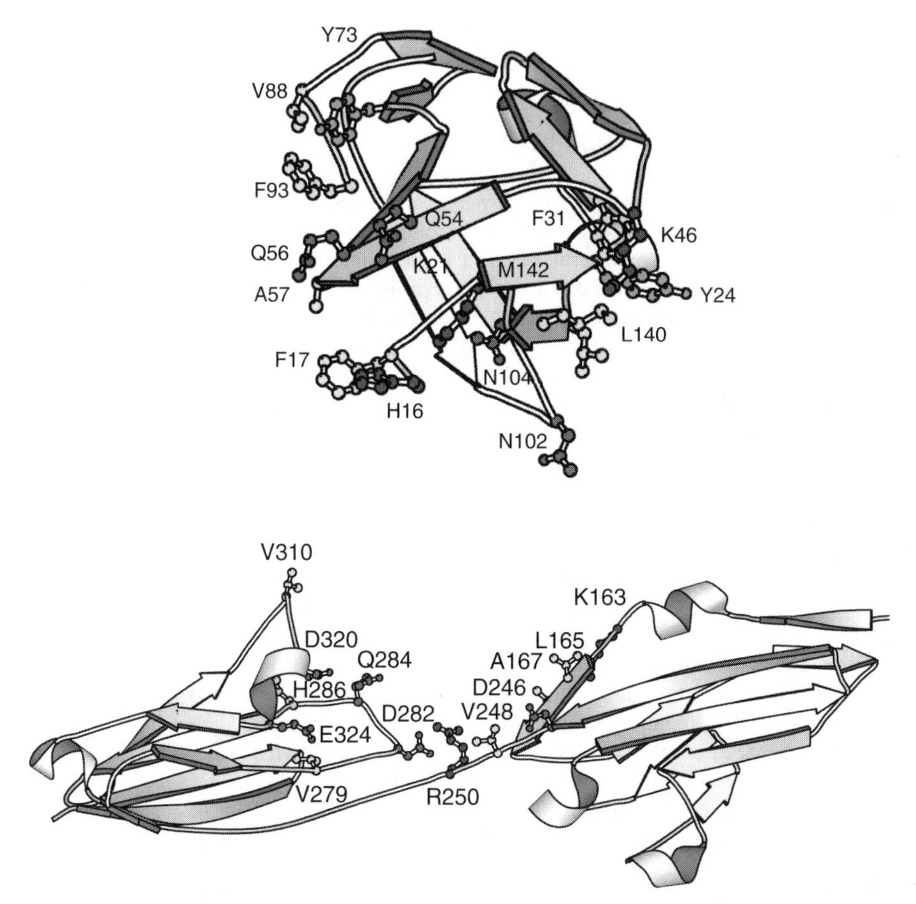

Figure 14.2 Molscript (Kraulis, 1991) figure showing fibroblast growth factor interaction with its receptor (code 1cvs; Plotnikov *et al.*, 1999). Residues at the interface are labelled. The two molecules have been pulled apart for clarity.

through subsequent modification. In addition, the gene-specified protein sequence can be shortened by proteolysis, or lengthened by addition of amino acids at either terminus.

Two common modifications, phosphorylation and glycosylation, are discussed in the context of the amino acids where they most often occur (tyrosine, serine, threonine and asparagine; see below). We direct the reader to the review by Krishna for more information on many other known types and specific examples (Krishna and Wold, 1993). The main conclusion is that modifications are highly specific, with specificity provided by primary, secondary and tertiary protein structure, although with detailed mechanisms being obscure. The biological function of the modified proteins is also summarized, from the reversible phosphorylation of serine, threonine and tyrosine residues that occurs in signalling through to the formation of disulphide bridges and other cross-links that stabilize tertiary structure, and on to the covalent attachment of lipids that allows anchorage to cell membranes. More detail on biological effects is given by Parekh and Rohlff (1997), especially where it concerns possible therapeutic applications. Many diseases arise by abnormalities in post-translational modification, and these are not necessarily apparent from genetic information alone.

14.3 AMINO ACID CLASSIFICATIONS

Humans have a natural tendency to classify, as it makes the world around us easier to understand. As amino acids often share common properties, several classifications have been proposed. This is useful, but a little bit dangerous if over-interpreted. Always remember that, for the reasons discussed above, it is very difficult to put all amino acids of the same type into an invariant group. A substitution in one context can be disastrous in another. For example, a cysteine involved in a disulphide bond would not be expected to be mutatable to any other amino acid (i.e. it is in a group on its own), one involved in binding to zinc could likely be substituted by histidine (group of two) and one buried in an intracellular protein core could probably mutate to any other hydrophobic amino acid (a group of 10 or more). We will discuss other examples below.

14.3.1 Mutation Matrices

One means of classifiying amino acids is a mutation matrix (or substitution or exchange matrix). This is a set of numbers that describe the propensities of exchanging one amino acid for another (for a comprehensive review and explanation see Durbin *et al.*, 1998). These are derived from large sets of aligned sequences by counting the number of times that a particular substitution occurs and comparing this to what would be expected by chance. High values indicate that a substitution is seen often in nature and so is favourable, and vice versa. The values in the matrix are usually calculated using some model of evolutionary time, to account for the fact that different pairs of sequences are at different evolutionary distances. Probably the best known matrices are the Point Accepted Mutation (PAM) matrices of Dayhoff *et al.* (Dayhoff *et al.*, 1978) and BLOSUM matrices (Henikoff and Henikoff, 1992).

Mutation matrices are very useful as rough guides for how good or bad a particular change will be. Another useful feature is that they can be calculated for different datasets to account for some of the protein features that effect amino acid properties, such as cellular locations (Jones *et al.*, 1994) or different evolutionary distances (e.g. orthologues or paralogues; Henikoff and Henikoff, 1992). Several mutation matrices are reproduced in Appendix II.

14.3.2 Classification by Physical, Chemical and Structural Properties

Although mutation matrices are very useful for protein sequence alignments, especially in the absence of known three-dimensional structures, they do not precisely describe the likelihood and effects of particular substitutions at particular sites in the sequence. Position-specific substitution matrices can be generated for the family of interest, such as the profile-HMM models generated by HMMER (Eddy, 1998) and provided by Pfam (Bateman *et al.*, 2000), and those generated by PSI-BLAST (Altschul *et al.*, 1997). However, these are automatic methods suited to database searching and identification of new members of a family, and as such do not really give any qualitative information about the chemistry involved at particular sites.

Taylor presented a classification that explains mutation data through correlation with the physical, chemical and structural properties of amino acids (Taylor, 1986). The major factor is the size of the side chain, closely followed by its hydrophobicity. Effects of differerent amino acids on protein structure can account for mutation data when these physico-chemical properties do not. For example, hydrophobicity and size differ widely between glycine, proline, aspartic acid and glutamic acid. However, they are still closely related in mutation matrices because they prefer sharply turning regions on the surface of the protein; the phi and psi bonds of glycine are unconstrained by any side chain, proline forces a sharp turn because its side chain is bonded to the backbone nitrogen as well as to carbon, and aspartate and glutamate prefer to expose their charged side chains to solvent.

The Taylor classification is normally displayed as a Venn diagram (Figure 14.3). The amino acids were positioned on this by multidimensional scaling of Dayhoff's mutation matrix, and then grouped by common physico-chemical properties. Size is subcategorized into small and tiny (with large included by implication). Affinity for water is described by several sets: polar and hydrophobic, which overlap, and charged, which is divided into positive and negative. Sets of aromatic and aliphatic amino acids are also marked. These properties were enough to distinguish between most amino acids. However, properties such as hydrogen-bonding ability and the previously mentioned propensity for sharply turning regions are not described well. Although these factors are less important on average, and would confuse the effects of more important properties if included on the diagram, the dangers of relying on simple classifications are apparent. This can be overcome somewhat by listing all amino acids which belong to each subset (defined as an intersection or union of the sets) in the diagram, for example 'small and non-polar', and including extra subsets to describe important additional properties. These subsets can be used to give qualitative descriptions of each position in a multiple alignment, by associating the positions with the smallest subset that includes all the amino acids found at that position.

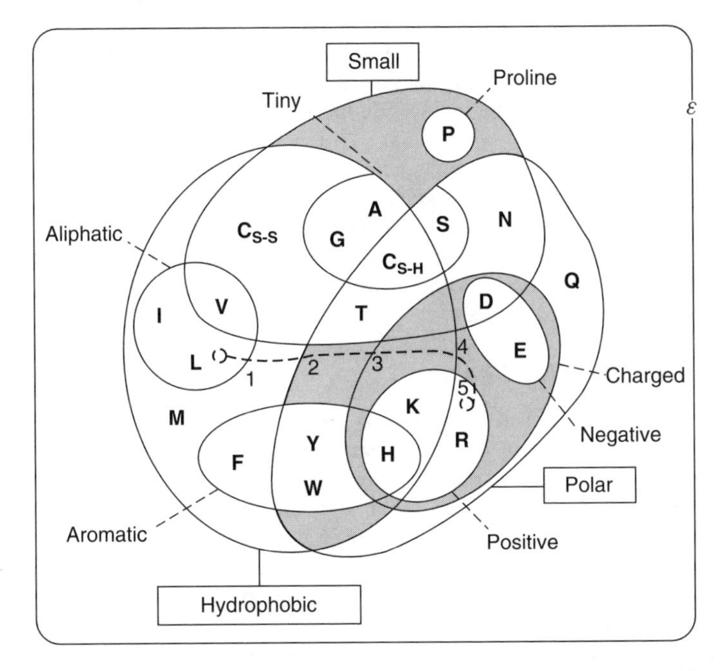

Figure 14.3 Venn diagram illustrating the properties of amino acids.

This may suggest alternative amino acids that could be engineered into the protein at each position.

14.4 PROPERTIES OF THE AMINO ACIDS

The sections that follow will first consider several major properties that are often used to group amino acids together. Note that amino acids can be in more than one group, and that sometimes properties as different as 'hydrophobic' and 'hydrophilic' can be applied to the same amino acids.

14.4.1 Hydrophobic Amino Acids

Probably the most common broad division of amino acids is into those that prefer to be in an aqueous environment (hydrophilic) and those that do not (hydrophobic). The latter can be divided according to whether they have *aliphatic* or *aromatic* side chains.

14.4.1.1 Aliphatic Side Chains

Strictly speaking aliphatic means that the side chain contains only hydrogen and carbon atoms. By this strict definition, the amino acids with aliphatic side chains are *alanine, isoleucine, leucine, proline* and *valine*. Alanine's side chain, being very short, means that it is not particularly hydrophobic and proline has an unusual geometry that gives it special roles in proteins as we shall discuss below. Although it also contains a sulphur atom, it is often convenient to consider *methionine* in the same category as isoleucine, leucine and valine. The unifying theme is that they contain largely non-reactive and flexible side chains that are ideally suited for packing in the protein interior.

Aliphatic side chains are very non-reactive, and are thus rarely involved directly in protein function, although they can play a role in substrate recognition. In particular, hydrophobic amino acids can be involved in binding/recognition of hydrophobic ligands such as lipids.

Several other amino acids also contain aliphatic regions. For example, arginine, lysine, glutamate and glutamine are *amphipathic*, meaning that they contain hydrophobic and polar areas. All contain two or more aliphatic carbons that connect the protein backbone to the non-aliphatic portion of the side chain. In some instances it is possible for such amino acids to play a dual role, with part of the side chain being buried in the protein and another being exposed to water.

14.4.1.2 Aromatic Side Chains

A side chain is aromatic when it contains an aromatic ring system. The strict definition has to do with the number of electrons contained within the ring. Generally, aromatic ring systems are planar and electrons are shared over the whole ring structure. *Phenylalanine and tryptophan* have very hydrophobic aromatic side chains, whereas *tyrosine* and *histidine* are less so. The latter two can often be found in positions that are somewhere between buried and exposed. The hydrophobic aromatic amino acids can sometimes substitute for aliphatic residues of a similar size, for example phenylalanine to leucine, but not tryptophan to valine.

Aromatic residues have also been proposed to participate in 'stacking' interactions (Hunter *et al.*, 1991) (Figure 14.4). Here, numerous aromatic rings are thought to stack

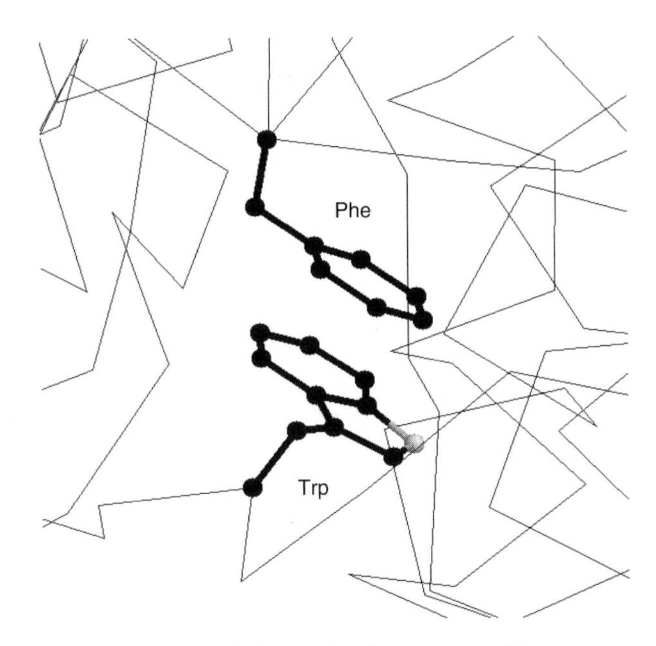

Figure 14.4 Example of aromatic stacking.

on top of each other such that their PI electron clouds are aligned. They can also play a role in binding to specific amino acids, such as proline. SH3 and WW domains, for example, use these residues to bind to their polyproline-containing interaction partners (Macias *et al.*, 2002). Owing to its unique chemical nature, histidine is frequently found in protein active sites as we shall see below.

14.4.2 Polar Amino Acids

Polar amino acids prefer to be surrounded by water. Those that are buried within the protein usually participate in hydrogen bonds with other side chains or the protein main-chain that essentially replace the water. Some of these carry a charge at typical biological pHs: *aspartate* and *glutamate* are negatively charged; *lysine* and *arginine* are positively charged. Other polar amino acids, *histidine, asparagine, glutamine, serine, threonine* and *tyrosine*, are neutral.

14.4.3 Small Amino Acids

The amino acids *alanine*, *cysteine*, *glycine*, *proline*, *serine* and *threonine* are often grouped together for the simple reason that they are all small in size. In some protein structural contexts, substitution of a small side chain for a large one can be disastrous.

14.5 AMINO ACID QUICK REFERENCE

In the sections that follow we discuss each amino acid in turn. For each we will briefly discuss general preferences for substitutions and important specific details regarding their possible structure and functional roles. More information is found on the WWW site that accompanies this chapter (www.russell.embl-heidelberg.de/aas). This website also features amino acid substitution matrices for transmembrane, extracellular and intracellular proteins. These can be used to numerically score an amino acid substitution, where unpreferred mutations are given negative scores, preferred substitutions are given positive scores and neutral substitutions are given zero scores.

14.5.1 Alanine (Ala, A)

14.5.1.1 Substitutions

Alanine can be substituted by other small amino acids.

14.5.1.2 Structure

Alanine is probably the dullest amino acid. It is not particularly hydrophobic and is non-polar. However, it contains a normal Cβ carbon, meaning that it is generally as hindered as other amino acids with respect to the conformations that the backbone can adopt. For this reason, it is not surprising to see alanine present in just about all non-critical protein contexts.

14.5.1.3 Function

The alanine side chain is very non-reactive, and is thus rarely directly involved in protein function, but it can play a role in substrate recognition or specificity, particularly in interactions with other non-reactive atoms such as carbon.

14.5.2 Isoleucine (Ile, I)

14.5.2.1 Substitutions

Isoleucine can be substituted by other hydrophobic, particularly aliphatic, amino acids.

14.5.2.2 Structure

Being hydrophobic, isoleucine prefers to be buried in protein hydrophobic cores. However, isoleucine has an additional property that is frequently overlooked. Like valine and threonine it is Cβ branched. Whereas most amino acids contain only one non-hydrogen substituent attached to their Cβ carbon, these three amino acids contain two. This means that there is a lot more bulkiness near to the protein backbone and this means that these amino acids are more restricted in the conformations the main chain can adopt. Perhaps the most pronounced effect of this is that it is more difficult for these amino acids to adopt an α-helical conformation, although it is easy and even preferred for them to lie within β-sheets.

14.5.2.3 Function

The isoleucine side chain is very non-reactive and is thus rarely directly involved in protein functions like catalysis, although it can play a role in substrate recognition. In particular, hydrophobic amino acids can be involved in binding/recognition of hydrophobic ligands such as lipids.

14.5.3 Leucine (Leu, L)

14.5.3.1 Substitutions

See Isoleucine.

14.5.3.2 Structure

Being hydrophobic, leucine prefers to be buried in protein hydrophobic cores. It also shows a preference for being within alpha helices more so than in beta strands.

14.5.3.3 Function

See Isoleucine.

14.5.4 Valine (Val, V)

14.5.4.1 Substitutions

See Isoleucine.

14.5.4.2 Structure

Being hydrophobic, valine prefers to be buried in protein hydrophobic cores. However, valine is also Cβ branched (see Isoleucine).

14.5.4.3 Function

See Isoleucine.

14.5.5 Methionine (Met, M)

14.5.5.1 Substitutions

See Isoleucine.

14.5.5.2 Structure

See Isoleucine.

14.5.5.3 Function

The methionine side chain is fairly non-reactive, and is thus rarely directly involved in protein function. Like other hydrophobic amino acids, it can play a role in binding/recognition of hydrophobic ligands such as lipids. However, unlike the proper aliphatic amino acids, methionine contains a sulphur atom, that can be involved in binding to atoms such as metals. However, whereas the sulphur atom in cysteine is connected to a hydrogen atom making it quite reactive, methionine's sulphur is connected to a methyl group. This means that the roles that methionine can play in protein function are much more limited.

14.5.6 Phenylalanine (Phe, F)

14.5.6.1 Substitutions

Phenylalanine can be substituted with other aromatic or hydrophobic amino acids. It particularly prefers to exchange with tyrosine, which differs only in that it contains an hydroxyl group in place of the ortho hydrogen on the benzene ring.

14.5.6.2 Structure

Phenylalanine prefers to be buried in protein hydrophobic cores. The aromatic side chain can also mean that phenylalanine is involved in stacking (Figure 14.4) interactions with other aromatic side chains.

14.5.6.3 Function

The phenylalanine side chain is fairly non-reactive, and is thus rarely directly involved in protein function, although it can play a role in substrate recognition (see Isoleucine). Aromatic residues can also be involved in interactions with non-protein ligands that themselves contain aromatic groups via stacking interactions (see above). They are also common in polyproline binding sites, for example in SH3 and WW domains (Macias *et al.*, 2002).

14.5.7. Tryptophan (Trp, W)

14.5.7.1 Substitutions

Trytophan can be replaced by other aromatic residues, but it is unique in terms of chemistry and size, meaning that often replacement by anything could be disastrous.

14.5.7.2 Structure

See Phenylalanine.

14.5.7.3 Function

As it contains a non-carbon atom (nitrogen) in the aromatic ring system, tryptophan is more reactive than phenylalanine although it is less reactive than tyrosine. Tryptophan can play a role in binding to non-protein atoms, but such instances are rare. See also Phenylalanine.

14.5.8 Tyrosine (Tyr, Y)

14.5.8.1 Substitutions

Tyrosine can be substituted by other aromatic amino acids. See Phenylalanine.

14.5.8.2 Structure

Being partially hydrophobic, tyrosine prefers to be buried in protein hydrophobic cores. The aromatic side chain can also mean that tyrosine is involved in stacking interactions with other aromatic side chains.

14.5.8.3 Function

Unlike the very similar phenylalanine, tyrosine contains a reactive hydroxyl group, thus making it much more likely to be involved in interactions with non-carbon atoms. See also Phenylalanine.

A common role for tyrosines (and serines and threonines) within intracellular proteins is phosphorylation. Protein kinases frequently attach phosphates to these three residues as part of a signal transduction process. Note that in this context, tyrosine will rarely substitute for serine or threonine since the enzymes that catalyse the reactions (i.e. the protein kinases) are highly specific (i.e. tyrosine kinases generally do not work on serines/threonines and vice versa (Hanks *et al.*, 1988)).

14.5.9 Histidine (His, H)

14.5.9.1 Substitutions

Histidine is generally considered to be a polar amino acid, however it is unique with regard to its chemical properties, which means that it does not substitute particularly well with any other amino acid.

14.5.9.2 Structure

Histidine has a pK_a near to that of physiological pH, meaning that it is relatively easy to move protons on and off of the side chain (i.e. changing the side chain from neutral to positive charge). This flexibility has two effects. The first is ambiguity about whether it prefers to be buried in the protein core or exposed to solvent. The second is that it is an ideal residue for protein functional centres (discussed below). It is false to presume that histidine is always protonated at typical pHs. The side chain has a pK_a of approximately 6.5, which means that only about 10% of molecules will be protonated. The precise pK_a depends on local environment.

14.5.9.3 Function

Histidines are the most common amino acids in protein active or binding sites. They are very common in metal binding sites (e.g. zinc), often acting together with cysteines or other amino acids (Figure 14.5; Wolfe *et al.*, 2001). In this context, it is common to see histidine replaced by cysteine.

The ease with which protons can be transferred on and off of histidines makes them ideal for charge relay systems such as those found within catalytic triads and in many cysteine and serine proteases (Figure 14.1). In this context, it is rare to see histidine exchanged for any amino acid at all.

14.5.10 Arginine (Arg, R)

14.5.10.1 Substitutions

Arginine is a positively-charged, polar amino acid. It thus most prefers to substitute for the other positively-charged amino acid, lysine, although in some circumstances it will also tolerate a change to other polar amino acids. Note that a change from arginine to lysine is not always neutral. In certain structural or functional contexts, such a mutation can be devastating to function (see below).

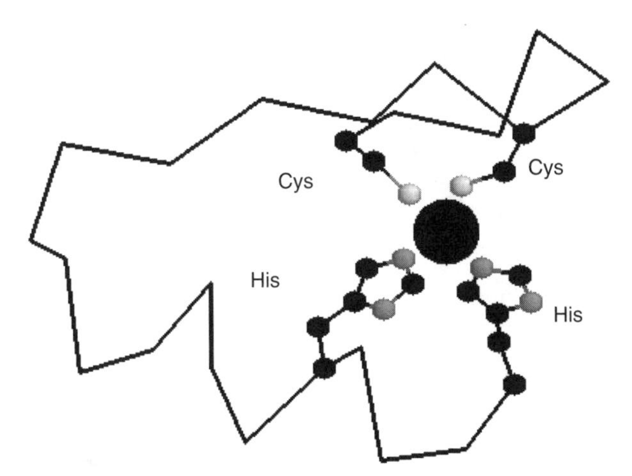

Figure 14.5 Example of a metal binding site coordinated by cysteine and histidine residues (code 1g2f; Wolfe *et al.*, 2001).

14.5.10.2 Structure

Arginine generally prefers to be on the surface of the protein, but its amphipathic nature can mean that part of the side chain is buried. Arginines are also frequently involved in salt-bridges where they pair with a negatively charged aspartate or glutamate to create stabilizing hydrogen bonds that can be important for protein stability (Figure 14.6).

14.5.10.3 Function

Arginines are quite frequent in protein active or binding sites. The positive charge means that they can interact with negatively-charged non-protein atoms (e.g. anions or carboxylate groups). Arginine contains a complex guanidinium group on its side chain that has a geometry and charge distribution that is ideal for binding negatively-charged groups on phosphates (it is able to form multiple hydrogen bonds). A good example can be found in the src homology 2 (SH2) domains (Figure 14.7; Waksman *et al.*, 1992). The two arginines shown in the figure make multiple hydrogen bonds with the phosphate. In this context arginine is not easily replaced by lysine. Although lysine can interact with phosphates, it contains only a single amino group, meaning it is more limited in the number of hydrogen bonds it can form. A change from arginine to lysine in some contexts can thus be disastrous (Copley and Barton, 1994).

14.5.11 Lysine (Lys, K)

14.5.11.1 Substitutions

Lysine can be substituted by arginine or other polar amino acids.

14.5.11.2 Structure

Lysine frequently plays an important role in structure. First, it can be considered to be somewhat amphipathic as the part of the side chain nearest to the backbone is long, carbon-containing and hydrophobic, whereas the end of the side chain is positively charged. For

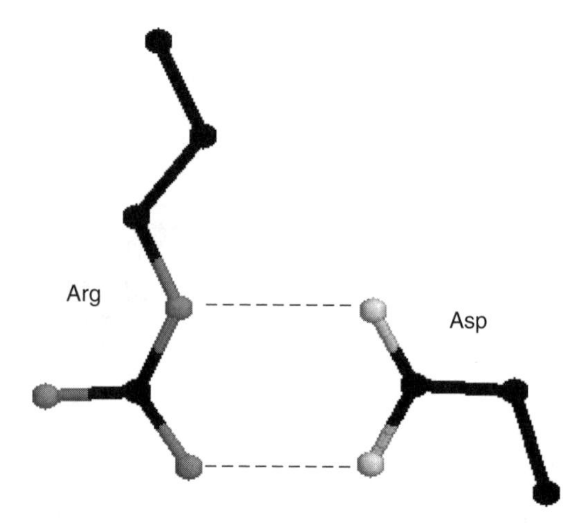

Figure 14.6 Example of a salt-bridge (code 1xel).

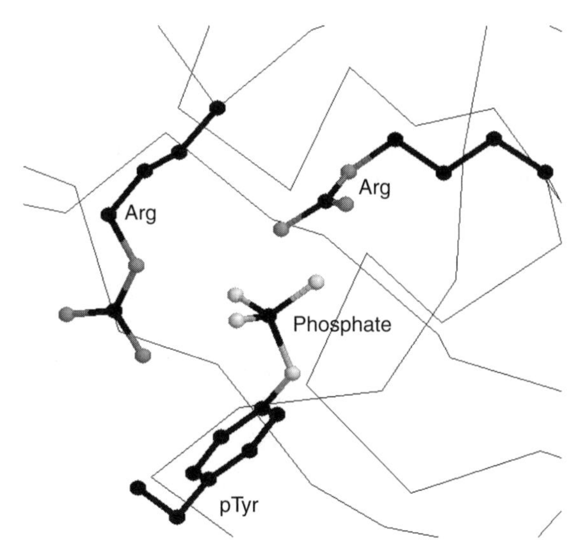

Figure 14.7 Interaction of arginine residues with phosphotyrosine in an SH2 domain (code 1sha; Waksman *et al.*, 1992).

this reason, one can find lysines where part of the side chain is buried and only the charged portion is on the outside of the protein. However, this is by no means always the case and generally lysines prefer to be on the outside of proteins. Lysines are also frequently involved in salt-bridges (see Arginine).

14.5.11.3 Function

Lysines are quite frequent in protein active or binding sites. Lysine contains a positively-charged amino group on its side chain that is sometimes involved in forming hydrogen bonds with negatively-charged non-protein atoms (e.g. anions or carboxylate groups).

14.5.12 Aspartate (Asp, D)

14.5.12.1 Substitutions

Aspartate can be substituted by glutamate or other polar amino acids, particularly asparagine, which differs only in that it contains an amino group in place of one of the oxygens found in aspartate (and thus also lacks a negative charge).

14.5.12.2 Structure

Being charged and polar, aspartates generally prefer to be on the surface of proteins, exposed to an aqueous environment. Aspartates (and glutamates) are frequently involved in salt-bridges (see Arginine).

14.5.12.3 Function

Aspartates are quite frequently involved in protein active or binding sites. The negative charge means that they can interact with positively-charged non-protein atoms, such as cations like zinc. Aspartate has a shorter side chain than the very similar glutamate meaning that is slightly more rigid within protein structures. This gives it a slightly stronger preference to be involved in protein active sites. Probably the most famous example of aspartate being involved in an active site is found within serine proteases such as trypsin, where it functions in the classical Asp-His-Ser catalytic triad (Figure 14.1). In this context, it is quite rare to see aspartate exchange for glutamate, although it is possible for glutamate to play a similar role.

14.5.13 Glutamate (Glu, E)

14.5.13.1 Substitutions

Substitution can be by aspartate or other polar amino acids, in particular glutamine, which is to glutamate what asparagine is to aspartate (see above).

14.5.13.2 Structure

See Aspartate.

14.5.13.3 Function

Glutamate, like aspartate, is quite frequently involved in protein active or binding sites. In certain cases, they can also perform a similar role to aspartate in the catalytic site of proteins such as proteases or lipases.

14.5.14 Asparagine (Asn, N)

14.5.14.1 Substitutions

Asparagine can be substituted by other polar amino acids, especially aspartate (see above).

14.5.14.2 Structure

Being polar asparagine prefers generally to be on the surface of proteins, exposed to an aqueous environment.

14.5.14.3 Function

Asparagines are quite frequently involved in protein active or binding sites. The polar side chain is good for interactions with other polar or charged atoms. Asparagine can play a similar role to aspartate in some proteins. Probably the best example is found in certain cysteine proteases, where it forms part of the Asn-His-Cys catalytic triad. In this context, it is quite rare to see asparagine exchange for glutamine.

Asparagine, when occurring in a particular motif (Asn-X-Ser/Thr) can be *N*-glycosylated (Gavel and von Heijne, 1990). Thus in this context it is impossible to substitute it with any amino acid at all.

14.5.15 Glutamine (Gln, Q)

14.5.15.1 Substitutions

Glutamine can be substituted by other polar amino acids, especially glutamate (see above).

14.5.15.2 Structure

See Asparagine.

14.5.15.3 Function

Glutamines are quite frequently involved in protein active or binding sites. The polar side chain is good for interactions with other polar or charged atoms.

14.5.16 Serine (Ser, S)

14.5.16.1 Substitutions

Serine can be substituted by other polar or small amino acids in particular threonine which differs only in that it has a methyl group in place of a hydrogen group found in serine.

14.5.16.2 Structure

Being a fairly indifferent amino acid, serine can reside both within the interior of a protein, or on the protein surface. Its small size means that it is relatively common within tight turns on the protein surface, where it is possible for the serine side chain hydroxyl oxygen to form a hydrogen bond with the protein backbone, effectively mimicking proline.

14.5.16.3 Function

Serines are quite common in protein functional centres. The hydroxyl group is fairly reactive, being able to form hydrogen bonds with a variety of polar substrates.

Perhaps the best known role for serine in protein active sites is exemplified by the classical Asp-His-Ser catalytic triad found in many hydrolases (e.g. proteases and lipases; Figure 14.1). Here, a serine, aided by a histidine and an aspartate act as a nucleophile to hydrolyse (effectively cut) other molecules. This three-dimensional 'motif' is found in many non-homologous (i.e. unrelated) proteins and is a classic example of molecular convergent evolution (Russell, 1998). In this context, it is rare for serine to exchange with threonine, but in some cases, the reactive serine can be replaced by cysteine, which can fulfil a similar role.

Intracellular serines can also be phosphorylated (see Tyrosine). Extracellular serines can also be *O*-glycosylated where a carbohydrate is attached to the side chain hydroxyl group (Gupta *et al.*, 1999).

14.5.17 Threonine (Thr, T)

14.5.17.1 Substitutions

Threonine can be substituted with other polar amino acids, particularly serine (see above).

14.5.17.2 Structure

Being a fairly indifferent amino acid, threonine can reside both within the interior of a protein or on the protein surface. Threonine is also $C\beta$ branched (see Isoleucine).

14.5.17.3 Function

Threonines are quite common in protein functional centres. The hydroxyl group is fairly reactive, being able to form hydrogen bonds with a variety of polar substrates. Intracellular threonines can also be phosphorylated (see Tyrosine) and in the extracellular environment they can be *O*-glycosylated (see Serine).

14.5.18 Cysteine (Cys, C)

14.5.18.1 Substitutions

In the case of cysteine there is no general preference for substitution with any other amino acid, although it can tolerate substitutions with other small amino acids. Cysteine has a role that is very dependent on cellular location, making substitution matrices dangerous to interpret (e.g. Barnes and Russell, 1999).

14.5.18.2 Structure

The role of cysteines in structure is very dependent on the cellular location of the protein in which they are contained. Within extracellular proteins, cysteines are frequently involved in disulphide bonds, where pairs of cysteines are oxidized to form a covalent bond. These bonds serve mostly to stabilize the protein structure and the structure of many extracellular proteins is almost entirely determined by the topology of multiple disulphide bonds (e.g. Figure 14.8).

The reducing environment inside cells makes the formation of disulphide bonds very unlikely. Indeed, instances of disulphide bonds in the intracellular environment are so rare that they almost always attract special attention. Disulphide bonds are also rare within the membrane, although membrane proteins may contain disulphide bonds within extracellular domains. Disulphide bonds are such that cysteines must be paired. If one half of a disulphide bond pair is lost, then the protein may not fold properly.

Figure 14.8 Example of a small, disulphide-rich protein (code 1tfx).

In the intracellular environment cysteines can still play a key structural role. Their sulphydryl side chain is excellent for binding to metals, such as zinc, meaning that cysteines (and other amino acids such as histidines) are very common in metal binding motifs such as zinc fingers (Figure 14.5). Outside of this context within the intracellular environment and when it is not involved in molecular function, cysteine is a neutral, small amino acid and prefers to substitute with other amino acids of the same type.

14.5.18.3 Function

Cysteines are also very common in protein active and binding sites. Binding to metals (see above) can also be important in enzymatic functions (e.g. metal proteases). Cysteine can also function as a nucleophile (i.e. the reactive centre of an enzyme). Probably the best known example of this occurs within the cysteine proteases, such as caspases or papains, where cysteine is the key catalytic residue, being helped by a histidine and an asparagine.

14.5.19 Glycine (Gly, G)

14.5.19.1 Substitutions

Glycine can be substituted by other small amino acids, but be warned that even apparently neutral mutations (e.g. to alanine) can be forbidden in certain contexts (see below).

14.5.19.2 Structure

Glycine is unique as it contains a hydrogen as its side chain (rather than a carbon as is the case for all other amino acids). This means that there is much more conformational flexibility in glycine and as a result of this it can reside in parts of protein structures that are forbidden to all other amino acids (e.g. tight turns in structures).

14.5.19.3 Function

The uniqueness of glycine also means that it can play a distinct functional role, such as using its backbone (without a side chain) to bind to phosphates (Schulze-Gahmen *et al.*, 1996). This means that if one sees a conserved glycine changing to any other amino acid, the change could have a drastic impact on function.

A good example is found among the protein kinases. Figure 14.9 shows a region around the ATP binding site in a protein kinase; the ATP is shown to the right of the figure and part of the protein to the left. The glycines in this loop are part of the classic 'Gly-X-Gly-X-X-Gly' motif present in the kinases (Hanks *et al.*, 1988). These three glycines are almost never mutated to other residues; only glycines can function to bind to the phosphates of the ATP molecule using their main chains.

14.5.20 Proline (Pro, P)

14.5.20.1 Substitutions

Proline can sometimes substitute for other small amino acids, although its unique properties mean that it does not often substitute well.

14.5.20.2 Structure

Proline is unique in that it is the only amino acid where the side chain is connected to the protein backbone twice, forming a five-membered ring. Strictly speaking, this makes proline an imino acid (since in its isolated form, it contains an NH^{2+} rather than an NH^{3+} group, but this is mostly just pedantic detail). This difference is very important as

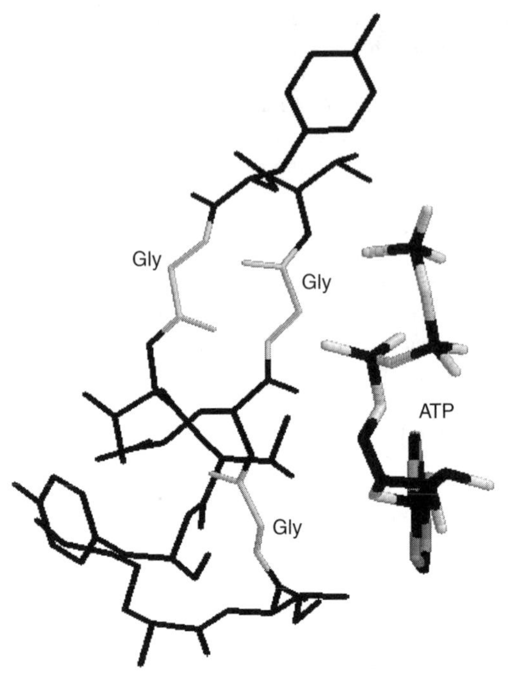

Figure 14.9 Glycine-rich phosphate binding loop in a protein kinase (code 1hck; Schulze-Gahmen *et al.*, 1996).

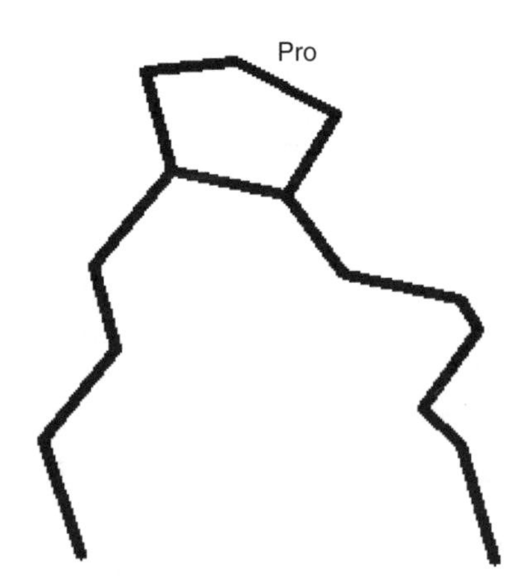

Figure 14.10 Example of proline in a tight protein turn (code 1ag6).

it means that proline is unable to occupy many of the main-chain conformations easily adopted by all other amino acids. In this sense, it can be considered to be an opposite of glycine, which can adopt many more main-chain conformations. For this reason proline is often found in very tight turns in protein structures (i.e. where the polypeptide chain must change direction; Figure 14.10). It can also function to introduce kinks into α-helices, since it is unable to adopt a normal helical conformation. Despite being aliphatic the preference for turn structure means that prolines are usually found on the protein surface.

14.5.20.3 Function

The proline side chain is very non-reactive. This, together with its difficulty in adopting many protein main-chain conformations means that it is very rarely involved in protein active or binding sites.

14.6 STUDIES OF HOW MUTATIONS AFFECT FUNCTION

Several studies have been carried out previously in an attempt to derive general principles about the relationship between mutations, structure, function and diseases. We review some of these below.

14.6.1 Single Nucleotide Polymorphisms (SNPs)

A SNP is a point mutation that is present at a measurable frequency in human populations. They can occur either in coding or non-coding DNA. Non-coding SNPs may have effects on important mechanisms such as transcription, translation and splicing. However, the effects of coding SNPs are easier to study and are potentially more damaging, and so they have received considerably more attention. They are also more relevant to this chapter. Coding SNPs can be divided into two main categories, synonymous (where there is no change in the amino acid coded for), and non-synonymous. Non-synonymous SNPs tend to occur at lower frequencies than synonymous SNPs. Minor allele frequencies also tend to be lower in non-synonymous SNPs. This is a strong indication that these replacement polymorphisms are deleterious (Cargill *et al.*, 1999).

To examine the phenotypic effects of coding SNPs, Sunyaev *et al.* (2000) studied the relationships between non-synonymous SNPs and protein structure and function. Three sets of SNP data were compared: disease causing susbtitutions, substitutions between orthologues and those represented by human alleles. Disease-causing mutations were more common in structurally and functionally important sites than were variations between orthologues, as might be expected. Allelic variations were also more common in these regions than were those between orthologues. Minor allele frequency and the level of occurrence in these regions were correlated, another indication of evolutionary selection of phenotype. The most damaging allelic variants affect protein stability, rather than binding, catalysis, allosteric response or post-translational modification (Sunyaev *et al.*, 2001). The expected increase in the number of known protein structures will allow other analyses and refinement of the details of the phenotypic effects of SNPs.

Wang and Moult (2001) developed a description of the possible effects of missense SNPs on protein structure and used it to compare disease-causing missense SNPs with a set from the general population. Five general classes of effect were considered: protein stability, ligand binding, catalysis, allosteric regulation and post-translational modification. The disease and population sets of SNPs contain those that can be mapped onto known protein structures, either directly or through homologues of known structure. Of the disease-causing SNPs, 90% were explained by the description, with the majority (83%) being attributed to effects on protein stability, as reported by Sunyaev *et al.* (2001). The 10% that are not explained by the description may cause disease by effects not easily identified by structure alone. Of the SNPs from the general population, 70% were predicted to have no effect. The remaining 30% may represent disease-causing SNPs previously unidentified as such, or molecular effects that have no significant phenotypic effect.

14.6.2 Site-directed Mutagenesis

Site-directed mutagenesis is a powerful tool for discovering the importance of an amino acid in the function of the protein. Gross changes in amino acid type can reveal sites that are important in maintaining the structure of the protein. Conversely, when investigating functionally interesting sites it is important to choose replacement residues that are unlikely to affect structure dramatically, for example by choosing ones of a similar size to the original. Peracchi (2001) reviews the use of site-directed mutagenesis to investigate mechanisms of enzyme catalysis, in particular those studies involving mutagenesis of general acids (proton donors), general bases (proton acceptors) and catalytic nucleophiles in active sites. These types of amino acid could be considered to be the most important to enzyme function as they directly participate in the formation or cleavage of covalent bonds. However, studies indicate that they are often important but not essential — rates are still higher than the uncatalysed reaction even when these residues are removed, because the protein is able to use an alternative mechanism of catalysis. Also, direct involvement in the formation and cleavage of bonds is only one of a combination of methods that an enzyme can use to catalyse a reaction. Transition states can be stabilized by complementary shape and electrostatics of the binding site of the enzyme and substrates can be precisely positioned, lowering the entropy of activation. These factors can also be studied by site-directed mutagenesis, with consideration of the physical and chemical properties of the amino acids again guiding the choice of replacements, along with knowledge of the structure of the protein.

14.6.3 Key Mutations in Evolution

Golding and Dean (1998) reviewed six studies that demonstrate the insight into molecular adaptation that is provided by combining knowledge of phylogenies, site-directed mutagenesis and protein structure. These studies emphasize the importance of protein structure when considering the effects of amino acid mutations.

Many changes can occur over many generations, with only a few being responsible for changes in function. For example, the sequences of lactate dehydrogenase (LDH) and malate dehydrogenase (MDH) from *Bacillus stearothermophillus* are only about 25% identical, but their tertiary structures are highly similar. Only one mutation, of uncharged glutamine 102 to positive arginine in the active site, is required to convert LDH into a highly specific MDH. The arginine is thought to interact with the carboxylate group which is the only difference between the substrate/products of the two enzymes (Figure 14.11; Wilks *et al.*, 1988).

Thus amino acid changes that appear to be radical or conservative from their scores in mutation matrices or amino acid properties may be the opposite when their effect on protein function is considered; glutamine to arginine has a score of 0 in the PAM250 matrix, meaning that it is neutral. The importance of the mutation at position 102 in LDH and MDH could not be predicted using this information alone.

Another study showed that phylogeny and site-directed mutagenesis can identify key amino acid changes that would likely be overlooked if only structure was considered; the reconstruction of an ancestral ribonuclease showed that the mutation that causes most of the five-fold loss in activity towards double-stranded RNA is of Gly38 to Asp, more than 5 Å from the active site (Golding and Dean, 1998).

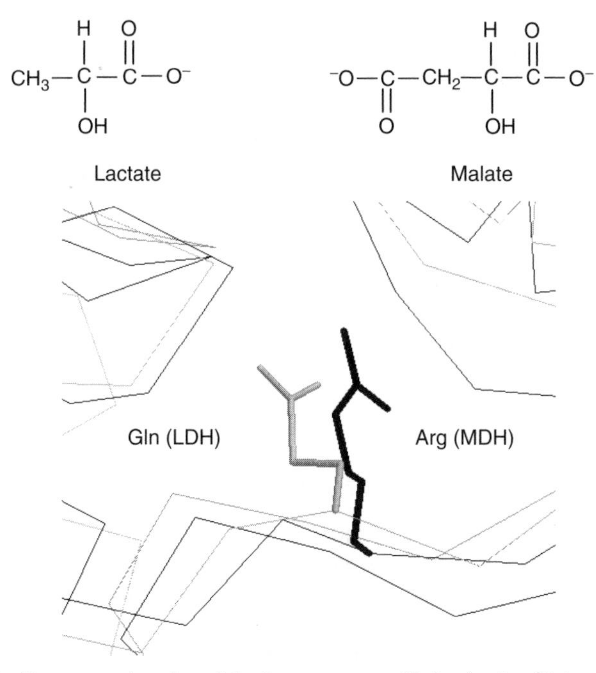

Figure 14.11 Lactate and malate dehydrogenase specificity (codes 9ltd and 2cmd; Wilks *et al.*, 1988).

A third study showed that knowledge of structure can be important in understanding the effects of mutations. Two different mutations in different locations in the haemoglobin genes of the bar-headed goose and Andean goose give both species a high affinity for oxygen. Structural studies showed that both changes remove an important van der Waals contact between subunits, shifting the equilibrium of the haemoglobin tetramer towards the high-affinity state. The important point in all these studies is that no single approach, such as phylogeny alone or structural studies alone, is enough to understand the effects of all amino acid mutations.

14.7 A SUMMARY OF THE THOUGHT PROCESS

It is our hope that this chapter has given the reader some guidelines for interpreting how a particular mutation might affect the structure and function of a protein. Our suggestion would be that you ask the following questions:

First about the protein:

1. What is the cellular environment?
2. What does it do? Is anything known about the amino acids involved in its function?
3. Is there a structure known or one for a homologue?
4. What protein family does it belong to?
5. Are any post-translational modifications expected?

Then about a particular amino acid:

1. Is the position conserved across orthologues? Across paralogues?
2. If a structure is known: is the amino acid on the surface? Buried in the core of the protein?
3. Is it directly involved in function or near (in sequence or space) to other amino acids that are?
4. Is it an amino acid that is likely to be critical for function? For structure?

Once these questions have been answered it should be possible to make a rational guess or interpretation of effects seen by an amino acid substitution and select logical amino acids for mutagenesis experiments.

REFERENCES

Altschul SF, Madden TL, Schaffer AA, Zhang J, Zhang Z, Miller W, *et al.* (1997). Gapped BLAST and PSI-BLAST: a new generation of protein database search programs. *Nucleic Acids Res* **25**: 3389–3402.

Ashburner M, Ball CA, Blake JA, Botstein D, Butler H, Cherry JM, *et al.* (2000). Gene ontology: tool for the unification of biology. The Gene Ontology Consortium. *Nature Genet* **25**: 25–29.

Barnes MR, Russell RB. (1999). A lipid-binding domain in Wnt: a case of mistaken identity? *Curr Biol* **9**: R717–R719.

Bateman A, Birney E, Durbin R, Eddy SR, Howe KL, Sonnhammer EL. (2000). The Pfam protein families database. *Nucleic Acids Res* **28**: 263–266.

Berman HM, Westbrook J, Feng Z, Gilliland G, Bhat TN, Weissig H, *et al.* (2000). The Protein Data Bank. *Nucleic Acids Res* **28**: 235–242.

Cargill M, Altshuler D, Ireland J, Sklar P, Ardlie K, Patil N, *et al.* (1999). Characterization of single-nucleotide polymorphisms in coding regions of human genes. *Nature Genet* **22**: 231–238.

Copley RR, Barton GJ. (1994). A structural analysis of phosphate and sulphate binding sites in proteins. Estimation of propensities for binding and conservation of phosphate binding sites. *J Mol Biol* **242**: 321–329.

Dayhoff MO, Schwartz RM, Orcutt BC. (1978). A model of evolutionary change in proteins. In Dayhoff MO. (Ed.), *Atlas of Protein Sequence and Structure*, Vol. 5. National Biomedical Research Foundation: Washington DC, pp. 345–352.

Durbin R, Eddy S, Krogh A, Mitchison G. (1998). *Biological Sequence Analysis. Probabalistic Models of Proteins and Nucleic Acids*. Cambridge University Press: Cambridge.

Eddy, SR (1998). Profile hidden Markov models. *Bioinformatics* **14**: 755–763.

Gavel Y, von Heijne G. (1990). Sequence differences between glycosylated and non-glycosylated Asn-X-Thr/Ser acceptor sites: implications for protein engineering. *Protein Eng* **3**: 433–442.

Gibson TJ, Spring J. (1998). Genetic redundancy in vertebrates: polyploidy and persistence of genes encoding multidomain proteins. *Trends Genet* **14**: 46–49; discussion 49–50.

Golding GB, Dean AM. (1998). The structural basis of molecular adaptation. *Mol Biol Evol* **15**: 355–369.

Gupta R, Birch H, Rapacki K, Brunak S, Hansen JE. (1999). O-GLYCBASE version 4.0: a revised database of O-glycosylated proteins. *Nucleic Acids Res* **27**: 370–372.

Hanks SK, Quinn AM, Hunter T. (1988). The protein kinase family: conserved features and deduced phylogeny of the catalytic domains. *Science* **241**: 42–52.

Henikoff S, Henikoff JG. (1992). Amino acid substitution matrices from protein blocks. *Proc Natl Acad Sci USA* **89**: 10915–10919.

Hunter CA, Singh J, Thornton JM. (1991). Pi–pi interactions: the geometry and energetics of phenylalanine–phenylalanine interactions in proteins. *J Mol Biol* **218**: 837–846.

Jeffrey PD, Russo AA, Polyak K, Gibbs E, Hurwitz J, Massague J, *et al.* (1995). Mechanism of CDK activation revealed by the structure of a cyclinA–CDK2 complex. *Nature* **376**: 313–320.

Jones DT, Taylor WR, Thornton JM. (1994). A mutation data matrix for transmembrane proteins. *FEBS Lett* **339**: 269–275.

Kraulis PJ. (1991). MOLSCRIPT: a program to produce both detailed and schematic plots of protein structures. *J Appl Crys* **24**: 946–950.

Krishna RG, Wold, F. (1993). Post-translational modification of proteins. *Adv Enzymol Relat Areas Mol Biol* **67**: 265–298.

Macias MJ, Wiesner S, Sudol M. (2002). WW and SH3 domains, two different scaffolds to recognize proline-rich ligands. *FEBS Lett* **513**: 30–37.

Parekh RB, Rohlff C. (1997). Post-translational modification of proteins and the discovery of new medicine. *Curr Opin Biotechnol* **8**: 718–723.

Peracchi A. (2001). Enzyme catalysis: removing chemically 'essential' residues by site-directed mutagenesis. *Trends Biochem Sci* **26**: 497–503.

Plotnikov AN, Schlessinger J, Hubbard SR, Mohammadi M. (1999). Structural basis for FGF receptor dimerization and activation. *Cell* **98**: 641–650.

Russell RB. (1998). Detection of protein three-dimensional side chain patterns: new examples of convergent evolution. *J Mol Biol* **279**: 1211–1227.

Sayle RA, Milner-White EJ. (1995). RASMOL: biomolecular graphics for all. *Trends Biochem Sci* **20**: 374.

Schoorlemmer J, Goldfarb M. (2001). Fibroblast growth factor homologous factors are intracellular signaling proteins. *Curr Biol* **11**: 793–797.

Schulze-Gahmen U, De Bondt HL, Kim SH. (1996). High-resolution crystal structures of human cyclin-dependent kinase 2 with and without ATP: bound waters and natural ligand as guides for inhibitor design. *J Med Chem* **39**: 4540–4546.

Sunyaev S, Lathe W III, Bork P. (2001). Integration of genome data and protein structures: prediction of protein folds, protein interactions and 'molecular phenotypes' of single nucleotide polymorphisms. *Curr Opin Struct Biol* **11**: 125–130.

Sunyaev S, Ramensky V, Bork P. (2000). Towards a structural basis of human non-synonymous single nucleotide polymorphisms. *Trends Genet* **16**: 198–200.

Taylor WR. (1986). The classification of amino acid conservation. *J Theor Biol* **119**: 205–218.

Waksman G, Kominos D, Robertson SC, Pant N, Baltimore D, Birge RB, *et al.* (1992). Crystal structure of the phosphotyrosine recognition domain SH2 of v-src complexed with tyrosine-phosphorylated peptides. *Nature* **358**: 646–653.

Wang Z, Moult J. (2001). SNPs, protein structure, and disease. *Hum Mutat* **17**: 263–70.

Wilks HM, Hart KW, Feeney R, Dunn CR, Muirhead H, Chia WN, *et al.* (1988). A specific, highly active malate dehydrogenase by redesign of a lactate dehydrogenase framework. *Science* **242**: 1541–1544.

Wilson CA, Kreychman J, Gerstein M. (2000). Assessing annotation transfer for genomics: quantifying the relations between protein sequence, structure and function through traditional and probabilistic scores. *J Mol Biol* **297**: 233–249.

Wolfe SA, Grant RA, Elrod-Erickson M, Pabo CO. (2001). Beyond the 'recognition code': structures of two Cys2His2 zinc finger/TATA box complexes. *Structure (Camb)* **9**: 717–723.

APPENDIX: TOOLS

Protein sequences

http://www.expasy.ch/
http://www.ncbi.nlm.nih.gov/

Amino acid properties

http://russell.embl-heidelberg.de/aas/

Domain assignment/sequence search tools

http://www.ebi.ac.uk/interpro/
http://www.sanger.ac.uk/Software/Pfam/
http://smart.embl-heidelberg.de/
http://www.ncbi.nlm.nih.gov/BLAST/
http://www.ncbi.nlm.nih.gov/COG/
http://www.cbs.dtu.dk/TargetP/

Protein structure

Databases of 3D structures of proteins
http://www.rcsb.org/pdb/
Structural classification of proteins
http://scop.mrc-lmb.cam.ac.uk/scop/

Protein function

http://www.geneontology.org/

SECTION 5

GENETICS/GENOMICS INTERFACES

CHAPTER 15

Gene Expression Informatics and Analysis

ANTOINE H. C. VAN KAMPEN*, JAN M. RUIJTER, BARBERA D. C. VAN SCHAIK, HUIB N. CARON, and ROGIER VERSTEEG

Academic Medical Center, University of Amsterdam
Meibergdreef 9, 1105 AZ Amsterdam
The Netherlands

* corresponding author

Bioinformatics for Geneticists. Edited by M.R. Barnes and I.C. Gray
 ISBNs: 0 470 84393 4; 0 470 84394 2 (PB)

15.1 INTRODUCTION

Unravelling the molecular mechanisms in living cells is one of the major challenges in current biology. Understanding these mechanisms will help us to recognize and finally treat a range of diseases such as cancer. Gene expression profiling provides one approach to study cellular processes at the gene level. There are many approaches for the measurement of gene expression, at a single gene level, technologies such as RT-PCR and Taqman provide detailed gene expression profiles across a defined range of tissues (Heid *et al.*, 1996; Riedy, *et al.*, 1995). At a genome-wide level, Serial Analysis of Gene Expression (SAGE) (Velculescu *et al.*, 1995) and DNA microarrays (Lockhart *et al.*, 1996; Schena *et al.*, 1995; Zammatteo *et al.*, 2000) enable the simultaneous measurement of the expression of thousands of genes in a single tissue.

The advances in physical transcript mapping afforded by the recently released draft sequence of the human genome (Lander *et al.*, 2001; Venter *et al.*, 2001) creates a new opportunity to combine genome-wide gene profiling and gene mapping efforts. Combination of gene expression data with gene positions will further unravel molecular mechanism in the cell. This chapter focuses on *in silico* sources of gene expression data, such as SAGE and the Human Transcriptome Map (HTM; Caron *et al.*, 2001), for the evaluation of gene expression across loci, specifically addressing the needs of positional cloning and cancer genetics.

Cancer results from changes in DNA sequence, which are reflected in altered amino acid sequences of the corresponding proteins or changes in protein expression levels, either of which ultimately changes cell function (King, 2000). These DNA changes can include (relatively) small mutations involving substitutions, insertions or deletions of bases, but also gross changes in DNA content per nucleus manifested as chromosome rearrangements or as gene amplifications.

DNA changes that eventually lead to cancer may be reflected in the expression levels of the corresponding genes or in the expression levels of genes that are directly or indirectly regulated by the mutated gene(s). Consequently, comparison and analysis of gene expression profiles of normal and tumour tissue at different stages of carcinogenesis helps to increase our knowledge of the molecular biology of cancer. Although the translation of expression profiles to relevant biological information is still one of the major challenges in biology and bioinformatics, integral gene expression analysis is already used extensively in cancer research (e.g. Alizadeh *et al.*, 2000; Ben-Dor *et al.*, 2000; Cole *et al.*, 1999; Golub *et al.*, 1999; Hastie *et al.*, 2000; Spieker *et al.*, 2001; Yeang *et al.*, 2001; Zhang *et al.*, 1997).

One problem that occurs in the comparison of gene expression profiles for normal and tumour tissues is the large number of genes that are differentially expressed. Not all these genes are interesting candidates for further investigation, most are not directly implicated in carcinogenesis, but instead they may be part of the multiple downstream pathways which are activated during carcinogenesis, including for example, responses of the cell to stress or apoptosis. Therefore, to facilitate the identification of candidate genes it could be useful to select those genes that are positioned at aberrant regions of chromosomes. Many such regions are already known for different types of cancer or they can easily be detected by screening tumour material (Mitelman *et al.*, 2001). For example, the embryonal tumour neuroblastoma shows common genetic aberrations such as the amplification of the MYCN oncogene (Schwab *et al.*, 1983) and loss of chromosome 1p (Brodeur *et al.*, 1977).

The Human Transcriptome Map (HTM) was specifically developed to enable the comparison of expression levels of genes in such regions. The HTM provides a clear example of a project that was initiated with a simple question: 'Is it possible to develop a tool that guides the identification of candidate genes from chromosomal regions known to be involved in neuroblastoma (or other cancers) from genome-wide gene expression profiles obtained with Serial Analysis of Gene Expression (SAGE)?'. To answer this question, the HTM integrates the position of human genes on chromosomes with genome-wide expression profiles provided by SAGE (Velculescu *et al.*, 1995).

Although the HTM seems a straightforward integration of a gene mapping database and SAGE expression profiles, the development of such an application is actually quite complex as will be shown in this chapter. Numerous aspects had to be considered during the development of the HTM such as the development of sequence analysis algorithms as part of the SAGE analysis, the application of statistical methods to analyse the data and the development of a relational database to enable the integration of data from different (public) resources.

HTM was initially developed for the selection of candidate genes but it also provides more fundamental insight into the organization of the human genome. Inspection of the expression profiles for all chromosomes reveals an intriguing pattern of domains of genes with an above-average expression in each tissue. These domains were named RIDGEs (Regions of Increased Gene Expression) and understanding them may further advance our knowledge of normal organization of the genome and of cancer.

In using computational tools such as HTM it is important to have a basic understanding of the underlying principles of the tool to avoid misinterpretation of results. In general, software applications are used as 'black boxes' that give answers to questions when data is put in. On the other hand, once underlying technologies and principles are understood, it is possible to identify new possibilities for application of tools or generation of ideas for the development of new tools. The use of public biological databases also requires caution since they may contain errors, ambiguous data or data may be missing (e.g. Karp, 1998; Karp *et al.*, 2001). Understanding the nature of the data contained in these databases will facilitate the interpretation of the results obtained. This chapter provides some examples of the issues and pitfalls that are involved in the construction and the application of gene expression analysis tools.

This chapter focuses specifically on issues related to the Human Transcriptome Map and therefore it necessarily concentrates on SAGE technology and bioinformatics of SAGE analysis. However, it will become clear that the overall approach, technologies and problems described during the development of the HTM are not specific for SAGE analysis but also apply to technologies such as DNA microarrays. In this chapter we will describe the SAGE and DNA microarray technologies and discuss some important differences between these two technologies. We will introduce the Cancer Genome Anatomy Project, which has made several tools and databases for gene expression analysis available via the internet. We will discuss the processing and statistical analysis of SAGE data and explain the data integration process that was required to construct the HTM.

Parts of Sections 15.4.2 and 15.6 are reprinted (abstracted/excerpted) with permission from Caron *et al.* (2001). (Copyright 2001 American Association for the Advancement of Science).

15.2 TECHNOLOGIES FOR THE MEASUREMENT OF GENE EXPRESSION

A range of methods are available to measure gene expression or changes in gene expression. In this section the SAGE and DNA microarray technologies are described since these are the primary methods used for genome-wide profiling.

15.2.1 Serial Analysis of Gene Expression (SAGE)

Serial Analysis of Gene Expression (SAGE; Velculescu *et al.*, 1995) is a technique used to construct quantitative genome-wide gene expression profiles (van Limpt *et al.*, 2000; Porter *et al.*, 2001; Scott and Chrast, 2001; Velculescu *et al.*, 2000). Three principles underlie the SAGE methodology (Figure 15.1):

(1) A short 10-base pair sequence tag contains sufficient information to uniquely identify a transcript provided that this tag is obtained from a unique position within each transcript (there are many more possible tags ($4^{10} = 1,048,576$) than human genes).

(2) Sequence tags can be linked together to form long serial molecules (concatemers) that can be cloned and sequenced.

(3) Counting of the number of times a particular tag is observed provides the expression level of the corresponding transcript.

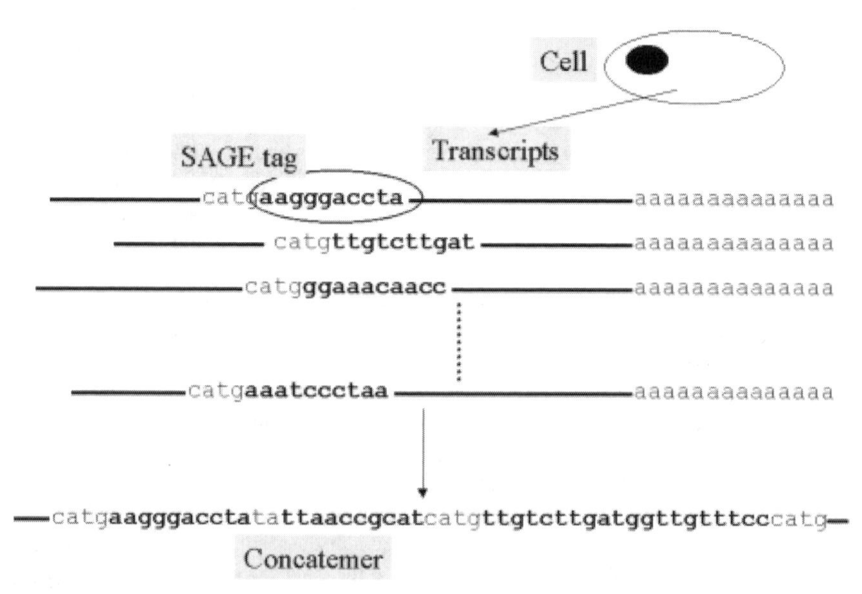

Figure 15.1 Serial Analysis of Gene Expression (SAGE; Velculescu *et al.*, 1995). mRNA is extracted from a cell tissue sample. Subsequently, a 10-base pair tag that is right to the most 3′ CATG site is extracted from each transcript by using the NlaIII restriction enzyme. These tags are then ligated to ditags, which are amplified and linked to form concatemers containing approximately 30 tags. These concatemers are then cloned and sequenced.

The sequenced concatemers consist of ditags and include approximately 30 to 40 tags. The sequenced concatemers are the starting point for data processing, which is explained in more detail in Section 15.4. The number of tags that is obtained in a SAGE experiment ranges from 10,000 to over 100,000. The frequency of a tag directly reflects the fraction of the corresponding transcript in the cell. In other words, if a particular tag is observed 25 times in a SAGE library that consists of 50,000 tags, then the number of corresponding transcripts is also 25 per 50,000 transcripts in the cell. In this sense, SAGE provides an 'absolute' expression level.

15.2.2 DNA Microarrays

The principle of a DNA microarray experiment is to hybridize labelled cDNA to DNA sequences that are immobilized on a solid surface in an ordered array. The labelled cDNA is often referred to as the target and the immobilized DNA sequences as the probe. A DNA microarray allows the detection and quantification of thousands of transcripts simultaneously. Two main types of DNA microarrays can be distinguished according to the arrayed material. The first type is the cDNA microarray in which the probes are usually products of the polymerase chain reaction generated from cDNA libraries or clone collections (Bowtell, 1999; Brown and Botstein, 1999; Schena *et al.*, 1995, 1996). These probes are spotted onto glass slides or nylon membranes at defined positions. The second type or arrays are the oligonucleotide arrays for which short 20–25mers are synthesized *in situ* by photolithography onto silicon wafers (GeneChip™ technology of Affymetrix (Lockhart *et al.*, 1996)). Alternatively, pre-synthesized oligonucleotides can be printed onto glass slides (Okamoto *et al.*, 2000, Zammatteo *et al.*, 2000). For target preparation, mRNA from cells or tissue is extracted, which is converted to cDNA and labelled. The target is then hybridized to the DNA probes on the array and detected by phospho-imaging or fluorescence scanning. In the case of fluorescence, two fluorescent dyes with different colours (Cy3 and Cy5) are used to label the cDNAs from two different cell populations. The resulting two targets are mixed and hybridized to the same array, which results in competitive binding of the target to the spotted probe sequences. Subsequently, the array is scanned using two different wavelengths, corresponding to the two dyes and the intensity of each spot in both channels is 'mixed' *in silico*. This results in an expression level, relative to the chosen control condition, for each gene that is represented on the array (see Chapter 9 Section 9.6, for some examples of output from oligonucleotide arrays).

15.2.3 Comparison of SAGE and DNA Microarrays

The SAGE and DNA microarray technologies differ in several important ways. SAGE measures expression levels that directly reflect the fraction of mRNAs in the cell, i.e. SAGE produces 'absolute' expression levels. In contrast, the DNA microarray technique measures expression levels relative to a control condition. Consequently, different SAGE libraries can be directly compared because the expression levels do not depend on the use of a reference mRNA or experimental conditions, while DNA microarray experiments can only be compared if they have been measured relative to the same control tissue under the same conditions. For the same reason, the gene expression levels within one SAGE library can be directly compared, while expression levels obtained for genes on one DNA microarray cannot be compared due to differences in labelling and hybridization efficiency of individual genes. Another difference between SAGE and DNA microarrays comprises the genes that can be measured in an experiment. In a DNA microarray experiment one only measures the genes for which the array contains probes, while SAGE in principle measures every mRNA in the sample. Consequently, SAGE is very suitable for discovering new genes, although low abundance transcripts are only likely to appear in large SAGE libraries. The DNA microarray is, however, very suitable for quickly screening cells or tissues for the expression of a pre-selected set of genes. A disadvantage of SAGE is that the extracted mRNA tags need to be identified *in silico* (Section 15.4) while for DNA microarrays it is already known which probes (genes) are on the array. Furthermore, the construction of a SAGE library requires much more effort than carrying out a DNA microarray experiment once the array has been printed.

15.3 THE CANCER GENOME ANATOMY PROJECT (CGAP)

The Cancer Genome Anatomy Project (CGAP; Lal *et al.*, 1999; Lash *et al.*, 2000; Riggins and Strausberg, 2001; Schaefer *et al.*, 2001; Strausberg *et al.*, 1997, 2000) is a project of the National Cancer Institute (NCI). Their main objective is to decipher the molecular mechanism of cancer. For this goal, information is gathered from different resources such as gene expression data, aberrations of chromosomes, gene variation and biochemical pathways. CGAP collaborates with the National Centre for Biotechnology Information (NCBI) to develop computational technologies for the management and analysis of these large amounts of data. All data and programs for analysis are made available via the internet (cgap.nci.nih.gov).

The HTM makes extensive use of two CGAP resources. Firstly, HTM includes the SAGE libraries that were constructed as part of the CGAP project. Secondly, HTM algorithms use the SAGEmap tag-to-gene mapping as a starting point for constructing an improved tag-to-gene mapping (Section 15.4.2). These mappings are used for SAGE tag identification. In addition, several other CGAP tools and databases are regularly used during the analysis of SAGE data. Therefore, this section provides a brief overview of the resources offered by CGAP.

The CGAP resource contains cDNA and SAGE libraries of normal cells and cancer cells in different stages. These libraries include the 3′ and 5′ clones of cDNAs from the dbEST database (Boguski *et al.*, 1993), the CGAP subset of dbEST, the Mammalian Gene Collection (MGC) subset of dbEST, randomly cloned cDNAs from the ORESTES (Open Reading Frame EST sequencing) project and SAGE libraries. The MGC is an NIH initiative that supports the production of cDNA libraries, clones and sequences (Strausberg *et al.*, 1999). The goal of the MGC is to provide a complete set of full-length (open reading frame) sequences and cDNA clones of expressed genes for human and mouse. The ORESTES project aims for the completion of gene annotation by sequencing randomly primed cDNAs (Pandey, 2001). CGAP also supports the generation of SAGE libraries and their sequencing to obtain gene expression profiles of normal, pre-cancer, and cancer cells, which resulted in high-quality SAGE gene expression profiles for a range of normal and tumour tissues. The generation of these profiles still continues. At present over 140 SAGE libraries are available from the SAGEmap database (Lal *et al.*, 1999; Lash *et al.*, 2000) including more than 5 million tags.

The CGAP Library Finder Tool retrieves any cDNA library from dbEST or SAGE libraries. The search can be narrowed to the CGAP, MGC or ORESTES subsets. A query first returns a single library or a list of libraries, each of which is linked to its own Library Info page where details of the library and its preparation can be found. The Library Finder

Tool allows the retrieval of libraries according to tissue type, tissue preparation, tissue histology, library protocol and library name.

CGAP offers a range of tools to examine gene expression data from their cDNA or SAGE collection. The Gene Library Summarizer (GLS) generates unique and non-unique genes expressed in a single cDNA library or library group. It then identifies the genes in each of these groups as known or unknown. The cDNA xProfiler is a tool that compares gene expression between two pools of libraries by counting the number of clones in the library. The Digital Gene Expression Displayer (DGED) is a tool that compares gene expression between two pools of libraries. In contrast to the cDNA xProfiler that counts clones, the DGED treats the presence of a gene in a library pool as a matter of degree. It compares the 'degree' of presence of a gene in pool A with its 'degree' of presence in pool B by using a chi-squared test. The SAGEmap xProfiler performs differential-type analyses on (pooled) SAGE libraries. Similar libraries can be placed into one of two groups based on their characteristics (e.g. normal colon and colon cancer). Comparisons are then made between the two groups using a statistical test developed specifically for SAGE data (Lash *et al*., 2000). The SAGEmap Virtual Northern (vNorthern) tool has been designed to accept mRNA or EST sequences as input. Possible tags are then extracted from this sequence and links provided to access the data from the various SAGE libraries currently represented on the SAGEmap website.

CGAP also provides access to the Mitelman database of chromosome aberrations in cancer (Mitelman *et al*., 2001). This database contains manually selected data from about 40,000 scientific articles and is organized as three distinct sub-databases. The sub-database 'Cases' contains the data that relates chromosomal aberrations to specific tumour characteristics in individual patient cases. The sub-database of 'Molecular Biology and Clinical Associations' contains no data from individual patient cases. Instead, the data is pulled from studies with distinct information about molecular biology or clinical associations. The molecular biology associations relate chromosomal aberrations and tumour histologies to genomic sequence data, while clinical associations relate chromosomal aberrations and tumour histologies to clinical variables such as prognosis, tumour grade and patient characteristics. The 'Reference' sub-database contains all the references culled from the literature.

Another tool to examine the chromosomes uses the CGAP FISH-mapped BACs, which are BAC clones that are mapped both cytogenetically by FISH and physically by STSs to the human genome. Genetic and physical SNP maps are available, which show the genetic and physical locations of confirmed, validated and predicted SNPs per individual chromosome.

CGAP also includes the graphical biochemical pathway maps from KEGG (Kanehisa and Goto, 2000) and BioCarta (www.biocarta.com). The entities on these maps are linked to the above-mentioned CGAP resources.

15.4 PROCESSING OF SAGE DATA

The processing of SAGE data generally consists of three steps. First a list of tags is compiled from the concatemer sequences. Secondly, the SAGE tags are identified and finally the expression levels can be compared statistically. The extraction of tags from the concatemer sequences is straightforward since each concatemer consists of ditags that are separated by the CATG sequence. Each ditag contains one tag in the $5' \rightarrow 3'$ (sense) direction and a second tag in the $3' \rightarrow 5'$ (complementary-reverse) orientation. The ditags are extracted from the concatemers and duplicate ditags are removed because they are most likely experimental artifacts (Velculescu *et al.*, 1995). The length of the resulting ditags must be between 20 and 24 bp. Shorter and longer ditags are discarded as experimental artefacts. Subsequently, from each extracted ditag the sense and complementary-reverse (which is converted to a sense tag) tags are extracted and added to the list of SAGE tags. The number of times that a tag occurs in this list directly reflects the expression level of the corresponding transcript.

As a result of this experimental procedure the association between tag and transcript from which the tag is extracted is lost. Consequently, after compiling the tag list (i.e. gene expression profile) each tag in this list has to be identified by matching it against a tag-to-gene map. This tag-to-gene mapping database must first be compiled by electronically extracting a tag from each mRNA/EST sequence in the GenBank database and subsequently storing the annotated tag in the tag-to-gene mapping. The compilation of this tag-to-gene mapping is one of the crucial steps in the SAGE analysis.

The CGAP SAGEmap tag-to-gene mapping (Lal *et al.*, 1999) is an example of such mapping. Typical entries in this tag-to-gene mapping look something like the following:

AAAAATACAA 5/EST/+3_label *43744 ESTs* AI093649, AI263776, N26090, N67808 (4 6)
TATTAGGATA 5/EST/+3_label *43744 ESTs* AI434789, AI813305, AW271602 (3 3)
AAAAAATACA 1/mRNA/+orient *1119 nuclear receptor subfamily 4, group A* D85245 (1 1)
AAAAAATACA 2/EST/+orient+3_label *107526 UDP-Gal:betaGlcNAc beta 1,4-galactosyltransferase, polypeptide 5* AA046634 (1 10)

Each entry (tag annotation) contains five attributes, i.e. the 10-bp tag (bold), the sequence type of the clones from which the tag was extracted (underlined), the UniGene cluster number and cluster name (italic), the accession codes of clones (comma delimited list) and two frequency numbers (between parentheses).

The sequence type provides information about the reliability of the determination of the 3′-end of the GenBank sequence. Since tags are only valid if extracted adjacent to the most 3′ CATG in the sequence, it is very important to establish whether the sequence indeed includes the 3′-end. The following sequence types are defined:

'1/mRNA/+orient' — Well-characterized mRNA or RefSeq sequence (Pruitt and Maglott, 2001).
'2/EST/+orient+3_label' — EST, with polyA signal and/or polyA tail, and labelled as 3′
'3/EST+orient' — EST, with polyA signal and/or polyA tail, but unlabelled
'4/EST+orient+5_label' — EST, with polyA signal and/or polyA tail, and labelled as 5′
'5/EST+3_label' — EST, without polyA signal or polyA tail, but labelled as 3′

The polyA signal and polyA tail both provide information about the 3′-end of the sequence. In the definition of these sequence types only the two most common polyA signals (ATTAAA and AATAAA) were considered. A polyA tail was defined as a stretch of 10 consecutive As at the end of the sequence of 10 consecutive Ts at the beginning of the sequence. Additional information to identify 3′-end sequences is obtained from the depositors of the cDNA sequences, which have assigned a label (3′ or 5′) to the GenBank sequence based on the cloning and sequencing procedures.

The frequency numbers provide information about the reliability and uniqueness of the tag. The first frequency number denotes the number of GenBank clones of this type, with this tag and this UniGene cluster assignment. The second frequency number denotes number of GenBank clones of this type with this tag in any UniGene cluster. In the example above, we see that the tag AAAAATACAA (5/EST+3_label) corresponds to four clones in UniGene cluster 43744. However, from the second frequency number it can be seen that this tag of this type is also extracted from two clones in one or two other UniGene clusters. Therefore, this tag is not unique for a gene or it may be an incorrect tag.

15.4.1 The Construction of Tag-to-Gene Mapping in HTM

To obtain a reliable mapping of gene expression profiles to chromosomes it is important to have a tag-to-gene mapping in which false positive tag identifications (tags that are extracted from the wrong position of the database sequence and therefore do not correspond to the experimentally determined tag of the gene) are removed. False positive tags would strongly compromise the genome-wide expression patterns. The CGAP SAGEmap tag-to-gene mapping contains many false positive tags because this mapping was designed to include all potential tags. To improve the quality of SAGE analysis, the Academic Medical Centre (AMC) tag-to-gene mapping process was constructed to exclude as many false positive tags as possible. The AMC tag-to-gene mapping basically comprises four steps:

1. Identification of the 3′-end of cDNA clones and the electronic extraction of tags.
2. Removal of erroneous tags that result due to EST sequence errors in the 10-bp tag.
3. Removal of erroneous tags that result due to EST sequence errors in the CATG sequence.
4. Identification of anti-sense tags.

Sequencing of cDNA clones occurs, by definition, from the 5′-end to the 3′-end of the sequence. The $5' \rightarrow 3'$ sequence is called the ‘sense’ sequence, while the $3' \rightarrow 5'$ sequence is called the ‘complementary-reverse’ sequence. This implies that the most likely orientation of sequences in a database of sequenced cDNA clones is either ‘sense’ or ‘complementary-reverse’. In the case of 3′-end sequences this will, respectively, show the polyA tail as an A-stretch at the end or as a T-stretch at the beginning of the sequence. However, two other possible sequence orientations (reverse or complement) occur in the GenBank database as a result of human errors in submitting or processing the sequence. The frequency of the four possible sequence orientations were analysed by using the 718,271 clones included in the CGAP SAGEmap tag-to-gene mapping of which 12,381 clones contain a stretch of >30 As or Ts at either end of the sequence. Of these clones, 11,476 (93%) end with >30 As (sense) or start with >30 Ts (complementary-reverse). Only 7% of the polyA tails are on the wrong side of the sequence and these clones could result from wrong sequence orientation in the database. Therefore, only the sense and complementary-reverse sequence orientations are considered in the subsequent electronic tag extraction procedures to build the AMC tag-to-gene map. The algorithms that were constructed to build the AMC tag-to-gene map used the cDNA clones (and UniGene cluster assignment) that are included in the CGAP SAGEmap tag-to-gene map. In addition, the sequence type that was assigned to each tag was used.

15.4.2 Identification of 3′-end cDNA Clones and Electronic Tag Extraction

The 3′-end of a processed gene transcript is characterized by a polyA tail and a polyA signal. Besides the two 'classical' polyA signals (AATAAA and ATTAAA), other polyA signals have been reported (Proudfoot, 1991; Sheets *et al.*, 1990; see Chapter 12 for more details). The clones included in the CGAP SAGEmap tag-to-gene map were analysed for the occurrence of 'alternative' polyadenylation signals. The clones containing either >30 As at the end or >30 Ts at the beginning of their sequence were selected. Polyadenylation signals are thought to occur within 50 to 100 bp from the polyA addition site (Salamov and Solovyev, 1997). Therefore the 150 nucleotides adjacent to the polyA or polyT stretch were analysed for the presence of the two classical polyadenylation signals, nine possible alternative polyA signals (AATTAA, AATAAC, AATAAT, AATACA, ACTAAA, AGTAAA, CATAAA, GATAAA, TATAAA) and six random hexamer sequences. The two classical polyA signals were found in 55.8 and 17.7% of those clones respectively, and showed a clear preference for occurrence within the first 50 nucleotides from the polyA tail. Four possible alternative polyA signals (AATTAA, AATAAT, CATAAA, AGTAAA) occur in these 50 nucleotides with a frequency ranging from 5.7 to 8.4%. The other five possible polyA signals and the six random hexamers showed no appreciable preference for occurring in the 3′-end of transcripts. Therefore, the sequence orientation algorithms that were developed were configured to search for the six most abundant polyA signals within 50 bp from the polyA site. The same frequency and position patterns for the six polyA signals were found in cDNA clones ending with at least 10 As or starting with at least 10 Ts. This indicates that the occurrence of stretches of 10 or more As or Ts at the end and the beginning of a cDNA sequence, respectively, is likely to represent a polyA tail.

The sequence types that are included in the CGAP SAGEmap tag-to-gene map provide additional information to identify the 3′-end clones. This sequence type was combined with the presence of one of the six polyA signals at either end of the clone sequence (within 50 bp) and/or a polyA tail (>10 As at the end or >10 Ts at the beginning) to select for reliable 3′-end clones. To minimize the risk of extracting erroneous tags (false positives) from GenBank sequences, only 'reliable 3′-end' clones were used for electronic tag extractions. When both strands of a cDNA encoded conflicting polyadenylation signals and/or polyA/polyT stretches, clones were not used for tag extraction.

15.4.3 Identification of 10-base Pair Tag Sequencing Errors

Single pass high throughput sequencing of EST libraries is one of the more error prone sequencing methods; therefore the chance of a sequence error is about 1% per base. Consequently, tags that are electronically extracted from database sequences may include sequencing errors. Therefore, the tags were checked for errors in the 10-bp sequence resulting from sequencing errors in ESTs. If it is assumed that sequencing errors are independent for each base and the error rate is 1%, then the probability of one error being present is only $10 \times 0.01 \times 0.99^9 = 0.091$. We designed algorithms that detected any combination of matching tags with maximal two-base substitutions, insertions or deletions because the chance that a tag will contain three errors is negligible (0.01%). To check for sequencing errors all EST clones in a UniGene cluster were compared pair-wise and checked for substitutions, insertions or deletions. If two tags were identical, except for one or two mismatches, a potential sequencing error in the tag might be involved. The tag corresponding to the largest number of clones was considered to be a correct tag. The tag with the potential sequencing error was removed when it was found in less than five ESTs/cDNAs and was five times less frequent then the correct tag. This ensured that variant tags resulting from frequent single nucleotide polymorphisms (SNPs) were not discarded in the AMC tag-to-gene mapping.

15.4.4 Identification of CATG Sequencing Errors

Sequence errors (Figure 15.2) in the most 3′ CATG sequence of an EST will result in skipping of the corresponding tag by the extraction algorithm and erroneous use of the next CATG for tag extraction. Also, an EST sequence error may create a new CATG distal

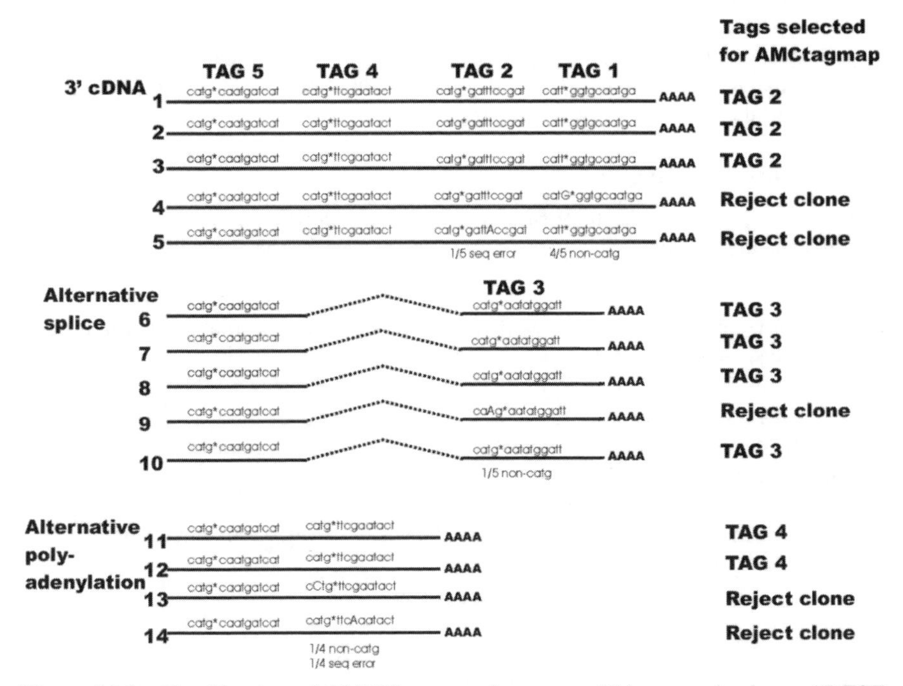

Figure 15.2 Identification of (CATG) sequencing errors. This example shows 15 EST clones (five 3′ cDNA clones, five 3′ cDNA clones of the alternatively spliced gene and five 3′ cDNA clones of the alternatively polyadenylated gene). TAG2 (GATTTCCGAT) is the correct tag for the first five clones. However, clone 4 is rejected because a CATG is created due to a sequencing error (T → G). If this clone was not rejected then TAG1 (GGTGCAATGA) would mistakenly be associated to this transcript. Clone 5 is rejected because TAG2 contains a sequencing error. Both sequencing errors are not considered to be SNPs because they only occur once in these five clones. TAG5 (AATATGGATT) is the correct tag for the alternatively spliced gene. Clone 9 is rejected because the CATG is destroyed due to a sequencing error (T → A). If this clone was not rejected then TAG5 would be mistakenly associated to this clone. In the case of the alternatively polyadenylated genes no clones are rejected because too few clones are available to make a decision. Consequently, TAG4 (TTCGAATACT) is extracted from clones 11 and 12, TAG4 (CAATGATCAT) from clone 13 (CATG was destroyed) and TAG6 (TTCAAATACT) from clones 14 and 15.

to the true most 3′ CATG. This also results in extraction of a false tag for an EST. An algorithm to remove these tags should preserve tags from alternatively spliced transcripts of the same gene. Each gene can have a series of tags belonging to alternatively spliced or alternatively polyadenylated transcripts. Furthermore, SNPs in the CATG sequence can cause extraction of alternative tags that are correct and should be preserved. Our algorithms were directed to the identification and removal of all tags that are caused by CATG sequence errors. The remaining tags were accepted as reliable tags.

15.4.5 Identification of Sense and Antisense Tags

One of the major problems with the UniGene clustering algorithm is that it can place overlapping genes encoded on opposite DNA strands in one UniGene cluster. In such cases, tag extraction routines may extract the tags from both genes. Therefore, algorithms to recognize oppositely oriented tags were designed. In such clusters, the orientation of the most frequent tag was considered as 'sense'. The antisense tags were marked and preserved in the AMC tag-to-gene mapping.

15.4.6 Comparison of SAGE Libraries

The HTM does not include statistical routines to establish whether two expression levels are significantly different. Therefore, once a candidate gene has been identified (based

TABLE 15.1 Public Resources (Software and Databases) Available for the (Statistical) Analysis of SAGE Data

Resource	Main Functionalities	Website
SAGE300 (Zhang *et al.*, 1997)	Tag extraction, tag identification, statistical comparison	www.sagenet.org
CGAP SAGEmap (Lal *et al.*, 1999)	Tag identification, statistical, xProfiler, Virtual Northern	www.ncbi.nlm.nih.gov/SAGE/
USAGE (van Kampen *et al.*, 2000)	Tag extraction, tag identification, statistical comparison, management of SAGE libraries (pool, merge, etc.)	www.cmbi.kun.nl/usage/
eSAGE (Margulies and Innis, 2000)	Tag extraction, statistical comparison, data management	ehm@umich.edu
Detecting sequencing errors (Colinge and Feger, 2001)	Detection of sequencing errors in SAGE libraries	georg.feger@serono.com
Audic and Claverie (1997)	Statistical comparison	igs-server.cnrs-mrs.fr/~audic/significance.html
SAGEstat (Kal *et al.*, 1999)	Statistical comparison	j.m.ruijter@amc.uva.nl or www.cmbi.kun.nl/usage/
POWER_SAGE (Man *et al.*, 2000)	Statistical comparison	michael.man@pfizer.com

on visual inspections of tag counts) one may calculate the statistical difference between the tag counts. Several statistical methods are available (see also Table 15.1) and are discussed in this section.

The aim of statistical comparison of two SAGE libraries is to reject the null hypothesis that the observed tag counts in both libraries are equal. Testing of this hypothesis is hampered by the fact that SAGE experiments are generally not repeated, and therefore, each SAGE library is only one measurement: the necessary information on biological variation and experimental precision is not available in the data. It is possible that all differences between two libraries are just the result of random sampling from the same population. Therefore, before starting a pair-wise comparison of specific tags in two libraries, the null hypothesis that the differences between libraries result from random sampling has to be rejected. In the context of SAGE research, only one reference to a test for this purpose has been published (Michiels *et al.*, 1999). This overall test is based on a simulation of a large number of possible distributions of two libraries within the pooled marginal totals of the observed SAGE libraries. By calculating the chi-squared statistic for each simulated pair of libraries, a distribution of this statistic under the null hypothesis can be constructed. From this simulated distribution and the chi-squared value of the observed libraries, one can then determine the probability of obtaining the observed tag distributions at random. Rejection of the null hypothesis that all differences between SAGE libraries are just the result of random sampling then opens the way for pair-wise comparisons.

15.4.7 Statistical Tests for Differences Between SAGE Libraries

Several statistical tests have been published for the pair-wise comparison of SAGE libraries. For all tests the null hypothesis states that there is no difference in tag numbers between the two libraries that are compared. It should be kept in mind that in most comparisons between specific tags in SAGE libraries, there is no *a-priori* knowledge about the direction of the effect. Therefore, all decision rules have to be formulated to result in a two-sided test. The significance level (α) can be set to 0.001 to safeguard against the rate of accumulation of false positives that may result from multiple testing (Bonferroni correction; Altman, 1991).

The different methods that can be used to test the difference between two SAGE libraries can be compared by considering the critical values. Critical values are defined as the highest or lowest number of tags that, given an observed number of tags in one library, needs to be found in the other library to result in a p-value below the significance level when the pair-wise test is carried out. They can be determined by repeatedly testing simulated tag numbers until the resulting p-value leads to rejection of the null hypothesis at the required level of significance.

In the original SAGE paper (Velculescu *et al.*, 1995), tag numbers in different libraries are compared pair-wise with a test based on a Monte Carlo simulation of tag counts. This approach is included in the SAGE software package SAGE300 (Zhang *et al.*, 1997). SAGE300 performs, in each pair-wise comparison, at least 100 with a maximum of 100,000 simulations to determine the chance of obtaining a difference in tag counts equal to or greater than the observed difference. This results in a one-sided p-value that has to be compared to $\alpha/2$. Since the Monte Carlo-based test of SAGE300 does not give the same p-value every time the same input is tested, each input is run six times and the mean p-value is used for the determination of the upper critical values that are given in Figure 15.3A. In this figure the critical values are given for two SAGE libraries of equal size (diamonds) and for two SAGE libraries of different size (squares). The critical

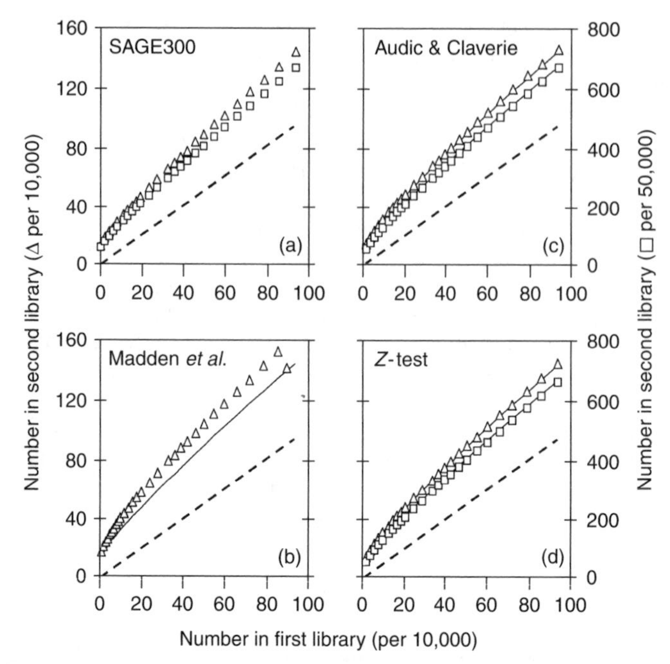

Figure 15.3 Comparison of the critical values of different tests for SAGE data. Critical values are defined as the numbers of tags that need to be found in the second SAGE library to be significantly different from the number of tags already found in the first SAGE library. Upper critical values for a 0.001 level of significance are given for (A) SAGE300 (Zhang *et al.*, 1997), and the tests of (B) Madden *et al.* (1997), (C) Audic and Claverie (1997) and (D) the Z-test of Kal *et al.* (1999). The critical values plotted in each graph are based on a first SAGE library with a total of 10,000 tags (reference values, plotted as a dotted continuous line on the x-axis) and a second library with a total of 10,000 tags (critical values plotted as triangles on the left y-axis) or a second library of 50,000 tags (critical values plotted as squares on the right y-axis). In B, C and D a plot of the critical values of SAGE300 (A) are added (thin lines) to facilitate comparison between tests. In B only critical values for a second library of 10,000 tags are given because Madden's test can only be used for libraries of similar size.

values of SAGE300 are copied as continuous lines into Figure 15.3B, C and D to facilitate comparison with other tests.

The test suggested by Madden *et al.* (1997) is based on only the number of observed specific tags in each SAGE library and the test statistic is calculated as:

$$Z = \frac{n_1 - n_2}{\sqrt{n_1} + \sqrt{n_2}} \tag{1}$$

with n_1 and n_2 as the number of specific tags in the first and second library, respectively. This test statistic is estimated to be normally distributed and can be compared to $Z_{\alpha/2}$. The test of Madden requires about 25% larger differences than SAGE300 to reach statistical significance and is, therefore, more conservative (Figure 15.3B). Only one set of critical values is given because this test can only be used for two libraries of similar size. However, the simple mathematics of this test (Eq. 1) are a point in its favour.

Audic and Claverie (1997) derived a new equation for the probability of finding n_2 or more tags in one library given the fact that n_1 tags have already been observed in the other library:

$$P(n_2|n_1) = \left(\frac{N_2}{N_1}\right)^{n_2} \frac{(n_1 + n_2)!}{n_1!n_2!(1 + N_2/N_1)^{(n_1+n_2+1)}} \tag{2}$$

with N_1 and N_2 as the total number of tags in the first and second library, respectively. A summation of this probability over all n from n_2 to infinity gives a one-sided p-value that can be compared to $\alpha/2$. The upper critical values for a significance level of 0.001 for Audic and Claverie's test are given in Figure 15.3C. For both the libraries of equal and different size these critical values are all within 1.5% of those of SAGE300.

The Z-test focuses on the proportions of specific tags in each library and is based on the normal approximation of the binomial distribution (Altman, 1991; Kal *et al.*, 1999). The test statistic Z is calculated as the difference in proportions divided by the standard error of this difference:

$$Z = \frac{p_1 - p_2}{\sqrt{p_0(1 - p_0)(1/N_1 + 1/N_2)}} \tag{3}$$

with $p_1 = n_1/N_1$ and $p_2 = n_2/N_2$. The proportion p_0, the expected proportion when the null hypothesis is true, is calculated as $p_0 = (n_1 + n_2)/(N_1 + N_2)$. Z is approximately normally distributed and can be compared to $Z_{\alpha/2}$. The critical values of the Z-test are given in Figure 15.3D and are also all within 1.5% of those of SAGE300.

The chi-squared test can be used for comparing SAGE libraries (Michiels *et al.*, 1999) after reorganizing the data in a 2 × 2 contingency table. However, this test is statistically equivalent to the Z-test on two proportions (Altman, 1991) and will give the same p-values and have the same critical values. Another test using 2 × 2 contingency tables is the Fischer exact test (Altman, 1991), which has also been applied to SAGE data (Man *et al.*, 2000). However, the sampling design required by this test does not apply to SAGE (Claverie, 1999; Conover, 1980) and moreover, for the large number of tags involved in SAGE, the chi-squared test is to be preferred. In the paper by Chen *et al.* (1998), a procedure based on Bayesian statistics is described to calculate the probability that the level of expression of a given mRNA is increased by at least x-fold between libraries. Although this procedure can be used to statistically judge differences in tag numbers, its approach is clearly different from the classical approach of hypothesis testing and results of these test procedures cannot be directly compared.

In conclusion, this comparison shows that SAGE300, Audic and Claverie's test (1997) and the Z-test, will all give the same test results when applied for pair-wise comparison of SAGE libraries whereas Madden's test will behave considerably more conservatively. In a Monte Carlo comparison of the chi-squared test, Fischer exact test and Audic and Claverie's test it was shown that the chi-squared test, which is equivalent to the Z-test, had the best power and robustness (Man *et al.*, 2000), especially at low expression levels.

15.4.8 Computational Resources for SAGE Analysis

Table 15.1 summarizes the public resources that are available for the analysis of SAGE data. The SAGE300 program (Zhang *et al.*, 1997) is probably the most commonly used application for SAGE analysis. To identify SAGE tags the SAGE300 program compiles a tag-to-gene map from human (EST) sequences in GenBank. A drawback of this method is that the orientation of the sequence is not checked before tag extraction and consequently, incorrect tags can result. SAGE300 also includes a Monte Carlo-based method for statistical comparison of SAGE libraries.

As part of CGAP the NCBI established the SAGEmap public database (Lal *et al.*, 1999), which includes SAGE libraries and a tag-to-gene mapping. SAGEmap also includes a 'reliable tag-to-gene map', which accounts for sequencing errors in GenBank sequences. These tag-to-gene maps can be downloaded and used in combination with applications such as Microsoft Access. Alternatively, the tag-to-gene maps are accessible online from the SAGEmap site but this only allows the analysis of one tag at a time. No full identification reports, i.e. for all tags in a SAGE tag list, can be generated as is possible with SAGE300, which unfortunately does not support the use of these tag-to-gene maps.

The USAGE application (van Kampen *et al.*, 2000) allows construction of tag-to-gene maps from the EMBL database for any organism. The program allows the extraction of tags from the sense and complement-reverse orientation of the sequence because the 3′-end of the clone is not determined prior to tag extraction. However, USAGE also includes both SAGEmap tag-to-gene maps and the AMC tag-to-gene map and allows the user to produce full tag identification reports. USAGE includes the Z-test for the statistical comparison of SAGE libraries (Kal *et al.*, 1999).

The eSAGE software (Margulies and Innis, 2000) is similar to USAGE. It includes the SAGEmap tag-to-gene mapping and performs statistical comparisons according to the test proposed by Claverie (1999). The input concatemers can contain any characters from the standard IUPAC code. In addition, eSAGE reads PHD files generated from phred-analysed sequence trace files (Ewing and Green, 1998; Ewing *et al.*, 1998) and uses the phred quality values for each base as a more accurate method of excluding low quality sequence data.

Colinge and Feger (2001) introduced a method to identify possible sequence errors in tags in SAGE libraries. This method in combination with an accurate tag-to-gene map can greatly enhance SAGE tag identification.

15.5 INTEGRATION OF BIOLOGICAL DATABASES FOR THE CONSTRUCTION OF THE HTM

To enable the mapping of gene expression profiles to chromosomes in the HTM, several public databases were integrated in a relational database. The HTM was constructed by mapping gene expression levels (SAGE tag counts) to gene positions as defined by the GeneMap99 database (Deloukas *et al.*, 1998). GeneMap99 gives the chromosomal position of 45,049 human expressed sequence tags (ESTs) and genes belonging to 24,106 UniGene clusters. The STS markers in GeneMap99 are assigned to a unique radiation hybrid code (RH-code), which is linked to the accession code of the corresponding clone in the rhdb_xrefs_human cross-reference file, which is part of the radiation hybrid database (RHdb; Rodriguez-Tome and Lijnzaad, 1997). This accession code is linked to the AMC tag-to-gene mapping to obtain the corresponding UniGene cluster and thereby the corresponding SAGE tags. The tags from the tag-to-gene mapping are linked to the expression levels in the selected SAGE libraries. If an accession code of an STS marker was not present in the cross-reference file then the UniGene cluster was retrieved instead of the accession code. The UniGene cluster was then used to retrieve the corresponding SAGE tags in the tag-to-gene map and the expression levels in the SAGE libraries.

15.5.1 The HTM Relational Database

E. F. Codd at IBM introduced the relational database in 1970 (Codd, 1970), since then this form of database has developed to fundamentally underpin most modern bioinformatics databases. A relational database is a collection of data items organized as a set of formally-described tables from which data can be accessed or reassembled in many different ways without having to re-organize the database tables (Ullman, 1988). Each table contains one or more data categories in columns. Each row contains a unique instance of data for the categories defined by the columns. It is important to carefully design the database model because a poorly designed database may be slow to query, hard to maintain and extend, and may contain inconsistent and redundant information.

15.5.2 Relational Database Design

Relational databases are a key concept in bioinformatics and so it is useful to take the HTM as an example of database design and construction. The integration of the aforementioned public databases and SAGE libraries into the HTM relational database is shown in an entity–relationship (ER) diagram (Figure 15.4). The ER diagram describes the HTM

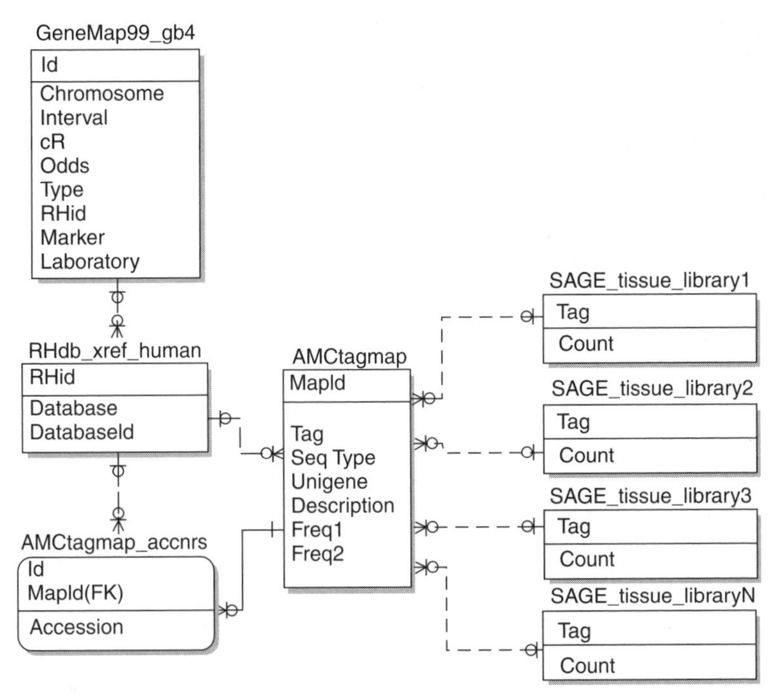

Figure 15.4 Relational model of database used in the HTM. Each table in the database (e.g. AMCtagmap) contains a number of attributes (e.g. Tag). The relationship between the tables are specified as 'zero-or-one to many' or as 'one to many'. For example, each tag in a SAGE library is linked to zero or more electronic tags in the 'AMCtagmap' table. Subsequently, each of these tags is linked via the 'RHdb_xref_human' table to the 'GeneMap99_gb 4' table to establish the mapping.

database tables and relationship between these tables. The relational database model was implemented by using the Postgresql relational database management system (RDBMS) (http://www.postgresql.org). A RDBMS tool allows the developer to:

1. Implement a database with tables, columns and indexes.
2. Define the so-called foreign keys, which specify relationships between rows of various tables.
3. Update the indexes automatically.
4. Interpret an SQL query and combine information from various tables.

Once the tables are implemented it is possible to upload data to the database or extract data from the database by using SQL (Structured Query Language). SQL is the standard user and application program interface to a relational database and is used both for interactive queries for information from a relational database and for gathering data for reports. The reader should be aware that despite first impressions, SQL is a very easy language to learn; 2 days' training can quickly enable a new user to perform complex database queries to integrate diverse forms of data. For example, the next SQL query returns all expression levels of genes mapped on chromosome 1 (compare the statements in this query with the ER diagram in Figure 15.4 to get an idea of what this query is doing):

```
SELECT gm.Chromosome, gm.cR, amc.Unigene, SUM(sage.Count)
FROM GeneMap99_gb4 AS gm JOIN RHdb_xrefs_human  AS rh
ON (gm.RHid = rh.RHid) JOIN AMCtagmap  AS amc
ON (rh.Databaseid = amc.Unigene) JOIN SAGE_tissue_library1
    AS sage
ON (amc.Tag = sage.Tag)
WHERE gm.chromosome = 'chr1'
AND rh.DatabaseName = 'UniGene'
GROUP BY gm.Chromosome, gm.cR, amc.Unigene
ORDER BY gm.cR
```

The relational database forms the core of HTM in which all required data to map expression profiles to chromosomal positions are stored. The SQL queries are part of the user-interface that is built on top of the relational database and which is introduced in the next section.

15.6 THE HUMAN TRANSCRIPTOME MAP

The Human Transcriptome Map (HTM; bioinfo.amc.uva.nl) is a database application that presents gene expression profiles for any chromosomal region in normal and pathological tissues (Caron *et al.*, 2001). The application can be used to search for genes that are over-expressed or silenced in cancer. The HTM provides three different ways to present gene expression profiles obtained with SAGE. The 'extended view' provides the most detailed level of information (Figure 15.5). In this view the expression profiles given for all SAGE tags that could be linked to the radiation hybrid map (RH-map) are shown. Different tags may correspond to a single gene as they may occur as a result of differential splicing or polyadenylation of the gene. In the 'concise view', no individual tags are included but information is presented at the gene (UniGene) level and consequently the tag counts for all tags belonging to the same genes are pooled, i.e. no distinction is

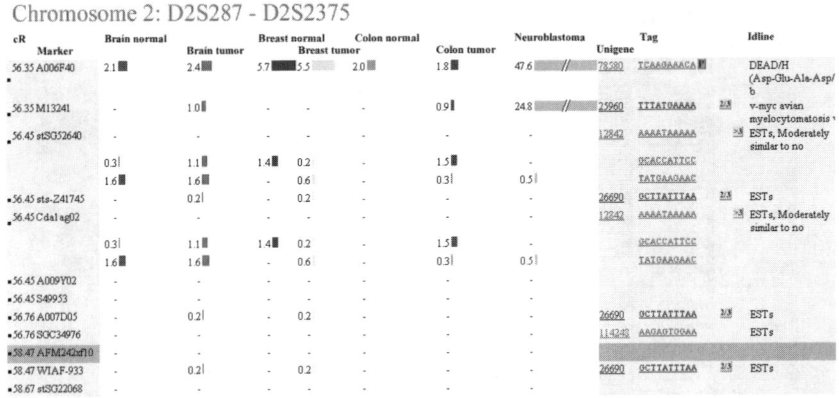

Chromosome 2: D2S287 - D2S2375

cR Marker	Brain normal	Brain tumor	Breast normal	Breast tumor	Colon normal	Colon tumor	Neuroblastoma	Unigene	Tag		Idline
56.35 A006F40	2.1	2.4	5.7	5.5	2.0	1.8	47.6 //	78580	TCAAGAAACA		DEAD/H (Asp-Glu-Ala-Asp/ b
56.35 M13241	-	1.0	-	-	-	0.9	24.8 //	25960	TTTATGAAAA	2/3	v-myc avian myelocytomatosis
56.45 stSG52640	-	-	-	-	-	-	-	12842	AAAATAAAAA	>3	ESTs, Moderately similar to no
	0.3	1.1	1.4	0.2	-	1.5	-		GCACCATTCC		
	1.6	1.6	-	0.6	-	0.3	0.5		TATGAAGAAC		
56.45 sts-Z41745	-	0.2	-	0.2	-	-	-	26690	GCTTATTTAA	2/3	ESTs
56.45 Cdal ag02	-	-	-	-	-	-	-	12842	AAAATAAAAA	>3	ESTs, Moderately similar to no
	0.3	1.1	1.4	0.2	-	1.5	-		GCACCATTCC		
	1.6	1.6	-	0.6	-	0.3	0.5		TATGAAGAAC		
56.45 A009Y02	-	-	-	-	-	-	-				
56.45 S49953	-	-	-	-	-	-	-				
56.76 A007D05	-	0.2	-	0.2	-	-	-	26690	GCTTATTTAA	2/3	ESTs
56.76 SGC34976	-	-	-	-	-	-	-	114242	AAGAGTGGAA		ESTs
58.47 AFM242xf10	-	-	-	-	-	-	-				
58.47 WIAF-933	-	0.2	-	0.2	-	-	-	26690	GCTTATTTAA	2/3	ESTs
58.67 stSG22068	-	-	-	-	-	-	-				

Figure 15.5 Extended view of a chromosome 2p region showing neuroblastoma-specific over-expression of the neighbouring genes N-myc (UniGene Hs.25960) and DDX-1 (UniGene Hs.78580). A small part of the interval D2S287 to D2S2375 is shown. The left-hand columns show the marker and centiRay position as defined on GeneMap99. The right-hand side shows the UniGene number, tag sequence and the description of the UniGene cluster. Expression levels in the libraries are normalized per 100,000 tags and shown by grey bars with a range from 0 to 15. Numbers give the counts per 100,000 tags. The tags are annotated by symbols (explained in the text). (*Reprinted with permission from Caron et al. (2001). Copyright 2001 American Association for the Advancement of Science*).

made between different gene variants. In both the concise and extended view, only a selected region between two framework markers of a chromosome is shown. In the 'whole chromosome view' the expression levels of all genes on a particular chromosome are displayed (Figure 15.6). Also in the whole chromosome view the tag counts for all tags belonging to the same gene are pooled to obtain an overall expression level. In this presentation each unit on the vertical axis represents one gene, i.e. the scale does not denote a genetic or physical distance. The RH-map contains errors (see Chapter 7) and, therefore, some genes map two or more times at slightly different positions. Genes that correspond to multiple markers on the RH-map are shown only on the HTM at the position of the highest LOD score. Only genes for which a tag was included in the AMC tag-to-gene map are displayed.

15.6.1 Annotation of the HTM

In the extended and concise view of the HTM, several annotation symbols are used.

15.6.1.1 Unreliable Tags

Two types of tags were considered unreliable for use in the HTM. They are marked as 'L', '2/3' or '>3' in a yellow box:

1. **Linker tags**. The SAGE technique may produce tags derived from linker oligo's used in library construction (V. E. Velculescu *et al.*, personal communication). These 73 linker tags are marked 'L' in a yellow box on the extended interval view, but their expression levels in the SAGE libraries are not shown.

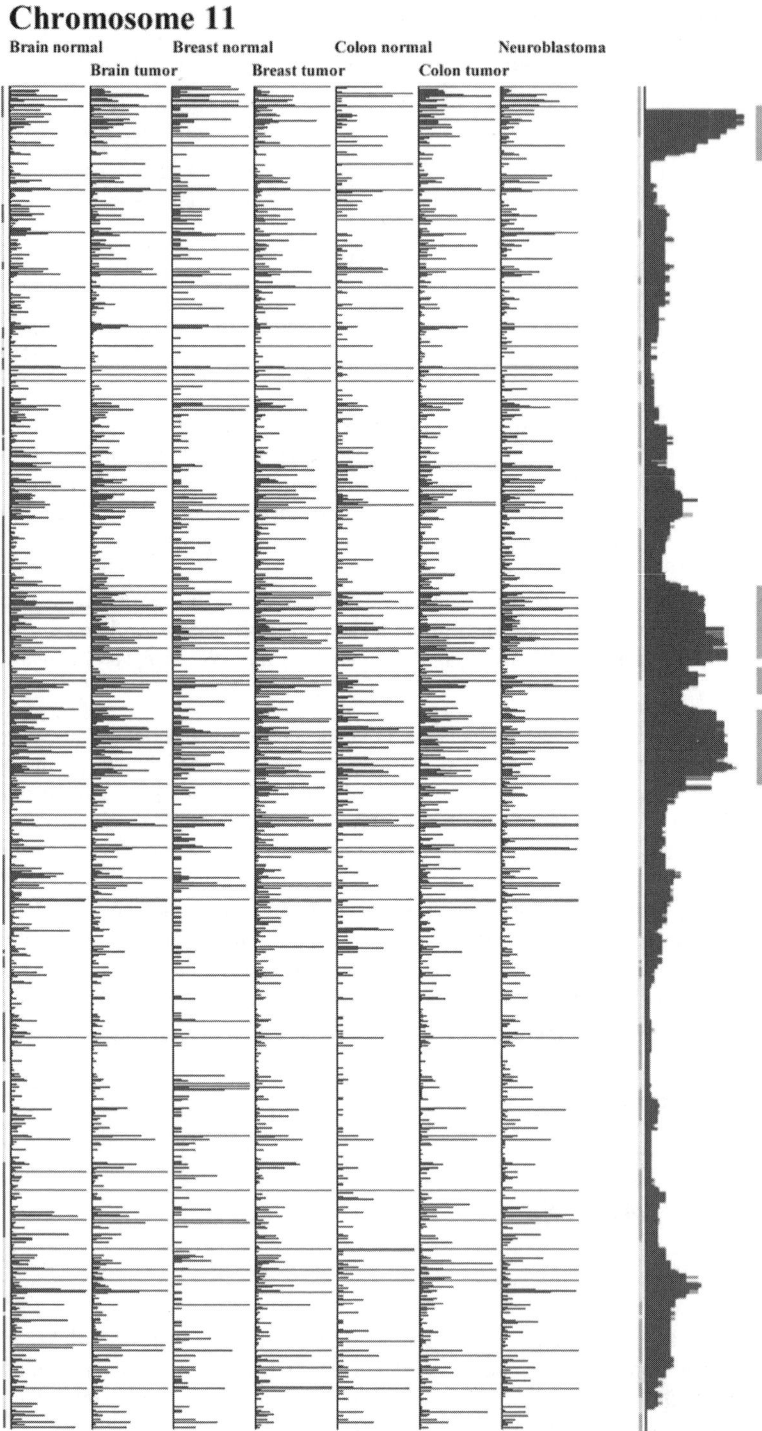
Chromosome 11
Brain normal
Brain tumor
Breast normal
Breast tumor
Colon normal
Colon tumor
Neuroblastoma

2. **Redundant tags**. Some tags are found for more than three UniGene clusters. This may be explained by coincidental limited sequence homologies between genes. Other redundant tags are derived from genes with a CATG close to the polyA tail. This generates tags with a strongly reduced sequence variability, as most of the tag consists of an A stretch. They are marked '>3' in a yellow box in the extended interval view and their expression levels in the SAGE libraries are *not* shown. Tags belonging to two or three UniGene clusters are marked in a yellow box with '2/3' respectively, and their expression levels in the libraries are shown.

15.6.1.2 Antisense Tags

In the extended interval view, tags with an antisense orientation are marked as 'AS' in a purple box. In the concise interval view, the cumulative expression levels for 'sense' and 'antisense' tags are shown as separate bars for each UniGene cluster. Antisense expression levels are not included in the whole chromosome views.

15.6.2 UniGene Clustering Errors

Hybrid UniGene clusters cause many problems, as they include ESTs from different genes. These genes, which usually have different map positions, each yield their own correct reliable tags. To identify the hybrid clusters the GenBank database (Genomes *Homo sapiens* section) was searched for the corresponding PAC sequenced in the Human Genome Project, as well as two adjacent PACs, for the markers mapped on GeneMap99. Tags from the gene corresponding to the marker are expected to be present on these PACs, whereas tags from a 'contaminating' gene in a hybrid cluster are not. The PACs were analysed for the presence of the 10-bp tag sequence plus adjacent CATG. When positive, the tag was marked on the extended interval view with a 'P' in a light green box. A one-nucleotide mismatch between tag and PAC sequence was accepted to cover SNPs or PAC sequencing errors (marked 'P' in a dark green box). When a PAC for a marker was known, but when the tag was not found in the sequence, the tag was marked 'P' in a red box. For all situations the expression level of the tag is shown in all views. This check is not yet available for all markers, but the progress in sequencing and annotation will provide this function for all UniGene clusters.

15.7 REGIONS OF INCREASED GENE EXPRESSION (RIDGES)

The Human Transcriptome Map provides an intriguing insight into the higher-order organization and regulation of expression in the human genome. From the whole chromosome views it is clear that there is a strong clustering of highly expressed genes in specific

---

Figure 15.6 Whole chromosome view of expression levels of the 1208 UniGene clusters mapped to chromosome 11 on the GB4 radiation hybrid map of GeneMap99. Each unit on the vertical axis represents one UniGene cluster. Expression is shown for SAGE libraries of 7 out of the 12 available tissue types. Expression levels in the libraries are normalized per 100.000 tags and tag counts from 0 to 15 are shown by horizontal blue bars while tag frequencies over 15 are shown as red bars (colors not shown in this figure). The section to the right represents a moving median with a window size of 39 UniGene clusters generated from the expression levels in 'all tissues'. The bars above the moving median indicate RIDGEs. (*Reprinted with permission from Caron et al. (2001). Copyright 2001 American Association for the Advancement of Science*).

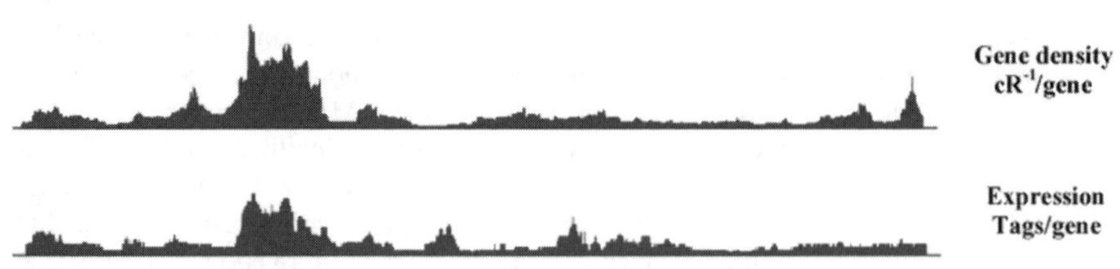

Figure 15.7 Comparison of median gene expression levels and gene density for chromosome 3. The lower diagram shows the expression levels as a moving median with a window size of 39 UniGene clusters. The upper diagram shows gene density. For each UniGene cluster, the average distance between adjacent clusters in a window of 39 adjacent UniGene clusters was calculated. The inverse of this value is shown (inverse centiRays per gene). (*Reprinted with permission from Caron et al. (2001). Copyright 2001 American Association for the Advancement of Science*).

domains, which were named Regions of Increased Gene Expression (RIDGEs) (Caron *et al.*, 2001). This is clearly demonstrated in Figure 15.6, which shows the whole chromosome view of expression levels of 1208 genes mapped to the RH-map of chromosome 11. Expression is shown for SAGE libraries of seven tissue types. To emphasize the RIDGEs more clearly, a moving median with a window size of 39 genes was calculated for 'all tissues', which pools all available SAGE libraries. From the resulting median values, RIDGEs were defined as regions in which at least 10 consecutive genes have a median expression level of at least four times the genomic median. Green bars in the resulting graph indicate the resulting RIDGEs. These RIDGEs were observed on most chromosomes. With the current definition, 27 RIDGEs could be identified (Caron *et al.*, 2001).

Analysis of RIDGEs for physical characteristics suggests that many of them have a high gene density. Figure 15.7 shows the correlation between RIDGEs and gene density (expressed as cR^{-1}/gene) for chromosome 3. This correlation between gene expression and density of mapped genes is found for most RIDGEs. Typical RIDGEs contain six to 30 mapped genes per centiRay, compared to one to two mapped genes per centiRay for weakly transcribed regions.

15.7.1 Statistical Evaluation of RIDGEs

To analyse whether the observed RIDGEs could be explained by the random variation in the distribution of expression levels of the 18,422 UniGene clusters in the HTM, a Monte Carlo simulation was performed. We permutated the genomic order of all 18,422 UniGene clusters in the Human Transcriptome Map and analysed 10,000 permutated datasets for the incidence of RIDGEs. The number of RIDGEs according to our definition was determined for each of the permutations. The observed number of RIDGEs in the Human Transcriptome Map (27) was about 38 standard deviations (0.7) higher than the average number of RIDGEs (0.4) observed in the permutations. The observed number of RIDGEs is therefore unlikely to result from random variation in the distribution of highly expressed genes over the genome.

15.8 DISCUSSION

This chapter has reviewed one possible approach to the analysis of gene expression data in which (statistical) data analysis, database technology, informatics and molecular biology play an important role. The HTM was designed to assist in the identification of genes that are involved in cancer; however it has a wider applicability to the study of any disease. In Chapter 9, approaches for the expression-based prioritization of positional gene candidates in disease loci were reviewed. The HTM could be a valuable tool for prioritizing such candidates. As the number of publicly available SAGE libraries increases their value will also increase (as every SAGE experiment can be directly compared).

Perhaps the most interesting aspect of the SAGE-based, Human Transcriptome Map is that it is somewhat different from other approaches and therefore it is complementary to microarray data. Integration of (public) databases using HTM, uncovered a previously unknown genomic phenomenon — regions of increased gene expression (RIDGEs). RIDGEs may provide more fundamental insight into the higher-order organization of the human genome. The biology of RIDGEs is not yet understood but they may play an important role in gene transcription and therefore, may be relevant to the study of carcinogenesis or any other disease which involves disregulation of gene expression.

RIDGEs would not have been revealed if DNA microarray data had been used. Since the overall expression profile for all chromosomes is similar in all tissues, the measurement of the expression of one tissue relative to a control tissue would reveal only genes that are differentially expressed between these tissues. Furthermore, as explained in Section 15.2, the expression levels of genes on one DNA microarray cannot be compared and therefore, these domains would not have this clear structure. However, DNA microarray data can be used to further understand the nature of RIDGEs. It can be envisioned that specific tumour samples have disturbed expression of entire transcriptional domains due to translocations. DNA microarrays are very suitable for measuring gene expression profiles for large numbers of (tumour) samples; integration of this data with the HTM would directly reveal whether gene expression in specific domains is turned on or off. Such experiments may further increase our knowledge about the organization of the genome with respect to gene expression.

The current HTM is not the end of gene expression analysis but can be regarded as the starting point of much more research that aims at understanding the biology of RIDGEs. This research includes the construction of a sequence-based HTM that is much more precise than the current map that is based on radiation hybrid data. Such a sequence-based map would allow a more precise definition of RIDGEs. Furthermore, this will allow the investigation of the correlation between RIDGEs and other domains such as gene density. To understand why many genes in RIDGEs are highly expressed in comparison to other regions one could search for regulatory sequences that are common for genes in such domains. Moreover, and maybe more interesting, is the hunt for regulatory sequences that turn complete domains of genes on and off. To enhance the search for regulatory sequences a comparison between the Human Transcriptome Map and a Mouse Transcriptome Map would be very valuable since conserved sequences can be identified (see Chapter 12 for an overview of some of the tools which may be suitable for such an analysis). For all this research much more bioinformatics and laboratory work is required. However, this will ultimately lead to a further understanding of the molecular biology of cancer and human disease.

REFERENCES

Alizadeh AA, Eisen MB, Davis RE, Ma C, Lossos IS, Rosenwald A, *et al.* (2000). Distinct types of diffuse large B-cell lymphoma identified by gene expression profiling. *Nature* **403**: 503–511.

Altman DG. (1991). *Practical Statistics for Medical Research*. Chapman-Hall: London.

Audic S, Claverie JM. (1997). The significance of digital gene expression profiles. *Genome Res* **7**: 986–995.

Ben-Dor A, Bruhn L, Friedman N, Nachman I, Schummer M, Yakhini Z. (2000). Tissue classification with gene expression profiles. *J Comput Biol* **7**: 559–583.

Boguski MS, Lowe TM, Tolstoshev CM. (1993). dbEST — database for 'expressed sequence tags'. *Nature Genet* **4**: 332–333.

Bowtell DD. (1999). Options available — from start to finish — for obtaining expression data by microarray. *Nature Genet* **21**: 25–32.

Brodeur GM, Sekhon G, Goldstein MN. (1977). Chromosomal aberrations in human neuroblastoma. *Cancer* **40**: 2256–2263.

Brown PO, Botstein, D. (1999). Exploring the new world of the genome with DNA microarrays. *Nature Genet* **21**: 33–37.

Caron H, van Schaik B, van der Mee M, Baas F, Riggins G, van Sluis P, *et al.* (2001). The human transcriptome map: clustering of highly expressed genes in chromosomal domains. *Science* **291**: 1289–1292.

Chen H, Centola M, Altschul SF, Metzger H. (1998). *J Exp Med* **188**: 1657–1668.

Claverie JM. (1999). Characterization of gene expression in resting and activated mast cells. *Hum Mol Genet* **8**: 1821–1832.

Codd EF. (1970). *Commun ACM* **13**: 377–387.

Cole KA, Krizman DB, Emmert-Buck MR. (1999). The genetics of cancer — a 3D model. *Nature Genet* **21**: 38–41.

Colinge J, Feger G. (2001). Detecting the impact of sequencing errors on SAGE data. *Bioinformatics* **17**: 840–842.

Conover WJ. (1980). *Practical Nonparametric Statistics*, John Wiley: New York.

Deloukas P, Schuler GD, Gyapay G, Beasley EM, Soderlund C, Rodriguez-Tome P, *et al.* (1998). A physical map of 30,000 human genes. *Science* **282**: 744–746.

Ewing B, Green P. (1998). Base-calling of automated sequencer traces using Phred. II. Error probabilities. *Genome Res* **8**: 186–194.

Ewing B, Hillier L, Wendl MC, Green P. (1998). Base-calling of automated sequencer traces using Phred. I. Accuracy assessment. *Genome Res* **8**: 175–185.

Golub TR, Slonim DK, Tamayo P, Huard C, Gaasenbeek M, Mesirov JP, *et al.* (1999). Molecular classification of cancer: class discovery and class prediction by gene expression monitoring. *Science* **286**: 531–537.

Hastie T, Tibshirani R, Eisen MB, Alizadeh A, Levy R, Staudt L, *et al.* (2000). 'Gene shaving' as a method for identifying distinct sets of genes with similar expression patterns. *Genome Biol* **1**: research 0003.1–0003.21.

Heid CA, Stevens J, Livak KJ, Williams PM. (1996). Real time quantitative PCR. *Genome Methods* **6**: 986–994.

Kal AJ, van Zonneveld AJ, Benes V, van den Berg M, Koerkamp MG, Albermann K, *et al.* (1999). Dynamics of gene expression revealed by comparison of serial analysis of gene expression transcript profiles from yeast grown on two different carbon sources. *Mol Biol Cell* **10**: 1859–1872.

van Kampen AH, van Schaik BD, Pauws E, Michiels EM, Ruijter JM, Caron HN, *et al.* (2000). USAGE: a web-based approach towards the analysis of SAGE data. Serial Analysis of Gene Expression. *Bioinformatics* **16**: 899–905.

Kanehisa M, Goto S. (2000). KEGG: Kyoto encyclopedia of genes and genomes. *Nucleic Acids Res* **28**: 27–30.

Karp PD. (1998). What we do not know about sequence analysis and sequence databases. *Bioinformatics* **14**: 753–754.

Karp PD, Paley S, Zhu J. (2001). Database verification studies of SWISS-PROT and GenBank. *Bioinformatics* **17**: 526–532; discussion 533–534.

King RJB. (2000). *Cancer Biology*. Pearson Education: Harlow, UK.

Lal A, Lash AE, Altschul SF, Velculescu V, Zhang L, McLendon RE, *et al.* (1999). A public database for gene expression in human cancers. *Cancer Res* **59**: 5403–5407.

Lander ES, Linton LM, Birren B, Nusbaum C, Zody MC, Baldwin J, *et al.* (2001). Initial sequencing and analysis of the human genome. *Nature* **409**: 860–921.

Lash AE, Tolstoshev CM, Wagner L, Schuler GD, Strausberg RL, Riggins GJ, *et al.* (2000). SAGEmap: a public gene expression resource. *Genome Res* **10**: 1051–1060.

van Limpt V, Chan A, Caron H, Sluis PV, Boon K, Hermus MC, *et al.* (2000). SAGE analysis of neuroblastoma reveals a high expression of the human homologue of the Drosophila Delta gene. *Med Pediatr Oncol* **35**: 554–558.

Lockhart DJ, Dong H, Byrne MC, Follettie MT, Gallo MV, Chee MS, *et al.* (1996). Expression monitoring by hybridization to high-density oligonucleotide arrays. *Nature Biotechnol* **14**: 1675–1680.

Madden SL, Galella EA, Zhu J, Bertelsen AH, Beaudry GA. (1997). SAGE transcript profiles for p53-dependent growth regulation. *Oncogene* **15**: 1079–1085.

Man MZ, Wang X, Wang Y. (2000). POWER_SAGE: comparing statistical tests for SAGE experiments. *Bioinformatics* **16**: 953–959.

Margulies EH, Innis JW. (2000). eSAGE: managing and analysing data generated with serial analysis of gene expression (SAGE). *Bioinformatics* **16**: 650–651.

Michiels EM, Oussoren E, Van Groenigen M, Pauws E, Bossuyt PM, Voute PA, *et al.* (1999). Genes differentially expressed in medulloblastoma and fetal brain. *Physiol Genomics* **1**: 83–91.

Mitelman F, Johansson B, Mertens F. (2001). http://cgap.nci.nih.gov/Chromosomes/Mitelman.

Okamoto H, Yonemori F, Wakitani K, Minowa T, Maeda K, Shinkai H. (2000). A cholesteryl ester transfer protein inhibitor attenuates atherosclerosis in rabbits. *Nature* **406**: 203–207.

Pandey A. (2001). Common standards for genomics and proteomics. *Trends Genet* **17**: 696.

Porter DA, Krop IE, Nasser S, Sgroi D, Kaelin CM, Marks JR, *et al.* (2001). A SAGE (serial analysis of gene expression) view of breast tumor progression. *Cancer Res* **61**: 5697–5702.

Proudfoot N. (1991). Poly(A) signals. *Cell* **64**: 671–674.

Pruitt KD, Maglott DR. (2001). RefSeq and LocusLink: NCBI gene-centered resources. *Nucleic Acids Res* **29**: 137–140.

Riedy MC, Timm EA Jr, Stewart CC. (1995). Quantitative RT-PCR for measuring gene expression. *Biotechniques* **18**: 70–74, 76.

Riggins GJ, Strausberg RL. (2001). Genome and genetic resources from the Cancer Genome Anatomy Project. *Hum Mol Genet* **10**: 663–667.

Rodriguez-Tome P, Lijnzaad P. (1997). The Radiation Hybrid Database. *Nucleic Acids Res* **25**: 81–84.

Salamov AA, Solovyev VV. (1997). Recognition of 3′-processing sites of human mRNA precursors. *Comput Appl Biosci* **13**: 23–28.

Schaefer C, Grouse L, Buetow K, Strausberg RL. (2001). A new cancer genome anatomy project web resource for the community. *Cancer J* **7**: 52–60.

Schena M, Shalon D, Davis RW, Brown PO. (1995). Quantitative monitoring of gene expression patterns with a complementary DNA microarray. *Science* **270**: 467.

Schena M, Shalon D, Heller R, Chai A, Brown PO, Davis RW. (1996). Parallel human genome analysis: microarray-based expression monitoring of 1000 genes. *Proc Natl Acad Sci USA* **93**: 10614–10619.

Schwab M, Alitalo K, Klempnauer KH, Varmus HE, Bishop JM, Gilbert F, *et al.* (1983). Amplified DNA with limited homology to myc cellular oncogene is shared by human neuroblastoma cell lines and a neuroblastoma tumour. *Nature* **305**: 245–248.

Scott HS, Chrast R. (2001). Global transcript expression profiling by Serial Analysis of Gene Expression (SAGE). *Genet Eng* **23**: 201–219.

Sheets MD, Ogg SC, Wickens MP. (1990). Point mutations in AAUAAA and the poly(A) addition site: effects on the accuracy and efficiency of cleavage and polyadenylation *in vitro*. *Nucleic Acids Res* **18**: 5799–5805.

Spieker N, van Sluis P, Beitsma M, Boon K, van Schaik BD, van Kampen AH, *et al.* (2001). The MEIS1 oncogene is highly expressed in neuroblastoma and amplified in cell line IMR32. *Genomics* **71**: 214–221.

Strausberg RL, Buetow KH, Emmert-Buck MR, Klausner RD. (2000). The cancer genome anatomy project: building an annotated gene index. *Trends Genet* **16**: 103–106.

Strausberg RL, Dahl CA, Klausner RD. (1997). New opportunities for uncovering the molecular basis of cancer. *Nature Genet* **15** (Special Issue): 415–416.

Strausberg RL, Feingold EA, Klausner RD, Collins FS. (1999). The mammalian gene collection. *Science* **286**: 455–457.

Ullman JD. (1988). *Principles of Database and Knowledge-Base Systems*. Computer Science Press: New York.

Velculescu VE, Vogelstein B, Kinzler KW. (2000). Analysing uncharted transcriptomes with SAGE. *Trends Genet* **16**: 423–425.

Velculescu VE, Zhang L, Vogelstein B, Kinzler KW. (1995). Serial analysis of gene expression. *Science* **270**: 484–487.

Venter JC, Adams MD, Myers EW, Li PW, Mural RJ, Sutton GG, *et al.* (2001). The sequence of the human genome. *Science* **291**: 1304–1351.

Yeang CH, Ramaswamy S, Tamayo P, Mukherjee S, Rifkin RM, Angelo M, *et al.* (2001). Molecular classification of multiple tumor types. *Bioinformatics*, **17** (Suppl. 1): S316–S322.

Zammatteo N, Jeanmart L, Hamels S, Courtois S, Louette P, Hevesi L, *et al.* (2000). Comparison between different strategies of covalent attachment of DNA to glass surfaces to build DNA microarrays. *Anal Biochem* **280**: 143–150.

Zhang L, Zhou W, Velculescu VE, Kern SE, Hruban RH, Hamilton SR, *et al.* (1997). Gene expression profiles in normal and cancer cells. *Science* **276**: 1268–1272.

CHAPTER 16

Proteomic Informatics

JÉRÔME WOJCIK and ALEXANDRE HAMBURGER

Hybrigenics
Paris, France

Bioinformatics for Geneticists. Edited by M.R. Barnes and I.C. Gray
 ISBNs: 0 470 84393 4; 0 470 84394 2 (PB)

16.1 INTRODUCTION

16.1.1 A Definition of Proteomics

As genomics is the study of the set of genes in genomes, proteomics deals with the analysis of the 'proteome', that is the product of translation of the transcriptome.

The completion of the sequencing of bacterial and higher eukaryotic organisms marks the beginning of the post-genomic era. As more and more raw data become available, new challenges arise, namely handling these data and making sense out of them. Proteomics is a way of giving relevant meaning to these data by redefining them in a higher-level, function-oriented context, closer to what we may broadly call 'biological function'.

16.1.2 Challenge Compared to Genomics: Identification of 'Function'

The term 'proteomics' yields a new conception of the functional assignment issue in biology. 'Prote-' indicates that function is sustained by proteins, not by genes, and '-omics' proposes that function is defined 'in context'. The function of a protein is not solely an individual property of the protein but is defined as a combination of its biochemical interactions with its partners and the environment in which it exists. Information on the scale of the whole cell is therefore needed to comprehensively understand the function of proteins.

Protein sequence information is often an endpoint for the geneticist, for example, an amino acid substitution may be defined by a SNP. But as a matter of fact, this is just one element of many that can tell us about the properties of a protein. Other meaningful information can tell us a great deal more about the nature of proteins, such as 3D structure, post-translational modifications, half-life, phenotypic role, enzymatic activity or quantity (abundance). These properties have also been proven to be tissue- and subcellular localization-specific. Beyond the properties of the protein itself, protein interactions are a rather novel data form that have been shown to be amenable to high-throughput analysis (which will be discussed shortly). These methods are powerful tools to define proteins and pathways in context on the cellular scale. Ultimately this is the objective of genetics and hence proteomics is a critical step in the progression from candidate gene to validated disease gene.

With the completion of many genome sequences, including human, the aforementioned issue of finding a relevant context to study biological data in is even more acutely felt. Many of the recent advances in proteomics have been made during the analysis of prokaryotic organisms. In this field more than any other, prokaryotes may point the way forward for analysis methods in higher eukaryotes, such as man, for these methods rely heavily on fully optimized and complete datasets, an ideal that we still struggle to achieve in studies of human material. We can safely assume that sequence data is not a sufficient and rich enough source of information to reach higher levels of understanding or meaningful definition of protein function. Indeed, a raw DNA sequence may be altered by several phenomena, making any assumption on function difficult. To name a few: alternative splicing may lead a single gene (or pre-mRNA) to produce many gene products (or mature mRNA) in eukaryotes. Further down the protein synthesis pathway, post-translational modifications may result in proteic cleavages, glycosylation, etc. The regulation of proteins is by itself an issue: post-transcriptional regulation of protein expression (changes in protein synthesis and degradation rates) induces no obvious correlation between protein and mRNA expression levels in humans (Anderson and Seilhamer, 1997) or in yeast (Gygi *et al.*, 1999); time and space regulations may sometimes be partially uncovered by sequence analysis (proteic translocation between subcellular compartments may be linked to the presence of peptide signals which are cleaved when the protein reaches a mature state) but the subcellular localization *per se*, turnover, dynamic behaviour or lifetime of a protein cannot be directly linked to sequence analysis alone.

16.2 PROTEOMIC INFORMATICS

From the term 'Proteomic Informatics', we have already given an overview of what 'proteomics' may be. As for 'Informatics', Luscombe (2001) defines Bioinformatics as 'conceptualizing biology in terms of molecules (in the sense of physical-chemistry) and then applying informatics techniques (derived from disciplines such as applied mathematics, computer science and statistics) to understand and organize the information associated with these molecules, on a large-scale'. As high-throughput methods for biological data generation have been developed, we need powerful automated tools for analysing and understanding them. This is the goal of proteomic informatics. Data may be seen as a dense, fuzzy cloud of points in a complex, multidimensional space. It is the role of Bioinformatics to find a relevant subspace and project our data in a meaningful and understandable way that will enable us to reap the rewards of our data while not losing valuable information. At first glance, Proteomic Informatics may be seen only as a tool for data handling and visualization but its purpose is actually two-fold. On one hand, data may be displayed in a comprehensive way through the efficient use of bioinformatics tools and stored in rich databases that keep track of experimental settings. On the other hand, algorithms may be developed and improved to extract new information. As would befit bioinformatics tools aimed at proteomics applications, they should be able to process high quantities of data and conceptualize them as integral parts of a cellular context; hence the need to develop algorithms allowing reconstruction or inference of cellular pathways and protein–protein interaction maps.

16.3 EXPERIMENTAL WORKFLOW: CLASSICAL PROTEOMICS

The most frequently used high-throughput technology designed to study the proteome is aimed at identifying and quantifying the expression levels of proteins localized in specific protein complexes. This method is sometimes referred to as 'Classical Proteomics', compared to 'Functional Proteomics' which concerns itself with the identification of interactions and cell processes. A typical approach consists in the separation of the various proteins of a cellular extract by gel electrophoresis followed by mass spectrometric analysis: comparison of the resulting experimental data with that available from sequence databases provides unique assignments for protein gel spots to their corresponding DNA sequences. Recent optimizations of the various steps provide one of the most powerful approaches in proteomics. The following section details the experimental workflow.

16.3.1 Proteome Purification

Sample preparation is the first and a crucial step in classical proteomics. The purer the sample, the more accurate the expression quantification and protein identification will be. Proteins can be extracted from whole cells (bacteria, yeasts...), tissues, or subcellular compartments (organelles). Purification methods include mainly centrifugation in density gradients, exclusion chromatography, affinity chromatography using for example peptide tags, antibodies (immuno-precipitation) or substrates (for reviews see Legrain *et al.* (2000) or Lee (2001)). A tandem affinity purification (TAP) involving a combination of two high-affinity tags linked to the protein of interest was also suggested as a general method for protein complex purification in mild conditions after expression in natural conditions (Rigaut *et al.*, 1999) and was recently comprehensively applied to the yeast proteome (Gavin *et al.*, 2002).

16.3.2 Proteome Separation: Electrophoresis

In the next step, the protein expression profile of the sample is typically deduced by 2D gel SDS-polyacrylamide gel electrophoresis (SDS-PAGE), a high-resolution technique for decomposing protein complexes of tenths of polypeptides (see Lee (2001) for review). Proteins are separated according to both isoelectric point (p*I*) and molecular weight (Mw), by a combination of isoelectric focusing and electrophoresis respectively. Spots are detected using colour stains, fluorescent dyes or radioactive labels (Figure 16.1).

Proteins can also be separated by classical 1D-PAGE but this requires reduction of the number of proteins in the cell extract, for instance by immuno-affinity purification (Ho *et al.*, 2002) or TAP (Gavin *et al.*, 2002).

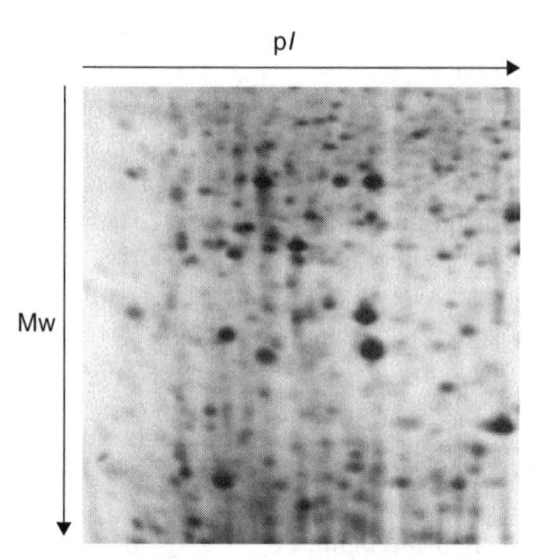

Figure 16.1 An example of 2D-PAGE. Proteins are identified by black spots after separation by electrical focusing (p*I*) and electrophoresis (Mw).

As SDS-PAGE becomes the most commonly used bidimensional protein separation method in proteomics, the technique is becoming standardized among different laboratories and databases of 2D gel images highlighting protein spots with appropriate links have been created for various proteomes (see Table 16.1).

16.3.3 Proteome Identification: Mass Spectrometry

Third, the separated protein spots on 2D gels are excised and digested in-gel with a protease (usually trypsin). The eluted peptides are then analysed by Mass Spectrometry (MS). Reaching a high level of sensitivity, automation and throughput for protein analysis, mass spectrometry has become one of the key technologies in the proteomics field.

Analysing femtomoles of protein materials is now routinely carried out using MALDI (Matrix-Assisted Laser Desorption/Ionization)/TOF (Time-Of-Flight)-based peptide mass fingerprinting, which provides a list of masses for the peptides contained in the digested 2D spot. Matching these against the list of calculated peptide masses from an appropriate protein sequence database characterizes the isolated protein (see for example, Houry *et al.*, 1999).When the mass fingerprint is not found in databases, Tandem Mass Spectrometry (or MS/MS) can be used to sequence the polypeptides, thus providing sequence tags that could allow protein identification by sequence similarity screening of classical bioinformatics databases (for example EMBL by using BLAST (Altschul *et al.*, 1997)). The combination of peptide mass fingerprinting followed by sequence tagging is a suite

TABLE 16.1 Main Online 2D-PAGE Proteomics Resources

Database	URL
Aarhus 2DPAGE database	biobase.dk/cgi-bin/celis
Aberdeen 2DPAGE	www.abdn.ac.uk/~mmb023/2dhome.htm
Argone protein mapping group	www.anl.gov/BIO/PMG/
Cyano2Dbase	www.kazusa.or.jp/cyano/cyano2D/
ES cell-2DPAGE	www.dur.ac.uk/~dbl0nh1/2DPAGE/
Harefield HSC 2DPAGE	www.harefield.nthames.nhs.uk/nhli/protein/
Maize Genome database	moulon.moulon.inra.fr
Maritime pine 2DPAGE	www.pierroton.inra.fr/genetics/2D/
Max-Planck Institut 2DPAGE	www.mpiib-berlin.mpg.de/2D-PAGE/
MDC Heart-2DPAGE	www.mdc-berlin.de/~emu/heart/
Parasite Host Cell Interaction 2DPAGE	www.gram.au.dk
Plant Plasma Membrane Database	sphinx.rug.ac.be:8080/ppmdb/index.html
SWISS-2DPAGE	www.expasy.ch/ch2d
SIENA-2DPAGE	www.bio-mol.unisi.it/2d/2d.html
SSI-2DPAGE	www.ssi.dk/en/forskning/tbimmun/tbhjemme.htm
TMIG 2DPAGE	proteome.tmig.or.jp/2D/
Université Paris 13 2DPAGE	www-smbh.univ-paris13.fr/lbtp/Biochemistry/ biochimie/bque.htm
2DWGDB (WebGel)	www-lmmb.ncifcrf.gov/2dwgDB
WU Inner Ear database	oto.wustl.edu/thc/innerear2d.htm
Yeast 2DPAGE	yeast-2dpage.gmm.gu.se/
Yeast Protein Map (YPM)	www.ibgc.u-bordeaux2.fr/YPM/

of powerful techniques used to analyse and identify proteins (Quadroni and James, 1999; Yates, 1998).

One step further, MS coupled with High Performance Liquid Chromatography (HPLC) techniques and/or combined with biochemical techniques (immunoprecipitation) can provide shotgun identification of proteins in complex biological mixtures in order to study protein–protein interaction, to locate and identify single protein or protein complexes from a subcellular fraction. For instance, using a combination of HPLC and ESI (Electrospray Ionization)-MS, it has been shown that a large transmembrane protein (the lactose permease) could be analysed and studied quickly and with high accuracy (Whitelegge *et al.*, 1999). High-throughput methods have also been designed to identify various post-translational modifications of proteins by mass spectrometry (Wilkins *et al.*, 1999).

16.3.4 Building Protein Expression 'Networks'

Proteome-wide characterization allows the production of global maps of differentially expressed proteins. By comparing several sets of expression patterns under different conditions (for instance, wild-type versus mutant or normal versus diseased) or at different time stages, one can deduce clusters of co-regulated proteins that could be interpreted as a protein expression 'network'. Such differential protein expression networks have been applied for instance to the elucidation of cell pathways, the characterization of cell types or the identification of pathogenic agents (for review see Legrain *et al.*, 2000). They are complementary to gene regulation networks produced by transcriptomics techniques (see Chapter 15).

Mass spectrometry also allows the identification of protein complexes, which could be conceptualized as clusters of the expression network. The technique was recently applied to detect yeast complexes on a proteome-wide scale (Gavin *et al.*, 2002; Ho *et al.*, 2002).

16.3.5 Analysing Protein Expression Data

Approaches to 2D gel image analysis may range from very basic to fairly complex. Several commercial 2D gel image analysis software packages are available that allow display, analysis and comparison of gel images, as well as determination, quantification and normalization of spots (Table 16.2). One can also use Flicker (Lemkin and Thornwall, 1999), a free web tool for comparing images from different internet sources. Given two gel images URL, Flicker loads the images and displays them in the web browser. They can be enhanced in various ways (spatial warping, pseudo 3-dimensional image sharpening...), while regions of interest can be 'landmarked' with several corresponding points in each gel image. One gel image is then warped to the geometry of the other and the two resulting images are compared visually in a third window (the 'flicker' window): as the two gels are rapidly alternated ('flickered'), the user can slide one gel past the other to visually align

TABLE 16.2 Some Gel Analysis Software

Software	Company	Reference
Melanie	Geneva Bioinformatics	www.expasy.ch/melanie
PDQuest	Bio-Rad	www.proteomeworks.bio-rad.com
Phoretix	Phoretix advanced	www.phoretix.com
Flicker		www.hi-beam.net

corresponding spots by matching local morphology. With such image analysis tools, an expert can locally visualize an expression network and formulate biological hypotheses. The next step is the automated numerization and database storage of protein expression patterns to allow high-throughput screening.

16.4 PROTEIN INTERACTION NETWORKS

If protein expression networks give information about co-regulation of proteins and their response to specific conditions, they are not completely informative about the biochemical function of gene products. Determining which other cell components interact with proteins addresses this issue. The function of a protein can be defined by the role it takes in cell pathways and the interactions in which it participates with other cell components (DNA, RNA, proteins, metabolites or lipids for instance). We distinguish here the interaction networks dealing only with proteins and produced by high-throughput experimental protocols from those containing heterogeneous factors (referred to as 'cell pathways'). The set of technologies used to produce interaction data on a large scale is referred to as 'Functional Proteomics'.

16.4.1 Experimental Technologies

Low-throughput technologies (co-immunoprecipitations, far-Western blots, 'pull-downs', etc, see Phizicky and Fields (1995) for review) are commonly used for studies on individual proteins. The study of interactions at the proteome level, however, requires high-throughput assays.

16.4.2 Yeast Two-Hybrid (Y2H)

The yeast two-hybrid system (Fields and Song, 1989) can detect interactions between two known proteins or polypeptides and can also search for unknown partners (prey) of a given protein (bait) (for review, see Vidal and Legrain, 1999). Yeast two-hybrid assay remains the main large-scale technology that is available to build protein interaction maps. Two strategies — namely the matrix approach and the fragment (or polypeptide) library screening approach — have been tested to find the most efficient way to explore proteomes for interactions (the interactome).

The matrix approach uses a collection of predefined open reading frames (ORFs), usually full-length proteins, as both bait and prey for interaction assays. Combinations of bait and prey can be assessed individually or after pooling cells expressing different bait or prey proteins. The intrinsic limitation of this strategy is that it tests only known proteins that are predefined. Y2H was first used to explore interactions among drosophila proteins involved in the control of cell cycle (Finley and Brent, 1994). Several studies have now been published for the yeast proteome, either comprehensive (Ito *et al.*, 2000, 2001, Uetz *et al.*, 2000) or using only a subset of specific baits (Newman *et al.*, 2000).

The alternative Y2H assay strategy uses exhaustive libraries to screen for the identification of new protein interacting partners. Applying this library screening approach to functionally related proteins results in connection of uncharacterized proteins to specific pathways. It can be also applied to whole cellular interactomes. Screening numerous randomly generated fragments contained in the libraries also permits the determination of interacting domains defined experimentally as the common sequence shared by the selected overlapping prey fragments (Rain *et al.*, 2001). This approach was first applied to determine protein networks for the T7 phage proteome which contains 55 proteins (Bartel *et al.*, 1996) and later applied to the yeast proteome focused on the RNA metabolism (Fromont-Racine *et al.*, 1997) and to the human gastric pathogen *Helicobacter pylori* (Rain *et al.*, 2001).

The two two-hybrid strategies are depicted in Figure 16.2. The pros and cons of each technology are discussed in a review (Legrain *et al.*, 2001). Table 16.3 draws an inventory of major two-hybrid large-scale assays performed so far.

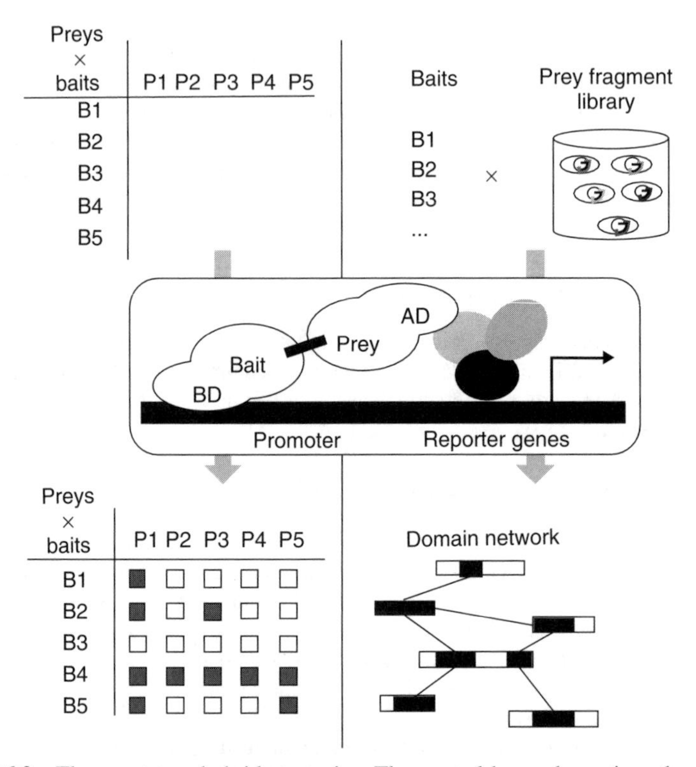

Figure 16.2 The yeast two-hybrid strategies. The central box schematizes the principle of the yeast two-hybrid assay: a protein domain that binds specifically to DNA sequences (BD) is fused to a polypeptide dubbed the 'bait' and a domain that recruits the transcription machinery (AD) is fused to a polypeptide dubbed the 'prey'. The basis of the assay is that transcription of a reporter gene will occur only if the bait and the prey polypeptides interact together. The matrix approach (first column) uses the same collection of proteins used as bait (B1–B5) and prey (P1–P5). The results can be drawn in a matrix where bait autoactivators (B4 for example) and 'sticky' prey proteins (P1 for example interacts with many proteins) are identified and discarded. The final result can be summarized as a list of interactions that can be heterodimers (B2–P3) or homodimers (B5–P5). The library screening approach identifies for each interacting prey protein the domain of interaction with a given bait. Sticky prey proteins are identified as fragments of proteins that are often selected regardless of the bait protein. An autoactivator bait can be used in the screening process with more stringent selective conditions.

TABLE 16.3 Key Figures in Large Scale Datasets for Protein–Protein Interaction Maps

Organism	Technology	Number of assays baits × preys	No. of interactions	Reference
Vaccinia virus	Protein array	Proteome × proteome	37	McCraith *et al.* (2000)
S. cerevisiae	Protein array	192 × proteome	281	Uetz *et al.* (2000)
	Pools of preys	Proteome × proteome	692	
S. cerevisiae	Pools of baits and preys	430 assays of pools (96 × 96)	175	Ito *et al.* (2000)
S. cerevisiae	Pools of baits and preys	3844 assays of pools (96 × 96)	841*	Ito *et al.* (2001)
S. cerevisiae	Protein array	162 × 162	213	Newman *et al.* (2000)
C. elegans	Protein array	29 × 29	8	Walhout *et al.* (2000)
	Library screening	27 × proteome	124	
HCV	Protein array	10 × proteome	0	Flajolet *et al.* (2000)
	Library screening	22 fragments × proteome	5	
S. cerevisiae	Library screening	15 × proteome	170	Fromont-Racine *et al.* (1997)
S. cerevisiae	Library screening	11 × proteome	113	Fromont-Racine *et al.* (2000)
H. pylori	Library screening	261 × proteome	1524	Rain *et al.* (2001)

* This number corresponds to highly significant interactions (more than three hits, see Ito *et al.*, 2001).

16.4.3 Other Technologies

Phage display technology is another assay used to screen a library of polypeptides for interaction with a target protein. Each polypeptide is expressed on the surface of a bacteriophage particle, as a fusion with a phage coat protein. This provides a physical link between the expressed polypeptide and its encoding gene. The phage-displayed polypeptide can be selected by binding to a target using affinity chromatography and further characterized by amplification and sequencing of the corresponding gene located within the phage particle. No protein–protein interaction map using phage display has been published so far either for an organism or an entire cell but the technology has a high-throughput potential (see for example Walter *et al.*, 2001). The technology is particularly suited for screening libraries of random polypeptide variants, such as antibody fragments and can be combined to the complementary yeast two-hybrid technology in order to obtain more relevant results (Tong *et al.*, 2001).

Protein microarrays are also emerging in order to study protein–protein interactions. Known proteins are precisely spotted on glass substrates and used to probe interactions with peptides (Lueking *et al.*, 1999) or proteins (Haab *et al.*, 2001). A similar method was also tested to screen for small molecules (MacBeath and Schreiber, 2000).

16.5 BUILDING PROTEIN INTERACTION NETWORKS

16.5.1 From Experimental Results to Graphs

When the two protein partners are identified, a graph can be built where the vertices are the proteins (bait or prey) and the edges are the protein interactions. This step is trivial when the two partners are known beforehand, for example in the two-hybrid matrix approach, but requires post-processing when a partner is screened against a library and has selected a target/prey. In the latter case, the prey gene must be sequenced and identified in sequence databases using tools such as BLAST (Altschul *et al.*, 1997). When several experimental protocols are combined, for instance phage display and yeast two-hybrid (Tong *et al.*, 2001), one can decide whether to consider the totality of the interactions or only those common to both techniques, depending on the desired trade-off between false negatives and false positives (see below).

Moreover, in the two-hybrid strategy using fragment libraries, the functionally interacting domains can be precisely mapped on proteins: the common sequence shared by the selected overlapping prey fragments experimentally defines the smallest docking site selected by the bait (Rain *et al.*, 2001). The interaction network can then also be represented as a graph where the vertices are protein domains instead of full-length proteins.

16.6 FALSE NEGATIVES AND FALSE POSITIVES

One major drawback of the high-throughput experimental technologies described above is the generation of potential false negatives and false positives, depending on the assay conditions.

False-negative interactions are biological interactions that are missed because of incorrect folding, inadequate subcellular localization, lack of specific post-translational modifications etc. In yeast two-hybrid assays, the matrix approach is prone to generate a high level of false negatives (see Table 16.3), because only two assays are performed for each pair of proteins (bait versus prey, and reciprocally), whereas the fragment library approach allows testing of millions of potential interactions simultaneously. For instance, the two exhaustive studies of the yeast proteome (Ito *et al.*, 2001; Uetz *et al.*, 2000) have failed to recapitulate as much as 90% of interactions previously described in the literature (Ito *et al.*, 2001). The intrinsic limitations of the matrix approach concerning the choice of selective conditions can also explain this high rate of false negatives (for review see Legrain *et al.*, 2001).

Conversely, searching for many potential interactions, especially when screening a random fragment library, increases the chance of selecting biologically non-significant interacting polypeptides, thus leading to false positives. First, some bait proteins might have a predisposition to activate the transcription of reporter genes without specific interaction with any prey protein. These *auto-activator* bait proteins may randomly select

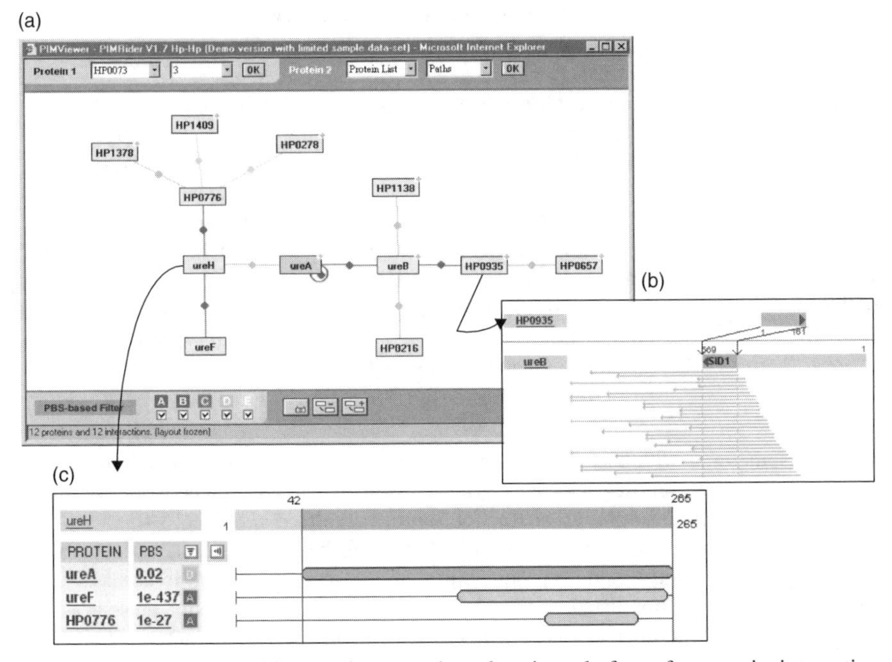

Figure 16.3 The PIMRider: an integrated exploration platform for protein interaction networks. The main window (a) displays the cell-wide protein interaction map as a graph and allows the biologist to navigate through the network, filter information depending on its reliability (PBS®) and focus on a particular pathway. Clicking on a specific interaction gives access to primary two-hybrid data (b) where interacting fragments and the computed Selected Interacting Domain (SID®) are positioned relative to the coding sequence of the two proteins. The biologist can make up his/her own mind about the interaction significance. All the interacting domains of one protein with its partners to formulate biological hypotheses, for instance about dominant negative interactors (c), can also be displayed.

prey proteins in addition to specific ones. Second, some chimeric prey proteins, dubbed *sticky* proteins, may similarly be non-specifically selected by many independent bait proteins. Discarding autoactivator bait proteins (that select many prey proteins) or sticky prey proteins (that are selected in many screens) leads to results with a reduced rate of false positives, although it may also mean a slightly increased number of false negatives (Ito *et al.*, 2000, 2001). Less stringent filtering was used for the interaction network of *H. pylori*, based on a fragment library approach (Rain *et al.*, 2001). A scoring scheme was designed that computes an E-value for each bait–prey interaction by comparing the observed pattern of selected prey fragments with the theoretical pattern that would be obtained by randomly picking fragments in the library. At the end, each interaction yields a reliability value (see Figure 16.3).

16.7 ANALYSING INTERACTION NETWORKS

The first protein interaction databases available on the internet provided a basic display of the alphabetical protein interaction list. An interaction is represented by its two protein

TABLE 16.4 Main Protein–Protein Interaction Databases

Database	URL	Reference
EcoCyc	ecocyc.org/ecocyc/ecocyc.html	Karp *et al.* (2000)
BIND	www.bind.ca	Bader and Hogue (2000)
Cellzome	yeast.cellzome.com	Gavin *et al.* (2002)
CuraGen portal	portal.curagen.com	Uetz *et al.* (2000)
DIP	dip.doe-mbi.ucla.edu	Xenarios *et al.* (2000)
FlyNets	gifts.univ-mrs.fr/FlyNets/	Sanchez *et al.* (1999)
Interact	bioinf.man.ac.uk/interactso.htm	Eilbeck *et al.* (1999)
MIPS	www.mips.biochem.mpg.de	Mewes *et al.* (2000)
PIM Rider	pim.hybrigenics.fr	Rain *et al.* (2001)
ProNet	pronet.doubletwist.com	

partners, sometimes with basic annotations or cross-references to other protein databases. Some websites also propose packages to graphically display interaction networks (Mrowka, 2001). The main protein–protein interaction sources are listed in Table 16.4.

However, a simple list of interactions poorly tackles the issue of result reproducibility. To evaluate false positives and reproducibility, access to primary data is necessary. For example, the interactions listed at the MIPS (Mewes *et al.*, 2000) only present a brief indication of the experimental source, such as 'two-hybrid' or 'co-immunoprecipitation', without any quality clue or reference to the source experiment or laboratory. Bioinformatics tools are now emerging to tackle this issue, such as the PIM Rider® (Rain *et al.*, 2001) which gives access to primary data (see Figure 16.3b).

Visualization software is in parallel being enriched with options to help the biologist in his/her discovery process. They let the user search for interaction paths between two given proteins, filter displayed interactions depending on their reliability value or simultaneously display all interacting domains identified in one specific protein (see Table 16.4 for examples, such as PIM Rider from Hybrigenics (Rain *et al.*, 2001), PIScout from LION Biosciences, or the visualisation tool of DIP (Xenarios *et al.*, 2000)).

16.8 CELL PATHWAYS

Cell pathways extend protein interaction networks by integrating interactions with lipids, small molecules (e.g. metabolites), RNA, DNA etc. They are mainly deduced from a compilation of literature resources, contrary to protein interaction networks that are technology-driven results.

16.8.1 Metabolic Pathways

The metabolism of living systems and their evolution have been investigated for a long time. The fluxes of metabolites inside a cell and the cascades of enzymatic reactions leading from one compound to another have been depicted in charts, that is, heterogeneous interaction networks mixing small molecules (metabolites) and proteins (enzymes). For example, Figure 16.4 illustrates the pyruvate metabolic pathway: the circles represent the small molecules that are the vertices of the metabolic network, whereas edges are catalytic reactions and are labelled with boxed enzymes. Several databases regroup information about these cell networks, especially for prokaryotic organisms (Kanehisa and Goto, 2000;

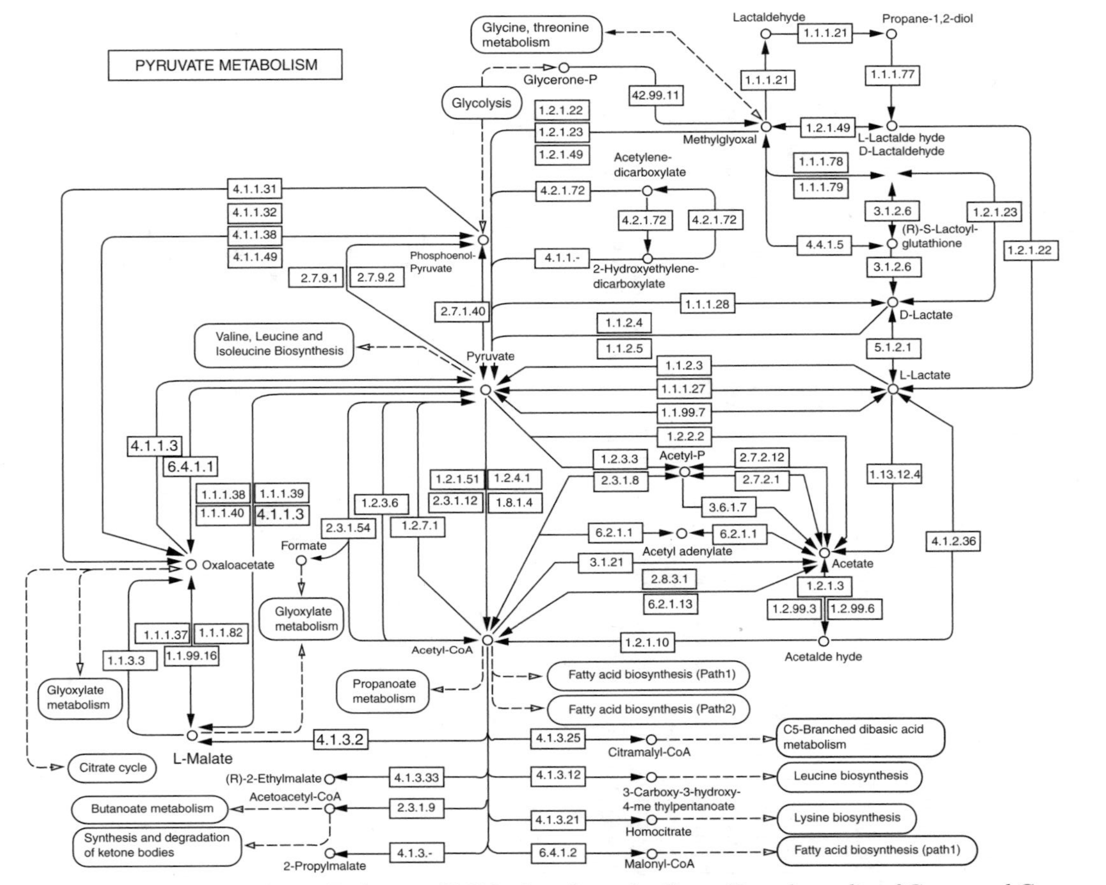

Figure 16.4 The pyruvate metabolism pathway. Reference 00620 taken from the *Kyoto Encyclopaedia of Genes and Genomes* (Kanehisa and Goto, 2000). Small circles represent molecule compounds and square boxes represent enzyme proteins, referenced by their EC numbers.

Karp *et al.*, 2000; Selkov *et al.*, 1998). Enzymes are referenced by their EC (Enzyme Commission) number, a system which overlays a functional hierarchy on enzymes (see Bairoch (2000) for a review).

16.8.2 Signal Transduction Networks

The signal transduction pathways are particular instances of internal cell pathways. They describe the cascades of molecular interactions from the reception of an extracellular signal (e.g. binding of a cytokine to its receptor) to the activation of transcription factors triggering the transcription of specific genes. The signal transduction networks are generally described in terms of physical interactions between proteins (e.g. binding or phosphorylation, etc; see Figure 16.5).

16.8.3 Gene Regulation Networks

Downstream of the signal transduction pathways a complex array of gene regulation networks takes place. The transcriptional regulatory networks mix heterogeneous physical interactions (protein–protein, protein–DNA, and protein–RNA) and genetic interactions (activation, inhibition, etc). Gene regulation networks are however still more studied at a higher level of abstraction (see Chapters 13 and 15).

Signal transduction and regulatory pathways have been constructed from individual experiments and stored in dedicated databases such as, SPAD http://www.grt.kyushu-u.ac.jp/spad/, TRANSFAC (Heinemeyer *et al.*, 1999), or MIPS (Mewes *et al.*, 2000).

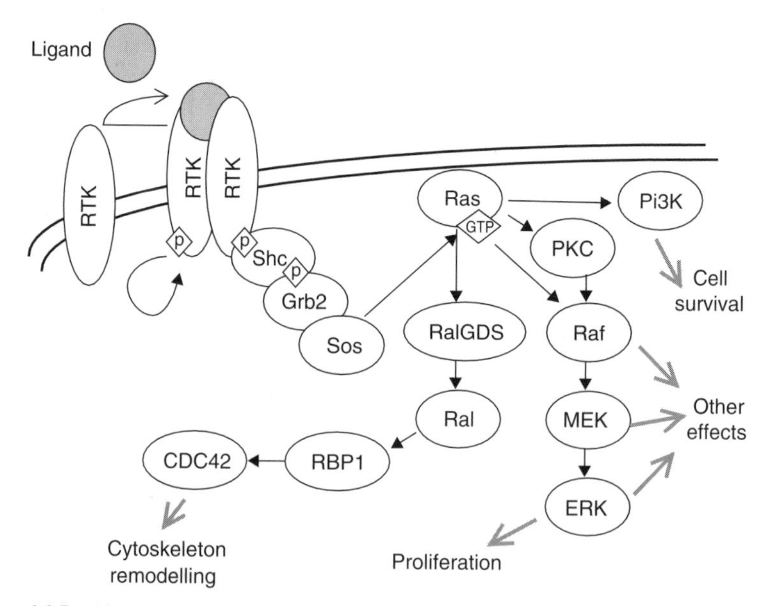

Figure 16.5 Signal transduction networks of TK receptors. The binding of a ligand to its tyrosine kinase receptor (RTK) provokes the dimerization of the receptor and the initialization of several intra-molecular signalling cascades, involving physical interactions and activation (black arrows: phosphorylation (P), GTP-binding (GTP), and others). One signal pathway triggers several biological effects (grey arrows).

These databases have allowed researchers to computationally predict regulatory networks, for example, Pilpel *et al.* (2001), computationally predicted an extensive transcriptional regulatory network in yeast by combinatorial analysis of promoter elements.

16.9 PREDICTION OF PROTEIN NETWORKS

16.9.1 Prediction of Functional Networks by Comparative Genomics

With the completion of many genome sequences, new techniques are emerging to predict the function of gene products by analysing the genes on a genome scale and comparing genomes between organisms. This new set of methods dubbed 'comparative genomics' has allowed the prediction of *functional* links between many proteins (see Eisenberg *et al.* (2000) for review).

Comparing genomes means comparing sequences of genes and establishing similarity links between genes means identifying orthologues, i.e. genes sharing the same function across organisms. In the following prediction method, the identification of orthology is often reduced to the detection of a significant sequence similarity, that is below a fixed E-value threshold, in a sequence similarity search (such as BLAST). Implications of this statement on prediction accuracy will be discussed below.

16.9.2 Gene Fusion Events

The gene fusion event method was first introduced by Marcotte *et al.* (1999a) and extended thereafter by other works (Enright *et al.*, 1999; Marcotte *et al.*, 1999b). The method is based on evolutionary interaction hypotheses. Basically, if two genes A and B participate in the same function, they are likely to be fused together during evolution to enhance the effective concentration of the fused gene product. Few mutations can then appear between the proteic domains from A and B. If genes A and B are once again separated, their products could still physically interact (Figure 16.6). Thus, if two separate genes in a given organism are fused together in another organism, they are likely to be functionally linked, that is to participate in the same structural complex, in the same biological pathway, in the same biological process or sometimes to physically interact (see examples in Figure 16.7). However, one cannot distinguish between these four kinds of functional links without extra information. The gene fusion event method is often referred to as the *Rosetta-stone* method (Marcotte *et al.*, 1999a) in reference to the Rosetta stone which allowed Champollion to make sense of hieroglyphs ('word fusion') by comparing them to Greek and Demotic (languages using 'unitary' words).

The gene fusion event method was applied to the prediction of the protein functional network of *Escherichia coli* by comparing its genome to a set of 22 genomes of archaeal, bacterial and eukaryotic species (Tsoka and Ouzounis, 2000). In terms of participation in fusion events, a three-fold preference was evidenced for metabolic enzymes compared with control sets. It is worth mentioning that 76% of the detected pairs of enzymes participating in fusion events are known to be subunits of an enzymatic complex in the EcoCyc database (Karp *et al.*, 2000; Table 16.4). The fusion event method thus seems to be able to detect physical interactions for metabolic enzymes.

16.9.3 Gene Neighbourhood

It was postulated for a long time that the way genes are organized in clusters in bacterial chromosomes is probably the result of an evolutionary constraint. The completion

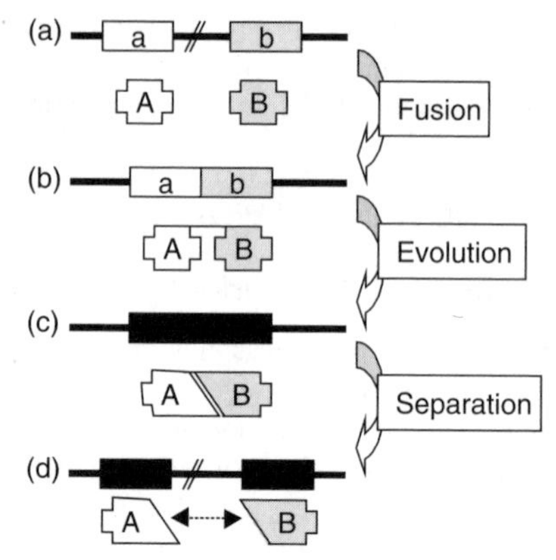

Figure 16.6 Underlying hypotheses of the gene fusion event method. This figure represents a model for the evolution of protein–protein interactions. If two genes a and b, originally separated in the genome (a), are fused together during evolution (b), the resulting chimeric protein A-B could mutate to develop intra-molecular contacts between A and B domains (c). Then, if the two initial genes are once again separated in genomes, the corresponding gene products A and B could still physically interact, or at least be functionally linked (d).

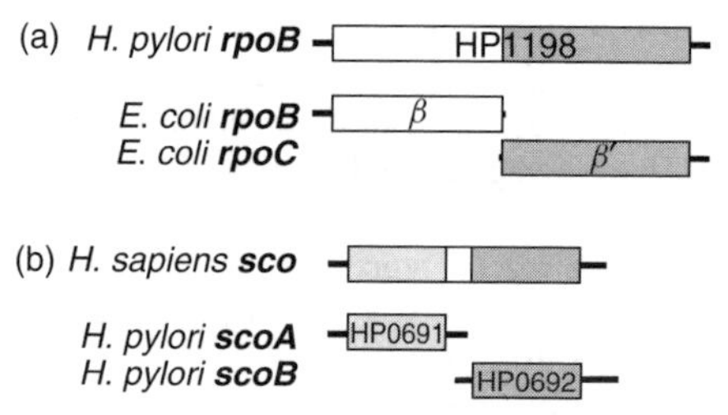

Figure 16.7 Examples of gene fusion events. (a) The β and β' subunits of the DNA-dependent RNA polymerase are encoded by two separate genes in most eubacteria and archea, but are fused together in a single gene in *Helicobacter pylori* (HP1198). These two subunits are known to be part of the RNA polymerase holoenzyme complex. (b) Similarly, the α and β subunits of the succinyl-CoA transferase in *H. pylori* (HP0691 and HP0692, respectively) are fused together in human and the corresponding gene products are predicted to physically interact in two-hybrid screens (Rain *et al.*, 2001).

of many genome sequences now allows testing of this hypothesis at a comprehensive level. Dandekar and co-workers first analysed three triplets of sequenced genomes to identify conserved gene pairs (Dandekar *et al.*, 1998). About 100 genes were found to be conserved as pairs, among them 75% of the encoded protein pairs physically interact. This suggests that conservation of gene order and physical interaction of encoded proteins are evolutionarily correlated.

Overbeek *et al.* (1999) extended this kind of analysis by building synteny groups, i.e. gene clusters across organisms, in order to infer functional links. They defined a gene cluster as a set of genes located on the same strand, and in which the maximal intergenic distance is 300 base pairs. If two genes X_A and Y_A in a given cluster of genome A have orthologues X_B and Y_B in a cluster of genome B, they are defined as functionally coupled. A coupling score is also derived depending on the number of organisms in which orthologous pairs are found and the phylogenetic distances between these organisms and A.

The use of this gene neighbourhood method is obviously more efficient for microbial genomes with their conserved gene organization. But it may also be extended for eukaryotes where operon-like cluster structures have been observed (Wu and Maniatis, 1999).

16.9.4 Phylogenetic Profiles

A phylogenetic profile is defined as the occurrence pattern of orthologues for a given gene in a set of reference genomes (Pellegrini *et al.*, 1999). It describes the absence or presence of a particular gene across this set of genomes (Figure 16.8). If two proteins have the same phylogenetic profile across these genomes (for instance P1 and P2, as well as P4 and P6 in Figure 16.8), it is assumed that they are functionally linked because they have probably co-evolved.

The major underlying hypothesis of the method is that orthologues, that is proteins having exactly the same function, are correctly identified. Moreover, all the reference genomes must be completely sequenced to avoid false-negative information. Note also that paradoxically if the identification of orthology heavily relies on sequence similarity, the phylogenetic profile method is referred to as a sequence-independent clustering algorithm, since proteins that are functionally linked in this way, i.e. that have the same phylogenetic profile, do not share sequence similarity in general.

16.9.5 Combination of Several Methods

Each of the previously described methods predicts functional links between proteins according to evolutionary and sequence-based hypotheses. Combining these approaches theoretically minimizes the false-positive prediction rate. Eisenberg and colleagues combined five types of protein–protein interaction links to build a functional linkage network for yeast, three of them are predictions from bioinformatics algorithms, two others are derived from experimental data (Marcotte *et al.*, 1999b):

Protein	*E. coli*	*H. pylori*	*S. aureus*	*S. cerevisiae*
P1	1	1	1	0
P2	1	1	1	0
P3	1	1	1	1
P4	1	0	1	1
P5	0	1	1	1
P6	1	0	1	1

Figure 16.8 Clustering by phylogenetic profiles. The presence or absence of six proteins labelled P1 to P6 is indicated by 1 or 0, respectively, in four genomes. Proteins with the same profiles are boxed.

- Links from the Rosetta-stone method
- Links from the phylogenetic profile method
- Links between yeast proteins that have *Escherichia coli* homologues linked in metabolic pathways, as defined in the EcoCyc database (Karp *et al.*, 2000)
- Links from known physical interactions in the DIP database (Xenarios *et al.*, 2000)
- Links between proteins whose mRNA levels are correlated in cell cycle microarray experiments (Spellman *et al.*, 1998)

The combination of these five networks represents over 93,000 pair-wise links between yeast proteins (about 30 links per protein), indicating a potentially high proportion of false positives. However, taking into account only 'highest confidence links', defined as links found by any two out of the three prediction methods or deduced from one of the two experimental techniques, reduces the number of links to 4130 (about 5%).

16.9.6 Inferences Across Organisms

Once a protein network is built for a given organism (by experimental or predictive methods) one might wonder how to transport it to other organisms. The classical inference mechanism involves two major steps:

1. A correspondence is established between proteomes, classically by identifying orthologues between organisms by sequence comparison.
2. The interaction links in the source protein network are transported to the target proteome along this correspondence.

The accuracy of these inference processes is highly dependent on the criteria chosen for orthology (i.e. conservation of function). Caveats of inferences will be further discussed in Section 16.10.

16.9.7 Protein Interaction Inferences

The inference process can be applied to all types of protein networks. It was recently tested on protein interaction methods (Wojcik and Schächter, 2001). An inference method similar to the one described above (correspondence according to sequence similarity on full-length sequences), referred to as the 'naive' method was assayed together with another method, dubbed the 'Interacting Domain Profile Pair' (IDPP) method, that combines sequence similarity searches with clustering based on interaction patterns and interaction domain information.

The principle of the IDPP method is illustrated by the prediction of a protein interaction network for *E. coli* from an experimental protein interaction map for *H. pylori* (Rain *et al.*, 2001) in Figure 16.9. From the 1524 interactions in the original *H. pylori* network, the IDPP method led to 881 interaction predictions, connecting 412 proteins of *E. coli* (9.6%). Compared to the naive method, the IDPP method yields 35 additional, highly domain-specific, predicted interactions. The use of sequence similarity searches restricted to interacting domains rather than full-length proteins increases the sensitivity of the method. Similarly, the use of interacting domain clusters instead of single interacting domain sequences allowed the detection of homologies at lower levels of sequence similarity (see Figure 16.10 for an example). Six-hundred and fifty-one interactions were predicted by the naive method but not by the IDPP method. Two hundred and fifty-two

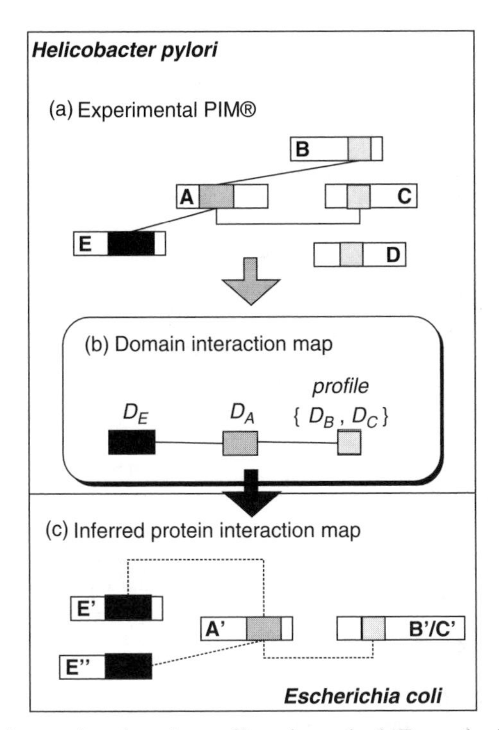

Figure 16.9 The interacting domain profile pair method. From the initial protein interaction map of *H. pylori* (a), an abstract domain cluster interaction map is derived (b). Domains are clustered together if (i) they share a significant sequence similarity and (ii) they share a common interaction property with a third partner (e.g. interacting domains of proteins B and C both interact with A). Each domain or profile of domains is then used as a probe to screen a library of *E. coli* protein sequences and domain cluster interactions are transferred (c).

of these 651 interactions were demonstrated to be false positives using the naive method since the prediction is achieved through sequence similarity of a region that does not contain the interacting domain. The 399 remaining interactions were obtained through sequence similarity that was significant when considering the whole protein but not when considering the shorter interacting domain and thus, might be considered as potential false positives.

16.10 ASSESSMENT AND VALIDATION OF PREDICTIONS

The methods described above predict protein networks. Each prediction method is based on a specific biological hypothesis and yields a set of given parameters both of which must be validated. The *validation* of bioinformatics predictions means the comparison of predicted results with the *state of the art* of biology. We distinguish here automated validation methods, that are systematic, reproducible, comparable and easy to perform but often yield weak biological confirmation, and manual validation methods, that are much more biologically informative but also more biased and laborious. We do not discuss here the validation of prediction methods *per se* but only the validation of predicted results.

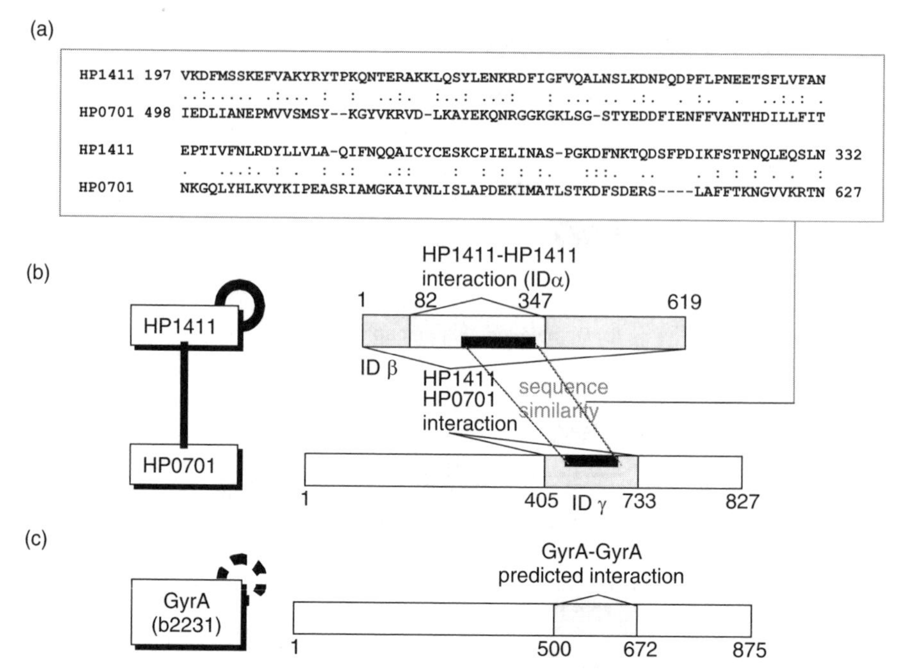

Figure 16.10 Prediction of GyrA homodimerization in *E. coli* by the IDPP method. In the *H. pylori* reference protein interaction map, the β interacting domain (ID) of HP1411 interacts with ID γ of HP0701 and HP1411 interacts with itself through ID α (b). When the IDPP method is applied, ID α and ID γ are clustered together since they both interact with the same region of HP1411 (b) and they share a sequence similarity (region 197–332 of HP1411 and region 498–627 of HP0701, 103 amino acid overlap, 32% of identity, (a)). This leads to the creation of a 'homodimer' profile pair connecting the α/γ domain profile with itself. When used as a probe to screen an *E. coli* protein sequence library, the α/γ domain profile selected a 172-amino acid-long domain on the GyrA protein, and GyrA was predicted to interact with itself through this domain (c). This prediction is confirmed by the literature: GyrA is known to form an A2–B2 complex with GyrB.

16.10.1 Automated Validations

The most widely used validation method is the 'keyword retrieval' technique. The principle is simple: if two proteins are linked together in the protein network, one compares their keywords according to a specific biological annotation and if they share similar keywords, the weight of the link is reinforced. The percentage of shared keywords at the network level is compared to a theoretical background noise to evaluate the global validity of the prediction. For instance, the keywords can be SWISS-PROT annotation keywords or functional categories (Jenssen *et al.*, 2001; Marcotte *et al.*, 1999b; Wojcik and Schächter, 2001). However, this validation method relies heavily on database annotations that are always reductive and sometimes false. For example, Marcotte *et al.* (1999a) noted that 'even truly related proteins show only a partial SWISS-PROT keyword overlap'. In this case they observed only a 35% overlap. Thus, this method, while significantly better than random noise, probably gives a poor biological validation. Cross-validating protein interaction predictions by comparing annotations of both partners is also very dependent on the existence, the format and the quality of these annotations.

The second idea is to consider that a prediction, here a functional link between proteins, made by several independent methods is more reliable (in fact the random background noise of co-occurring independent facts is lower). This was used for instance to define high-confidence links in protein networks (Marcotte *et al.*, 1999b) or to assess interaction predictions against physical location of genes in prokaryotic genomes (Wojcik and Schächter, 2001). In that case, there is a caveat to assess the real independence of prediction methods since the majority of them are sequence based. Basically, the more independent the prediction methods, the more relevant (in terms of false positives) the overlapping results will be.

Finally, predicted protein–protein links can be evaluated by checking their existence in dedicated databases, such as MIPS (Mewes *et al.*, 2000), DIP (Xenarios *et al.*, 2000) or OMIM (Hamosh *et al.*, 2000). For instance these databases can be used to validate networks predicted from literature mining (Jenssen *et al.*, 2001). These manually curated databases however regroup heterogeneous information and one must be cautious about data source quality. Predictions can also be compared to other types of data, such as gene clusters deduced from microarray data (Jenssen *et al.*, 2001). In both cases, the significance of the predictions is evaluated by calculating the fold improvement over a virtual random experiment and/or the correlation between the two datasets.

16.10.2 Manual Validations

Using manual validation, each predicted interaction link between two proteins of a network is assessed by manually comparing the annotations in public databases, by checking literature references of each protein partner. This method is obviously low-throughput and by essence biased, but can lead to interesting conclusions about protein network quality.

It was for the first time applied to the assessment of inferred protein interactions from *H. pylori* to *E. coli* (J. Wojcik *et al.*, unpublished data). The inference process is based on clustering and a definition of orthology restricted to the interacting protein domains (Wojcik and Schächter, 2001). The true positive prediction rate was evaluated to be at least 12%, i.e. at least 12% of the 1280 predicted interactions make biological sense according to biological curators. Three main causes were identified to explain predictions that are not confirmed by the literature: (i) predictions are true positives but are not yet referenced in the literature; (ii) one of the protein functions in the source interaction was completely lost during evolution (the corresponding gene has only paralogues in *E. coli*); or (iii) the source interaction is a false-positive result. The comparison of these exact but not statistically significant results with those obtained by automated validation by keyword retrieval (Wojcik and Schächter, 2001) emphasizes the need to have real and exhaustive reference datasets in order to validate predictions.

16.10.3 Literature Mining

The literature mining method, sometimes called 'Information Retrieval', can be viewed both as an assessment method to predict protein networks and as a prediction method *per se*. Assuming that the major part of current biology knowledge is contained in scientific literature, the parsing of titles, headings, abstracts and/or full texts of articles should enable us to extract links between genes or proteins and then build networks. Several techniques exist to perform this parsing, including linguistic methods that tag parts of words (e.g.

Ono *et al.*, 2001) or statistical methods that estimate discriminating word distributions (e.g. Marcotte *et al.*, 2001). One major issue in these studies is the establishment of an unambiguous nomenclature for gene or gene product names. Gene name dictionaries can be created from various nomenclature databases such as HUGO, LocusLink or OMIM, but problems remain due to insufficient synonym definition, synonym variations and gene families with fuzzy naming conventions.

A recent work aimed to analyse over 10 million MEDLINE records to detect and count human gene symbols or names co-occurring in titles or abstract. This resulted in a protein interaction network containing about 140,000 interactions connecting 7512 human genes (Jenssen *et al.*, 2001). This is the largest protein network predicted from literature mining so far. For now mining literature is more profitably used to help the scientist by screening abstracts and reducing the number of articles to read. This is used to enrich the Database of Interacting Proteins (DIP) (Marcotte *et al.*, 2001).

16.11 EXPLOITING PROTEIN NETWORKS

Once a protein network is experimentally built or predicted or inferred by bioinformatics algorithms, it represents a valuable source of information to understand molecular mechanisms on the scale of a whole cell, either by assigning function to gene products in context (local analysis of the protein map) or by analysing the global network shape and suggesting biological hypotheses.

16.11.1 Functional Assignments: the 'Guilt-By-Association' Rule

The first attempts to assign function used 'guilt-by-association' methods to annotate proteins on the basis of the annotations of their interacting partners or, more generally, of the proteins sharing a common property in a given cluster (Mayer and Hieter, 2000).

For example, a set of yeast protein interactions described in the literature or revealed by large-scale two-hybrid screens was analysed through a clustering method (Schwikowski *et al.*, 2000) based on cellular role and subcellular localization annotations from the Yeast Proteome Database (Costanzo *et al.*, 2000). The function of an uncharacterized protein is assigned on the basis of the known functions of its interacting partners. A function was assigned to 29 proteins (out of 554) that have two or more interacting proteins with at least one common function.

However, 'guilt-by-association' functional assignments must be used with caution. First the predictions are highly dependent on the database function annotations which are often reductive (only one keyword) and sometimes false. Poorly defined annotations can gather different concepts and induce biologically non-significant clustering. The assignments also obviously depend on the quality of the source protein network. If there are too few connections or, on the contrary, if there are too many false-positive connections to a protein node, the guilt-by-association would lead to erroneous conclusions. This point is especially crucial with two-hybrid interaction data, for which false positives represent highly connected nodes in the network.

Last, but not least, a major hurdle in this kind of automated function annotation method, common to all bioinformatics prediction algorithms, is the absence of an independent reference dataset and validation methods. For instance, the 29-function assignments made in the former study were compared with the corresponding high confidence links obtained in the study of Marcotte *et al.* (1999b) which were themselves partially predicted from interactions listed at MIPS, one of the yeast protein interaction databases used in the original study (Schwikowski *et al.*, 2000). This exemplifies the fact that predictions must be used with caution: the oversight of the initial study hypothesis and the deficiency in independent data sources could lead to biased conclusions.

Bioinformatics clustering of protein interactions still represents a powerful annotation tool which will become more and more useful as the interaction data accumulate and their quality improves. However, in order to be used successfully for appropriate functional annotation, the data needs to be stored in elaborate structures that allow each individual scientist to test their own hypothesis against complex heterogeneous primary data and then to design further experiments to validate the functional assignment.

16.12 DEDUCING PREDICTION RULES FROM NETWORKS

Given a protein interaction network and assuming that it is complete enough and has a low rate of false positives, one can deduce from the list of protein–protein interactions some biological information at a molecular level. We give here two examples of statements deduced from the analysis of comprehensive interaction maps that could *a posteriori* be used to predict protein interactions.

16.12.1 Domain–Domain Interactions

Two independent groups have analysed the available protein–protein interaction network of *Saccharomyces cerevisiae* in terms of domain–domain interactions (Ito *et al.*, 2000, 2001; Mewes *et al.*, 2000; Uetz *et al.*, 2000; Xenarios *et al.*, 2000). The first group (Park *et al.*, 2001) considered protein structural domains from the SCOP classification (scop.mrc-lmb.cam.ac.uk/scop/; Murzin *et al.*, 1995) and the second group (Sprinzak and Margalit, 2001) studied motifs from the InterPro database (www.ebi.ac.uk/interpro/; Apweiler *et al.*, 2001). The basic idea is to count the co-occurrence of pairs of domains in interacting proteins, and to compare it to a theoretical background, in order to use over-represented domain pairs as predictors.

16.12.2 Correlated Mutations

As stated previously, the interactions in which a protein participates define its function. The specificity of these interactions is essential for the protein function to some extent. Thus, if the protein evolves and some point mutations occur at the interaction interface, 'complementary' mutations should also occur on protein partners to guarantee the interaction specificity. This hypothesis was developed by Pazos *et al.* (1997) who showed that correlated mutations in interacting domain pairs occur favourably close to the structural protein–protein interface. They proposed the use of this information to help to discriminate between several docking propositions when the 3D structure of both protein partners is known.

16.12.3 Analysis of the Shape of Protein Networks

Rather than focusing on a specific protein node in a protein network, one can analyse the whole interaction map to deduce biological hypotheses on the cellular scale. Jeong and co-workers published such an analysis of the public yeast protein interaction map (Jeong *et al.*, 2001). They showed that this network forms a scale-free network: the probability that a given protein interacts with k partners follows a power law. This kind of structure is particularly tolerant to random attacks on one hand, and fragile against attacks targeted on the most connected nodes on the other hand (Albert *et al.*, 2000). Similar non-homogeneous network structures were also evidenced for metabolic networks (Jeong *et al.*, 2000) and another protein interaction map in bacteria (Rain *et al.*, 2001). The authors established a positive correlation between connectivity and lethality: highly connected proteins are three times more likely to be essential, i.e. the yeast cell dies if the corresponding gene is deleted. This correlation has been attributed to evolutionary selection.

Although the existence of such a correlation makes biological sense, one should probably wonder about the relative weight of technological bias in establishing it. Jeong's work indeed rests mainly on interaction data produced by one systematic two-hybrid system in yeast. The technology is prone to induce false negatives and false positives, as illustrated and commented on in a more recent similar study (Ito *et al.*, 2001). The corresponding protein interaction network which contains 1870 proteins (31% of the whole yeast proteome), is not complete. Its shape would probably be different if all 'real' interactions were known. Proteins that exhibit few interacting partners in this network could actually represent highly connected nodes. Conversely, false positives in the two-hybrid system are likely to result in highly-connected nodes of the network: so-called 'sticky prey' proteins bind 'by chance' to many independent bait proteins. The correlation between lethality and centrality in networks evidenced by Jeong and co-workers, could actually be much stronger if genes that are both non-essential and highly connected on the one hand and genes that are both essential and poorly connected on the other hand, proved to be the consequences of a technological bias in data.

16.12.4 Precautions for Protein Networks

To conclude, both local 'guilt-by-association' functional assignment rules and global network analysis methods are fragile against poor interaction data quality or incompleteness. They can thus hardly produce reliable 'local' conclusions and the fact that the conclusion appears biologically meaningful is not evidence of the validity of the demonstration *per se*.

It is, moreover, imperative to assess the technological data bias prior to analysing networks and formulating biological conclusions. Ideally, the false-negative rate should be minimized by building comprehensive networks and false positives should be filtered out by independent bioinformatics or experimental validations. Meanwhile, both should be assessed using technology-specific reliability score assignments (Rain *et al.*, 2001).

16.13 CONCLUSION

Since proteins and RNA sustain function rather than genes, and since function can no longer be considered as an individual property of each molecular actor taken independently from others, proteomics has appeared as the post-genomic method of choice. High-throughput experimental technologies are now routinely used to produce protein expression and interaction networks. When combined with complementary literature data, these networks become cellular pathways that are key elements for understanding the cell functions in context. Proteomic informatics enables the massive production of these data by storing them in dedicated databases allowing quality control, and by proposing adapted mining and visualization tools. Bioinformatics algorithms further allow prediction of protein networks by comparing genome sequences or by inferring networks across organisms. Even if one still lacks independent reference datasets and validation methods to precisely evaluate the efficiency of such algorithms; they will probably soon be the main tools for looking for associations between heterogeneous biological data.

While the sheer amount of data made available by various means of analysis is a challenge in itself, its heterogeneity should be one of today's main concerns: not only does it make any hypothesis difficult to test extensively, it is only by cross-referencing independent data-sources that we will be able to develop a consistent corpus of knowledge and extract from it an adequate validation set for reliable comparison and accurate evaluation of existent and as yet undiscovered analysis methods.

ACKNOWLEDGEMENTS

We thank P. Durand, V. Schächter, Y. Chemama and P. Legrain for stimulating discussions and meticulous reading of the manuscript.

REFERENCES

Albert R, Jeong H, Barabasi AL. (2000). *Nature* **406**: 378–382.

Altschul SF, Madden TL, Schaffer AA, Zhang J, Zhang Z, Miller W, *et al.* (1997). *Nucleic Acids Res* **25**: 3389–3402.

Anderson L, Seilhamer J. (1997). *Electrophoresis* **18**: 533–537.

Apweiler R, Attwood TK, Bairoch A, Bateman A, Birney E, Biswas M, *et al.* (2001). *Nucleic Acids Res* **29**: 37–40.

Bader GD, Hogue CW. (2000). *Bioinformatics* **16**: 465–477.

Bairoch A. (2000). *Nucleic Acids Res* **28**: 304–305.

Bartel PL, Roecklein JA, SenGupta D, Fields S. (1996). *Nature Genet* **12**: 72–77.

Costanzo MC, Hogan JD, Cusick ME, Davis BP, Fancher AM, Hodges PE, *et al.* (2000). *Nucleic Acids Res* **28**: 73–76.

Dandekar T, Snel B, Huynen M, Bork P. (1998). *Trends Biochem Sci* **23**: 324–328.

Eilbeck K, Brass A, Paton N, Hodgman C. (1999). In *Seventh International Conference of Intelligent Systems for Molecular Biology*, pp. 87–94.

Eisenberg D, Marcotte EM, Xenarios I, Yeates TO. (2000). *Nature* **405**: 823–826.

Enright AJ, Iliopoulos I, Kyrpides NC, Ouzounis CA. (1999). *Nature* **402**: 86–90.

Fields S, Song O. (1989). *Nature* **340**: 245–246.

Finley RL Jr, Brent R. (1994). *Proc Natl Acad Sci USA* **91**: 12980–12984.

Flajolet M, Rotondo G, Daviet L, Bergametti F, Inchauspé G, Tiollais P, *et al.* (2000). *Gene* **242**, 369–379.

Fromont-Racine M, Mayes AE, Brunet-Simon A, Rain JC, Colley A, Dix I, *et al.* (2000). *Yeast* **17**: 95–110.

Fromont-Racine M, Rain JC, Legrain P. (1997). *Nature Genet* **16**: 277–282.

Gavin AC, Bösche M, Krause R, Grandi P, Marzioch M, Bauer A, *et al.* (2002). *Nature* **415**: 141–147.

Gygi SP, Rochon Y, Franza BR, Aebersold R. (1999). *Mol Cell Biol* **19**: 1720–1730.

Haab BB, Dunham MJ, Brown PO. (2001). *Genome Biol* **2**:RESEARCH0004.

Hamosh A, Scott AF, Amberger J, Valle D, McKusick VA. (2000). *Hum Mutat* **15**: 57–61.

Heinemeyer T, Chen X, Karas H, Kel AE, Kel OV, Liebich I, *et al.* (1999). *Nucleic Acids Res* **27**: 318–322.

Ho Y, Gruhler A, Heilbut A, Bader GD, Moore L, Adams SL, *et al.* (2002). *Nature* **415**: 180–183.

Houry WA, Frishman D, Eckerskorn C, Lottspeich F, Hartl FU. (1999). *Nature* **402**: 147–54.

Ito T, Chiba T, Ozawa R, Yoshida M, Hattori M, Sakaki Y. (2001). *Proc Natl Acad Sci USA* **98**: 4569–4574.

Ito T, Tashiro K, Muta S, Ozawa R, Chiba T, Nishizawa M, *et al*. (2000). *Proc Natl Acad Sci USA* **97**: 1143–1147.

Jenssen TK, Laegreid A, Komorowski J, Hovig E. (2001). *Nature Genet* **28**: 21–28.

Jeong H, Mason SP, Barabasi A-L, Oltvai ZN. (2001). *Nature* **411**: 41–42.

Jeong H, Tombor B, Albert R, Oltvai ZN, Barabasi A-L. (2000). *Nature* **407**: 651–654.

Kanehisa M, Goto S. (2000). *Nucleic Acids Res* **28**: 27–30.

Karp PD, Riley M, Saier M, Paulsen IT, Paley SM, Pellegrini-Toole A. (2000). *Nucleic Acids Res* **28**: 56–59.

Lee KH. (2001). *Trends Biotechnol* **19**: 217–222.

Legrain P, Jestin J-L, Schächter V. (2000). *Curr Opin Biotechnol* **11**: 402–407.

Legrain P, Wojcik J, Gauthier JM. (2001). *Trends Genet* **17**: 346–352.

Lemkin PF, Thornwall G. (1999). *Mol Biotechnol* **12**: 159–172.

Lueking A, Horn M, Eickhoff H, Bussow K, Lehrach H, Walter G. (1999). *Anal Biochem* **270**: 103–111.

MacBeath G, Schreiber SL. (2000). *Science* **289**: 1760–1763.

Marcotte EM, Pellegrini M, Ng HL, Rice DW, Yeates TO, Eisenberg D. (1999a). *Science* **285**: 751–753.

Marcotte EM, Pellegrini M, Thompson MJ, Yeates TO, Eisenberg D. (1999b). *Nature* **402**: 83–86.

Marcotte EM, Xenarios I, Eisenberg D. (2001). *Bioinformatics* **17**: 359–363.

Mayer ML, Hieter P. (2000). *Nature Biotechnol* **18**: 1242–1243.

McCraith S, Holtzman T, Moss B, Fields S. (2000). *Proc Natl Acad Sci USA* **97**: 4879–4884.

Mewes HW, Frishman D, Gruber C, Geier B, Haase D, Kaps A, *et al*. (2000). *Nucleic Acids Res* **28**: 37–40.

Mrowka R. (2001). *Bioinformatics* **17**: 669–671.

Murzin AG, Brenner SE, Hubbard T, Chothia C. (1995). *J Mol Biol* **247**: 536–540.

Newman JR, Wolf E, Kim PS. (2000). *Proc Natl Acad Sci USA* **97**: 13203–13208.

Ono T, Hishigaki H, Tanigami A, Takagi T. (2001). *Bioinformatics* **17**: 155–161.

Overbeek R, Fonstein M, D'Souza M, Pusch GD, Maltsev N. (1999). *Proc Natl Acad Sci USA* **96**: 2896–2901.

Park J, Lappe M, Teichmann SA. (2001). *J Mol Biol* **307**: 929–938.

Pazos F, Helmer-Citterich M, Ausiello G, Valencia A. (1997). *J Mol Biol* **271**: 511–523.

Pellegrini M, Marcotte EM, Thompson MJ, Eisenberg D, Yeates TO. (1999). *Proc Natl Acad Sci USA* **96**: 4285–4288.

Phizicky EM, Fields S. (1995). *Microbiol Rev* **59**: 94–123.

Pilpel Y, Sudarsanam P, Church GM. (2001). *Nature Genet* **29**: 153–159.

Quadroni M, James P. (1999). *Electrophoresis* **20**: 664–677.

Rain JC, Selig L, de Reuse H, Battaglia V, Reverdy C, Simon S, *et al*. (2001). *Nature* **409**: 211–216.

Rigaut G, Shevchenko A, Rutz B, Wilm M, Mann M, Seraphin B. (1999). *Nature Biotechnol* **17**: 1030–1032.

Sanchez C, Lachaize C, Janody F, Bellon B, Roder L, Euzenat J, *et al*. (1999). *Nucleic Acids Res* **27**: 89–94.

Schwikowski B, Uetz P, Fields S. (2000). *Nature Biotechnol* **18**: 1257–1261.

Selkov E Jr, Grechkin Y, Mikhailova N, Selkov E. (1998). *Nucleic Acids Res* **26**: 43–45.
Spellman PT, Sherlock G, Zhang MQ, Iyer VR, Anders K, Eisen MB, *et al.* (1998). *Mol Biol Cell* **9**: 3273–3297.
Sprinzak E, Margalit H. (2001). *J Mol Biol* **311**: 681–692.
Tong AH, Drees B, Nardelli G, Bader GD, Brannetti B, Castagnoli L, *et al.* (2001). *Science* **13**: 13.
Tsoka S, Ouzounis CA. (2000). *Nature Genet* **26**: 141–142.
Uetz P, Giot L, Cagney G, Mansfield TA, Judson RS, Knight JR, *et al.* (2000). *Nature* **403**: 623–627.
Vidal M, Legrain P. (1999). *Nucleic Acids Res* **27**: 919–929.
Walhout AJ, Sordella R, Lu X, Hartley JL, Temple GF, Brasch MA, *et al.* (2000). *Science* **287**: 116–122.
Walter G, Konthur Z, Lehrach H. (2001). *Comb Chem High Throughput Screen* **4**: 193–205.
Whitelegge JP, le Coutre J, Lee JC, Engel CK, Prive GG, Faull KF, *et al.* (1999). *Proc Natl Acad Sci USA* **96**: 10695–10698.
Wilkins MR, Gasteiger E, Gooley AA, Herbert BR, Molloy MP, Binz PA, *et al.* (1999). *J Mol Biol* **289**: 645–67.
Wojcik J, Schächter V. (2001). *Bioinformatics* **17**: S296–S305.
Wu Q, Maniatis T. (1999). *Cell* **97**: 779–790.
Xenarios I, Rice DW, Salwinski L, Baron MK, Marcotte EM, Eisenberg D. (2000). *Nucleic Acids Res* **28**: 289–291.
Yates JR III. (1998). *J Mass Spect* **33**: 1–19.

CHAPTER 17

Concluding Remarks: Final Thoughts and Future Trends

MICHAEL R. BARNES[1] and IAN C. GRAY[2]

[1]*Genetic Bioinformatics and* [2]*Discovery Genetics*
Genetics Research Division
GlaxoSmithKline Pharmaceuticals, Harlow, Essex, UK

The sequencing of the human genome is complete. This is an obvious milestone for all fields of biology, none more so than genetics. As we have seen throughout this book, the availability of a complete genome makes the study of the genetics of an organism much less haphazard, and bioinformatics is an essential enabling skill for the geneticist to make the most of the genome. In the pre-genome era, geneticists probed the genome like early explorers penetrating a dark continent ripe for exploration. Relying on only the most basic data they painstakingly reconstructed genes and methodically drafted maps to find disease alleles. Now, in the post-genome era, instead of stars and a compass the genetic explorers have the equivalent of a global positioning satellite system and a detailed A–Z directory of genes. With all this technology it might be hard to imagine how genetics can now fail to locate disease genes, but failure will still be a frequent outcome.

Why? Firstly, we may be looking for something that does not exist or is too small to detect using existing methodology. Complex human disease is a product of both environment and genes, but the environment is often overlooked as a source of disease, particularly in the current era of high-profile genetics. The contribution of a single gene to a multifactorial disease or trait may be vanishingly small and consequently even large studies may have insufficient power to detect it. Secondly our directory of genes may not be as comprehensive as we think, with significant weaknesses in certain areas, for example the assignment of function to poorly understood regulatory motifs and the degree and nature of inter-individual genome diversity. Thirdly our maps are not yet completely

Bioinformatics for Geneticists. Edited by M.R. Barnes and I.C. Gray
 ISBNs: 0 470 84393 4; 0 470 84394 2 (PB)

error free. Bioinformatics cannot help with the first problem directly, although novel statistical methods may improve the chances of identifying small genetic effects and will form part of a continually evolving software suite for genetic analysis of complex traits. As more pieces of the genetic and environmental jigsaw puzzle are put into place for each complex trait, it should become progressively easier to position the remaining pieces to give a more complete picture. Although bioinformatics may be perceived as playing a secondary role in developing techniques for improved statistical analysis of complex trait data, it is the key to providing the equally important solutions required for a truly complete characterization of the genome coupled with unimpeachable data integrity.

17.1 HOW MANY GENES?

The biggest revelation of the human genome sequencing project was that humans appear to have fewer genes than we had expected. Estimates of the total number of human genes were widely anticipated to reach the 100,000 gene mark (Aparicio, 2000). As sequencing progressed these estimates were downgraded to 60–70,000 and finally as the first draft appeared estimates were consolidated to a mere 35,000 genes (Ewing and Green, 2000). If this figure is to be believed, then humans have only seven times as many genes as yeast, ~2.5 times as many as the fly *Drosophila melanogaster* and less than twice as many as the nematode worm *Caenorhabditis elegans*. This figure may increase as understanding of the genome and gene prediction increases, although it seems unlikely that the number will rise beyond 50,000.

This smaller than expected number of genes might be viewed as good news for geneticists — fewer genes to screen for disease association. But fewer genes does not necessarily equate to reduced complexity. Complexity can manifest at many levels, including splicing, gene regulation, post-transcriptional editing and post-translational modification. In Chapter 12, we described the *Drosophila* DSCAM gene which has 115 exons which are alternatively spliced to code for 38,016 related but distinct protein isoforms (Schmucker *et al*., 2000). This remarkable gene gives us a hint that many of the gene models described so far in humans could under-represent the true diversity of the human gene repertoire. Instead it may be wise to view every gene transcript as a unit specific to a particular tissue, time or cellular condition. Alterations in any of these conditions could direct the expression of an alternative transcript.

It may also be pertinent to question the definition of a gene. Traditionally a gene is viewed as a protein-coding unit. Transcripts which do not obviously code for a protein are often dismissed as 'regulatory RNA' — a virtual dumping ground for transcripts which we are just beginning to understand (see Szymanski and Barciszewski, 2002). This situation is exacerbated by the wealth of data generated by genomics; for example a very large number of ESTs and cDNAs show no *in silico* evidence of splicing (i.e. by each end aligning either side of an intron in a genomic sequence). There are a number of explanations for the existence of such transcripts. They could be derived from a real gene but simply do not span an intron and therefore show no evidence of splicing; alternatively they could be *in vitro* artefacts generated during the construction of cDNA libraries or *in vivo* artefacts generated from cryptic promoters or pseudogenes.

This highlights one of the biggest challenges for the bioinformatic interpretation of the human genome — data overload. Gene prediction and annotation tools generally disregard unspliced ESTs as supporting evidence for the existence of a gene. This is a necessary precaution to avoid over-prediction of genes across the genome; tools designed to analyse whole genomes have to sacrifice sensitivity to avoid extensive over-prediction of genes and to maintain the performance of genome analysis pipelines, but where geneticists seek to identify all candidate genes in a defined locus, it may be prudent to evaluate equivocal information such as unspliced ESTs in a more thorough fashion. This can be achieved easily with genome browser tools such as Ensembl and the UCSC human genome browser which present all available data across a locus. However, it is wise to proceed with caution when planning experimental work based on ambiguous data derived from *in silico* sources in order to avoid frustration as well as wasted time and resources. Simple, rapidly executed experiments to provide supporting evidence for the *in silico* observation should be the first step.

17.2 MAPPING THE GENOME AND GAINING A VIEW OF THE FULL DEPTH OF HUMAN VARIATION

Our incomplete understanding of genes and genome organization may not necessarily be a big problem for genetics. Experimental frameworks can be primarily focused on the physical and genetic composition of a region, in terms of genetic markers, recombination frequency and other characteristics, rather than its perceived functional content. 'Phenotype-driven' family-based whole genome linkage scans to identify genes responsible for monogenic traits illustrate one such approach. Use of linkage disequilibrium (LD) to identify genomic regions of genetic association is a second example, and is more appropriate for complex traits. This approach assumes little about the function of a marker or gene, but can allow mapping of a genetic association to a very small region (typically 10–100 kb) following the construction of detailed population-based LD maps. Completion of an LD map of the entire human genome will in itself be a highly significant milestone for genetics. Already provisional LD maps of chromosomes 21, 22 and 19 have been published (Dawson *et al.*, 2002; Patil *et al.*, 2001; Michael Phillips personal communication). A whole genome LD map generated by many of the former members of TSC should be made publicly available in late 2003. This will finally make comprehensive SNP-based whole genome association scans a realistic possibility; selecting SNPs which tag all of the major haplotype blocks across the genome will shift the emphasis toward good experimental design and away from conjecture when initiating genetic association studies.

However, evolution toward a whole-genome haplotype-based approach to genetic studies will present considerable challenges. For example, although all of the available evidence suggests that the majority of haplotypes in any given genomic region are common to multiple ethnic groups (Gabriel *et al.*, 2002), haplotype frequencies may vary considerably between groups. Thus markers that tag common haplotypes in one ethnic group may not identify the most common haplotypes in other groups. Furthermore, approaches based on attempts to associate common haplotypes with a disease state are broadly reliant on the veracity of the 'common disease caused by common variants' hypothesis (see Pritchard, 2001). A low frequency haplotype which is associated with disease may evade detection, and a rare predisposing SNP occurring on a common haplotypic background may not be detected due to insufficient statistical power. Only empirical data gathered over the next few years will reveal the true scale of such issues. A further consideration is the increase in throughput and reduction in cost required to render the necessary scale of genotyping for population-based association studies, which are likely to require several million data points per genome-wide experiment, feasible. However significant investment in this area has led to promising improvements across a range of genotyping platforms over the last few years and we expect this trend to continue.

17.3 HOLISTIC ANALYSIS OF COMPLEX TRAITS

One of the weaknesses of genetic association studies is the difficulty in drawing a firm conclusion regarding the robustness of the finding from the statistical evidence for association between a given gene and trait, particularly if the level of significance is marginal. A key future application of bioinformatics is likely to be the drawing together of diverse threads of data from a number of sources in a more holistic approach toward the analysis of complex traits. The output from human linkage and population-based association studies can be combined with animal model quantitative trait loci, phenotypic data from systematic gene knock-out and transgenic mouse approaches, genome-wide expression data from microarrays, proteomic profiles and other sources, to provide a substantial body of evidence relating to the gene or locus in question. This will require the development of both new interfaces for the integration of disparate datasets and sophisticated global analysis software.

17.4 A FINAL WORD ON BIOINFORMATICS

It is always difficult to present a rapidly moving field such as bioinformatics in a book. Despite the best efforts of the authors, editors and publisher, by the time this book reaches the reader many of the tools described in the preceding chapters will have evolved to offer yet more functionality and utility. Keeping abreast of new developments in bioinformatics is as important an activity as using the data themselves. Current awareness of the field is essential to ensure that all of the relevant available data are captured, maximizing research efficiency. Finally, the best approach to becoming proficient in the use of software tools is often trial and error, and bioinformatics is no exception; trial and error *in silico* can obviate the far less desirable prospect of trial and error in the laboratory, so do not be afraid to experiment with bioinformatics applications — see what the human genome can yield in your hands. Good luck!

ACKNOWLEDGEMENTS

MRB and ICG would like to acknowledge the efforts of all of the authors who have contributed to this volume. This book has taken shape after many discussions with many of our collegues at GSK and in the wider scientific community. The first drafts were moulded into final chapters with the assistance of several willing proof readers, particularly Christopher Southan, Aruna Bansal, Ralph McGinnis and Mary Plumpton. We would also like to express our gratitude to Joan Marsh, Layla Paggett, Amie Tibble and Monica Twine at John Wiley for able assistance in the preparation of the manuscript. Finally this volume would not have been possible without the support and encouragement of Robin Dement and Ian Purvis at GSK.

REFERENCES

Aparicio SA. (2000). How to count human genes. *Nature Genet* **25**: 129–130.

Dawson E, Abecasis GR, Bumpstead S, Chen Y, Hunt S, Beare DM, *et al.* (2002). A first-generation linkage disequilibrium map of human chromosome 22. *Nature* **418**: 544–548.

Ewing B, Green P. (2000). Analysis of expressed sequence tags indicates 35,000 human genes. *Nature Genet* **25**: 232–234.

Gabriel SB, Schaffner SF, Nguyen H, Moore JM, Roy J, Blumenstiel B, *et al.* (2002). The structure of haplotype blocks in the human genome. *Science* **296**: 2225–2229.

Patil N, Berno AJ, Hinds DA, Barrett WA, Doshi JM, Hacker CR, *et al.* (2001). Blocks of limited haplotype diversity revealed by high-resolution scanning of human chromosome 21. *Science* **294**: 1719–1723.

Pritchard JK. (2001). Are rare variants responsible for susceptibility to complex diseases? *Am J Hum Genet* **69**: 124–137.

Schmucker D, Clemens JC, Shu H, Worby CA, Xiao J, Muda M, *et al.* (2000). Drosophila DSCAM is an axon guidance receptor exhibiting extraordinary molecular diversity. *Cell* **101**: 671–684.

Szymanski M, Barciszewski J. (2002). Beyond the proteome: non-coding regulatory RNAs. *Genome Biol* **3** (reviews 5): 1–8.

APPENDIX I

IA IUPAC Nucleotide Ambiguity Codes

IUPAC Code	Meaning	Complement
A	A	T
C	C	G
G	G	C
T/U	T	A
M	A or C	K
R	A or G	Y
W	A or T	W
S	C or G	S
Y	C or T	R
K	G or T	M
V	A or C or G	B
H	A or C or T	D
D	A or G or T	H
B	C or G or T	V
N	G or A or T or C	N

IB IUPAC Amino Acid Codes

IUPAC Amino Acid Code	Three Letter Code	Amino Acid
A	Ala	Alanine
C	Cys	Cysteine
D	Asp	Aspartate
E	Glu	Glutamate
F	Phe	Phenylalanine
G	Gly	Glycine
H	His	Histidine
I	Ile	Isoleucine
K	Lys	Lysine
L	Leu	Leucine
M	Met	Methionine
N	Asn	Asparagine
P	Pro	Proline
Q	Gln	Glutamine
R	Arg	Arginine

Bioinformatics for Geneticists. Edited by M.R. Barnes and I.C. Gray
 ISBNs: 0 470 84393 4; 0 470 84394 2 (PB)

IUPAC Amino Acid Code	Three Letter Code	Amino Acid
S	Ser	Serine
T	Thr	Threonine
V	Val	Valine
W	Trp	Tryptophan
Y	Tyr	Tyrosine

IC Human Codon Usage Table

Second Codon					
First Codon	U	C	A	G	Last Codon
U	Phe	Ser	Tyr	Cys	**U**
	Phe	Ser	Tyr	Cys	**C**
	Leu	Ser	**Stop**	**Stop**	**A**
	Leu	Ser	**Stop**	Trp	**G**
C	Leu	Pro	His	Arg	**U**
	Leu	Pro	His	Arg	**C**
	Leu	Pro	Gln	Arg	**A**
	Leu	Pro	Gln	Arg	**G**
A	Ile	Thr	Asn	Ser	**U**
	Ile	Thr	Asn	Ser	**C**
	Ile	Thr	Lys	Arg	**A**
	Met	Thr	Lys	Arg	**G**
G	Val	Ala	Asp	Gly	**U**
	Val	Ala	Asp	Gly	**C**
	Val	Ala	Glu	Gly	**A**
	Val	Ala	Glu	Gly	**G**

APPENDIX II

Amino Acid Substitution Matrices

More information on these matrices is available on the following www site (www.russell.embl-heidelberg.de/aas).

Bioinformatics for Geneticists. Edited by M.R. Barnes and I.C. Gray
 ISBNs: 0 470 84393 4; 0 470 84394 2 (PB)

IIA — All Protein Types

	ALA	ARG	ASN	ASP	CYS	GLN	GLU	GLY	HIS	ILE	LEU	LYS	MET	PHE	PRO	SER	THR	TRP	TYR	VAL
ALA		Arg (-2)	Asn (0)	Asp (0)	Cys (-2)	Gln (0)	Glu (0)	Gly (1)	His (-1)	Ile (-1)	Leu (-2)	Lys (-1)	Met (-1)	Phe (-3)	Pro (1)	Ser (1)	Thr (1)	Trp (-6)	Tyr (-3)	Val (0)
ARG	Ala (-2)		Asn (0)	Asp (-1)	Cys (-4)	Gln (1)	Glu (-1)	Gly (-3)	His (2)	Ile (-2)	Leu (-3)	Lys (3)	Met (0)	Phe (-4)	Pro (0)	Ser (0)	Thr (-1)	Trp (2)	Tyr (-4)	Val (-2)
ASN	Ala (0)	Arg (0)		Asp (2)	Cys (-4)	Gln (1)	Glu (1)	Gly (0)	His (2)	Ile (-2)	Leu (-3)	Lys (1)	Met (-2)	Phe (-3)	Pro (0)	Ser (1)	Thr (0)	Trp (-4)	Tyr (-2)	Val (-2)
ASP	Ala (0)	Arg (-1)	Asn (2)		Cys (-5)	Gln (2)	Glu (3)	Gly (1)	His (1)	Ile (-2)	Leu (-4)	Lys (0)	Met (-3)	Phe (-6)	Pro (-1)	Ser (0)	Thr (0)	Trp (-7)	Tyr (-4)	Val (-2)
CYS	Ala (-2)	Arg (-4)	Asn (-4)	Asp (-5)		Gln (-5)	Glu (-5)	Gly (-3)	His (-3)	Ile (-2)	Leu (-6)	Lys (-5)	Met (-5)	Phe (-4)	Pro (-3)	Ser (0)	Thr (-2)	Trp (-8)	Tyr (0)	Val (-2)
GLN	Ala (0)	Arg (1)	Asn (1)	Asp (2)	Cys (-5)		Glu (2)	Gly (-1)	His (3)	Ile (-2)	Leu (-2)	Lys (1)	Met (-1)	Phe (-5)	Pro (0)	Ser (-1)	Thr (-1)	Trp (-5)	Tyr (-4)	Val (-2)
GLU	Ala (0)	Arg (-1)	Asn (1)	Asp (3)	Cys (-5)	Gln (2)		Gly (0)	His (1)	Ile (-2)	Leu (-3)	Lys (0)	Met (-2)	Phe (-5)	Pro (-1)	Ser (0)	Thr (0)	Trp (-7)	Tyr (-4)	Val (-2)
GLY	Ala (1)	Arg (-3)	Asn (0)	Asp (1)	Cys (-3)	Gln (-1)	Glu (0)		His (-2)	Ile (-3)	Leu (-4)	Lys (-2)	Met (-3)	Phe (-5)	Pro (0)	Ser (1)	Thr (0)	Trp (-7)	Tyr (-5)	Val (-1)
HIS	Ala (-1)	Arg (2)	Asn (2)	Asp (1)	Cys (-3)	Gln (3)	Glu (1)	Gly (-2)		Ile (-2)	Leu (-2)	Lys (0)	Met (-2)	Phe (-2)	Pro (0)	Ser (-1)	Thr (-1)	Trp (-3)	Tyr (0)	Val (-2)
ILE	Ala (-1)	Arg (-2)	Asn (-2)	Asp (-2)	Cys (-2)	Gln (-2)	Glu (-2)	Gly (-3)	His (-2)		Leu (2)	Lys (-2)	Met (2)	Phe (1)	Pro (-2)	Ser (-1)	Thr (0)	Trp (-5)	Tyr (-1)	Val (4)
LEU	Ala (-2)	Arg (-3)	Asn (-3)	Asp (-4)	Cys (-6)	Gln (-2)	Glu (-3)	Gly (-4)	His (-2)	Ile (2)		Lys (-3)	Met (4)	Phe (2)	Pro (-3)	Ser (-3)	Thr (-2)	Trp (-2)	Tyr (-1)	Val (2)
LYS	Ala (-1)	Arg (3)	Asn (1)	Asp (0)	Cys (-5)	Gln (1)	Glu (0)	Gly (-2)	His (0)	Ile (-2)	Leu (-3)		Met (0)	Phe (-5)	Pro (-1)	Ser (0)	Thr (0)	Trp (-3)	Tyr (-4)	Val (-2)
MET	Ala (-1)	Arg (0)	Asn (-2)	Asp (-3)	Cys (-5)	Gln (-1)	Glu (-2)	Gly (-3)	His (-2)	Ile (2)	Leu (4)	Lys (0)		Phe (0)	Pro (-2)	Ser (-2)	Thr (-1)	Trp (-4)	Tyr (-2)	Val (2)
PHE	Ala (-3)	Arg (-4)	Asn (-3)	Asp (-6)	Cys (-4)	Gln (-5)	Glu (-5)	Gly (-5)	His (-2)	Ile (1)	Leu (2)	Lys (-5)	Met (0)		Pro (-5)	Ser (-3)	Thr (-3)	Trp (0)	Tyr (7)	Val (-1)
PRO	Ala (1)	Arg (0)	Asn (0)	Asp (-1)	Cys (-3)	Gln (0)	Glu (-1)	Gly (0)	His (0)	Ile (-2)	Leu (-3)	Lys (-1)	Met (-2)	Phe (-5)		Ser (1)	Thr (0)	Trp (-6)	Tyr (-5)	Val (-1)
SER	Ala (1)	Arg (0)	Asn (1)	Asp (0)	Cys (0)	Gln (-1)	Glu (0)	Gly (1)	His (-1)	Ile (-1)	Leu (-3)	Lys (0)	Met (-2)	Phe (-3)	Pro (1)		Thr (1)	Trp (-2)	Tyr (-3)	Val (-1)
THR	Ala (1)	Arg (-1)	Asn (0)	Asp (0)	Cys (-2)	Gln (-1)	Glu (0)	Gly (0)	His (-1)	Ile (0)	Leu (-2)	Lys (0)	Met (-1)	Phe (-3)	Pro (0)	Ser (1)		Trp (-5)	Tyr (-3)	Val (0)
TRP	Ala (-6)	Arg (2)	Asn (-4)	Asp (-7)	Cys (-8)	Gln (-5)	Glu (-7)	Gly (-7)	His (-3)	Ile (-5)	Leu (-2)	Lys (-3)	Met (-4)	Phe (0)	Pro (-6)	Ser (-2)	Thr (-5)		Tyr (0)	Val (-6)
TYR	Ala (-3)	Arg (-4)	Asn (-2)	Asp (-4)	Cys (0)	Gln (-4)	Glu (-4)	Gly (-5)	His (0)	Ile (-1)	Leu (-1)	Lys (-4)	Met (-2)	Phe (7)	Pro (-5)	Ser (-3)	Thr (-3)	Trp (0)		Val (-2)
VAL	Ala (0)	Arg (-2)	Asn (-2)	Asp (-2)	Cys (-2)	Gln (-2)	Glu (-2)	Gly (-1)	His (-2)	Ile (4)	Leu (2)	Lys (-2)	Met (2)	Phe (-1)	Pro (-1)	Ser (-1)	Thr (0)	Trp (-6)	Tyr (-2)	

IIB Extracellular Proteins

	ALA	ARG	ASN	ASP	CYS	GLN	GLU	GLY	HIS	ILE	LEU	LYS	MET	PHE	PRO	SER	THR	TRP	TYR	VAL
ALA		Arg (0)	Asn (0)	Asp (-1)	Cys (-4)	Gln (0)	Glu (0)	Gly (0)	His (0)	Ile (0)	Leu (0)	Lys (0)	Met (0)	Phe (-1)	Pro (0)	Ser (0)	Thr (0)	Trp (-2)	Tyr (-1)	Val (0)
ARG	Ala (0)		Asn (0)	Asp (0)	Cys (-5)	Gln (0)	Glu (0)	Gly (0)	His (0)	Ile (0)	Leu (-1)	Lys (1)	Met (0)	Phe (-1)	Pro (0)	Ser (0)	Thr (0)	Trp (-1)	Tyr (0)	Val (0)
ASN	Ala (0)	Arg (0)		Asp (1)	Cys (-6)	Gln (0)	Glu (0)	Gly (0)	His (0)	Ile (-1)	Leu (-2)	Lys (0)	Met (-1)	Phe (-2)	Pro (0)	Ser (0)	Thr (0)	Trp (-3)	Tyr (-1)	Val (-1)
ASP	Ala (-1)	Arg (0)	Asn (1)		Cys (-7)	Gln (0)	Glu (0)	Gly (0)	His (0)	Ile (-2)	Leu (-2)	Lys (0)	Met (-2)	Phe (-2)	Pro (0)	Ser (0)	Thr (0)	Trp (-3)	Tyr (-2)	Val (-1)
CYS	Ala (-4)	Arg (-5)	Asn (-6)	Asp (-7)		Gln (-5)	Glu (-6)	Gly (-6)	His (-5)	Ile (-5)	Leu (-5)	Lys (-6)	Met (-5)	Phe (-5)	Pro (-6)	Ser (-5)	Thr (-5)	Trp (-5)	Tyr (-4)	Val (-4)
GLN	Ala (0)	Arg (0)	Asn (0)	Asp (0)	Cys (-5)		Glu (0)	Gly (0)	His (0)	Ile (-1)	Leu (-1)	Lys (0)	Met (0)	Phe (-2)	Pro (0)	Ser (0)	Thr (0)	Trp (-1)	Tyr (-1)	Val (0)
GLU	Ala (0)	Arg (0)	Asn (0)	Asp (0)	Cys (-6)	Gln (0)		Gly (-1)	His (0)	Ile (-1)	Leu (-1)	Lys (0)	Met (0)	Phe (-2)	Pro (0)	Ser (0)	Thr (0)	Trp (-1)	Tyr (-1)	Val (0)
GLY	Ala (0)	Arg (0)	Asn (0)	Asp (0)	Cys (-6)	Gln (0)	Glu (-1)		His (0)	Ile (-2)	Leu (-2)	Lys (-1)	Met (-2)	Phe (-3)	Pro (0)	Ser (0)	Thr (0)	Trp (-2)	Tyr (-2)	Val (-2)
HIS	Ala (0)	Arg (0)	Asn (0)	Asp (0)	Cys (-5)	Gln (0)	Glu (0)	Gly (0)		Ile (-1)	Leu (-1)	Lys (0)	Met (-1)	Phe (-1)	Pro (0)	Ser (0)	Thr (0)	Trp (-1)	Tyr (0)	Val (-1)
ILE	Ala (0)	Arg (0)	Asn (-1)	Asp (-2)	Cys (-5)	Gln (-1)	Glu (-1)	Gly (-2)	His (-1)		Leu (1)	Lys (0)	Met (0)	Phe (0)	Pro (-1)	Ser (-1)	Thr (0)	Trp (-1)	Tyr (0)	Val (2)
LEU	Ala (0)	Arg (-1)	Asn (-2)	Asp (-2)	Cys (-5)	Gln (-1)	Glu (-1)	Gly (-2)	His (-1)	Ile (1)		Lys (-1)	Met (1)	Phe (0)	Pro (0)	Ser (-1)	Thr (0)	Trp (-2)	Tyr (-1)	Val (1)
LYS	Ala (0)	Arg (1)	Asn (0)	Asp (0)	Cys (-6)	Gln (0)	Glu (0)	Gly (-1)	His (0)	Ile (0)	Leu (-1)		Met (-1)	Phe (-2)	Pro (0)	Ser (0)	Thr (0)	Trp (-2)	Tyr (-1)	Val (0)
MET	Ala (0)	Arg (0)	Asn (-1)	Asp (-2)	Cys (-5)	Gln (0)	Glu (0)	Gly (-2)	His (-1)	Ile (0)	Leu (1)	Lys (-1)		Phe (0)	Pro (-1)	Ser (-1)	Thr (0)	Trp (-1)	Tyr (-1)	Val (0)
PHE	Ala (-1)	Arg (-1)	Asn (-2)	Asp (-2)	Cys (-5)	Gln (-2)	Glu (-2)	Gly (-3)	His (-1)	Ile (0)	Leu (0)	Lys (-2)	Met (0)		Pro (-2)	Ser (-2)	Thr (-1)	Trp (1)	Tyr (2)	Val (0)
PRO	Ala (0)	Arg (0)	Asn (0)	Asp (0)	Cys (-6)	Gln (0)	Glu (0)	Gly (0)	His (0)	Ile (-1)	Leu (0)	Lys (0)	Met (-1)	Phe (-2)		Ser (0)	Thr (0)	Trp (-3)	Tyr (-1)	Val (0)
SER	Ala (0)	Arg (0)	Asn (0)	Asp (0)	Cys (-5)	Gln (0)	Glu (0)	Gly (0)	His (0)	Ile (-1)	Leu (-1)	Lys (0)	Met (-1)	Phe (-2)	Pro (0)		Thr (1)	Trp (-1)	Tyr (-1)	Val (-1)
THR	Ala (0)	Arg (0)	Asn (0)	Asp (0)	Cys (-5)	Gln (0)	Glu (0)	Gly (0)	His (0)	Ile (0)	Leu (0)	Lys (0)	Met (0)	Phe (-1)	Pro (0)	Ser (1)		Trp (-1)	Tyr (-1)	Val (0)
TRP	Ala (-2)	Arg (-1)	Asn (-3)	Asp (-3)	Cys (-5)	Gln (-1)	Glu (-1)	Gly (-2)	His (-1)	Ile (-1)	Leu (-2)	Lys (-2)	Met (-1)	Phe (1)	Pro (-3)	Ser (-1)	Thr (-1)		Tyr (1)	Val (-1)
TYR	Ala (-1)	Arg (0)	Asn (-1)	Asp (-2)	Cys (-4)	Gln (-1)	Glu (-1)	Gly (-2)	His (0)	Ile (0)	Leu (-1)	Lys (-1)	Met (-1)	Phe (2)	Pro (-1)	Ser (-1)	Thr (-1)	Trp (1)		Val (0)
VAL	Ala (0)	Arg (0)	Asn (-1)	Asp (-1)	Cys (-4)	Gln (0)	Glu (0)	Gly (-2)	His (-1)	Ile (2)	Leu (1)	Lys (0)	Met (0)	Phe (0)	Pro (0)	Ser (-1)	Thr (0)	Trp (-1)	Tyr (0)	

IIC Intracellular Proteins

	ALA	ARG	ASN	ASP	CYS	GLN	GLU	GLY	HIS	ILE	LEU	LYS	MET	PHE	PRO	SER	THR	TRP	TYR	VAL
ALA		Arg (0)	Asn (-1)	Asp (-1)	Cys (0)	Gln (0)	Glu (0)	Gly (0)	His (-1)	Ile (0)	Leu (0)	Lys (0)	Met (0)	Phe (-1)	Pro (0)	Ser (0)	Thr (0)	Trp (-2)	Tyr (-1)	Val (0)
ARG	Ala (0)		Asn (0)	Asp (0)	Cys (-1)	Gln (0)	Glu (0)	Gly (0)	His (0)	Ile (-1)	Leu (-1)	Lys (1)	Met (0)	Phe (-2)	Pro (0)	Ser (0)	Thr (0)	Trp (-1)	Tyr (-1)	Val (-1)
ASN	Ala (-1)	Arg (0)		Asp (1)	Cys (-1)	Gln (0)	Glu (0)	Gly (0)	His (0)	Ile (-2)	Leu (-2)	Lys (0)	Met (-1)	Phe (-2)	Pro (-1)	Ser (0)	Thr (0)	Trp (-2)	Tyr (-1)	Val (-2)
ASP	Ala (-1)	Arg (0)	Asn (1)		Cys (-2)	Gln (0)	Glu (1)	Gly (0)	His (0)	Ile (-3)	Leu (-3)	Lys (0)	Met (-2)	Phe (-3)	Pro (0)	Ser (0)	Thr (0)	Trp (-2)	Tyr (-2)	Val (-2)
CYS	Ala (0)	Arg (-1)	Asn (-1)	Asp (-2)		Gln (-2)	Glu (-2)	Gly (-1)	His (0)	Ile (0)	Leu (0)	Lys (-1)	Met (0)	Phe (0)	Pro (-2)	Ser (0)	Thr (0)	Trp (-1)	Tyr (0)	Val (0)
GLN	Ala (0)	Arg (0)	Asn (0)	Asp (0)	Cys (-2)		Glu (1)	Gly (0)	His (0)	Ile (-2)	Leu (-1)	Lys (0)	Met (0)	Phe (-2)	Pro (0)	Ser (0)	Thr (0)	Trp (-2)	Tyr (-1)	Val (-1)
GLU	Ala (0)	Arg (0)	Asn (0)	Asp (1)	Cys (-2)	Gln (1)		Gly (-1)	His (0)	Ile (-2)	Leu (-2)	Lys (0)	Met (-1)	Phe (-2)	Pro (0)	Ser (0)	Thr (0)	Trp (-2)	Tyr (-1)	Val (-1)
GLY	Ala (0)	Arg (0)	Asn (0)	Asp (0)	Cys (-1)	Gln (0)	Glu (-1)		His (-1)	Ile (-3)	Leu (-3)	Lys (0)	Met (-2)	Phe (-3)	Pro (0)	Ser (0)	Thr (-1)	Trp (-2)	Tyr (-2)	Val (-2)
HIS	Ala (-1)	Arg (0)	Asn (0)	Asp (0)	Cys (0)	Gln (0)	Glu (0)	Gly (-1)		Ile (-2)	Leu (-1)	Lys (0)	Met (-1)	Phe (-1)	Pro (-1)	Ser (0)	Thr (0)	Trp (0)	Tyr (1)	Val (-1)
ILE	Ala (0)	Arg (-1)	Asn (-2)	Asp (-3)	Cys (0)	Gln (-2)	Glu (-2)	Gly (-3)	His (-2)		Leu (2)	Lys (-1)	Met (1)	Phe (0)	Pro (-2)	Ser (-2)	Thr (0)	Trp (-1)	Tyr (0)	Val (2)
LEU	Ala (0)	Arg (-1)	Asn (-2)	Asp (-3)	Cys (0)	Gln (-1)	Glu (-2)	Gly (-3)	His (-1)	Ile (2)		Lys (-1)	Met (2)	Phe (1)	Pro (-2)	Ser (-2)	Thr (-1)	Trp (0)	Tyr (0)	Val (1)
LYS	Ala (0)	Arg (1)	Asn (0)	Asp (0)	Cys (-1)	Gln (0)	Glu (0)	Gly (0)	His (0)	Ile (-1)	Leu (-1)		Met (0)	Phe (-2)	Pro (0)	Ser (0)	Thr (0)	Trp (-1)	Tyr (-1)	Val (-1)
MET	Ala (0)	Arg (0)	Asn (-1)	Asp (-2)	Cys (0)	Gln (0)	Glu (-1)	Gly (-2)	His (-1)	Ile (1)	Leu (2)	Lys (0)		Phe (1)	Pro (-1)	Ser (-1)	Thr (0)	Trp (0)	Tyr (0)	Val (0)
PHE	Ala (-1)	Arg (-2)	Asn (-2)	Asp (-3)	Cys (0)	Gln (-2)	Glu (-2)	Gly (-3)	His (-1)	Ile (0)	Leu (1)	Lys (-2)	Met (1)		Pro (-2)	Ser (-2)	Thr (-1)	Trp (1)	Tyr (2)	Val (0)
PRO	Ala (0)	Arg (0)	Asn (-1)	Asp (0)	Cys (-2)	Gln (0)	Glu (0)	Gly (0)	His (-1)	Ile (-2)	Leu (-2)	Lys (0)	Met (-1)	Phe (-2)		Ser (0)	Thr (0)	Trp (-2)	Tyr (-1)	Val (-1)
SER	Ala (0)	Arg (0)	Asn (0)	Asp (0)	Cys (0)	Gln (0)	Glu (0)	Gly (0)	His (0)	Ile (-2)	Leu (-2)	Lys (0)	Met (-1)	Phe (-2)	Pro (0)		Thr (0)	Trp (-2)	Tyr (-1)	Val (-1)
THR	Ala (0)	Arg (0)	Asn (0)	Asp (0)	Cys (0)	Gln (0)	Glu (0)	Gly (-1)	His (0)	Ile (0)	Leu (-1)	Lys (0)	Met (0)	Phe (-1)	Pro (0)	Ser (0)		Trp (-2)	Tyr (-1)	Val (0)
TRP	Ala (-2)	Arg (-1)	Asn (-2)	Asp (-2)	Cys (-1)	Gln (-2)	Glu (-2)	Gly (-2)	His (0)	Ile (-1)	Leu (0)	Lys (-1)	Met (0)	Phe (1)	Pro (-2)	Ser (-2)	Thr (-2)		Tyr (2)	Val (-1)
TYR	Ala (-1)	Arg (-1)	Asn (-1)	Asp (-2)	Cys (0)	Gln (-1)	Glu (-1)	Gly (-2)	His (1)	Ile (0)	Leu (0)	Lys (-1)	Met (0)	Phe (2)	Pro (-1)	Ser (-1)	Thr (-1)	Trp (2)		Val (0)
VAL	Ala (0)	Arg (-1)	Asn (-2)	Asp (-2)	Cys (0)	Gln (-1)	Glu (-1)	Gly (-2)	His (-1)	Ile (2)	Leu (1)	Lys (-1)	Met (0)	Phe (0)	Pro (-1)	Ser (-1)	Thr (0)	Trp (-1)	Tyr (0)	

IID Membrane Proteins

	ALA	ARG	ASN	ASP	CYS	GLN	GLU	GLY	HIS	ILE	LEU	LYS	MET	PHE	PRO	SER	THR	TRP	TYR	VAL
ALA		Arg (-1)	Asn (-1)	Asp (0)	Cys (0)	Gln (-2)	Glu (0)	Gly (1)	His (-3)	Ile (0)	Leu (-2)	Lys (-2)	Met (-1)	Phe (-2)	Pro (0)	Ser (2)	Thr (1)	Trp (-4)	Tyr (-3)	Val (0)
ARG	Ala (-1)		Asn (2)	Asp (1)	Cys (-1)	Gln (6)	Glu (2)	Gly (0)	His (5)	Ile (-3)	Leu (-3)	Lys (9)	Met (0)	Phe (-4)	Pro (-3)	Ser (-1)	Thr (-1)	Trp (5)	Tyr (-1)	Val (-2)
ASN	Ala (-1)	Arg (2)		Asp (6)	Cys (-1)	Gln (3)	Glu (1)	Gly (-2)	His (3)	Ile (-3)	Leu (-4)	Lys (5)	Met (-2)	Phe (-4)	Pro (-2)	Ser (2)	Thr (1)	Trp (-3)	Tyr (-1)	Val (-3)
ASP	Ala (0)	Arg (1)	Asn (6)		Cys (-3)	Gln (2)	Glu (8)	Gly (2)	His (3)	Ile (-3)	Leu (-5)	Lys (3)	Met (-3)	Phe (-6)	Pro (-2)	Ser (0)	Thr (0)	Trp (-4)	Tyr (-2)	Val (-3)
CYS	Ala (0)	Arg (-1)	Asn (-1)	Asp (-3)		Gln (-3)	Glu (-3)	Gly (-1)	His (-1)	Ile (-1)	Leu (-1)	Lys (-3)	Met (-1)	Phe (1)	Pro (-3)	Ser (2)	Thr (0)	Trp (1)	Tyr (3)	Val (0)
GLN	Ala (-2)	Arg (5)	Asn (3)	Asp (2)	Cys (-3)		Glu (7)	Gly (-2)	His (7)	Ile (-4)	Leu (-2)	Lys (6)	Met (-2)	Phe (-4)	Pro (0)	Ser (-1)	Thr (-2)	Trp (0)	Tyr (0)	Val (-4)
GLU	Ala (0)	Arg (2)	Asn (1)	Asp (8)	Cys (-3)	Gln (7)		Gly (3)	His (2)	Ile (-4)	Leu (-5)	Lys (1)	Met (-3)	Phe (-6)	Pro (-3)	Ser (0)	Thr (-1)	Trp (-3)	Tyr (-5)	Val (-2)
GLY	Ala (1)	Arg (0)	Asn (-2)	Asp (3)	Cys (-1)	Gln (-1)	Glu (3)		His (-3)	Ile (-2)	Leu (-4)	Lys (-1)	Met (-3)	Phe (-5)	Pro (-1)	Ser (1)	Thr (0)	Trp (-2)	Tyr (-5)	Val (-1)
HIS	Ala (-3)	Arg (5)	Asn (3)	Asp (3)	Cys (-1)	Gln (7)	Glu (2)	Gly (-3)		Ile (-4)	Leu (-4)	Lys (4)	Met (-3)	Phe (-3)	Pro (-4)	Ser (-2)	Thr (-2)	Trp (-1)	Tyr (6)	Val (-4)
ILE	Ala (0)	Arg (-3)	Asn (-3)	Asp (-3)	Cys (-1)	Gln (-4)	Glu (-4)	Gly (-2)	His (-4)		Leu (1)	Lys (-4)	Met (1)	Phe (-1)	Pro (-3)	Ser (-1)	Thr (0)	Trp (-3)	Tyr (-4)	Val (2)
LEU	Ala (-2)	Arg (-3)	Asn (-4)	Asp (-5)	Cys (-1)	Gln (-2)	Glu (-5)	Gly (-4)	His (-4)	Ile (1)		Lys (-4)	Met (1)	Phe (1)	Pro (-1)	Ser (-2)	Thr (-1)	Trp (-2)	Tyr (-3)	Val (0)
LYS	Ala (-2)	Arg (9)	Asn (5)	Asp (3)	Cys (-3)	Gln (6)	Glu (1)	Gly (-1)	His (4)	Ile (-4)	Leu (-4)		Met (-1)	Phe (-5)	Pro (-4)	Ser (-1)	Thr (-2)	Trp (3)	Tyr (1)	Val (-4)
MET	Ala (-1)	Arg (0)	Asn (-2)	Asp (-3)	Cys (-1)	Gln (-2)	Glu (-3)	Gly (-3)	His (-3)	Ile (1)	Leu (1)	Lys (-1)		Phe (0)	Pro (-3)	Ser (-2)	Thr (0)	Trp (-2)	Tyr (-3)	Val (1)
PHE	Ala (-2)	Arg (-4)	Asn (-4)	Asp (-6)	Cys (1)	Gln (-4)	Glu (-6)	Gly (-5)	His (-3)	Ile (-1)	Leu (1)	Lys (-5)	Met (0)		Pro (-4)	Ser (-1)	Thr (-2)	Trp (-3)	Tyr (2)	Val (-1)
PRO	Ala (0)	Arg (-3)	Asn (-2)	Asp (-2)	Cys (-4)	Gln (0)	Glu (-3)	Gly (-1)	His (-4)	Ile (-3)	Leu (-1)	Lys (-4)	Met (-3)	Phe (-4)		Ser (-1)	Thr (-1)	Trp (-6)	Tyr (-5)	Val (-3)
SER	Ala (2)	Arg (-1)	Asn (2)	Asp (0)	Cys (1)	Gln (-1)	Glu (0)	Gly (1)	His (-2)	Ile (-1)	Leu (-2)	Lys (-1)	Met (-2)	Phe (-1)	Pro (-1)		Thr (2)	Trp (-3)	Tyr (0)	Val (-1)
THR	Ala (1)	Arg (-1)	Asn (1)	Asp (0)	Cys (0)	Gln (-2)	Glu (-1)	Gly (0)	His (-2)	Ile (0)	Leu (-1)	Lys (-2)	Met (0)	Phe (-2)	Pro (-1)	Ser (2)		Trp (-4)	Tyr (-3)	Val (0)
TRP	Ala (-3)	Arg (5)	Asn (-3)	Asp (-4)	Cys (1)	Gln (0)	Glu (-3)	Gly (-2)	His (-1)	Ile (-3)	Leu (-2)	Lys (3)	Met (-2)	Phe (-3)	Pro (-6)	Ser (-3)	Thr (-4)		Tyr (-3)	Val (-2)
TYR	Ala (-3)	Arg (-1)	Asn (-1)	Asp (-2)	Cys (3)	Gln (0)	Glu (-5)	Gly (-5)	His (6)	Ile (-4)	Leu (-2)	Lys (1)	Met (-3)	Phe (2)	Pro (-5)	Ser (0)	Thr (-3)	Trp (-3)		Val (-4)
VAL	Ala (0)	Arg (-2)	Asn (-3)	Asp (-3)	Cys (0)	Gln (-4)	Glu (-2)	Gly (-1)	His (-4)	Ile (2)	Leu (0)	Lys (-4)	Met (1)	Phe (-1)	Pro (-3)	Ser (-1)	Thr (0)	Trp (-2)	Tyr (-4)	

GLOSSARY OF TERMS AND ABBREVIATIONS

BLAST Basic Local Alignment Search Tool — a tool for identifying sequences in a database that match a given query sequence. Statistical analysis is applied to judge the significance of each match. Matching sequences may be homologous to, or related to, the query sequence. There are several versions of BLAST:

— **BLASTP** compares an amino acid query sequence against a protein sequence database

— **BLASTN** compares a nucleotide query sequence against a nucleotide sequence database

— **BLASTX** compares a nucleotide query sequence translated in all reading frames against a protein sequence database

— **TBLASTN** compares a protein query sequence against a nucleotide sequence database dynamically translated in all reading frames

— **TBLASTX** compares the six-frame translations of a nucleotide query sequence against the six-frame translations of a nucleotide sequence database.

BLAT BLAST-Like Alignment Tool. BLAT might superficially appear to be like BLAST, also being a tool for detecting subsequences that match a given query sequence, however BLAT and BLAST have a number of differences. BLAT was developed at the UCSC; it searches the human genome by keeping an index of the entire genome in memory. The index consists of all non-overlapping 11-mers except for repeat sequences. A BLAT search of the human genome will quickly find sequences of 95% and greater similarity of length 40 bases or more. It may miss more divergent or shorter sequence alignments (see the UCSC FAQ for more details on this tool — http://genome.ucsc.edu/FAQ.html).

CDS Coding sequence.

Contig Map A map depicting the relative order of overlapping (contiguous) clones representing a complete genomic or chromosomal segment.

DAS (Distributed Annotation System) DAS is a protocol for browsing and sharing genome sequence annotations across the Internet, allowing users to search and compare annotations from several sources. Ensembl provides a DAS reference server giving access to a wide range of specialist annotations of the human genome (see http://www.ensembl.org/das/ for more detail).

Data Mining The ability to query very large databases in order to satisfy a hypothesis ("top-down" data mining); or to interrogate a database in order to generate new hypotheses based on rigorous statistical correlations ("bottom-up" data mining).

Bioinformatics for Geneticists. Edited by M.R. Barnes and I.C. Gray
 ISBNs: 0 470 84393 4; 0 470 84394 2 (PB)

Domain (protein) A region of special biological interest within a single protein sequence. However, a domain may also be defined as a region within the three-dimensional structure of a protein that may encompass regions of several distinct protein sequences that accomplishes a specific function. A domain class is a group of domains that share a common set of well-defined properties or characteristics.

Electronic PCR (ePCR) An electronic process analogous to lab based PCR. Two primers are used to map a sequence feature (e.g. a SNP). To validate the position both primers must map in the same vicinity spanning a defined distance, effectively producing an electronic PCR product.

Expressed Sequence Tag (EST) A short sequence read from an expressed gene derived from a cDNA library. Databases storing large numbers of ESTs can be used to gauge the relative abundance of different transcripts in cDNA libraries and the tissues from which they are derived. An EST can also act as a physical tag for the identification, cloning and full length sequencing of the corresponding cDNA or gene.

FASTA format FASTA format, originally devised for Lipman & Pearson's FASTA (Fast-All) sequence alignment algorithm, is one of the simplest and most widely accepted formats for sequences, taking the form of a simple header preceded by a ">" sign and sequence on the following line, e.g.

```
>sequence_id
gataggctgagcgatgcgatgctagctagctagc
```

Golden Path The golden path is a term applied to the first and subsequent assemblies of the human genome.

Hidden Markov model (HMM) A joint statistical model for an ordered sequence of variables. The result of stochastically perturbing the variables in a Markov chain (the original variables are thus "hidden"), where the Markov chain has discrete variables which select the "state" of the HMM at each step. The perturbed values can be continuous and are the "outputs" of the HMM. A Hidden Markov Model is equivalently a coupled mixture model where the joint distribution over states is a Markov chain. Hidden Markov models are valuable in bioinformatics because they allow a search or alignment algorithm to be trained using unaligned or unweighted input sequences; and because they allow position-dependent scoring parameters such as gap penalties, thus more accurately modelling the consequences of evolutionary events on sequence families.

Homology (strict) Two or more biological species, systems or molecules that share a common evolutionary ancestor. (general) Two or more gene or protein sequences that share a significant degree of similarity, typically measured by the amount of identity (in the case of DNA), or conservative replacements (in the case of protein), that they register along their lengths. Sequence "homology" searches are typically performed with a query DNA or protein sequence to identify known genes or gene products that share significant similarity and hence might inform on the ancestry, heritage and possible function of the query gene.

***in silico* (biology)** (Lit. computer mediated). The use of computers to simulate, process, or analyse a biological experiment.

NCBI National Center for Biotechnology Information, Washington, D.C., USA.

Open reading frame (ORF) Any stretch of DNA that potentially encodes a protein. Open reading frames start with a start codon, and end with a termination codon. No termination codons may be present internally. The identification of an ORF is the first indication that a segment of DNA may be part of a functional gene.

Ortholog/Paralog Paralogs are genes related by duplication within a genome. Orthologs retain the same function in the course of evolution, whereas paralogs evolve new functions, even if these are related to the original one.

PERL PERL is the short form acronym for Practical Extraction and Report Language. Perl is relatively straightforward up to a certain level — this has encouraged its development as the primary language of biological computing.

Relational Database A database that follows E. F. Coddís 11 rules, a series of mathematical and logical steps for the organization and systemization of data into a software system that allows easy retrieval, updating, and expansion. A relational database management system (RDBMS) stores data in a database consisting of one or more tables of rows and columns. The rows correspond to a record (tuple); the columns correspond to attributes (fields) in the record. RDBMSs use Structured Query Language (SQL) for data definition, data management, and data access and retrieval. Relational and object-relational databases are used extensively in bioinformatics to store sequence and other biological data.

Secondary structure (protein) The organization of the peptide backbone of a protein that occurs as a result of hydrogen bonds e.g. alpha helix, Beta pleated sheet.

Sequence Tagged Site (STS) A unique sequence from a known chromosomal location that can be amplified by PCR. STSs act as physical markers for genomic mapping and cloning.

Single Nucleotide polymorphism (SNP) A DNA sequence variation resulting from substitution of one nucleotide for another.

SQL Structured Query Language. A type of programming language used to construct database queries and perform updates and other maintenance of relational databases, SQL is not a fully-fledged language that can create standalone applications, but it is powerful enough to create interactive routines in other database programs.

Substitution matrix A model of protein evolution at the sequence level resulting in the development of a set of widely used substitution matrices. These are frequently called Dayhoff, MDM (Mutation Data Matrix), BLOSUM or PAM (Percent Accepted Mutation) matrices. They are derived from global alignments of closely related sequences. Matrices for greater evolutionary distances are extrapolated from those for lesser ones.

Tertiary structure (protein) Folding of a protein chain via interactions of its side-chain molecules including formation of disulphide bonds between cysteine residues.

UCSC University of California Santa Cruz

UTR Untranslated region. The non coding region of an mRNA transcript flanking either side of the open reading frame.

INDEX

Note: page numbers in *italics* refer to figures and tables